Adult Haliplidae of the World

World Catalogue of Insects

VOLUME 18

The titles published in this series are listed at *brill.com/wci*

Adult Haliplidae of the World

An Illustrated Diagnostic Guide

By

Bernhard van Vondel

BRILL

ISSN 1398-8700
ISBN 978-90-04-75966-4
ISBN 978-90-04-75967-1 (PDF)
DOI https://doi.org/10.1163/9789004759671

Library of Congress Control Number: 2026938109

Brill and Koninklijke Brill BV are part of De Gruyter Brill.
www.degruyterbrill.com

Questions about General Product Safety Regulation: productsafety@degruyterbrill.com

Cover illustration: photo of Peltodytes roughleyi Vondel by Bernhard van Vondel.

Contents

Foreword

It is a special pleasure for me for several reasons to write a foreword for Bernhard van Vondel's book Adult Haliplidae of the World. First of all, I started my scientific career with my Diploma thesis on the larva of *Haliplus lineatocollis*, a common species of the group. Moreover, since the Handbook of Zoology Coleoptera project started in 2002, I had a very enjoyable and efficient co-operation with Bernhard van Vondel, who is the worldwide leading expert of the family. A third reason is that Haliplidae is a fascinating group, even though it is hardly known to the general public. The beetles are small, with an inconspicuous colouration, aquatic, and relatively cryptic in their environment, usually staying hidden among filamentous algae or Characeae.

Haliplidae belong to Adephaga, the second largest suborder of Coleoptera, with a total of ca. 50.000 described species, and mainly characterized by predacious habits of larvae and adults. In contrast to the typical adephagan groups like the species-rich ground beetles (ca. 40.000 spp.) and diving beetles (ca. 4.000 spp.), or also the highly specialized whirligig beetles (Gyrinidae) (ca. 1000 spp.), Haliplidae (ca. 250 spp.) mainly feed on algae as adults and exclusively as larvae, and are thus a highly unusual exception in the suborder. Despite the superficially inconspicuous habitus, haliplids display a distinctive apomorphy, greatly enlarged metacoxal plates, which almost cover the entire ventral abdomen and serve as an additional space for breathing air. The larvae are a unique case within Adephaga and beetles generally, with mandibles suitable for sucking out contents of single cells of algae, unique climbing adaptations of the fore legs, and dorsal tracheal gills or microtracheal gills unlike all other breathing organs occurring in other aquatic groups of beetles. Haliplidae are also a key taxon of Adephaga, with the systematic position disputed for a long time and arguably not fully resolved yet. Analyses of morphological data suggested either a placement as the second branch in Adephaga after Gyrinidae, or alternatively a sister group relationship to the aquatic Dytiscoidea. The latter option was finally supported by transcriptomic analyses.

Adult Haliplidae of the World will certainly be tremendously useful for everybody collecting or investigating Haliplidae. The keys to the genera and species of the family are very useful for identifying specimens from all corners of the planet. An outstanding feature of the book is that it is magnificently illustrated, with numerous excellent morphological line drawings, all of them prepared by the author himself. In addition, habitus photos are provided, and also distribution maps. Precise diagnoses are given for all species, with additional information on the biology, habitats, immature stages, and distribution.

The rich information presented in the book is certainly an example that should be followed by enthusiasts of other groups of Coleoptera.

All in all, I can only praise this work of Bernhard van Vondel, who has dedicated decades of study to an often underestimated but nevertheless fascinating group of Coleoptera.

Rolf Georg Beutel

Acknowledgements

The Uyttenboogaart-Eliasen Foundation is greatly acknowledged for the financial support of this project and my research in general. I am very grateful to Dr. Rolf Beutel for his very valuable cooperation in this and many earlier projects. I received some photographs for which I am very gratefull (see under credits). I am also grateful to publishers and authors for their permission to use drawings and photographs (see credits).

Credits

All drawings are made by the author and most are published earlier. Most of the photographs are made by the author, some are with permission taken from other publications and some specially made on request.

For papers published in Entomologische Berichten Amsterdam, The Coleopterists Bulletin and ZooKeys the copyright belongs to the author(s).

Manfred Jäch gave permission to use my own pictures published in Koleopterologische Rundschau Vol. 67 (1997): 119–120 (Mazzoldi, P. & B.J. van Vondel), Vol. 78 (2008) : 69–194 (B.J. van Vondel) and Vol. 86 (2016): 51–59 (B.J. van Vondel); Water Beetles of China 1 (1995): 111–154, Vol. 3 (2003): 289–294; and Water Beetles of New Caledonia 1 (2010): 237–242 (L. Hendrich & B.J. van Vondel).

Keith Maguire gave permission to use my own figures published in Records of the South Australian Museum Vol. 28: 61–101.

Tanja Cowell (De Gruyter Brill) gave permission to use my own pictures published in Tijdschrift voor Entomologie Vol 134 (1991): 75–144, Vol 135 (1992): 275–297, Vol 136 (1993): 289–316, Vol 149 (2006): 227–273 (Vondel, B.J. van, M. Holmen & P. Petrov), Vol 151 (2008): 141–143, Vol 152 (2009): 333–338, Vol 153 (2010): 239–314, Vol 155 (2012): 57–66 (Vondel, B.J. van & J. Bergsten), Vol 157 (2014): 137–144. Vol 161 (2018): 3–10, Vol. 163 (2021): 101–298, Vol 164 (2021): 77–90 (Vondel, B.J. van & K. Matsumoto), Vol 166 (2023): 117–124.

All photographs are made by the author with the following exceptions:

Haliplus (Australiplus) sindus (courtesy of Chris Watts Adelaide, Australia)
Haliplus (Haliplus) kirgisiensis (courtesy of Jaakko Mattila & Pekka Malinen, Finish Museum of Natural History, Helsinki, Finland)
Haliplus (Haliplus) morii (courtesy of Hiroyuki Yoshitomi, Matsuyama, Japan)
Haliplus (Liaphlus) brasiliensis (courtesy of Michael Balke and Katja Neven, Zoologische Staatssammlung München, Germany)
Haliplus (Liaphlus) diopus (courtesy of Christophe Rivier and Antoine Mantilleri , Musée National d'Histoire Naturelle Paris, France)
Haliplus (Liaphlus) ferrugineus (courtesy of Chris Watts, Adelaide, Australia)
Haliplus (Liaphlus) figuratus (courtesy of Jaakko Mattila & Pekka Malinen, Finish Museum of Natural History, Helsinki, Finland)
Haliplus (Liaphlus) fiorentini (courtesy of Cesar Benetti, Manaus, Brasil)
Haliplus (Liaphlus) perroti (courtesy of Christophe Rivier and Antoine Mantilleri , Musée National d'Histoire Naturelle Paris, France)

Haliplus (*Liaphlus*) *peruanus* (courtesy of Michael Balke and Katja Neven, Zoologische Staatssammlung München, Germany)
Haliplus (*Liaphlus*) *regili* (courtesy of Cesar Benetti, Manaus, Brasil)
Haliplus (*Paraliaphlus*) *alberti* (courtesy of Yoandri Megna, Zoologische Staatssammlung München, Germany)

Tables, Figures and Plates

Tables

Figures

Plates

Introduction

This book deals with the waterbeetle-family Haliplidae (Coleoptera), also known as crawling waterbeetles. The adults of all species worldwide are treated, while the larvae are treated in general. At present 233 species are known (Vondel 2005, 2007, 2013, 2016a, 2023). This number concerns extant species. The very few extinct species are not treated here.

General information is given on morphology, zoogeography and biology. Keys to genera and species and descriptions, drawings and photographs of all species are included. Distribution maps are presented for each species.

No revisional work has been included.

The most recent publications covering this area are from Zimmermann (1924) and Scholz (1929). More recent works are dealing with the worlds faunal regions (Vondel, 1991, 1992a, 1993a, b, 1995, 2006, 2008, 2010c, 2023), a limited part of Europe (Freude & Hoch, 1971; Hoch, 1960; Holmen, 1987; Lindroth, 1960, Vondel 1997) or one country (Balfour-Browne, 1953; Brakman, 1966; Everts, 1898/1922; Foster, 1981; Guignot, 1931–33; Hansen, 1930; Vondel, 1985, 1992b, 1995a, b, 2003a, b, 2009, 2010); Hendrich & Vondel, 2010; Short et al. 2010; Jia & Vondel, 2011; Vondel & Bergsten, 2012; Vondel et al. 2006; Lee et al. 2015; Jia et al. 2024.

 | HTTPS://DOI.ORG/10.1163/9789004759671-002

Diagnostic Characters of the Family

Adults: The Haliplidae are among other water beetles and also among all other beetles easily regognisable by the strongly enlarged hind-coxal plates, which at least reach to the fifth sternite (Fig. 10). The length ranges from 2 to 5 mm. The colour ranges from completely yellow to red-brown, with or without dark maculation.

In the water these beetles can be distinguished from the other waterbeetles by the way they swim. They move their legs alternately (that's why they are called "crawling water beetles"), while representatives of the water beetle family Dytiscidae, with which some species could be confused, move their legs at the same time.

Immature stages: Haliplidae have five immature stages: egg, first instar larva (instar I), second instar larva (instar II), third instar larva (instar III) and pupa.

Eggs are nearly round to oval, about 0.35–0.45 long and without distinct surface-structure. The chance to meet the eggs during fieldwork is very low.

The larvae are very slender, the length ranges from about 2 mm (first instars) to about 14 mm (third instars, Figs. 3–6). The colour ranges from almost white to strongly blackened by external growth of micro-organisms. The legs have five segments (coxa, trochanter, femur, tibia, tarsus), while each tarsus bears only one claw. The mandibles have a suction-channel. The length of the pupa ranges from 2.0–5.0 mm (Fig. 7). The immature stages of Haliplidae are not further treated here.

 | HTTPS://DOI.ORG/10.1163/9789004759671-003

Biogeography

The family Haliplidae is represented in all major faunal regions by 233 described species. Faunal regions as treated here are defined as in Fig. 1. The highest number of species is known from the temperate and subtropical zones of the northern hemisphere. In the tropical zones the known number of species is rather low, most of these species are seldomly collected (Zimmermann, 1920, 1924; Vondel, 1992b, 1993, 2008, 2010; Hendrich & Vondel, 2010).

The genus *Haliplus* is worldwide distributed. Of the remaining Palaearctic genera *Peltodytes* occurs in most parts of the world, except the neotropical and australasian regions, and is well developed in North America, while *Brychius* is restricted to the holarctic region. The genus *Phalilus* only occurs in the Australasian region.

Geographical distribution is given at the level of nations, and provinces of Australia, Canada, China, Russia and the United States. Moreover, individual islands like those of Indonesia and Japan are occasionally separated.

The delimitation of the Faunal Regions and the subdivision of Russia follow Löbl & Löbl (2017) (Figs. 1 and 2). For the distributionmaps (Plates 260–281) a worldmap from www.freeworldmaps.net is used as a base.

 | HTTPS://DOI.ORG/10.1163/9789004759671-004

1

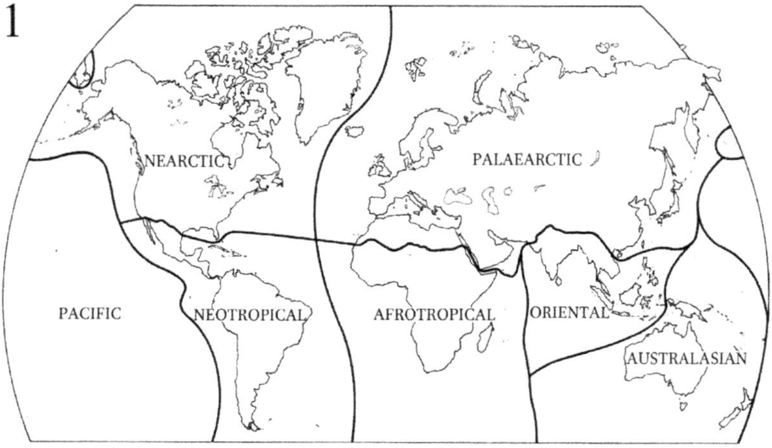

2

FIGURES 1, 2 Faunal regions. 1, Main Worlds Faunal Regions. 2, Faunal regions of Russia

Biology

Habitat: Haliplidae can be found in a large variety of freshwater habitats, some species even in brackish water. Most species live in stagnant or slowly-running water of pools, ponds, lakes, marshes, ditches, canals, brooklets, brooks, rivers etc. Some species, such as *Brychius*-species need clean oxygen-rich running water. Usually the waterbodies have a permanent character, but also semi-permanent or temporary waters are inhabited. As they are poor swimmers they are usually found between the vegetation along the shores. Most species prefer waters with a well developed vegetation of filamentous algae or Characeans. Acid conditions are usually avoided by most species. Some species are equiped to survive in poly-productive waters and does not seem to be very sensitive to pollution.

Ecological information on Middle European species is given by Koch (1989).

Mode of life: Haliplidae live most of the year in the water and can be found there, dependent on climatical conditions, throughout the year. They crawl over en between waterplants, usually near the bank and less in open water. Haliplidae are not excellent swimmers, the beetles move their hindlegs one after another, giving the impression of walking through the water. The larvae are not capable to swim. The beetles have to come to the surface to renew their airsupply under the elytra and coxal plates; the tip of the abdomen is pierced through the watersurface. Probably *Brychius*-species living in clean oxygen-rich water stay under water for a long time, because under these circumstances the airbubble works like a physical gill. In this way the beetle can survive in fast running water without getting catched by the current.

With a few exceptions Haliplidae have wings and in general they are supposed to be capable to fly. In the temporate zones there are, however, only few observations of flying-activities. In the tropics Haliplidae are often attracted to light. Jackson did research on the flight-capacity of European species (1952, 1956, a, b, 1973).

Life-cycle: Usually mating takes place in spring or summer. During copulation the male holds the female with his fore- and mid-legs. The hindlegs are moved alternately along the metacoxal plates, obviously to refresh the water near the airsupply for a better function of the physical gill. From time to time a certain vibration in the body and the hind-legs may be observed, but its function is unknown. After mating *Haliplus*-species place their eggs in the stems of algae or seed-plants, for which they use present holes or damages or holes that they create themselves by biting. *Peltodytes*-species attach the eggs to the stems. Untill now it is not possible to identify these eggs on species-level. After hatching, 9–16 days later, the instar I larvae begin to eat filamentous algae (the *Haliplus ruficollis*-related species) or characeans (most of the *H. confinis* and *H. fulvus*-related species). One group of larvae, which possess fore-legs with thumb-like tibial lobes hardly move forward, but draw the algal threads to the mandibles by means of the fore-legs. The other group, lacking the tibial lobes, climb along the stems of mainly Charophyta. The sharp mandibles pierce into the cells and suck the contents out of it. After 6–10 days moulting takes place. The instar II larvae live another 7–13 days before moulting to instar III. The life of instar III is very variable in length (Seeger, 1971a–c). In Michigan (USA) Hickman (1931a, b) has observed pupation after 10 days, shortly followed by emergence. In other cases the larvae hibernated and pupated next year. In my own rearing-experiments some larvae hibernated twice before emergence. When the larva is ready for pupation it goes ashore and makes itself a small cavity in the soil or the sand not far from the water. It may last more than a month before actual pupation takes place. In spring the adult, except the eyes still uncoloured, will emerge and enter the water after a while. It may last some days before it has his final coloration. Possibly these young adults will just reproduce next year. Sometimes even two generations a year seem to be possible. Further biological notes are given by Brocher (1922). Falkenström (1926) studied the life cycle of *Haliplus immaculatus*. Seeger (1971a–c) did extensive research on morphology, bionomics, ethology, autecology, zoogeography and habitat distribution of a number of Haliplidae-species. The biology of Nearctic species is extensively studied by Matheson (1912) and Hickman (1930, 1931a, b).

Food: The larvae are herbivorous. Some species eat all kinds of filamentous algae, other species eat characeans and may be in some cases seed-plants are used as foodplant.

In Figs 3–6 some examples are shown of third instar larvae. Fig. 7 shows an example of the pupa.

The actual knowledge about immature stages of Haliplidae is presented by Vondel (2025).

The adults are omnivorous. Besides the vegetable food, usually as the larvae, all kind of animal food is taken. Seeger (1971a) examined the contents of the gut of several species and discovered a large differentiation in the diet of the separate species. Especial the *H.ruficollis*-related species are mainly herbivorous, while some other species are merely carnivorous. The animal food may consist of insect-eggs, larvae of Chironomidae and other water-insects, worms, smaller and larger crustaceans etc. In my rearing-experiments with 16 Middle-European *Haliplus*-species all ate from freshly killed *Asellus aquaticus*. From died *Haliplus*-specimens the soft parts are immediately eaten by the remaining beetles.

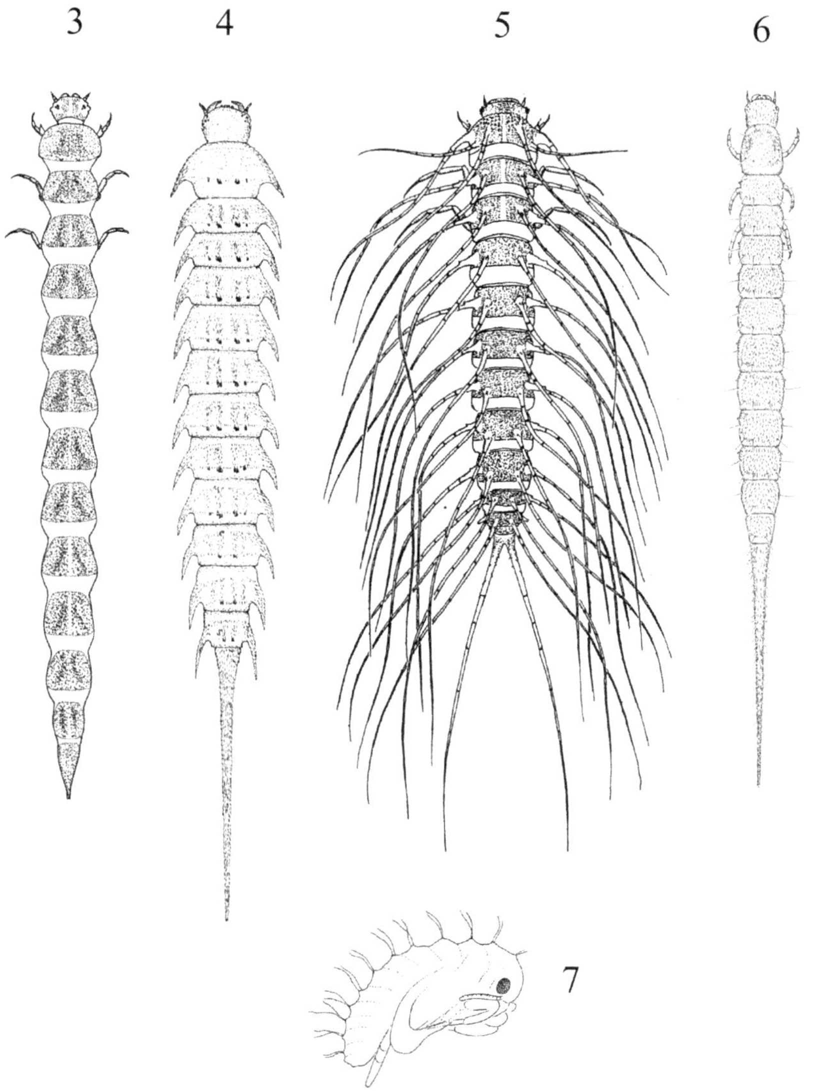

FIGURES 3–7 Immature stages of Haliplidae. 3–6, 3rd instar larva: 3, *Brychius elevatus*; 4, *Haliplus maculatus*; 5, *Peltodytes caesus*; 6, *Phalilus pinderi*. 7, pupa of *Haliplus immaculatus*

Collecting

Haliplidae can be collected in a large variety of waterbodies as pools, ditches, lakes, marshes, brooks and rivers by using a waternet or a kitchen-sieve. The beetles usually hide between the vegetation. After sweeping with the net through the waterplants, especially close to the shores and banks, the contents of the net can best be thrown in a white tray or on a cloth or sheet. The beetles will move shortly after that, but larvae will keep hiding for a long time and patience is needed. The best way to collect larvae is to throw the contents of the net after sweeping through the waterplants and especially the algae, in a wide-mashed sieve above a clean white tray with a little clean water. After a while and even after days the larvae will crawl out and fall into the tray. In wintertime larvae and also beetles can be collected by sieving the soil along the shore. In warmer countries the beetles will also come at light.

Identification of adults in the field is in most cases impossible. Conservation is possible by killing them with aethyl-acetate, in boiling water or in alcohol. The beetles and larvae may stay in 80 % alcohol untill further preparation or examination is needed. Extraction of genitaliae is easy, but nice mounting becomes difficult after a stay in alcohol.

During fieldwork larvae are seldomly recognised, because of their slender and small body, their slow movement and their perfect camouflage between the filamentous algae they are often living from (Figs 3–6). The pupae are about oval, in lateral view contracted ventrally, dorsally provided with long setae. Length ranges from 2–5 mm (Fig. 7). Colour usually white to yellow-white. During fieldwork pupae may be met, when the soil along waterbodies is sieved.

 | HTTPS://DOI.ORG/10.1163/9789004759671-006

Rearing

Most of the Central European species have been successfully reared (Falkenström, 1926; Beier, 1929; Seeger, 1971a; Vondel, 1986b). Matheson (1912) and Hickman (1937) reared some Nearctic species. Recently Japanese scientists were successful in rearing some Japanese species (Kitano &Sano 2011, Yamasaki et al. 2020).

The methods are fully described by Falkenström (1926), Hickman (1930), Seeger (1971a) and Vondel (1997).

One method is to use white trays of about 20 × 30 × 6 cm, filled with a thin layer of sand and a few cm of water. Suitable food is added, such as filamentous algae for larvae and adults and freshly killed small animals like Asellus or insect-larvae for the adults. Pieces of seed-plants like Elodea are necessary to oviposit on or in. Adults of one species, males and females are put into such a tray. The tray has to be covered by a glass pane to prevent beetles to walk or fly into another tray with possibly other species. The algal food has to be checked from time to time, but the animal food seems only to be attractive for a short time. The remains must be removed to keep the water clean.

The best thing is to check regularly if eggs are laid on or in the stems of the plants. If eggs are found they best can be done in small cups with a little water and some filamentous algae. When enough eggs are hatched the process of moulting to instar II or III larvae can be documented by preserving in alcohol some larvae of the first, second and third instar.

To allow instar III larvae to pupate soil is needed. The little cup with larvae in the water can be placed in a larger cup filled with soil or sand. A little branch in the water allows the larvae to crawl out of the water and to reach the soil for pupation. After emerging from the pupa the fresh beetle will enter again the water. This whole process may last some months, but can also last some years in case the larvae hibernates twice.

Attention has to be drawn upon some possible problems. When using water or plants out of the field there is a chance that eggs or larvae from other species than you are planning to rear are introduced into your trays, probably causing confusion. The use of "clean" material like tap-water and self-raised plants make results more reliable.

 | HTTPS://DOI.ORG/10.1163/9789004759671-007

Preparation of Specimens

Adult Haliplidae have to be clean to recognise possible micropunctuation. Unless they are permanently preserved in alcohol they are best mounted, with the legs spread out, on cards with water-soluble glue, as it is often necessary to examin the ventral side. Gluing the beetle with one side to the top of a small card, giving view on at least half of the prosternal and metasternal apophysis is also possible. It is advisable to extract the aedaegus of male specimens before mounting and to glue the aedaegus to the same card as the beetle.

HTTPS://DOI.ORG/10.1163/9789004759671-008

Taxonomy

Genera and subgenera

At present the family Haliplidae has 4 genera *Haliplus* Latreille, 1802; *Brychius* Thomson, 1859; *Peltodytes* Régimbart, 1878 and *Phalilus* Guignot, 1935.

The genus *Haliplus* is divided into 11 subgenera: *Algophilus* Zimmermann, 1924; *Apteraliplus* Chandler, 1943; *Australiplus* Vondel, 2019; *Borealiplus* Vondel, 2019; *Haliaplus* Ádám, 1993; *Haliplidius* Guignot, 1928; *Haliplus* s. str.; *Liaphlus* Guignot, 1928; *Neohaliplus* Netolitzky, 1911; *Nipponiplus* Vondel, 2019; *Paraliaphlus* Guignot, 1930.

8

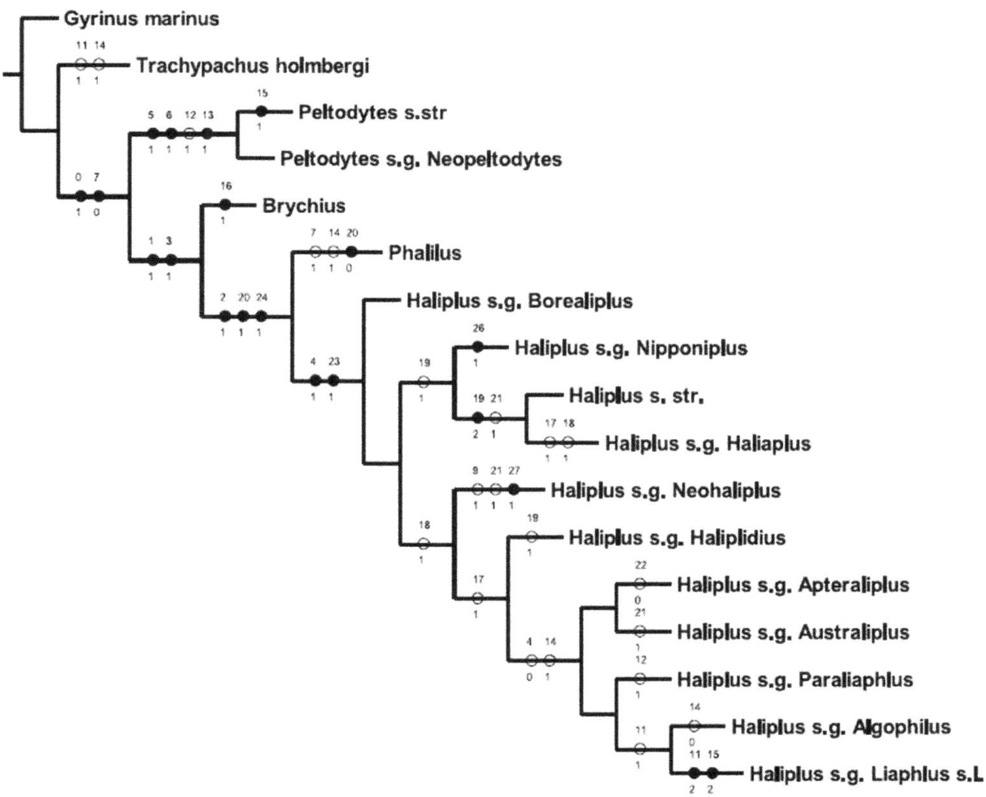

FIGURE 8 Cladogram of Haliplidae (Vondel, 2019)

Species-list with Faunal Region (PL = Palaearctic, NA = Nearctic, NT = Neotropic, AF = Afrotropic, OR = Oreintal, AU = Australasian)

Family Haliplidae Aubé, 1836
= Haliplides Aubé, 1836
= Brychiini Ádám, 1996
= Peltodytinae Böving & Craighead, 1931
= Haliplinae Böving & Craighead, 1931

1 Genus *Brychius* Thomson, 1859 NA PL

elevatus (Panzer, 1793) PL
= *Dytiscus elevatus* Panzer, 1793
= *Brychius cristatus* J. Sahlberg, 1874
= *Brychius intermedius* G. Müller, 1908
= *Brychius elevatus* ssp. *norvegicus* Munster, 1922
= *Brychius obenbergeri* Říha, 1952
= *Brychius rossicus* Semenov, 1898
glabratus (Villa & Villa, 1835) PL
= *Haliplus glabratus* Villa & Villa, 1835
= *Haliplus aequatus* Aubé, 1836
hornii Crotch, 1873 NA
= *albertanus* Carr, 1928
hungerfordi Spangler, 1954 NA
pacificus Carr, 1928 NA

2 Genus *Haliplus* Latreille, 1802 AF AU NA NT OR PL

= *Cnemidotus* Illiger, 1802
= *Hoplitus* Clairville, 1806

3 Subgenus *Algophilus* Zimmermann, 1924 AF

lathridioides Zimmermann, 1924 AF

 | HTTPS://DOI.ORG/10.1163/9789004759671-010

4 Subgenus *Apteraliplus* Chandler, 1943 NA

parvulus (Roberts, 1913) NA
= *Brychius parvulus* Roberts, 1913
= *Haliplus wallisi* Hatch, 1944

5 Subgenus *Australiplus* Vondel, 2019 AU

sindus Watts, 1988 AU

6 Subgenus *Borealiplus* Vondel, 2019 NA, NT

borealis LeConte, 1850 NA
oblongus Zimmermann, 1921 NT

7 Subgenus *Haliaplus* Ádám, 1993 PL

= *Hoplites* Kinel, 1929
laminatus (Schaller, 1783) PL
= *Dytiscus laminatus* Schaller, 1783
= *Haliplus ater* W. Redtenbacher, 1842
= *Haliplus cinereus* Aubé, 1836
= *Haliplus laminatus* ♀ var. *punctulatus* Penecke, 1901
= *Haliplus lombardus* Fiori, 1904

8 Subgenus *Haliplidius* Guignot, 1928 NT PL

confinis Stephens, 1828 PL
= *Haliplus confinis* var. *concolor* Lentz, 1879
= *Haliplus confinis* var. *halberti* Bullock, 1928
= *Haliplus confinis* var. *nebulosus* Stephens, 1828
= *Haliplus lineatus* Aubé, 1836
= *Haliplus confinis* var. *pallens* Fowler, 1887
obliquus (Fabricius, 1787) PL
= *Dytiscus obliquus* Fabricius, 1787
= *Haliplus albinus* Reitter, 1911

= *Dytiscus amoenus* Olivier, 1795
= *Haliplus albinus* var. *muelleri* Reitter, 1912
rejseki Šťastný & Boukal, 2003 PL
subseriatus Zimmermann, 1921 NT
varius Nicolai, 1822 PL
= *Haliplus maritimus* Fairmaire & Laboulbène, 1855
= *Haliplus pictus* Mannerheim, 1844

9 Subgenus *Haliplus* Latreille, 1802 NA NT PL

= *Haliplinus* Guignot, 1939
aliae Vondel, 2003 PL
apicalis Thomson, 1868 NA PL
=? *Haliplus brevis* Stephens, 1828
= *Haliplus striatus* Sharp, 1869
= *Haliplus strigatus* Roberts, 1913
blanchardi Roberts, 1913 NA
distinctus Wallis, 1933 NA
dorsomaculatus Zimmermann, 1924 NA
= *Haliplus allisonae* Brigham, 1977
falli Mank, 1940 NA
fluviatilis Aubé, 1836 PL
= *Haliplus fluviatilis* var. *maculatus* Gozis, 1915
= *Haliplus fluviatilis* var. *mannerheimi* Gozis, 1915
fulvicollis Erichson, 1837 PL
= *Haliplus jakowlewi* Semenov, 1898
furcatus Seidlitz, 1887 PL
= *Haliplus fluviatilis* var. *mannerheimii* Seidlitz, 1887
fuscicornis Holmen, Vondel & Petrov, 2006 PL
harminae Vondel, 1990 PL
heydeni Wehncke, 1875 PL
= *Haliplus foveostriatus* Thomson, 1884
= *Haliplus multipunctatus* Wehncke, 1875
= *Haliplus ruficollis* var. *pedemontanus* Fiori, 1904
= *Haliplus fulvicollis* var. *romanus* Fiori, 1904
immaculatus Gerhardt, 1877 PL
immaculicollis Harris, 1828 NA NT
= *Haliplus americanus* Aubé, 1838

= *Haliplus pallidus* Roberts, 1913
= *Haliplus robertsi* Zimmermann, 1924
interjectus Lindberg, 1937 PL
kamiyai Nakane, 1963 PL
kirgisiensis Holmen & Vondel, 2006 PL
lineolatus Mannerheim, 1844 PL
= *Haliplus browneanus* Sharp, 1913
= *Haliplus brownei* Sharp, 1913
= *Haliplus nomax* F. Balfour-Browne, 1911
= *Haliplus nomax* var. *laevigatus* Falkenström, 1936
= *Haliplus schaumi* Solsky, 1867
= *Haliplus transversus* Thomson, 1870
longulus LeConte, 1850 NA
= *Haliplus hoppingi* Wallis, 1933
morii Hayashi, Iwata & Yoshitomi, 2023 PL
ruficollis (De Geer, 1774) PL
= *Dytiscus ruficollis* De Geer, 1774
= *Haliplus affinis* Stephens, 1828
= *Dytiscus impressus* Fabricius, 1787
= *Dytiscus marginepunctatus* Panzer, 1793
= *Haliplus fluviatilis* var. *maculatus* Seidlitz, 1887
= *Haliplus fluviatilis* ab. *seidlitzi* Csiki, 1946
= *Haliplus rubicundus* Babington, 1836
samojedorum J. Sahlberg, 1880 PL
sibiricus Motschulsky, 1860 PL
= *Haliplus alsaticus* Scriba, 1905
= *Haliplus borealis* Gerhardt, 1877
= *Haliplus holsaticus* Reitter, 1908
= *Haliplus lindbergi* Falkenström, 1939
= *Haliplus sahlbergi* Falkenström, 1939
= *Haliplus wehnckei* Gerhardt, 1877
simplex Clark, 1863 PL
= *Haliplus minutus* Takizawa, 1931
= *Haliplus medvedevi* Gramma, 1980
stagninus Leech, 1949 NA
steppensis Guignot, 1954 PL
turkmenicus Vondel, 2006 PL
zacharenkoi Gramma & Prysnyi, 1973 PL

10 Subgenus *Liaphlus* Guignot, 1928 AF AU NA NT OR PL

= *Haliaplus* Ádám, 1993
= *Hoplites* Kinel, 1929
abbreviatus Wehncke, 1880 PL
= *Haliplus gafnyi* Vondel, 1991
= *Haliplus villiersi* Legros, 1972
= *Haliplus kulleri* Vondel, 1988
= *Haliplus jaechi* Vondel, 1991
= *Haliplus ortali* Vondel, 1991
africanus Aubé, 1838 AF
alastairi Watts, 1988 AU
alluaudi Régimbart, 1903 AF
andalusicus Wehncke, 1872 PL
angusi Vondel, 1991 PL
angustifrons Régimbart, 1892 OR PL
= *Haliplus shanicus* Guignot, 1956
= *Haliplus manipurensis* Vazirani, 1966
= *kotoshonis* Vondel, 1991 [misidentification]
annulatus Roberts, 1913 NA NT
= *Haliplus minor* Zimmermann, 1924
= *ohioensis* Wallis, 1933
arrowi Guignot, 1936 OR PL
astrakhanus Vondel, 1991 PL
australis Clark, 1862 AU
bachmanni Vidal Sarmiento & Grosso, 1970 NT
basinotatus Zimmermann, 1924 PL
= *Haliplus basinotatus latiusculus* Nakane, 1985
bonariensis Steinheil, 1869 NT
brasiliensis Zimmermann, 1924 NT
camposi Guignot, 1948 NT
carinatus Guignot, 1936 NT
chinensis Falkenström, 1932 PL
colombiensis Vondel & Spangler, 2008 NT
columbiensis Wallis, 1933 NA
concolor LeConte, 1852 NA NT
confluentus Roberts, 1913 NA NT
costaricanus Vondel & Spangler, 2008 NT
coxaspinipes Vondel, 2010 AF
crassus Chapin, 1930 NT

cribrarius LeConte, 1850 NA
cubensis Chapin, 1930 NT
= *Haliplus bierigi* Guignot, 1936
cylindricus Roberts, 1913 NA
= *ungularis* Wallis, 1933
dalmatinus G. Müller, 1900 PL
= *Haliplus dalmatinus* var. *weberi* G. Müller 1900
diopus Guignot, 1955 OR
discessus Guignot, 1936 AF
= *Haliplus natalensis* var. *discessus* Guignot, 1936
drechseli Vondel & Spangler, 2008 NT
ebolovensis Guignot, 1955 AF
elsaltous Vondel & Spangler, 2008 NT
excoffieri Vondel, 1991 PL
eximius Clark, 1863 OR PL
= *Haliplus emmerichi* Falkenström, 1936
= *Haliplus hiogoensis* Kano & Kamiya, 1931
= *Haliplus modestus* Zimmermann, 1924
exsecratus Guignot, 1936 AF
ferruginipes Régimbart, 1892 AU
= *Haliplus nicholasi* Watts, 1988
figuratus J. Sahlberg, 1908 AF
fiorentini Benetti & Hamada, 2017 NT
flavicollis Sturm, 1834 PL
= *Haliplus flavicollis* var. *expallidus* Zaitzev, 1908
= *Haliplus flavicollis* var. *pallida* Semenov, 1904
fulvus (Fabricius, 1801) NA PL
= *Dytiscus fulvus* Fabricius, 1801
= *Haliplus fulvus* var. *carlittensis* Régimbart, 1901
= *Dytiscus interpunctatus* Marsham, 1802
= *Haliplus lapponum* Thomson, 1856
= *Haliplus lapponum* var. *niger* Seidlitz, 1887
= *Haliplus pyraeneus* Delarouzée, 1857
= *Haliplus salinarius* Wallis, 1933
= *Haliplus fulvus sparreschneideri* Munster, 1922
= *Haliplus subguttatus* Roberts, 1913
= *Haliplus unicolor* Fiori, 1914
= *Haliplus fulvus* var. *unicolor* Munster, 1922
fuscipennis Germain, 1855 NT
gamma Vondel & Bergsten, 2012 AF

garambanus Guignot, 1958 AF
genieri Vondel, 2010 AF
gracilis Roberts, 1913 NA
gravidoides Vondel & Spangler, 2008 NT
gravidus Aubé, 1838 NT
= *Haliplus brandeni* Wehncke, 1883
= *Haliplus obconicus* Régimbart, 1889
= *Haliplus robustus* Sharp, 1877
guignoti Vondel, 2010 AF
guttatus Aubé, 1836 PL
= *Haliplus pelopis* J. Sahlberg, 1908
halsei Watts & McRae, 2010 AU
havaniensis Wehncke, 1880 NA NT
heppneri Vondel & Spangler, 2008 NT
incrassatus Régimbart, 1899 AF
= *Haliplus aspilus* Guignot, 1957
indicus Régimbart, 1899 OR
= *Haliplus pulchellus* var. *indicus* Régimbart, 1899
= *Haliplus agarwali* Vazirani, 1984
indistinctus Zimmermann, 1928 NT
insularis Guignot, 1960 AF
javanicus Vondel, 1993 OR
kantneri Vondel, 2010 AF
kapuri Vazirani, 1975 OR
kotoshonis Kano & Kamiya, 1931 OR, PL
= *diruptus* J. Balfour-Browne, 1946
= *davidi* Vondel, 1991
lamottei Legros, 1954 AF
langleyi Vondel & Spangler, 2008 NT
leechi Wallis, 1933 NA
= *Haliplus leechi carteri* Leech, 1949
= *Haliplus salmo* Wallis, 1933
lewisii Crotch, 1873 NA NT
maculatus Motschulsky, 1860 PL
= *Haliplus syriacus* Wehncke, 1880
= *Haliplus zimmermanni* Gschwendtner, 1921
maculicollis Zimmermann, 1924 NT
maculipennis Schaum, 1864 AF
= *Haliplus abyssinicus* Régimbart, 1895
= *Haliplus maculipennis capensis* Guéorguiev, 1967

madagascariensis Vondel & Bergsten, 2012 AF
megapunctatus Vondel & Spangler, 2008 NT
mesoamericanus Vondel & Spangler, 2008 NT
methneri Zimmermann, 1926 AF
mexicanus Vondel & Spangler, 2008 NT
mimulus Guignot, 1956 AF
minimus Vondel & Spangler, 2008 NT
mucronatus Stephens, 1828 PL
= *Haliplus badius* Aubé, 1836
= *Haliplus parallelus* Babington, 1836
= *Haliplus siculus* Wehncke, 1883
nanus Guignot, 1936 NT
napolovi Vondel, 2003 OR
natalensis Wehncke, 1880 AF
= *Haliplus nigrosignatus* Régimbart, 1895
nedungaduensis Vondel, 1993 OR
nieseri Vondel & Spangler, 2008 NT
nigerianus Vondel, 2010 AF
nitens LeConte, 1850 NA
= *Haliplus canadensis* Wallis, 1933
novateutonianus Vondel, 2016 NT
oklahomensis Wallis, 1933 NA NT
ornatipennis Zimmermann, 1921 NT
ovalis Sharp, 1884 PL
panamanus Chapin, 1930 NT
= *Haliplus soekhnandanae* Makhan, 1992
perroti Guignot, 1950 OR PL
peruanus Zimmermann, 1924 NT
philippinus Chapin, 1930 OR
pilbaraensis Watts & McRae, 2010 AU
pruthii Vazirani, 1966 OR
pulchellus Clark, 1863 OR PL
= *Haliplus oceanicus* Régimbart, 1886
regili Benetti & Hamada, 2017 NT
rubidus Perris, 1857 PL
= *Haliplus perforatus* Schaum, 1859
rufescens Régimbart, 1894 AF
samosirensis Vondel, 1993 OR
sharpi Wehncke, 1880 PL
= *Haliplus amrishi* Makhan, 1999

= *Haliplus tsukushiensis* Yoshimura, 1932
= *Haliplus holmeni* Vondel, 1991
signatipennis Régimbart, 1892 AU
signatus Sharp, 1882 NT
solitarius Sharp, 1882 NT
srilankanus Vondel, 1993 OR
stepheni Watts, 1988 AU
sulawesicus Vondel, 2008 OR
tantoyucanus Vondel & Spangler, 2008 NT
tanzanianus Vondel, 2010 AF
testaceus Zimmermann, 1924 NT
testudo Clark, 1862 AU
= *Haliplus nigrolineatus* Wehncke, 1883
thoracicus Zimmermann, 1923 NT
timmsi Vondel, 1995 AU
tocumenus Vondel & Spangler, 2008 NT
triplehorni Vondel & Spangler, 2008 NT
tumidus LeConte, 1880 NA NT
= *Haliplus curtulus* Sharp, 1887
unicarinatus Vondel & Spangler, 2008 NT
uniformis Zimmermann, 1920 AF
= *Haliplus concolor* Régimbart, 1894
valdiviensis Moroni, 1980 NT
vancouverensis Matheson, 1912 NA
varicator Guignot, 1954 AF
variegatus Sturm, 1834 PL
= *Haliplus variegatus galilea* J. Sahlberg, 1913
= *Haliplus leopardinus* J. Sahlberg, 1900
= *Haliplus variegatus* var. *obsoletus* Westhoff, 1881
= *Haliplus subnubilus* Babington, 1836
= *Haliplus transvolgensis* Semenov, 1904
venustus Régimbart, 1894 AF
volpei Vondel, 2009 OR
watsoni Vondel, 2010 AF
wattsi Vondel, 1995 AU
youngi Vondel & Spangler, 2008 NT

11 Subgenus *Neohaliplus* Netolitzky, 1911 AF AU PL

= *Protohaliplus* Scholz, 1929
lineatocollis (Marsham, 1802) AF PL
= *Dytiscus lineatocollis* Marsham, 1802
= *Dytiscus bistriolatus* Duftschmid, 1805
= *Haliplus lineatocollis arabicus* Zaitsev, 1953
= *Haliplus lineatocollis* var. *beckmanni* Ahlwarth, 1913
= *Haliplus lineatocollis* var. *directus* Gozis, 1915
= *Haliplus lineatocollis* var. *intemeratus* Gozis, 1915
= *Haliplus lineatocollis* var. *pallidus* J. Sahlberg, 1903
= *Haliplus lineatocollis* var. *pici* Régimbart, 1895
= *Haliplus lundbladi* Falkenström, 1939b
= *Haliplus melanocephalus* Stephens, 1828
= *Haliplus suffusus* Wollaston, 1863
= ? *Dyticus thoracicus* Geoffroy, 1785
= *Haliplus transversalis* Gautier des Cottes, 1861
= *Haliplus trimaculatus* Drapiez, 1820
ruficeps, Chevrolat, 1861 PL

12 Subgenus *Nipponiplus* Vondel, 2019 PL

japonicus Sharp, 1873 PL
= *Haliplus hummeli* Falkenström, 1932
= *Haliplus brevior* Nakane, 1963
= *Haliplus rishwani* Makhan, 1999
latreillei Jia & Vondel, 2011 PL
regimbarti Zaitzev, 1908 PL
= *Haliplus brevis* Wehncke, 1880
= *Haliplus sauteri* Zimmermann, 1924

13 Subgenus *Paraliaphlus* Guignot 1930 NA NT

alberti Megna, Hendrich & Balke, 2018 NT
apostolicus Wallis, 1933 NA
connexus Matheson, 1912 NA
deceptus Matheson, 1912 NA NT

= *Haliplus suturalis* Roberts, 1913
= *Haliplus variomaculatus* Brigham & Sanderson, 1973
fasciatus Aubé, 1838 NA
grandis Vondel & Spangler, 2008 NT
kenneri Vondel, 2014 NT
leopardus Roberts, 1913 NA
lucigelinae Vondel & Alarie, 2016 NA
mimeticus Matheson, 1912 NA NT
= *Haliplus eremicus* Wells, 1989
= *Haliplus rug*osus Roberts, 1913
mutchleri Wallis, 1933 NA
pantherinus Aubé, 1838 NA
pseudofasciatus Wallis, 1933 NA
punctatus Aubé, 1838 NA
tortilipenis Brigham & Sanderson, 1972 NA
triopsis Say, 1823 NA

14 Genus *Peltodytes* Régimbart, 1879 AF NA NT OR PL

= *Cnemidotus Illiger sensu* Erichson, 1832

15 Subgenus *Neopeltodytes* Satô 1963 NA NT

bradleyi Young, 1961 NA
callosus (LeConte, 1852) NA NT
= *Cnemidotus callosus* LeConte, 1852
darlingtoni Young, 1961 NT
dietrichi Young, 1961 NA
dunavani Young, 1961 NA
duodecimpunctatus (Say, 1823) NA
= *Haliplus duodecimpunctatus* Say 1823
= *Peltodytes lengi* Roberts, 1913
edentulus (LeConte, 1863) NA
= *Cnemidotus edentulus* LeConte, 1863
festivus (Wehncke, 1876) NA
= *Cnemidotus festivus* Wehncke, 1876
floridensis Matheson, 1912 NA
litoralis Matheson, 1912 NA

muticus (LeConte, 1863) NA
= *Cnemidotus muticus* LeConte, 1863
oppositus Roberts, 1913 NA
ovalis Zimmermann 1924 NT
pedunculatus (Blatchley, 1910) NA
= *Cnemidotus pedunculatus* Blatchley, 1910
roughleyi Vondel, 2021 NA
sexmaculatus Roberts, 1913 NA
shermani Roberts, 1913 NA
simplex (LeConte, 1852) NA NT
= *Cnemidotus simplex* LeConte, 1852
= *Peltodytes dispersus* Roberts, 1913
= *Cnemidotus mexicanus* Wehncke, 1883
tamaulipensis Young, 1964 NT
tortulosus Roberts, 1913 NA
= *Peltodytes obesus* Peschet, 1923

16 Subgenus *Peltodytes* Régimbart

caesus (Duftschmid, 1805) PL
= *Dytiscus caesus* Duftschmid, 1805
= ? *Dyticus curculinus* O.F. Müller, 1776
= *Cnemidotus delhermi* Mazeret, 1923
= *Cnemidotus caesus* var. *levantinus* J. Sahlberg, 1903
= *Haliplus quadrimaculatus* Drapiez, 1820
= *Haliplus rotundatus* Gistel, 1857
congoensis Zimmermann, 1924 AF
coomani Peschet, 1923 OR PL
= *Peltodytes wui* Gschwendtner, 1934
dauricus Zimmermann, 1924 PL
ghanaensis Vondel, 2010 AF
intermedius (Sharp, 1873) PL
= *Cnemidotus intermedius* Sharp, 1873
nodieri Guignot, 1936 AF
pekinensis Vondel, 1992 PL
quadratus Régimbart, 1895 AF
rotundatus (Aubé, 1836) PL
= *Cnemidotus rotundatus* Aubé, 1836
= *Cnemidotus conifer* Seidlitz, 1887

sinensis (Hope, 1845) OR PL
= *Peltodytes aschnae* Makhan, 1999
= *Peltodytes koreanus* Takizawa, 1931
= *Haliplus variabilis* Clark, 1863
speratus Régimbart, 1906 AF
= *alluaudi* Guignot, 1936
sumatrensis Régimbart, 1885 OR
ugandanus Vondel, 2010 AF

17 Genus *Phalilus* Guignot, 1935 AU

bistriatus (Wehncke, 1880) AU
fortescueensis (Watts & McRae, 2010) AU
fuscatus (Clark, 1862) AU
gibbus (Clark, 1862) AU
hydei (Vondel, 1995) AU
oberthuri (Guignot, 1935) AU
pinderi (Watts & McRae, 2010) AU
storeyi (Vondel, 1995) AU

Phylogeny

The phylogenetic relationships of Haliplidae with other groups of Adephaga was studied several times (Beutel & Roughley, 1988; Beutel, 1997; Beutel et al., 2008; Lawrence et al., 2011; Baca et al. 2017 and others). The Haliplidae (all or not together with Amphizoidae, Aspyditidae, Dytiscidae, Meruidae, Noteridae, Trachypachidae) form a sistergroup of Gyrinidae. The phylogeny of the genera, subgenera and species of Haliplidae was studied by Beutel & Ruhnau (1990) and Vondel (2019) (Fig. 8).

9

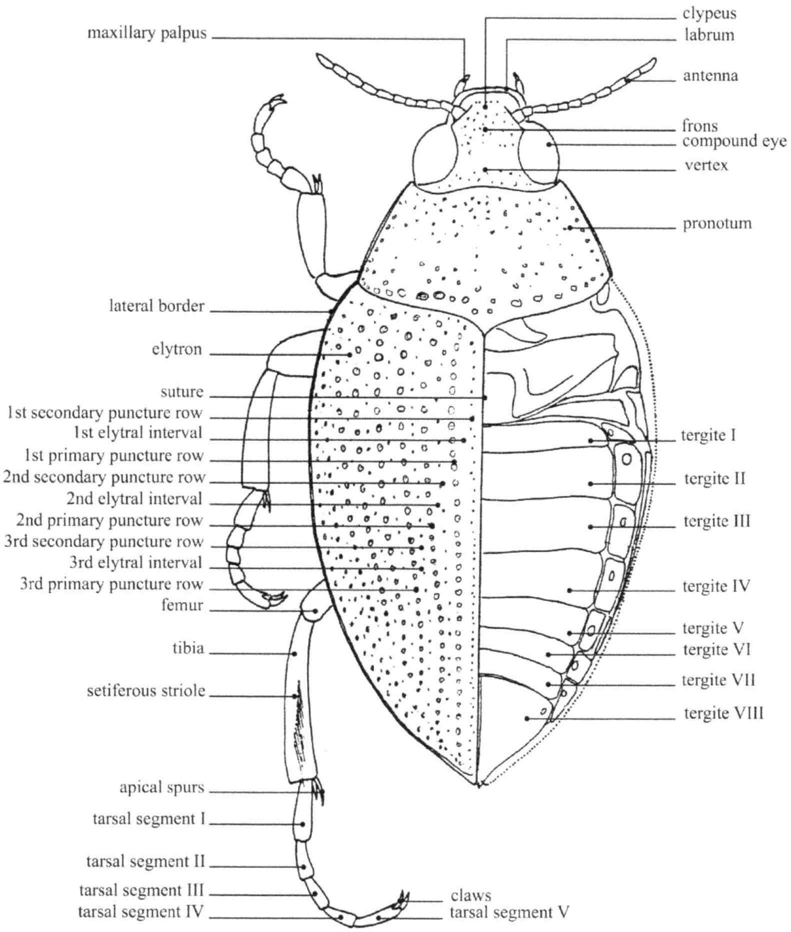

FIGURE 9 *Haliplus flavicollis*, dorsal side, right wing and elytron removed

DNA Barcoding

For a number of Haliplidae species genetic data are available in BOLDSYSTEMS, Genbank (part of International Nucleotide Sequence Database Collaboration) and other (regional) data banks. Researchers can explore these data centers for specific information.

 HTTPS://DOI.ORG/10.1163/9789004759671-012

Morphology of the Adults

The morphology of adult Haliplidae is recently treated by Holmen (1987) and Vondel (1997, 2016b). The morphological term are explained in Figs 9–24.

Body: Usually torpedo-like, widely or narrowly oval, sometimes parallel-sided. Sides of pronotum and elytra usually continuously curved, sometimes both sides making a distinct angle to each other. Yellow, brown or rust-coloured, with or without distinct or vague maculation. Body-length 1.5–5.0 mm.

Head: Yellow to yellow-brown, yellow-red, brown or even almost black. Eyes large, weakly to more strongly protruding, in lateral view oval, not incised. Labrum transverse and short, anteriorly concave, straight or convex and with a fringe of hairs, punctured or nearly unpunctured or partly microreticulated dorsally. Clypeus hardly or not longer than labrum, usually slightly wider than labrum, anteriorly straight or concave and densely punctured. Clypeus fixed posteriorly to frons without any distinct suture. Frons and vertex fused without any visible suture. A fold above each antennal base. Frons and vertex weakly and sparsely to strongly and densely punctured, punctures on vertex usually stronger and darkened. Ventral side of head, behind the mouthparts, composed of a gula medially and the genae laterally. Sides of head behind the eyes with one or two vertical furrows (genal lines), some Australian species lack those genal lines. Antennae moderately long, 11-segmented and filiform, basal two segments wider and usually darker than the following segments, distal segments usually longer and wider than proximal segments; a case of teratology is described by Vondel (1986a). Labium with a large mentum, which partly surrounds a small ligula. Mentum with a rounded, teeth-like or in two teeth devided extension in the middle. Labial palps composed of three segments, the penultimate of which often bears a tooth on the innerside, and a distinct palpiger attached to the ligula. Maxilla with a curved lacinia having long spines along the inner edge, a long two-segmented galea and a four-segmented palpus on a palpiger. Mandibles slightly asymmetrical, stout and strongly curved, ending in a distal tooth, a retinaculum and often some additional teeth along the inner edge.

Pronotum: Yellow to yellow-brown or yellow-red, without or with different forms of maculation. Well developed, sides usually converging anteriorly, sometimes parallel in posterior part. Posterior margin curved, backwardly projecting in the middle, covering the scutellum, posterior corners usually not produced. Anterior pronotal margin usually curved, slightly protruding in the middle and in the anterior corners. Pronotal sides usually with a distinct lateral border. Pronotum often with two basal plicae which may enclose a transverse

 | HTTPS://DOI.ORG/10.1163/9789004759671-013

10

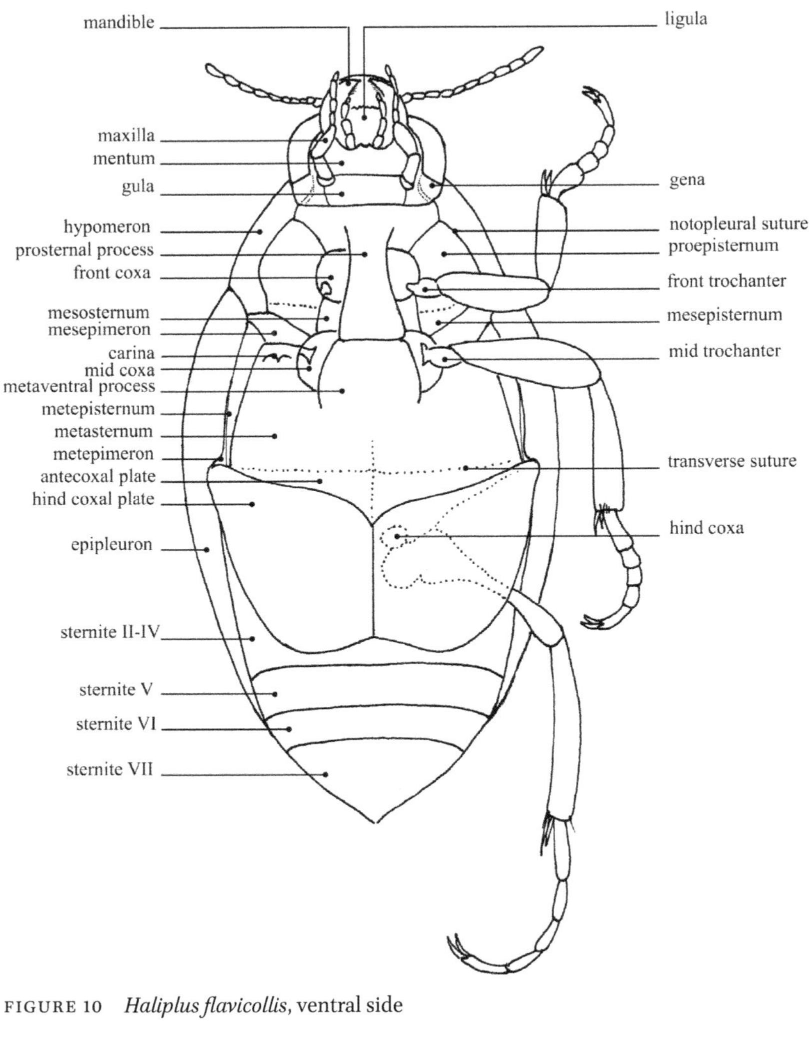

FIGURE 10 *Haliplus flavicollis*, ventral side

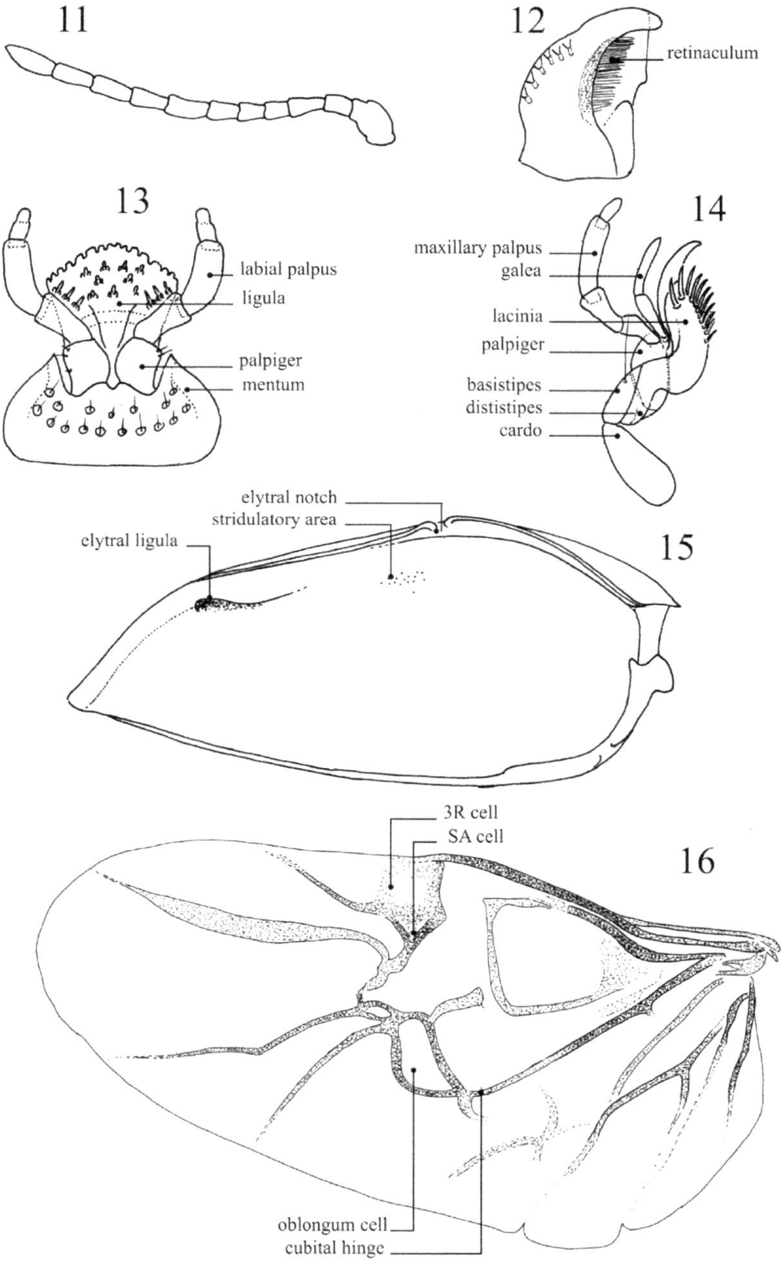

FIGURES 11–16 *Haliplus spec.* 11–15, H. *flavicollis*: 11, antenna; 12, mandible; 13, labium; 14, maxilla; 15, elytron in ventral view. 16, *Haliplus sibiricus*, wing

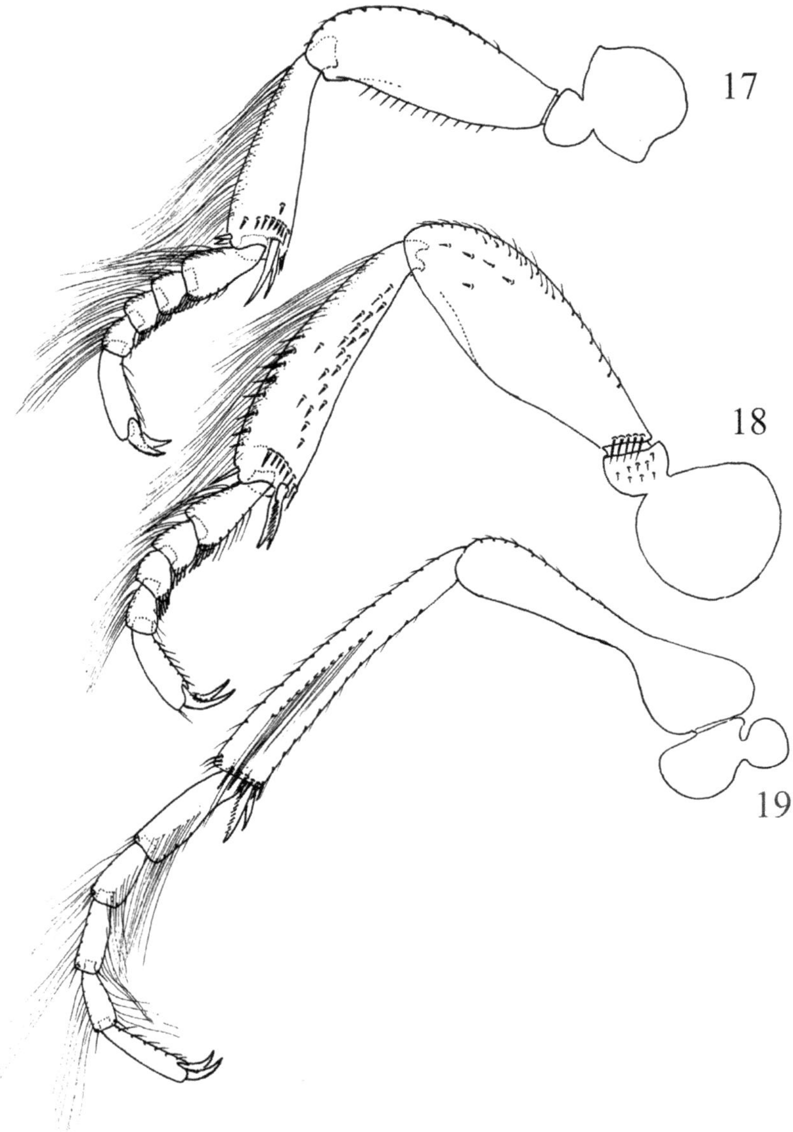

FIGURES 17–19 *Haliplus flavicollis*. 17, foreleg; 18, midleg; 19, hindleg

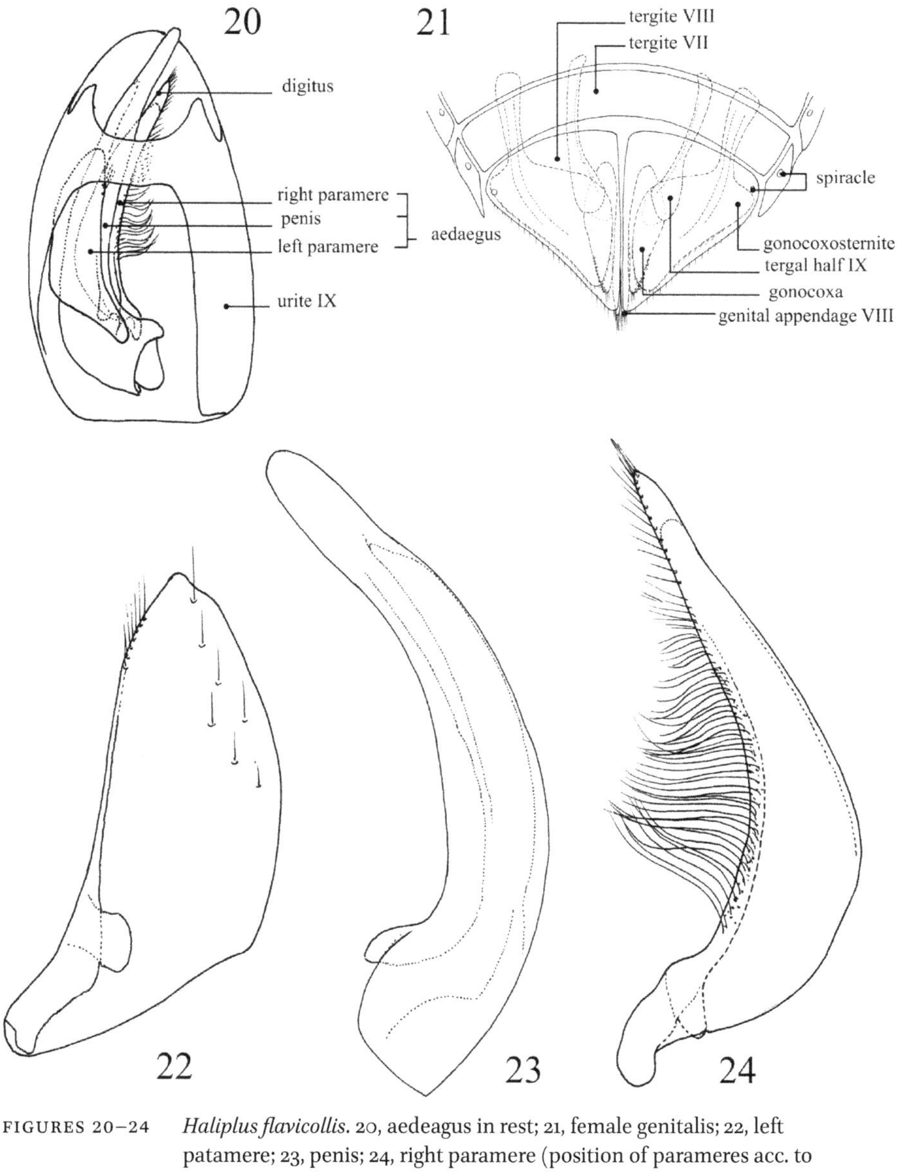

FIGURES 20–24 *Haliplus flavicollis*. 20, aedeagus in rest; 21, female genitalis; 22, left patamere; 23, penis; 24, right paramere (position of parameres acc. to Vondel, 1997)

impression. Surface more or less densely punctured, often strong widened punctures along the base. Prosternum transverse, laterally reaching the hypomera in anterior part, strongly elevated in the middle, provided with a large apophysis (or prosternal process), which is truncate posteriorly or parallel or widened anteriorly in some non-European species, flat or single or double grooved ventrally, and reaches the metaventral apophysis (or metasternal process). The proepisterna are visible from below, reaching the procoxal cavities. Notopreural suture present. Mesothorax small, its dorsal surface is completely covered by the elytra and the pronotum. Mesosternum small, partly covered by the prosternal apophysis. Mesepimera and mesepisterna visible from below. Metathorax large, dorsally covered by the elytra. Metasternum large, transverse, with a row of strong punctures along the often hardly visible suture between the antecoxal plate and the anterior part of the metasternum. Metasternum separated laterally from the epipleurae by the metepisterna and metepimera, in the median part produced anteriorly as the metaventral apophysis.

Elytra: Yellow to yellow-brown or yellow-red without or with a reduced or very extensive maculation. Well developed and covering the entire abdomen. At rest meeting along the entire elytral suture. Each elytron with 10 or more rows of primary punctures or in some non-European species punctures very irregular and not in distinct rows. Along the suture and on the intervals between primary puncture-rows rows of secondary punctures, which are usually weaker but sometimes as strong as primary punctures. Punctures often with a not always clearly visible hair and often darkened. In some species it might be possible that the primary punctures are connected with glands secreting poisonous or distasteful substance (Young, 1961). In the females and sometimes also in the males of many species micropunctuation is present on the entire or parts of the surface. Elytra with a lateral border throughout their length, which may be serrate anteriorly and/or posteriorly in some species.

Elytra bent down- and inwards at the sides, continuing ventrally as the elytral epipleura, which does not reach the apical point. Each elytron has a characteristic notch to receive the hind coxae, connecting the air-storage beneath the hind coxal plates and that beneath the elytra. Ventral side of each elytron with a lateral ligula, this extension fits into a depression of the dorsal side of the abdomen, fixing the elytron when the insect is not flying. A few species are known to produce sounds by stridulation by rubbing the stridulatory apparatus on the ventral side of the elytra along the edge of the abdomen (Beier, 1929).

Hindwings: Membranous, hyaline and well-developed, but absent in the subgenera *Algophilus* and *Apteraliplus*. According to the nomenclature of Ward (1979) the venation is of an Adephagan type in which the M4 originates anterior to the middle of the Oblongum Cell. A few other characteristics of the haliplid wing should be mentioned: Oblongum Cell broadly oval or in shape of a parallelogram, M4 originates close to M3, the 3R Cell larger than the SA Cell, a wedge-shaped cell between 1A1 and 1A2 absent, subcubital binding patch of spicules absent. Wing-folding also of an Adephagan type, with the cubital hinge not very far removed from the Oblongum Cell. The folding itself takes place by the aid of tergal binding patches.

Legs: Yellow to yellow-brown or yellow-red, usually darkened towards the coxae or sometimes completely darkened, often covered with a distinct micro-punctuation. Fairly long and slender. Fore- and midlegs with coxae as a ball joint, small globular trochanters, rather thick femora with variable rows of spines and setae, tibiae which are distally widened and provided with two strong spurs and a row of spines and usually a row of long swimming hairs, five tarsal segments of which the first one is usually longer and the last one much longer than the other segments, two claws of equal or different shape. Both fore- and midlegs are freely movable. Hindlegs with small coxal ball joints, larger globular trochanters, femora which are strongly narrowed in the middle and provided with only few setae and in distal part a short row of setae, tibia which are provided with rows of seate or swmimming hairs and in some groups with a setiferous striole (a row of closely set punctures) with swimming hairs on the dorsal side and two long spurs and a row of spines on distal end, five tarsal segment of which the first one and the last one are longer than the other segments, two claws. The coxal cavity and most of the femur of the hindleg is covered by a large plate (metacoxal plate or lobe). Although the hindlegs seem to be limited in their movement by the metacoxal plates, they can still be moved in various directions. In cross-section the parts of the legs are sub-cylindrical. The space enclosed by the metacoxal plates serves as an airstorage and is connected with the space enclosed by the elytra through the anterolateral corners of the coxae. A case of teratology is described by Vondel (1986a).

Legs are often occupied by Protozoa like Ciliata as is less often the case on other bodyparts.

Abdomen: Abdomen dorsally with 8 visible, partly membranous tergites (segments I–VIII), and with a number of more or less sclerotized pleurites. Laterally with 7 laterosternites. In the laterosternites and in tergite VIII the spiracles of the tracheal system are situated. Tergites and laterosternites are totally covered by the elytra. Ventrally 6 sternites are exposed, sternite VII sometimes with a median furrow or ridge, pointed apically. Sternites II–IV fused without any visible suture and immovably fixed to hind coxae along their anterior margin. Tergite IX, gonocoxosternites and sternites VIII–IX invaginated, more or less reduced, or modified to function as parts of the reprocuctive system.

Surface sculpture: In most species body allmost completely without micro-reticulation, except parts of the coxae, margins of the sternites, often on the legs and sometimes the apical margin of the labrum. Punctures of various size are present on most parts of the body (visible at 10×). Micro-punctuation (visible at 30× or more) is in many species present on the elytra, but often restricted to the females. The dorsal surface appears glabrous, but especial in the apical and lateral parts rather long to very short setae may be present. A few species has the body nearly completely covered with a dense and relatively strong micropunctuation.

Female genitalia: Externally comprised by the invaginated segments VIII and IX. Ventral parts of abdominal segment VIII give rise to the gonocoxosternites, formed by a pair of triangular sclerites, each with an anteriorly directed strut. Sclerotized parts of tergite IX present as two anteriorly directed struts ("valvifers"), attached to the paired gonocoxae ("genital valves"), these are flattened and usually triangular. Gonocoxosternites, parts of tergite IX, gonocoxae and the genital appendages VIII ("vulvosclerites") comprise the sclerotized parts of the ovipositor. Further details are given by Burmeister (1976, 1980).

Male genitalia: Externally comprised by the invaginated segment IX and perhaps non-sclerotized remnants of post-genital segments. Urite IX (genital plate or gonosomite) probably consists of sternite IX and tergite IX which are fused into a sheath partly enclosing the aedaegus at rest. The aedaegus consists of two lateral parts, the parameres and a median part, the penis; parameres proximally movably attached to the penis. Aedaegus not bilaterally symmetrical. Left paramere usually short, with sparse or more dense setation, in some species with a digitus, a solid distal extension. Right paramere usually longer than the left one, in most species with fringes of long hairs and in some species a digitus. Penis asymmetrical and laterally compressed with a tongue-shaped or pointed, sometimes strongly twisted, apex; sperm-groove, in which a tongue-like lobe, on the convex dorsal side.

When at rest in the abdomen, the aedaegus lies almost horizontally on its right side. During copulation the organ is turned downwards.

Identification Keys to the Worlds Haliplidae

For practical reasons identifications keys are given for each genus or subgenus and divided by Faunal region. This division is illustrated in Table 1.

TABLE 1 Review of the worlds Haliplidae, the faunal regions where they occur. The numbers correspond with separate identification keys. When a specimen has to be identified, first decide to which genus or subgenus it belongs, from which faunal region it originates and then chose the identification key to the species. Sometimes more genera or subgenera are combined in one key.

		World	Palae-arctic	Oriental	Austral asian	Afro-tropical	Ne-arctic	Neo-tropical
Brychius		1	2				12	
Haliplus	*Algophilus*	1				1		
	Apteraliplus	1					1	
	Australiplus	1			9			
	Borealiplus	1					14	17
	Haliaplus	1	3					
	Haliplidius	1	4					17
	Haliplus s.str	1	5	5			13	
	Liaphlus	1	3	8	9	10	14	17
	Neohaliplus	1	6			6		
	Nipponiplus	1	5					
	Paraliaphlus	1					15	17
Peltodytes	*Neopeltodytes*	1					16	16
	Peltodytes s.str	1	7	7		11		
Phalilus		1			9			

1 Identification Key to the Genera and Subgenera of Haliplidae

1 Last segment of palpi as long or longer than penultimate segment (Pl. 212: 4, 5). Elytral suture with fine line on posterior half (Pl. 212:1). Metacoxal lobes reaching to sixth abdominal sternite, usually margined laterally (Pl. 212:1). Head with one genal line behind the eyes (Pl. 219:3) .... *Peltodytes*>2

HTTPS://DOI.ORG/10.1163/9789004759671-014

– Last segment of palpi shorter than penultimate segment (Pl.1: 4, 5). Elytral suture without fine line (Pl. 1:1). Metacoxal plates not reaching sixth abdominal sternite, not margined laterally (Fig. 10) 3
2 Metaventral process with a central impression (Pl.212:6) .... subgenus *Peltodytes* s. str.
– Metaventral process without a central impression (Pl. 192:6)...................... .... subgenus *Neopeltodytes*
3 Epipleura broad, extending almost to the tip of elytra (Pl. 1:8) ... *Brychius*
– Epipleura usually ending before base of last abdominal segment (Fig. 10) 4
4 Head without genal lines behind the eyes (Pl. 231:3). Males with two first tarsal segments of fore- and midlegs widened and ventrally with adhesive setae (Pl. 231:5) (only Australasian) .... *Phalilus*
– Head with two genal lines behind the eyes (Pl. 171:4). Males with three first tarsal segments of fore- and midlegs widened and ventrally with adhesive setae (Pl. 26:8, 9) .... *Haliplus*>5
5 Dorsally and ventrally with strong micropunctation clearly visible at 30× magnification between the small primary punctures. Elytral epipleura without strong punctures .... *Haliplus* subgenus *Haliplidius*
– Elytra smooth or with a less stronger micropuncturation (visible at 50×) between the strong primary puncture-rows. Epipleura with coarse punctures 6
6 Pronotal base with plicae (Pl. 18:1) 7
– Pronotal base without plicae (Pl. 45:1) 10
7 Pronotal plicae long and curved (Pl. 171:1) *Haliplus* subgenus *Neohaliplus*
– Pronotal plicae about straigt and not more than the length of the pronotum 8
8 Prosternal process rounded posteriorly (Pl. 7:5) .... *Haliplus* subgenus *Apteraliplus* one species *Haliplus* (*Apteraliplus*) *parvulus*
– Prosternal process posteriorly about straight (Pl. 17:7) 9
9 In males left paramere rectangular (Pl. 173:7) .... *Haliplus* subgenus *Nipponiplus*
– In males left paramere triangular (Pl. 18:9) .... *Haliplus* s. str.
10 Pronotum narrowed in posterior half (Pl. 6:1) .... *Haliplus* subgenus *Algophilus* .... one species *Haliplus* (*Algophilus*) *lathridioides*
– Pronotum not narrowed in posterior half (Pl. 11:1) 11
11 Penultimate segment of labial triangular (Pl. 9:5)... .... *Haliplus* subgenus *Borealiplus*

– Penultimate segment of labial pali about parallel-sided (Fig. 13) 12

12 Setiferous striole on dorsal face of hind tibia long, running to about the apex of the tibia (Pl. 11:8) .... *Haliplus* subgenus *Haliaplus*, only one species *Haliplus* (*Haliplus*) *laminatus*

– Setiferous striole on dorsal face of hind tibia at most over half of the tibia, ending far from the apex of the tibia (Pl. 86:4) 13

13 Lateral margin of pronotum serrate (Pl. 8:1) .... *Haliplus* subgenus *Australiplus*, ... only one species *Haliplus* (*Australiplus*) *sindus*

– Lateral margin of pronotum not serrate (Pl. 86:1) 14

14 Metaventral process laterally with sharp plica from anterior to posterior end (Pl. 191:5) .... *Haliplus* subgenus *Paraliaphlus*

– Metaventral process flat or with various impressions, but not laterally with complete sharp plicae (Pl. 85:5) .... *Haliplus* subgenus *Liaphlus*

2 Key to the Palaearctic Species of *Brychius*

1 Elytra with longitudinal ridges (Pl. 1:1) .... *Brychius elevatus*

– Elytra without longitudinal ridges (Pl. 2:1) .... *Brychius glabratus*

3 Key to the Palaearctic Species of *Haliplus* Subgenera *Haliaplus* and *Liaphlus*

1 Distance between eyes about eye-width (Pl. 139:1) 2

– Distance between eyes at least 1.2 × eyewidth (Pl. 45:1) 4

2 Prosternal process furrowed on both sides (Pl. 139:5) .... *Haliplus* (*Liaphlus*) *pulchellus*

– Prosternal process strongly impressed in middle, at least in apical half (Pl. 113:5) 3

3 Elytra with basal black band, eighth puncture-row with about nine small dark blotches (Pl. 113:1) .... *Haliplus* (*Liaphlus*) *maculipennis*

– Elytra without basal black band, eighth puncture-row not with dense row of dark blotches (Pl. 106:1) .... *Haliplus* (*Liaphlus*) *kotoshonis*

4 Head relatively wide, distance between eyes at least 2.3 × eye-width Pl. 121:1). Metasternal process flat in middle (Pl. 121:5). Elytra without dark blotches .... *Haliplus* (*Liaphlus*) *mucronatus*

– Head narrower. Metasternal process at least with small pit or groove in middle (Pl. 49:5) 5
5 Sides of prosternal process before coxae extending forwards as complete plicae to anterior edge of prosternum (Pl. 86:5) 8
– Sides of prosternal process before coxae not completely extending forwards as plicae to anterior edge of prosternum (Pl. 85:5) 6
6 Plicae on sides of prosternal process widely interrupted before coxae (Pl. 49:5). Longer apical spur of hind tibia usually as long as or longer than first tarsal segment (Pl. 49:4). Elytra mostly with darker blotches (Pl. 49:1) .... *Haliplus* (*Liaphlus*) *andalusicus*
– Plicae on sides of prosternal process completely absent before coxae (Pl. 85:5). Elytra without clear blotches. Apical spurs of hind tibia not as long as first tarsal segment (Pl. 85:4) 7
7 Punctures of primary elytral rows wide in basal area, distance between rows not more than diameter of punctures (Pl. 141:1). Distance between eyes 1.7–1.9 × eye-width .... *Haliplus* (*Liaphlus*) *rubidus*
– Punctures of primary elytral rows rather small, distance between rows more than diameter of punctures (Pl. 85:1). Distance between eyes 1.4–1.6 × eye-width .... *Haliplus* (*Liaphlus*) *flavicollis*
8 Setiferous striole on dorsal face of hind tibia stretched over apical half (Pl. 11:8). Elytra usually with more or less confluent and interrupted dark lines following rows of dark punctures, but distinct blotches usually absent. Basal elytral punctures coalescing to form short grooves (Pl. 11:1.) Male: first tarsal segment of midlegs strongly dilated ventrally (Pl. 11:5). Female: elytra almost completely covered with micro-punctures..................... *Haliplus* (*Haliaplus*) *laminatus*
– Setiferous striole on dorsal face of hind tibia usually not longer than 1/3 × tibia length, not reaching apical point. Male: first 3 tarsal segments of midlegs weakly dilated ventrally. Female: elytra without clear micropunctuation 9
9 Elytra without dark blotches. Pronotal and elytral punctures black (Pl. 80:1) .... *Haliplus* (*Liaphlus*) *eximius*
– Elytra with dark blotches, or when these blotches are strongly reduced or absent punctures on disc of pronotum are not black 10
10 Elytra with basal transverse black band as dark as sutural maculation (Pl. 144:1) 25
– Elytra without basal black band. Sometimes some transverse darkening separated from basal margin or a brown basal band, clearly less dark than sutural maculation 11

11 Widened basal punctures on pronotum twice the width of elytral primary punctures. Body-shape parallel or subparallel. Prosternal process narrow, length 2.6 × width, slightly narrowed near coxae. Elytra with mostly vague blotches, one large blotch on middle of suture and three smaller ones on each elytron (Pl. 95:1) .... *Haliplus* (*Liaphlus*) *guttatus*
– Basal punctures on pronotum usually not twice as wide as elytral primary punctures. Body not clearly parallel 12
12 Length 3.6–4.5 mm 21
– Length 2.6–3.6 mm 13
13 Elytra with longitudinal distinct blotches in even intervals, blotches not connected to darkened suture (also check link via 17) 14
– Elytral blotches different 17
14 Metasternal process flat anteriorly (Pl. 54:5). Vertex not darkened (Pl. 54:1) 15
– Metasternal process grooved anteriorly (Pl. 73:5). Vertex darkened (Pl. 73:1) 16
15 Setiferous striole weak, about 1/6 × tibia length (Pl. 54:4). First elytral primary puncture-row with about 28 punctures. Pronotum anteriorly partly bordered (Pl. 54:1). Males: left paramere with or without solid digitus (Pl. 54: 7, 8) .... *Haliplus* (*Liaphlus*) *astrakhanus*
– Setiferous striole strong, about 1/3 × tibialength (Pl. 11:4). First elytral primary puncture-row usually with more than 28 punctures. Pronotum anteriorly not bordered (Pl. 111:1). Males: left paramere without solid digitus (Pl. 111:7) .... specimens of *Haliplus* (*Liaphlus*) *maculatus*
16 Distance between eyes less than 1.7 × eyewidth. Males: top of penis short (Pl. 73:8) .... specimens of *Haliplus* (*Liaphlus*) *dalmatinus*
– Distance between eyes more than 1.7 × eyewidth. Males: top of penis long, usually slightly widened apically (Pl. 165:8) .... specimens of *Haliplus* (*Liaphlus*) *variegatus*
17 Elytra with large maculation, discal blotch and blotch in apical part connected to darkened suture (Pl. 106:1) 18
– Elytra with usually vague maculation on first and even intervals, or only darkening on discal parts of puncture-rows, blotches on apical part not connected to suture (Pl. 45:1) .... *Haliplus* (*Liaphlus*) *abbreviatus*
18 Length of prosternal process 2.3 × width (Pl. 50:5) .... *Haliplus* (*Liaphlus*) *angusi*
– Length of prosternal process less than 2.3 × width (Pl. 106:5) 19
19 Distance between eyes more than 1.7 × eyewidth (Pl. 165:1) .... specimens of *Haliplus* (*Liaphlus*) *variegatus*

– Distance between eyes less than 1.7 × eyewidth (Pl. 106:1) .... specimens of *Haliplus* (*Liaphlus*) *kotoshonis*

20 Distance between eyes less than 1.4 × eyewidth 21

– Distance between eyes more than 1.4 × eyewidth 22

21 Black suture not reaching first secondary puncture-row in basal part. Body strongly tapering apically (Pl. 79:1). Penis hump-backed before top (Pl. 79:10) .... *Haliplus* (*Liaphlus*) *excoffieri*

– Black suture reaching first secondary puncture-row in basal part. Body more oval (Pl. 62:1). Penis gradually curved to top (Pl. 62:8) .... *Haliplus* (*Liaphlus*) *chinensis*

22 Elytral discal blotch connected to suture (Pl. 132:1) .... *Haliplus* (*Liaphlus*) *ovalis*

– Elytral discal blotch not connected to suture 23

23 Length of hind tarsus including claws more than 1.3 × tibia length. Prosternal process slightly narrowed near coxae, narrowed part 0.6 to 0.7 × widest part (Pl. 73:5). Punctures of setiferous striole not connected in a groove (Pl. 73:4). Male: digitus of right paramere 1/5 × length of basal part (Pl. 73:9) .... large specimens of *Haliplus* (*Liaphlus*) *dalmatinus*

– Length of hind tarsus less than 1.3 × tibia length. Prosternal process strongly narrowed near coxae, narrowed part 1/2 × width of widest part 24

24 Distance between eyes less than 1.6 × eyewidth. Setiferous striole on dorsal side of hind tibia not strongly grooved, separate points usually clearly visible. Male: apical part of penis very narrow and pointed (Pl. 111:8)......... .... *Haliplus* (*Liaphlus*) *maculatus*

– Distance between eyes more than 1.6 × eyewidth. Setiferous striole on dorsal side of hind tibia strongly grooved, separate points not or hardly visible (Pl. 86:4). Male: apical part of penis wide and rounded at the top (Pl. 86:8) .... *Haliplus* (*Liaphlus*) *fulvus*

25 Length 3.5–4.1 mm. Black sutural stripe in anterior half at most reaching first secondary puncture-row (Pl. 57:1) .... *Haliplus* (*Liaphlus*) *basinotatus*

– Length at most 3.6 mm. Black sutural stripe reaching first primary puncture-row (Pl. 144:1) .... *Haliplus* (*Liaphlus*) *sharpi*

4 Key to the Palaearctic Species of *Haliplus* Subgenus *Haliplidius*

1 Basal pronotal plicae distinct (Pl. 12:1) 2

– Basal pronotal plicae weak or absent (Pl. 13:1) 3

2 Sides of prosternal apophysis distinctly beaded (Pl. 12:6). Body broad. Black coloration of elytral striae distinctly interrupted (Pl. 12:1). Median lobe (penis) with subapical basal-directed denticle (Pl. 12:9).......................

.... *Haliplus* (*Haliplidius*) *confinis*
– Sides of prosternal apophysis not beaded (Pl. 14:5). Body narrow, elongate. Black coloration of elytral striae almost continuous, very distinct (Pl. 14:1). Median lobe (penis) not with subapical basal-directed denticles (Pl. 14:8) .... *Haliplus* (*Haliplidius*) *rejseki*
3 Anterior and posterior margin of pronotum with a distinct broad black band (Pl. 16:1). Apex of femora black (Pl. 16:8) ....*Haliplus* (*Haliplidius*) *varius*
– Posterior margin of pronotum narrowly blackened, anterior margin at most slightly darkened (Pl. 13:1). Femur without black apex. Pronotum without basal plicae .... *Haliplus* (*Haliplidius*) *obliquus*

5 Key to the Palaearctic Species of *Haliplus* Subgenera *Haliplus* s. str. and *Nipponiplus*

1 Prosternal process with transverse rim basally (Pl. 38:8) 2
– Prosternal process without transverse rim basally 5
2 Male protarsal claws wider, of different width (Pl. 38:9). Female elytra strongly punctured. .... *Haliplus* (*Haliplus*) *samojedorum* (part)
– Male protarsal claws of equal width. Female elytra without micropunctation 3
3 Prosternal process with distinct ridge medially (Pl. 32:6). Elytral dark pattern strongly developed (Pl. 32:1) .... *Haliplus* (*Haliplus*) *kamiyai*
– Prosternal process somewhat elevated medially, but without a distinct ridge. Elytral dark pattern more reduced 4
4 Metasternal process almost flat, with at most a weak depression on each side (Pl. 173:4) .... *Haliplus* (*Nipponiplus*) *japonicus*
– Metasternal process with a deep depression on each side (Pl. 175:4) .... *Haliplus* (*Nipponiplus*) *regimbarti*
5 Metasternal process with an impression on each side (Pl. 174:4) 6
– Metasternal process usually with medial impression about or flat 14
6 Pronotum without basal plicae (Pl. 174:1) 7
– Pronotum with (sometimes very short) basal plicae (Pl. 18:1) 8
7 Head darkened (Pl. 36:1). In males left paramere triangular (Pl. 36:7) .... *Haliplus* (*Haliplus*) *morii*
– Head not darkened (Pl. 174:1). In males left paramere rectangular (Pl. 174:7) .... *Haliplus* (*Nipponiplus*) *latreillei*

8 Prosternal process very narrow between front coxae, not clearly furrowed along the sides or basally (Pl. 24:6). Male with apex of penis very narrowly rounded (Pl. 24:10). Female elytra never micropunctured .... *Haliplus* (*Haliplus*) *fulvicollis*

– Prosternal process wider between front coxae, at least weakly furrowed basally and often also along the sides. Apex of penis obliquely truncate. Female elytra sometimes micropunctured 9

9 Shape of body narrow, sometimes with sides parallel in the middle 10

– Shape of body rather wide, not parallel sided. Penis as in Pl. 33:10 (females not found yet) .... *Haliplus* (*Haliplus*) *kirgisiensis*

10 Elytral dark lines more or less confluent into a common elytral spot behind the middle, connected to the suture, otherwise maculation hardly developed (Pl. 38:1). Prosternal process less conctricted between front coxae (Pl. 38:6). Penis as in Pl. 38:11. Female elytra completely and strongly micropunctured .... *Haliplus* (*Haliplus*) *samojedorum*(part)

– Elytral dark pattern different. Prosternal process more constricted between front coxae. Penis different. Female elytra with strong or weak or absent micropunctation 11

11 Elytra with strongly interrupted dark lines, often strongly maculated (Pl. 25:1). Penis as in Pl. 25:10. Female elytra at most weakly micropunctured. .... *Haliplus* (*Haliplus*) *furcatus*

– Elytra with more or less interrupted or continuous dark lines without maculation 12

12 Elytral dark lines continuous and strong (Pl. 18:1). Female elytra strongly micropunctured .... *Haliplus* (*Haliplus*) *apicalis*

– Elytral dark lines not continuous. Elytra not strongly micropunctured 13

13 Primary puncture rows not or vaguely darkened (Pl. 17:1). Metasternal process with strong impression on each side (Pl. 17:7). Female elytra without micropunctures .... *Haliplus* (*Haliplus*) *aliae*

– Elytral puncture rows with interrupted although not strong lines (Pl. 42:1). Metasternal process with weak impression on each side (Pl. 42:4). Female elytra completely but weakly micropunctured .... *Haliplus* (*Haliplus*) *steppensis*

14 First three tarsal segments of fore– and midlegs widened and ventrally provided with closely set sucker–hairs: males: 15

– First three tarsal segments of fore– and midlegs not widened, ventrally at most with few hairs, in most cases elytra with micropunctation: females: 29

15 First tarsal segment of midlegs strongly excised (Pl. 34:9) .... *Haliplus* (*Haliplus*) *lineolatus*

– First tarsal segment of midlegs not strongly excised, at most slightly concave 16
16 Pronotum without basal plicae or at most with one or two punctures suggesting a plica (Pl. 44:1) 17
– Pronotum with clear basal plicae 19
17 Elytral puncture rows with dark lines 18
– Elytra with vague dark marks, or with discal mark behind the middle, dark lines hardly developed, harminae. (Pl. 27:1) .... *Haliplus* (*Haliplus*) *harminae*
18 Head darkened (Pl. 36:1) .... *Haliplus* (*Haliplus*) *morii*
– Head not darkened (Pl. 44:1) .... *Haliplus* (*Haliplus*) *zacharenkoi*
19 Body oblong, sides paralel. Usually one central mark behind the middle connected to the suture (Pl. 38:1) *Haliplus* (*Haliplus*) *samojedorum* (part)
– Body wide, sides not parallel, elytra usually without vague marks 20
20 First tarsal segment of fore-legs ventrally dilated (Pl. 29:8) 21
– First tarsal segments of fore-legs ventrally simple, not specifically modified 23
21 First tarsal segment of fore-legs ventrally with a longitudinal row of short spines (Pl. 31:8), black elytral stripes widely interrupted (Pl. 31:1) .... *Haliplus* (*Haliplus*) *interjectus*
– First tarsal segment of fore-legs ventrally with a sharp ridge, black elytral stripes hardly interrupted.(Pl. 26:8, 29:8) 22
22 Claws of forelegs almost equal in length (Pl. 26:8). Antennae darkened apically (Pl. 26:2) .... *Haliplus* (*Haliplus*) *fuscicornis*
– Claws of forelegs clearly unequal in length and curvature (Pl. 29:8) .... *Haliplus* (*Haliplus*) *immaculatus*
23 Claws of fore–legs clearly unequal in length and shape 24
– Claws of fore–legs equal or subequal in length and shape 27
24 Body rather wide. Pronotal plicae short and usually curved 25
– Body rather narrow. Pronotal plicae longer and usually straight. Elytral dark lines usually hardly interrupted in the apical half, not confluent, no spotted appearance (Pl. 39:1). Transverse row of punctures on abdominal sternite VI usually widely interrupted in the middle .... *Haliplus* (*Haliplus*) *sibiricus*
25 Sutural secondary punctures dense and in two irregular rows in basal part (Pl. 40:1). Aedaegus as in Pl. 40: 6–8 .... *Haliplus* (*Haliplus*) *simplex*
– Sutural secondary punctures sparse in basal part (Pl. 43:1) 26

26 Dark lines on elytral primary puncture rows interrupted, not confluent, no spotted appearance (Pl. 43:1). Transverse row of punctures on abdominal sternite VI very weak in the middle .... *Haliplus* (*Haliplus*) *turkmenicus*
– Dark lines on elytral primary puncture rows interrupted and partly confluent, giving the elytra a pronouncedly spotted appearance (Pl. 37:1). Transverse row of punctures on abdominal sternite VI not widely interrupted in the middle .... *Haliplus* (*Haliplus*) *ruficollis*
27 Pronotal plicae very short, at most consisting of one or two punctures suggesting a plica. Elytral dark lines more or less continuous in apical half (Pl. 44:1) .... *Haliplus* (*Haliplus*) *zacharenkoi*
– Pronotal plicae clearly developed 28
28 Prosternal process clearly grooved in the anterior part (Pl. 28:6). Body usually widest near the shoulders, strongly tapering backwards. Elytral dark lines widely interrupted and often connected by blotches of which the largest one is situated on the middle of the suture. Basal primary punctures on elytra groovelike connected (Pl. 28:1) .... *Haliplus* (*Haliplus*) *heydeni*
– Prosternal process usually flat or slightly convex, sometimes slightly grooved (Pl. 23: 6). Body slender, shoulders less pronounced, tapering backwards less strongly. Elytral dark lines not connected by blotches, but locally widened. Generally light yellow or yellow (Pl. 23:1) .... *Haliplus* (*Haliplus*) *fluviatilis*
29 Pronotum without plicae. Antennae not darkened in apical half 30
– Pronotum with plicae or antennae darkened in apical half 32
30 Elytra with with dark blotches, no clear dark lines (Pl. 27:1). Female elytra without micropunctures .... *Haliplus* (*Haliplus*) *harminae*
– Elytra with more or less interrupted dark lines, no dark blotches 31
31 Sutural row of secondary punctures sparse and regular. Dark lines on primary puncture rows only interrupted in basal third (Pl. 44:1) .... *Haliplus* (*Haliplus*) *zacharenkoi*
– Sutural row of secondary punctures irregular and more dense. Dark lines on primary puncture rows interrupted in apical half and often with connecting marks (Pl. 40:1) .... *Haliplus* (*Haliplus*) *simplex*
32 Elytra with micropunctation, sometimes restricted to the apex (if there are only few micropunctures in the apical area) 33
– Elytra without micropunctation 41
33 Elytral apex broadly rounded or truncate, lateral border wide posteriorly. Micropunctation usually also visible on the anterior half of the elytra. Elytral dark lines usually narrow, interrupted in the basal part, often

continuous in the apical half, no connecting blotches. Body rather wide, subparallel (Pl. 34:1) .... *Haliplus* (*Haliplus*) *lineolatus*
– Elytral apex more narrowly rounded or truncate, lateral border rather narrow posteriorly 34
34 Segment 5 of front tarsi usually with 3 ventral spines (Pl. 31:8), segment 5 of mid tarsi usually with 4 ventral spines, the spines are rather long and the proximal ones are more closely set than the distal ones. Body fairly short, with rather evenly rounded sides. Elytral dark lines usually strongly interrupted, giving the elytra a pale, spotted appearance (Pl. 31:1).............. .... *Haliplus* (*Haliplus*) *interjectus*
– Segment 5 of front and mid tarsi usually with more spines, these are rather evenly dispersed and usually shorter 35
35 Antennae darkened apically (Pl. 26:1) .... *Haliplus* (*Haliplus*) *fuscicornis*
– Antennae not clearly darkened apically 36
36 Metasternal process usually flat in anterior part. Prosternal process usually flat or slightly convex, but sometimes weakly grooved 37
– Metasternal process usually grooved in anterior part. Prosternal process usually grooved 38
37 Body oblong, shoulders not pronounced. Usually light yellow to yellow. Dark elytral lines not connected by blotches, but locally widened, usually causing a double V–pattern (Pl. 23:1) .... *Haliplus* (*Haliplus*) *fluviatilis*
– Body wide, not parallel. Dark elytral lines not confluent, not showing a clear double V–pattern. (Pl. 43:1) .... *Haliplus* (*Haliplus*) *turkmenicus*
38 First elytral puncture row with 30–35 punctures 39
– First elytral puncture row usually with 40 or more punctures 40
39 Sutural secondary punctures sparse in basal part. Elytral dark lines more or less continuous in apical half (Pl. 44:1) .... *Haliplus* (*Haliplus*) *zacharenkoi*
– Sutural secondary punctures dense and in two irregular rows in basal part. Elytral dark lines interrupted in basal and in apical part (Pl. 40:1)..... .... *Haliplus* (*Haliplus*) *simplex*
40 Body usually with pronounced shoulders, widest before the middle. Pronotal plicae short and usually curved (pl. 37:1). Prosternal process only weakly grooved in the anterior half (Pl. 37:6). Dark elytral lines often interrupted and confluent (Pl. 37:1). Transverse row of punctures on the sixth abdominal sternite not widely interrupted in the middle .... *Haliplus* (*Haliplus*) *ruficollis*
– Body widest in the middle. Pronotal plicae long, usually more than 1/3x the length of the pronotum, and straight. Dark elytral lines interrupted in the basal half, often almost continuous in the apical half, usually

not confluent (Pl. 39:1). Prosternal process usually grooved completely (Pl. 39:6). Transverse row of punctures on the sixth abdominal sternite widely interrupted in the middle. Basal punctures on elytra sometimes connected to form transverse grooves (Pl. 39:1) .... *Haliplus* (*Haliplus*) *sibiricus*

Dark elytral lines mostly continuous, not confluent. Shoulders not pronounced. Body subparallel (Pl. 29:1). Transverse row of punctures on the sixth abdominal sternite usually widely interrupted .... *Haliplus* (*Haliplus*) *immaculatus*

– Dark elytral lines often interrupted and confluent. Shoulders pronounced, body widest before the middle (Pl. 28:1). Transverse row of punctures on the sixth abdominal sternite not widely interrupted. First elytral primary puncture row usually with less than 40 punctures. Basal punctures on elytra connected groove–like (Pl. 28:1) .... *Haliplus* (*Haliplus*) *heydeni*

6 Key to the Species of *Haliplus* Subgenus *Neohaliplus*

1 Length more than 2.2mm. General appearance yellow-red with usually dark maculation on elytra and pronotum. Basal punctures on elytra confluent, forming grooves (Pl. 171:1). Prosternal process usually strongly impressed in the posterior part. Metasternal process with strongly punctured grooves on both sides (Pl. 171:7). Males with many widened bristles on top of the left paramere (Pl. 171:9) .... *Haliplus* (*Neohaliplus*) *lineatocollis*

– Length less than 2.2mm. General appearance pale yellow with weak maculation on the elytra. Basal punctures of elytra small, separate, not forming grooves (Pl. 172:1). Prosternal process narrow, not or hardly impressed in the middle. Metasternal process flat, without grooves or strongly impressed rows of confluent punctures (Pl. 172:6). Males without widened bristles on top of the left paramere (Pl. 172:8) .... *Haliplus* (*Neohaliplus*) *ruficeps*

7 Key to the Palaearctic and Oriental Species of *Peltodytes*

1 Margins of pronotum narrow, not thicker than antenna, hind corners hardly protruding. Elytra apically not incided (Pl. 212:1) 2

– Margins of pronotum broad, distinctly thicker than antenna, hind corners strongly protruding. Elytra apically incised (Pl. 224:1) 7

2 Hind margin of metacoxal plates with a weak blunt tooth (Pl. 212:9) 3

– Hind margin of metacoxal plates with a sharp long tooth (Pl. 217:6) 4

3 Fourth elytral puncture-row complete or at most slightly interrupted in the middle. Base of pronotum strongly impressed, in the impression a transverse row of strong darkened punctures. Not with basal marks on pronotum, larger than darkened punctures. Vertex at most with two vague marks (Pl. 212:1). Males: surface of apical parts of parameres hairy (Pl. 212:14) .... *Peltodytes* (*Peltodytes*) *caesus*

– Fourth elytral puncture-row only clearly present in basal part, represented by 3–5 punctures. Base of pronotum hardly impressed, on both sides of impression two joined strong punctures and about three small ones, connected by a dark mark, vertex with two small distinct marks (Pl. 222:1). Males: surface of apical parts of parameres areolate (Pl. 222:11) .... *Peltodytes* (*Peltodytes*) *sinensis*

4 Antennae slender, segments longer than wide (Pl. 215:1) 5

– Antennae short, segments 2–9 about as long as wide (Pl. 221:1). Males: surface of apical parts of parameres hairy (Pl. 221:13) .... *Peltodytes* (*Peltodytes*) *rotundatus*

5 Punctures on elytral epipleura black. Prosternal process clearly bordered in anterior half (Pl. 219:5). Males: surface of apical parts of parameres aeolate (Pl. 219:12) 6

– Punctures on elytral epipleura not black. Prosternal process usually not completely and clearly bordered in anterior half (Pl. 215:4). Surface of apical parts of parameres hairy (Pl. 215:11) .... *Peltodytes* (*Peltodytes*) *dauricus*

6 Pronotum without black punctures in posterior corners between margin and widened basal punctures (Pl. 219:1). Proepisternum without strong dark punctures, when very weak punctures present these are not darkened. Teeth of coxal plates usually long but with a blunt point, especially in lateral view (Pl. 219: 7, 8). Head with a black collar behind eyes (Pl. 219:3) .... *Peltodytes* (*Peltodytes*) *pekinensis*

– Pronotum with black punctures in posterior corners between margin and widened basal punctures (Pl. 217:1). Proepisternum with strong darkened punctures. Teeth of coxal plates long and sharp (Pl. 217:6). Head without a black collar directly behind eyes ...... *Peltodytes* (*Peltodytes*) *intermedius*

7 Elytra with large dark marks and usually with black band connecting basal punctures. Two clear sharp dark spots on vertex (Pl. 224:1) .... *Peltodytes* (*Peltodytes*) *sumatrensis*

– Elytra with at most a few small marks and base usually not black. Vertex with at most two vague spots (Pl. 214:1) ... *Peltodytes* (*Peltodytes*) *coomani*

8 Key to the Oriental Species of *Haliplus* Subgenus *Liaphlus*

1 Prosternal process with two longitudinal grooves (Pl. 139:5) 2
– Prosternal process flat or impressed in the middle (Pl. 86:5) 11
2 Elytra maculate 3
– Elytra only narrowly darkened along suture (Pl. 125:1)..................................
.... *Haliplus* (*Liaphlus*) *nedungaduensis*
3 Margins of elytra serrate in basal part (Pl.150:1) 4
– Margins of elytra not clearly serrate in basal part 7
4 Distance between eyes 0.9–1.0 × width of one eye (Pl. 139:1) ....
.... specimens of *Haliplus* (*Liaphlus*) *pulchellus*
– Distance between eyes 1.1–1.5 × width of one eye 5
5 Metasternal process not impressed in the middle (Pl. 150:5)........................
.... *Haliplus* (*Liaphlus*) *sulawesicus*
– Metasternal process impressed in the middle 6
6 Metasternal process with transverse impression, in which two pits can be recognised (Pl. 100:5) .... *Haliplus* (*Liaphlus*) *indicus*
– Metasternal process with two separated pits (Pl. 74:5) ....
.... *Haliplus* (*Liaphlus*) *diopus*
7 Metasternal process with one pit in the middle (Pl. 123:5) ....
.... *Haliplus* (*Liaphlus*) *napolovi*
– Metasternal process with two pits 8
8 Mesepimeron with about 4 strong punctures (Pl. 136:5). Length less than 2.6 mm .... *Haliplus* (*Liaphlus*) *philippinus*
– Mesepimeron without or rarely with one or two punctures. Length more than 2.6 mm 9
9 Pronotum usually not continuously margined, sometimes traces of a margin in posterior part, posterior corners usually clearly protruding. Basal black band on elytra reaching fifth primary puncture-row (Pl. 139:1)
.... specimens of *Haliplus* (*Liaphlus*) *pulchellus*
– Pronotum usually finely margined except near anterior corner, posterior corners not or hardly protruding. Basal dark band on elytra reaching at least sixth primary puncture row 10
10 Elytral basal band brown, reaching margins when seen from above, black marks very irregular. Pronotum slightly darker in the middle (Pl. 143:1). Length 3.3–3.5 mm .... *Haliplus* (*Liaphlus*) *samosirensis*
– Elytral basal band black, reaching sixth primary puncture-row, black marks well defined. Pronotum clearly darkened in the middle (Pl. 103:1). Length 2.8–3.0 mm .... *Haliplus* (*Liaphlus*) *javanicus*

11 Elytra without any markings .... *Haliplus* (*Liaphlus*) *eximius*
– Elytra with markings 12
12 Two longitudinal markings on disc of pronotum (Pl. 138:1) ....
.... *Haliplus* (*Liaphlus*) *pruthii*
– At most small markings on disc of pronotum 13
13 Elytral markings not connected with darkened suture (Pl. 105:1) 14
– Elytral markings connected with darkened suture (Pl. 106:1) 15
14 Pronotum with two small marks in the middle. Narrowly darkened along elytral suture, long dark marks on sixth interval, about 30 punctures in first primary puncture row (Pl. 134:1) .... *Haliplus* (*Liaphlus*) *perroti*
– Pronotum without marks. Elytral suture weakly darkened, about 23 punctures in first primary puncture row (Pl. 105:1) ... *Haliplus* (*Liaphlus*) *kapuri*
15 Prosternal process strongly grooved and with a deep pit in apical part (Pl. 167:1). Elytral base usually darkened 16
– Prosternal process flat or weakly impressed in apical part (Pl. 106:5). Elytral base usually not darkened (Pl. 106:1) ....
.... *Haliplus* (*Liaphlus*) *kotoshonis*
16 Mesepimeron with about three strong punctures near coxae. Fourth secondary elytral puncture row restricted to basal half, sixth row restricted to basal 1/3, eighth row absent .... *Haliplus* (*Liaphlus*) *volpei*
– Mesepimeron at most with a few smal punctures. Fourth, sixth and eighth secondary puncture row more or less continuous, although sometimes sparse 17
17 Distance between eyes 0.8–1.0 × width of one eye. Base of pronotum usually wider than elytral base, posterior corners rounded. Pronotum short, width at base 2.1–2.3 × length in the middle, strong and in basal corners coarse punctures. Margins thicker than elytral margin near shoulders. Basal dark band usually reaching to fifth puncture-row. Anterior edge of prosternum only margined near prosternal process. About 38 sutural punctures, about 32 punctures in first primary row (Pl. 51:1) ....
.... *Haliplus* (*Liaphlus*) *angustifrons*
– Distance between eyes 1.0–1.1 × width of one eye. Width of pronotum at base about 2.0 × length in the middle, margins about as thick as elytral margins near shoulders 18
18 Pronotal posterior corners clearly protruding (Pl.148:1). Anterior edge of prosternum completely margined. About 33 punctures in first primary row. Suture darkened to first primary puncture-row (Pl. 148:1) ....
.... *Haliplus* (*Liaphlus*) *srilankanus*

– Pronotal posterior corners not protruding. Anterior edge of prosternum only margined near prosternal process. About 24 punctures in first primary row. Suture darkened to first secondary puncture-row (Pl. 53:1) .... *Haliplus (Liaphlus) arrowi*

9 Key to the Australasian Haliplidae

1 Pronotal base usually with longitudinal plicae (Pl. 231:1). Between plicae strongly impressed. Prosternal process posteriorly strongly impressed in middle (Pl. 231:5) 2

– Pronotal base without longitudinal plicae (Pl. 55:1). Prosternal process not strongly impressed in middle (Pl. 55:8) 10

2 Last abdominal sternite with strong longitudinal ridge or keel (males) (Pl. 231:8, 9) or with a weak keel (females) (Pl. 231:11, 12) 3

Last abdominal sternite without ridge or keel 4

3 Pronotal plicae absent or hardly recognizable, base of pronotum strongly and coarsely punctured (Pl. 233:1). Last sternite with clearly visible ridge in posterior half, which is hardly visible in lateral view (Pl. 233: 7, 8) .... *Phalilus storey*

Pronotal plicae long and strongly bent, base of pronotum moderately punctured (Pl. 231:1). Last sternite with strong ridge, which is also clearly visible in lateral view (males) (Pl. 231:8, 9) or with weak ridges (females) (Pl. 231:11, 12) .... *Phalilus oberthuri*

4 Pronotal plicae long, base of elytral puncture-row 5 strongly grooved (Pl. 232:1) 5

– Pronotal plicae short and straight, base of elytral puncture-row 5 at most slightly grooved (Pl. 226:1) 8

5 Elytra with strong dark/light colour pattern (Pl. 232:1) .... *Phalilus pinderi*

– Elytra uniformly coloured or with indistinct colour pattern 6

6 Elytral apex with strong dent (Pl. 227:7). Row of strong comb-like spines on outside of protibia 7

– Elytral apex at most with weak dent (Pl. 226:1). Row of spines on outside of protibia not strong. Pronotal basal plicae curved and strongly bent inwards before base. Pronotum distinctly darkened along base and along anterior edge. Elytron with distinct interrupted dark lines on primary puncture-rows (Pl. 226:1). Metasternal process in middle without dense hairy punctures (Pl. 226:5). Male: protarsal claws equal in length (Pl. 226:4); left paramere with solid digitus (sometimes wide and then not always clearly recognizable) with hairs on top (Pl. 226:8, 11); penis wider in middle or nearly parallel, inner side abruptly bent before top (Pl. 226:9, 12) .... *Phalilus bistriatus*

7 Pronotal basal plicae straight at base and anteriorly hardly curved. Pronotum basally and anteriorly at most vaguely darkened. Elytron with vague, hardly interrupted, dark lines on primary puncture-rows (Pl. 230:1). Metasternal process in middle with dense hairy punctures (Pl. 230:5). Male: protarsal claws clearly unequal in length (Pl. 230:4); left paramere without recognizable solid digitus and without hairs on apex (Pl. 230:8); penis narrow, tip slightly widened, inner side gradually curved to top (Pl. 230:9) .... *Phalilus hydei*

– Pronotal basal plicae weakly curved at base. Pronotum not darkened. Elytra without maculation or dark lines (Pl. 227:1). Metasternal process in the middle without hairy punctures (Pl. 227:5). Male: protarsal claws clearly unequal in length; left paramere without solid digitus and with few hairs (Pl. 227:8); penis straight on the inside and abruptly bent at the truncate tip (Pl. 227:9) .... *Phalilus fortescueensis*

8 Females .... *Phalilus fuscatus* or *P. gibbus*

– Males 9

9 Penis very narrow (Pl. 228:9) .... *Phalilus fuscatus*

– Penis broad (Pl. 229:9) .... *Phalilus gibbus*

10 Pronotum with longitudinal dark mark in middle (Pl. 145:1) 11

– Pronotum with dark mark at most anteriorly 12

11 Secondary punctures on elytron weak and sparse. Anterior elytral margin at most weakly serrate (Pl. 145:1). Specimens from New Guinea .... *Haliplus* (*Liaphlus*) *signatipennis*

– Secondary punctures on elytron very strong and dense, almost as strong as primary punctures. Anterior elytral margin clearly serrate (Pl. 150:1). Specimens from Seram .... *Haliplus* (*Liaphlus*) *sulawesicus*

12 Uniformly yellow-brown. Pronotum with lateral margins serrate (Pl. 8:1). Prosternal process impressed in middle (Pl. 8:5) .... *Haliplus* (*Australiplus*) *sindus*

– Elytron usually with dark marks or stripes, but at least with darkened punctures. Pronotum with lateral margins not serrate. Prosternal process not impressed in middle 13

13 Elytral puncture-row 5 basally ending in deep transverse impression .. 14

– Elytral puncture-row 5 basally at most with few punctures standing close together or more or less confluent 16

14 Elytron with dark band along base reaching to puncture-row. Pronotum with or without anterior dark mark (Pl. 149:1) .... *Haliplus* (*Liaphlus*) *stepheni*

– Elytral base at most with dark mark near puncture-row 4. Pronotum without marks 15

15 Metasternal process flat, at most with some strong punctures on both sides close to each other (Pl. 47:5). Elytron with secondary punctures near base of primary row 5 strong, basal punctures of primary row 6 strong. Elytron usually with well defined maculation (Pl. 47:1) .... *Haliplus* (*Liaphlus*) *alastairi*

– Metasternal process with clear pit on both sides (Pl. 156:5). Elytron with secondary punctures near base of primary row 5 weak. Basal punctures of primary row 6 weak. Elytron usually with interrupted dark lines, which are rarely connected by vague marks (Pl. 156:1) .... *Haliplus* (*Liaphlus*) *timmsi*

16 Elytron with extensive maculation, base broadly darkened to puncture-row (Pl. 82:1) .... *Haliplus* (*Liaphlus*) *ferruginipes*

– Elytron without extensive maculation, base not distinctly darkened ... 17

17 Proepisternum (Fig. 10) strongly and usually densely, sometimes even coarsely punctured 18

– Proepisternum (Fig. 10) at most weakly punctured in anterior part 19

18 Prosternal process usually slightly but gradually widening anteriorly, not clearly narrowed before coxae (Pl. 169:5). Last sternite with strong punctures on most of its surface. Male: left paramere with small solid digitus (Pl. 169:7) .... *Haliplus* (*Liaphlus*) *wattsi*

– Prosternal process slightly narrowed before coxae, more or less sinuate, anteriorly about as wide as posteriorly (Pl. 137:5). Last sternite with moderately strong to weak punctures on about apical half. Male: left paramere without solid digitus (Pl. 137:7) ...... *Haliplus* (*Liaphlus*) *pilbaraensis*

19 Elytra with continuous or hardly interrupted dark stripes on primary puncture-rows. Row of secondary punctures along suture dense and in regular row, interval 2 with about 6–12 punctures and interval 4+6 only anteriorly with a few punctures Pl. 154:1). Metasternal process with clear lateral impressions (Pl. 154:5). Male: left paramere without solid digitus (Pl. 154:7) .... *Haliplus* (*Liaphlus*) *testudo*

– Elytra usually without dark stripes on primary puncture-rows 20

20 Elytral shoulders and apex dentate. Row of secondary punctures along suture very dense and in irregular row, elytral interval 3 with complete dense row of secondary punctures (Pl. 55:1). Metasternal process with lateral rows of strong punctures, which usually lie in a slight impression (Pl. 55:5). Male: left paramere with small solid digitus (Pl. 55:7) .... *Haliplus* (*Liaphlus*) *australis*

– Elytral shoulders and apex smooth. Row of secondary punctures along suture not dense and not irregular, elytral interval 3 with very sparse row of secondary punctures (Pl. 96:1). Metasternal process with two plicae from anterior edge backwards to two weak impressions (Pl. 96:5). Male: left paramere without solid digitus (Pl. 96:7) .... *Haliplus* (*Liaphlus*) *halsei*

10 Key to the Afrotropical Species of *Haliplus* Subgenus *Liaphlus*

1 Elytra immaculate. At most some vague marks present around punctures 2
– Elytra distinctly maculate 9
2 Apical margin of elytra serrate (Pl. 99:1) 3
– Apical margin of elytra not serrate (Pl. 121:1) 6
3 Shoulder margin serrate (Pl. 99:1) .... *Haliplus* (*Liaphlus*) *incrassatus*
– Shoulder margin not serrate (Pl. 161:1) 4
4 Elytral punctures, secondary and primary, not standing in distinct rows. Secondary puncture rows dense (Pl. 46:1). Metaventral process not impressed in the middle (Pl. 46:5). Body parallel-sided 5
– Elytral punctures more or less in rows Secondary puncture rows very sparse (161:1). Metaventral process impressed in the middle (Pl. 161:5). Body oval .... *Haliplus* (*Liaphlus*) *uniformis*
5 Metacoxal lobes rounded posteriorly (Pl. 46:7) .... *Haliplus* (*Liaphlus*) *africanus*
– Metacoxal lobes posteriorly with long sharp spine (Pl. 68:7) .... *Haliplus* (*Liaphlus*) *coxaspinipes*
6 Total length less than 3.7 mm. Width between the eyes less than 1.4 × width of one eye 8
– Total length more than 3.9 mm. Width between the eyes more than 1.7 × width of one eye 7
7 Elytral punctures not very strong. Width between the eyes more than 2.2 × width of one eye (Pl. 121:1). Prosternal process not wide posteriorly and laterally margined in anterior part (Pl. 121:5) .... *Haliplus* (*Liaphlus*) *mucronatus*
– Elytral punctures of first four rows very strong in anterior part. Width between the eyes less than 2.0× width of one eye (Pl. 141:1). Prosternal process very wide posteriorly and laterally not margined in anterior part (Pl. 141:5) .... *Haliplus* (*Liaphlus*) *rubidus*

8 Colour yellowish. Prosternal process posteriorly about as wide as anteriorly, weakly narrowed in anterior part, narrowed part about 0.7× width at posterior end, posterior part slightly widening or nearly parallel (Pl. 81:5) .... *Haliplus* (*Liaphlus*) *exsecratus*
– Colour reddish. Prosternal process posteriorly less wide than anteriorly, clearly narrowed in anterior part, narrowed part about 0.6× width at posterior end, posterior part strongly widening posteriorly (Pl. 142:5) .... *Haliplus* (*Liaphlus*) *rufescens*
9 Apical margin of elytra not serrate 10
– Apical margin of elytra serrate 21
10 Prosternal process impressed in the middle (Pl. 113:5) 11
– Prosternal process not impressed in the middle (Pl. 83:3) 13
11 Elytral base with distinct dark band, eighth primary puncture row with small dark marks (Pl. 113:1) .... *Haliplus* (*Liaphlus*) *maculipennis*
– Elytral base not darkened, eighth primary puncture row not with dark marks (Pl. 49:1) 12
12 Total length less than 3.2 mm (Pl. 49:1) ... *Haliplus* (*Liaphlus*) *andalusicus*
– Total length more than 3.4 mm (Pl. 95:1) ...... *Haliplus* (*Liaphlus*) *guttatus*
13 Pronotum immaculate 14
– Pronotum maculate 19
14 Metaventral process impressed in the middle 16
– Metaventral process with two impressions 15
15 Elytral maculation connected to suture (Pl. 83:1) .... *Haliplus* (*Liaphlus*) *figuratus*
– Elytral maculation not connected to suture (Pl. 88:1) .... *Haliplus* (*Liaphlus*) *gamma*
16 Prosternal process grooved on each side (Pl. 164:5) .... *Haliplus* (*Liaphlus*) *varicator*
– Prosternal process at most slightly grooved in the middle 17
17 Total length less than 3.5 mm 18
– Total length more than 3.5 mm (Pl. 86:1) .... *Haliplus* (*Liaphlus*) *fulvus*
18 Elytral maculation well developed. At least an anchor like mark on the disk, which is usually connected to the suture (Pl. 165:1) .... *Haliplus* (*Liaphlus*) *variegatus*
– Elytral maculation weak and consisting of separate small marks, which may be almost absent (Pl. 45:1) .... *Haliplus* (*Liaphlus*) *abbreviatus*
19 Metaventral process impressed in the middle (Pl. 77:5) 20
– Metaventral process with 2 impressions (Pl. 89:5) .... *Haliplus* (*Liaphlus*) *garambanus*

20 Pronotal mark widest anteriorly (Pl. 77:1). Prosternal process about parallel, not or hardly impressed in anterior part (Pl. 77:5). Elytral colour yellowish. Tip of penis narrowing to the top (Pl. 77:8) .... *Haliplus* (*Liaphlus*) *ebolovensis*
– Pronotal mark widest posteriorly (Pl. 107:1). Prosternal process widest anteriorly, anterior part strongly impressed (Pl. 107:5). Elytral colour reddish. Tip of penis not or hardly narrowing to the top (Pl. 107:8) .... *Haliplus* (*Liaphlus*) *lamottei*
21 Metaventral process with punctured grooves along lateral margins, flat in the middle (Pl. 48:5) .... *Haliplus* (*Liaphlus*) *alluaudi*
– Metaventral process with one or two pits or oval impressions 22
22 Elytral maculation not connected to suture (Pl. 102:1) 23
– Elytral maculation connected to suture (Pl. 114:1) 24
23 Lateral margins of pronotum convex (Pl. 102:1). Prosternal process about parallel, metaventral process with separate impressions (Pl. 102:5) .... *Haliplus* (*Liaphlus*) *insularis*
– Lateral margins of pronotum concave (Pl. 152:1). Prosternal process strongly narrowed before coxae, metaventral process with two pits close together in a common impression (Pl. 152:5) .... *Haliplus* (*Liaphlus*) *tanzanianus*
24 Pronotum immaculate 25
– Pronotum maculate 26
25 Elytra only with discal dark mark in posterior half (Pl. 114:1) .... *Haliplus* (*Liaphlus*) *madagascariensis*
– Elytra with maculation on disc, along base and on intervals (Pl. 127:1) .... *Haliplus* (*Liaphlus*) *nigerianus*
26 Pronotum with central anterior mark (Pl. 168:1). Male: right paramere with very long parallel apical digitus (Pl. 168:9) .... *Haliplus* (*Liaphlus*) *watsoni*
– Pronotum with central anterior and posterior mark, which are often confluent (Pl. 124:1). Male: right paramere with short digitus (Pl. 124:9) 27
27 Primary and secondary punctures nearly equal in strength (Pl. 124:1, 3). Metaventral process with a deep central impression, in which usually two pits close to each other, distance less than 0.2 times width of metaventral process (Pl. 124:5) .... *Haliplus* (*Liaphlus*) *natalensis*
– Primary and secondary punctures clearly different in strength. Metaventral process with two pits less close to each other, distance about 0.3 times width of metaventral process, pits sometimes in a common impression 28

28 Shoulders rasp-like, pronotal margins concave, pronotal base strongly punctured and impressed (Pl. 75:1) .... *Haliplus* (*Liaphlus*) *discessus*
– Shoulders usually not rasp-like. Pronotal margins about straight, pronotal base not strongly depressed 29
29 Primary and secondary elytral punctures almost similar (Pl. 94:1). Body parallel .... *Haliplus* (*Liaphlus*) *guignoti*
– Primary and secondary elytral punctures clearly different, secondary punctures much weaker. Body more oval 30
30 Prosternal process with two longitudinal rows of strong punctures, not or hardly grooved (Pl. 104:5). Elytral puncture rows usually weak and sparse (Pl. 104:1, 3). Males: penis gradually bent, with weak angle at 1/3 from the top (Pl. 104:8) .... *Haliplus* (*Liaphlus*) *kantneri*
– Prosternal process with two usually strong longitudinal impression in posterior part. Elytral puncure rows moderately strong and dense 31
31 Prosternal process anteriorly more than 1.7× width posteriorly (Pl. 90:5) .... *Haliplus* (*Liaphlus*) *genieri*
– Prosternal process anteriorly less than 1.3× width posteriorly 32
32 Elytral puncture row 8 and 9 stronger than 7 or 6, impressed behind shoulders (Pl. 117:1). Male: penis abruptly bent before the top, bent tip about 0.2× total length of penis (Pl. 117:8) ... *Haliplus* (*Liaphlus*) *methneri*
– Elytral puncture row 8 and 9 not or hardly stronger and not more impressed than row 7 or 6. Male: penis gradually rounded to the top or the bent tip is about 0.4× length of penis 33
33 Elytral secondary punctures in interval 4, 6 and 8 sparse, in 6 and/or 8 sometimes absent; basal primary punctures of row 2 to 4 strong (Pl. 119:1). Males: penis gradually bent to the top (Pl. 119:8) .... *Haliplus* (*Liaphlus*) *mimulus*
– Elytral secondary punctures in interval 4 only basally, absent in interval 6 and 8; basal primary punctures of row 2 to 4 small (Pl. 166:1). Male: penis abruptly bent before top, bent tip 0.4 × total length of penis (Pl. 166:8) .... *Haliplus* (*Liaphlus*) *venustus*

11 Key to the North African and Afrotropical Species of *Peltodytes*

1 Metaventral process with complete plica on each side (Pl. 212:7) 2
– Metaventral process without or with anteriorly short plica on each side (Pl. 218:5) 3
2 Coxal plate with long sharp posterior tooth (Pl. 221:8) .... *Peltodytes* (*Peltodytes*) *rotundatus*

– Coxal plate with blunt posterior tooth (Pl. 212:9) .... *Peltodytes* (*Peltodytes*) *caesus*

3 Prosternal process longitudinally only impressed in the middle, sometimes traces of impressions along margin of posterior half. Metacoxal lobes laterally not margined (Pl. 220:8) *Peltodytes* (*Peltodytes*) *quadratus*

– Prosternal process in posterior part longitudinally impressed along the sides and impressed or strongly punctured in the middle. Metacoxal lobes laterally margined (Pl. 218:7) 4

4 Pronotal base without puncture row between plicae (Pl. 218:1) 5

– Pronotal base with row of punctures, which may be interrupted (Pl. 225:1) 6

5 Pronotal posterior corners not strongly protruding, base between plicae weakly impressed (Pl. 216:1). Elytral discal dark mark confluent with sutural darkening, base around scutellum weakly impressed, second and third punctures in basal 1/3 of row 5 and 6 a little larger than in row 1 to 3 (Pl. 216:1) .... *Peltodytes* (*Peltodytes*) *ghanaensis*

– Pronotal posterior corners strongly protruding, base between plicae strongly impressed (Pl. 218:1). Elytral discal mark usually not confluent with sutural darkening, base around scutellum strongly impressed; second and third punctures in basal 1/3 of row 5 and 6 much larger than in row 1 to 3 and irregular placed (Pl. 218:1) .... *Peltodytes* (*Peltodytes*) *nodieri*

6 Elytral discal mark connected to the sutural darkening (Pl. 225:1). Metacoxal lobes posteriorly rounded or very slightly angulate (Pl. 225:7). Male: penis narrowed before the top (Pl. 225:9) .... *Peltodytes* (*Peltodytes*) *ugandanus*

– Elytral discal marks usually not connected to the sutural darkening (Pl. 223:1). Metacoxal lobes posteriorly clearly angulate (Pl. 223:7). Male: top of penis rounded (Pl. 223:9) 7

7 Width of prosternal process posteriorly 1.1–1.2 × width anteriorly (Pl. 223:5). Metaventral process anteriorly with small plica on each side (Pl. 223:5). Elytral basal punctures in a strong common impressionn (Pl. 223:1). Pronotal margin convex, complete row of punctures between basal plicae (Pl. 223:1) .... *Peltodytes* (*Peltodytes*) *speratus*

– Width of prosternal process posteriorly 1.3–1.4 × width anteriorly (Pl. 213:5). Metaventral process anteriorly without plica on each side (Pl. 213:5). Elytral basal punctures in a weak common impression. Pronotal margin straight, row of punctures between basal plicae interrupted in the middle (Pl. 213:1) .... *Peltodytes* (*Peltodytes*) *congoensis*

12 Key to the Nearctic Species of *Brychius*

1 Prosternal process constricted at the level of procoxae, gradually widening posteriorly to apex making the anterior lobe distinctly wedge-shaped (Pl. 5:5). Last abdominal segment with blunt point (Pl. 5:7). Hood lobe of aedaeagus truncated, a deep distinct channel separating it from the median lobe (Pl. 5:9) .... *Brychius pacificus*

– Prosternal process constricted at the level of procoxae, nearly parallel to apex (Pl. 3:5). Last abdominal segment with long sharp point (Pl. 3:7). Hood lobe of aedaeagus rounded, not distinctly separated from the median lobe (Pl. 3:9) 2

2 Humeral margin of elytra hooked and with serration; margin there clearly widened (Pl.4:1). Margin of pronotum posteriorly hardly concave, disc of pronotum strongly impressed (Pl. 4:1). Large specimens, total length 3.7–4.4 mm. Aedaeagus with apex of median lobe long and slender, narrowing at the tip, hood lobe pointed (Pl. 4:9) .... *Brychius hungerfordi*

– Humeral margin of elytra weakly hooked, without serration (Pl. 3:1). Margin of pronotum posteriorly concave, disc of pronotum weakly impressed (Pl. 3:1). Total length 3.0–3.8 mm. Aedaeagus with curvature of the median lobe not pronounced, apex rounded and the hood lobe small, laying almost against the median lobe (Pl. 3:9) .... *Brychius hornii*

13 Key to the Nearctic and Neotropical Species of *Haliplus* s. str.

1 Apices of elytra strongly sinuate (Pl. 19:1) . *Haliplus* (*Haliplus*) *blanchardi*

– Apices of elytra not strongly sinuate, rounded or almost subtruncate and feebly sinuate 2

2 Basal pronotal plicae shorter, less than one quarter the length measured from base of plicae along the plicae to the anterior margin of the pronotum (Pl. 30:1) .... *Haliplus* (*Haliplus*) *immaculicollis*

– Basal pronotal plicae longer, more than one–quarter the length measured from base of plicae to anterior margin of pronotum (Pl. 18:1) 3

3 Elytra with blackish uninterrupted longitudinal lines formed by the closely arranged blackened punctures, but without blotches (Pl. 18:1) 4

– Elytra with more or less definite blotches, never with uninterrupted lines (Pl. 21:1) 5

4 Robust, average length 3.4 mm. 6th elytral interval with secondary punctures (Pl. 41:1). Penis almost L-shaped, with distinct subapical angle dorsally (Pl. 41:8). Elytra of female without micropunctation .... *Haliplus* (*Haliplus*) *stagninus*

– Smaller, average length 3.0 mm. 6th elytral interval without secondary punctures (Pl. 18:1). Penis almost L-shaped, but smoothly curved dorsally (Pl. 18:10). Elytra of females with clear micropunctation .... *Haliplus* (*Haliplus*) *apicalis*

5 Prosternal process channelled (Pl. 21:5). Penis broad, with prominent dorsal lobe; right paramere with variable setae (Pl. 21:8) 6

– Prosternal process not channelled, but grooved along sides (Pl. 25:5). Penis elongate, slender, without prominent dorsal lobe (Pl. 35:8); right paramere with setae rather sparsely placed along ventral edge (Pl. 35:9) 7

6 Metaventral process at most weakly impressed at sides (Pl. 21:5). Penis with very prominent dorsal lobe (Pl. 21:8), right paramere with trumpet- or hammer-like ended setae on distal half (Pl. 21: 9, 10). Elytra of female without micropunctation .... *Haliplus* (*Haliplus*) *dorsomaculatus*

– Metaventral process clearly longitudinally impressed at sides (Pl. 20:5). Penis with dorsal lobe present, but much less prominent (Pl. 20:8); right paramere with pointed setae (Pl. 20:9). Elytra of female with dense micropunctation .... *Haliplus* (*Haliplus*) *distinctus*

7 Pronotum more narrowed anteriorly, at base about 1.7 × width at anterior edge (Pl. 35:1). Last sternite without micropunctation. Elytra of females without micropunctation .... *Haliplus* (*Haliplus*) *longulus*

– Pronotum less narrowed anteriorly, at base about 1.5 × width at anterior edge (Pl. 22:1). Last sternite with strong micropunctation. Elytra of females with strong and dense micropunctation .... *Haliplus* (*Haliplus*) *falli*

14 Key to the Nearctic Species of *Haliplus* Subgenus *Liaphlus* and *Borealiplus*

1 Penultimate segment of labial palpi triangular (Pl. 9:5) .... *Haliplus* (*Borealiplus*) *borealis*

– Penultimate segment of labial palpi about parallel-sided 2

2 Length more than 3.2 mm. Prosternal process flat, not evidently grooved on each side 9

– Length less than 3.2 mm. Prosternal process grooved on each side 3

3 Elytral apex serrate (Pl. 65:1) 4

– Elytral apex not serrate (Pl. 97:1) 7

4 Elytra not maculated (Pl. 65:1) .... *Haliplus* (*Liaphlus*) *concolor*

– Elytra maculated 5

5 Shoulders raspy (Pl. 159:1) .... *Haliplus* (*Liaphlus*) *tumidus*

– Shoulders not raspy 6

6 Length less than 2.4 mm, maculation not in lines (Pl. 110:1), sternites very weakly punctured, 7th sternite very weakly punctured apically .... *Haliplus* (*Liaphlus*) *lewisii*

– Length more than 2.4 mm, maculation often in dark lines on puncture rows which may be vaguely connected (Pl. 130:1), sternites moderately punctured, 7th sternite densely punctured apically .... *Haliplus* (*Liaphlus*) *oklahomensis*

7 Metaventral impression on each side round (Pl. 97:5) .... *Haliplus* (*Liaphlus*) *havaniensis*

– Metaventral impressions extended posteriorly (Pl. 66:5) 8

8 Length more than 2.7 mm. Shoulders serrate (Pl. 66:5) .... *Haliplus* (*Liaphlus*) *confluentus*

– Length less than 2.7 mm. Shoulders not serrate (Pl. 52:1) .... *Haliplus* (*Liaphlus*) *annulatus*

9 Metaventral process impressed in posterior part (in lateral view) (Pl. 70:5) 10

– Metaventral process not impressed in posterior part (in lateral view) (Pl. 163:6) 12

10 Subsutural rows of interstrial punctures single and relatively sparse (Pl. 91:1). Form elongate. Prosternal process very feebly or not at all margined at apex in lateral view (Pl. 91:5). Metaventral process not tumid between coxae, moderately deeply depressed behind, margined (Pl. 91:5). Pronotum without dark blotch at the anterior margin (Pl. 91:1) .... *Haliplus* (*Liaphlus*) *gracilis*

– Subsutural rows of interstrial punctures partly double and dense (Pl. 70: 1, 3). Form usually broader. Prosternal process evidently margined at apex in lateral view (Pl. 70:6). Metaventral process feebly to strongly tumid between coxae, not or very briefly margined (Pl. 70:5). Pronotum with a dark blotch at the anterior margin (Pl. 70:1) 11

11 Length 4.4–4.9 mm; punctation of elytral striae coarse (Pl. 70:1) .... *Haliplus* (*Liaphlus*) *cribrarius*

– Length less than 4.0 mm; punctation of elytral striae finer (Pl. 128:1) .... *Haliplus* (*Liaphlus*) *nitens*

12 Elytra not or hardly maculated, dark lines on most parts of puncture rows 13

– Elytra maculated, no continuous dark lines on puncture rows 14

13 Elytral punctures not darkened (Pl. 163:1) .... *Haliplus* (*Liaphlus*) *vancouverensis*

– Elytral punctures darkened (Pl. 72:1) .... *Haliplus* (*Liaphlus*) *cylindricus*

14 Pronotal basal punctures concolorous or almost concolorous with pronotum (Pl. 64:1) .... *Haliplus* (*Liaphlus*) *columbiensis*
– Pronotal basal punctures distinctly darkened 15
15 Pronotum with anterior wide dark mark (Pl. 109:1). Discal elytral punctures of 2nd and 3rd row stronger than basal pronotal punctures opposite these rows. Secondary punctures on 1st interval more or less double (Pl. 109:1) .... *Haliplus* (*Liaphlus*) *leechi*
– Pronotum anteriorly with at most a weak darkening (Pl. 86:1). Discal elytral punctures of 2nd and 3rd rows about as strong as basal pronotal punctures opposite these rows. Secondary punctures in 1st interval regular and at least basally in a single row (Pl. 86:1) .... *Haliplus* (*Liaphlus*) *fulvus*

15 Key to the Nearctic Species of *Haliplus* Subgenus *Paraliaphlus*

Includes *Haliplus kenneri* from Baja California and *Haliplus alberti* from Cuba.

1 Anterior margin of pronotum beaded at sides, anteriorly without median spot (Pl. 180:1) 2
– Anterior margin of pronotum not beaded at sides, frequently with anterior median spot (Pl. 183:1) 5
2 Elytral sutural darkening reaching to first primary puncture row. Elytral apex weakly serrate or undulate (Pl. 180:1) 3
– Elytral sutural darkening not reaching to first primary puncture row. Elytral apex clearly serrate (Pl. 178:1) 4
3 Hind tibia with distinct setiferous striole on dorsal side (Pl. 180:4) .... *Haliplus* (*Paraliaphlus*) *fasciatus*
– Hind tibia without or with very short setiferous striole on dorsal side (Pl. 188:1) .... *Haliplus* (*Paraliaphlus*) *pseudofasciatus*
4 Elytral apex serrate up to posterior tip (Pl. 178:1) .... *Haliplus* (*Paraliaphlus*) *connexus*
– Serration on elytral apex getting weak near posterior tip (Pl. 190:1) .... *Haliplus* (*Paraliaphlus*) *tortilipenis*
5 Elytra almost unicoloured or with weak maculation (maculation and ground colour not strongly differing) 6
– Elytra with strong maculation (maculation and ground colour strongly differing) 7
6 Elytra unicoloured. Elytral apex with clear sharp serration (Pl. 182:1) .... *Haliplus* (*Paraliaphlus*) *kenneri*

– Elytra almost unicoloured or with clear maculation, which however differs not much with ground colour. Elytral apex not serrate, but slightly undulate (Pl. 185:1) .... *Haliplus* (*Paraliaphlus*) *mimeticus*

7 Trochanter of midlegs with strong punctures (Pl. 183:3) .... *Haliplus* (*Paraliaphlus*) *leopardus*

– Trochanter of midlegs not with strong punctures 8

8 Carina on anterior part of metepisternum strongly darkened, clearly darker than surrounding area (Fig. 10) .... *Haliplus* (*Paraliaphlus*) *pantherinus*

– Carina on anterior part of metepisternum not distinctly darker than surrounding area 9

9 Prosternal process clearly narrowed in anterior part (Pl. 177:5) 10

– Prosternal process anteriorly not or hardly narrower than posteriorly (Pl. 189:5) 11

10 Length more than 3.5 mm (Pl. 177:1) .. *Haliplus* (*Paraliaphlus*) *apostolicus*

– Length less than 3.4 mm (Pl. 184:1) ..... *Haliplus* (*Paraliaphlus*) *lucigelinae*

11 Elytral sutural darkening broad, (almost) reaching to first primary row (Pl. 189:1) 12

– Elytral sutural darkening narrow, at most reaching to secondary puncture row (Pl. 181:1) 15

12 Prosternal process clearly wider anteriorly than posteriorly (Pl. 189:1).... 13

– Prosternal process anteriorly at most a little wider, but usually as wide as posteriorly (Pl. 191:5) 14

13 Pronotum with dark mark on anterior part (Pl. 189:1). Penis with short wide bent apex (Pl. 189:8) .... *Haliplus* (*Paraliaphlus*) *punctatus*

– Pronotum without dark mark (Pl. 176:1). Penis with abruptly bent top (Pl. 176:8) .... *Haliplus* (*Paraliaphlus*) *alberti*

14 Prosternal process narrowed before coxae (Pl. 191:5). Elytral ground colour usually yellowish brown. Male: protarsal claws about half the length of last tarsal segment. Tip of penis abruptly bent and very narrow (Pl. 191:8) .... *Haliplus* (*Paraliaphlus*) *triopsis*

– Prosternal process more parallel to slightly widening anteriorly (Pl. 186:5). Elytral ground colour usually more reddish brown. Male: protarsal claws longer, about 3/4 × length of last tarsal segment. Penis with wide bent apex (Pl. 186:8) .... *Haliplus* (*Paraliaphlus*) *mutchleri*

15 Length less than 3.6 mm, elytral base roughly punctured (Pl. 179:1) .... *Haliplus* (*Paraliaphlus*) *deceptus*

– Length more than 4.0 mm, elytral base not roughly punctured (Pl. 181:1) .... *Haliplus* (*Paraliaphlus*) *grandis*

16 Key to the Nearctic and Neotropical Species of *Peltodytes* Subgenus *Neopeltodytes*

1 Posterior femora entirely black or brown or yellow (Pl. 192:5) 2
– Posterior femora black, ringed with yellow before the femoral apex (Pl. 197:6) 18
2 Posterior femora pale, femoral apex only darker (Pl. 211:5) 19
– Posterior femora entirely black or brown 3
3 Elytra without subhumeral spot, or black wipe (Pl. 193:3) 4
– Elytra with a subhumeral spot, or black wipe(Pl. 192:3) 10
4 Elytra without distinct callosity, or tubercle 6
– Elytra with a distinct callosity (Pl. 193:4) 5
5 Elytral punctures with black ring (Pl. 206:1, 4). Tubercle strongly developed on two puncture rows and with sharp ridges (Pl. 206:1, 6) .... *Peltodytes* (*Neopeltodytes*) *roughleyi*
– Elytral punctures simply (Pl. 193:1, 5). Tubercle single and between puncture rows, top of tubercle smooth (Pl. 193:1, 4) .... *Peltodytes* (*Neopeltodytes*) *callosus*
6 Median elytral spots coalescent with suture (Pl. 202:1) .... *Peltodytes* (*Neopeltodytes*) *muticus*
– Median elytral spots not coalescent with suture 7
7 Elytral base broadly darkened to first punctures. (Pl. 207: 1, 4). Prosternal process with clear anterior transverse ridge (Pl. 207:6) .... *Peltodytes* (*Neopeltodytes*) *sexmaculatus* (partim)
– Elytral base usually not darkened to first punctures, but at most weakly darkened or with a small dark base. Prosternal process with or without anterior transverse ridge 8
8 Metacoxal plates posteriorly distinctly angulate (Pl. 204:9) .... *Peltodytes* (*Neopeltodytes*) *ovalis*
– Metacoxal plates posteriorly at most weakly angulate (Pl. 210:9) 9
9 Anterior apex of prosternal process with a transverse ridge, which slightly continues backwards along lateral margins. Metaventral process without sharp plicae on each side (Pl. 210:6) .... *Peltodytes* (*Neopeltodytes*) *tamaulipensis*
– Anterior apex of prosternal process usually not with a clear transverse ridge, but with few strong punctures in a row, which sometimes seems to form a ridge (Pl. 209: 6, 8). Metaventral process with sharp plicae on each side (Pl. 209:6) .... *Peltodytes* (*Neopeltodytes*) *simplex*

10 Sutural black band uneven, noticeably wider basally than apically (Pl. 192:1). Subhumeral spot on row 9–10 (Pl. 192:3) 12
– Sutural black band even and narrow, not wider basally than apically; sutural blotch not connected to suture. Subhumeral spot on row 8–10 ... 11
11 Pronotum without basal plicae (Pl. 207:1) .... *Peltodytes* (*Neopeltodytes*) *sexmaculatus* (partim)
– Pronotum with basal plicae (Pl. 199:1) *Peltodytes* (*Neopeltodytes*) *festivus*
12 Last abdominal segment shining, polished 13
– Last abdominal segment dull, rugose (sometimes also in *P. floridensis*) (Pl. 208:14) 17
13 Sutural stripe generally broad at base, reaching the first discal stria (Pl. 200:1). Aedaeagus with tooth on inner margin (Pl. 200:11) 15
– Sutural stripe generally narrower at base, not reaching discal stria (Pl. 192;1) 14
14 Outer margin of aedaeagus undulate (Pl. 192:11) .... *Peltodytes* (*Neopeltodytes*) *bradleyi*
– Outer margin of aedaeagus not undulate (Pl. 205:11). Elytral spots small, distinct, median can be confluent with sutural darkening (Pl. 205:1) .... *Peltodytes* (*Neopeltodytes*) *pedunculatus*
15 Elytra with 3–7 punctures between the 1st and 2nd discal stria in front of median dark spot; spots large, confluent, median distinctly coalescent with the sutural margin (Pl. 200:1). Penis strongly but regularly bent at the apex, parameres with fringes of hair (Pl. 200:10–12) .... *Peltodytes* (*Neopeltodytes*) *floridensis*
– Elytra usually with at most 1 or 2 punctures between the 1st and the 2nd discal stria in front of the median dark spot 16
16 Metaventral process with marginal grooves nearly absent in anterior part, more densely but weakly punctured (Pl. 203:6). Penis with backwards directed point before the top (Pl. 203:11) .... some specimens of *Peltodytes* (*Neopeltodytes*) *oppositus*
– Metaventral process with deep margined grooves in anterior part, with few strong punctures (Pl. 196:6). Penis rounded and narrowed before the top, parameres without fringes of hair (Pl. 196:10–12) .... *Peltodytes* (*Neopeltodytes*) *dunavani*
17 Posterior coxal plates usually rounded at apex, scarcely perceptibly angulate (Pl. 208:10); 1st and 2nd joints of middle tarsi in male strongly produced (Pl. 208:8). Metaventral process on both sides strongly impressed and coarsely punctured (Pl. 208:6) ... *Peltodytes* (*Neopeltodytes*) *shermani*

– Posterior coxal plates usually angulate (Pl. 203:9), 1st and 2nd joints of middle tarsi in male thickened, but scarcely produced. Metaventral process flat, coarsely punctured, weak longitudinal groove on each side (Pl. 203:6) .... most specimens of *Peltodytes* (*Neopeltodytes*) *oppositus*

18 Base of head with black collar (Pl. 198:1). Elytra without subhumeral spot (Pl. 198:3). Metaventral process with isolated transverse row of few strong punctures (Pl. 198:6) .... *Peltodytes* (*Neopeltodytes*) *edentulus*

– Base of head without black collar (Pl. 197:1). Elytra with or without subhumeral spot (Pl. 197:3, 4). Metaventral process with strong scattered punctures (Pl. 197:7) .... *Peltodytes* (*Neopeltodytes*) *duodecimpunctatus*

19 Length less than 4.0 mm 20

– Length 4.0–5.0 mm, dark spots indistinct, base of pronotum and base of elytra depressed, hump–backed (Pl. 211:1). Apices of metacoxal plates rounded (Pl. 211:9). Penis rounded, not narrowed at the apex (Pl. 211:11) .... *Peltodytes* (*Neopeltodytes*) *tortulosus*

20 Elytral suture narrowly darkened (Pl. 195:1) 21

– Elytral suture in anterior half darkened to first puncture row (Pl. 194:1) .... *Peltodytes* (*Neopeltodytes*) *darlingtoni*

21 Hind femur yellow (Pl. 201:5). Pronotal basal dark marks small, elytral puncture rows not very strong or coarse, except on dark spots no dark stripes on puncture rows (Pl. 201:1). Penis subapically nearly as wide as subbasally (Pl. 201:11) .... *Peltodytes* (*Neopeltodytes*) *litoralis*

– Hind femur light brown (Pl. 195:5). Pronotal basal dark marks large and sometimes connected; elytral puncture rows strong and coarse in basal area; dark stripes on large parts of first puncture rows (Pl. 195:1). Penis subapically 7/5 as wide as near the base (Pl. 195:11) .... *Peltodytes* (*Neopeltodytes*) *dietrichi*

17 Key to the Neotropical Species of *Haliplus* (All Subgenera)

1 Metaventral process with at each side a sharp plica running to anterior edge (note that *Haliplus* (*Paraliaphlus*) *grandis* might have this character!) (Pl. 185:5) 2

– Metaventral process without plicea to anterior edge (Pl. 61:1) 3

2 Elytral apex with undulate or with weak serration (Pl. 185:1) .... *Haliplus* (*Paraliaphlus*) *mimeticus*

– Elytral apex with clear dense serration (Pl. 182:1) .... *Haliplus* (*Paraliaphlus*) *kenneri*

3 Body completely covered with strong micropunctures, elytra not maculated (Pl. 15:1) .... *Haliplus* (*Haliplidius*) *subseriatus*
– Body not completely covered with micropunctures, when females with fine micropunctures then elytra maculated 4
4 Eighth elytral puncture row carinate in preapical part (Pl. 61:1) 5
– Eighth elytral puncture row not carinate in preapical part 6
5 Seventh elytral puncture row carinate in preapical part (Pl. 61:1) .... *Haliplus* (*Liaphlus*) *carinatus*
– Seventh elytral puncture row not carinate in preapical part (Pl. 160:1) .... *Haliplus* (*Liaphlus*) *unicarinatus*
6 Elytra without maculation 5
– Elytra with maculation 8
7 Metaventral process with two round pits basally (Pl. 65:5) .... *Haliplus* (*Liaphlus*) *concolor*
– Metaventral process with long grooves (Pl. 153:5)
8 Elytral punctures not darkened (Pl. 153:1) ... *Haliplus* (*Liaphlus*) *testaceus*
– Elytral punctures at least for the greater part darkened 9
9 Metaventral process with two long almost parallel grooves (Pl. 58:5) .... *Haliplus* (*Liaphlus*) *bonariensis*
– Metaventral process with two short diverging grooves (Pl. 87:7) .... *Haliplus* (*Liaphlus*) *fuscipennis*
10 Prosternal process with length less than 1.5 × width at base (Pl. 69:5) ... 11
– Prosternal process with length more than 1.5 × width at base (Pl. 116:5) 12
11 Metaventral process with narrow marginal rim and on ach side a weak groove starting in the middle of the lateral rim running in medial-anterior direction (Pl. 84:5) .... *Haliplus* (*Liaphlus*) *fiorentini*
– Metaventral process with wide marginal rim (Pl. 69:5) .... *Haliplus* (*Liaphlus*) *crassus*
12 Pronotal base with plicae opposite 4th or 5th elytral puncture row (Pl. 56:1) 13
– Pronotal base without plicae 17
13 Pronotum without maculation (Pl. 30:1). Prosternal process posteriorly impressed in the middle (Pl. 30:5) .... *Haliplus* (*Haliplus*) *immaculicollis*
– Pronotum with maculation. Prosternal process posteriorly not impressed in the middle 14
14 Pronotum with central mark from anterior to posterior edge, in posterior half with light central line (Pl. 116:1) *Haliplus* (*Liaphlus*) *mesoamericanus*
– Pronotum not with central mark reaching from anterior to posterior edge, but only along anterior or posterior edge 15

15 Metaventral process with 2 round impressions on each side (Pl. 56:5) .... *Haliplus* (*Liaphlus*) *bachmanni*
– Metaventral process not with 2 round impressions on each side 16
16 Pronotum laterally margined (Pl. 60:1) .... *Haliplus* (*Liaphlus*) *camposi*
– Pronotum laterally not margined (Pl. 157:1) .... *Haliplus* (*Liaphlus*) *tocumenus*
17 Metaventral process only impressed in the middle (Pl. 147:5) .... *Haliplus* (*Liaphlus*) *solitarius*
– Metaventral process not only impressed in the middle 18
18 Prosternal process strongly narrowed in anterior half, anteriorly narrower than posteriorly 19
– Prosternal process at most weakly narrowed in anterior half 22
19 Pronotum with longitudinal central mark (Pl. 155:1) .... *Haliplus* (*Liaphlus*) *thoracicus*
– Pronotum without longitudinal central mark 20
20 Length less than 3.0 mm (Pl. 10:1) .... *Haliplus* (*Borealiplus*) *oblongus*
– Length more than 3.0 mm 21
21 Metaventral process very wide with two strongly diverging grooves (Pl. 155:5) .... *Haliplus* (*Liaphlus*) *megapunctatus*
– Metaventral process narrow, grooves not strongly diverging (Pl. 162:5) .... *Haliplus* (*Liaphlus*) *valdiviensis*
22 Elytral puncture rows 4 to 6 basally impressed to about 1/4 of elytral length (Pl. 112:1) .... *Haliplus* (*Liaphlus*) *maculicollis*
– At most fifth elytral puncture row basally impressed 23
23 Fifth elytral puncture row basally not in a common impression (check both elytra) 30
– Fifth elytral puncture row basally in a common impression (Pl. 101:1) . 24
24 Basal impression of fifth elytral puncture row extended along base. Preapical elytral punctures of row 8 with very strong punctures (Pl. 151:1) .... *Haliplus* (*Liaphlus*) *tantoyucanus*
– Basal impression of fifth elytral puncture row not extended along base. Punctures of elytral row 8 weak in preapical part 25
25 Metaventral process with 2 more or less confluent pits on each side (Pl. 101:5). Pronotum not clearly maculate, at most with small anterior mark and small vague marks opposite fifth elytral puncture row (Pl. 101:1) .... *Haliplus* (*Liaphlus*) *indistinctus*
– Metaventral process with a single groove on each side 26
26 Pronotum with broad longitudinal central mark from anterior to posterior edge (Pl. 118:1) .... sometimes *Haliplus* (*Liaphlus*) *mexicanus*

– Pronotum at most with mark along anterior edge or base slightly darkened 27
27 Metaventral process with strong nearly round pit on each side and a number of smaller pits (Pl. 63:5) .... *Haliplus* (*Liaphlus*) *colombiensis*
– Metaventral process with groove on each side 28
28 Length less than 3.5 mm 29
– Length more than 3.5 mm (Pl. 181:1) .... sometimes *Haliplus* (*Paraliaphlus*) *grandis*
29 Shoulders raspy (Pl. 159:1). Metaventral process with impression on each side, at most weak groove towards anterior edge (Pl. 159:5) .... sometimes *Haliplus* (*Liaphlus*) *tumidus*
– Shoulders not raspy. Metaventral process with continuous groove on each side (Pl. 179:1) .... sometimes *Haliplus* (*Paraliaphlus*) *deceptus*
30 Metaventral process with 2 isolated round pits on each side (Pl. 98:5) . 31
– Metaventral process not with 2 round pits on each side, but at most on each side with 2 impressions which are more or less confluent 33
31 Elytral suture darkened (Pl. 108:1) 32
– Elytral suture not darkened (Pl. 98:1) .... *Haliplus* (*Liaphlus*) *heppneri*
32 Secondary punctures on elytral row 3–7 in single row (Pl. 133:1) .... *Haliplus* (*Liaphlus*) *panamanus*
– Secondary punctures on elytral row 3–7 very dense and in more than one irregular row (Pl. 108:1) .... sometimes *Haliplus* (*Liaphlus*) *langleyi*
33 Metaventral process with single round impression on each side or an impression with less stronger backwards projected groove or puncture row (Pl. 170:5) 34
– Metaventral process with long posteriorly also strong groove on each side, in which sometimes 2 single impressions can be recognized or with sharp marginal plica (Pl. 120:5) 44
34 Metaventral impressions clearly round (Pl. 170:5) 35
– Metaventral impressions extended posteriorly (Pl. 159:5) 38
35 Body wide, length less than 1.6 × width. Males: tip of penis gradually narrowed 36
– Body less wide, length more than 1.7 × width 37
36 Length more than 2.9 mm. Shoulders densely raspy. Anterior elytral margin with about 20 teeth. Secondary elytral punctures, except along suture, usually weak and sparse. Pronotum laterally not clearly margined (Pl. 170:1) .... *Haliplus* (*Liaphlus*) *youngi*
– Length less than 2.9 mm. Shoulders at most weakly raspy. Anterior elytral margin with about 5 teeth. Secondary elytral punctures usually strong and dense (Pl. 71:1) .... *Haliplus* (*Liaphlus*) *cubensis*

37 Length more than 2.5 mm (Pl. 97:1) .... *Haliplus* (*Liaphlus*) *havaniensis*
– Length less than 2.5 mm (Pl. 122:1) .... *Haliplus* (*Liaphlus*) *nanus*
38 Prosternal process gradually widening anteriorly (Pl. 159:5) 39
– Prosternal process about parallel or narrowed near coxa (Pl. 181:5) 41
39 Elytral base darkened from suture to at least fifth puncture row (Pl. 146:1). Prosternal process usually clearly narrowed posteriorly (Pl. 146:5) 40
– Elytral base not darkened, shoulders raspy. Pronotum generally without mark, if so elytral punctures outside maculation distinctly darkened. Elytral apical margin distinctly serrate (Pl. 159:1). Sutural apex of coxal plates dentate. Claws of midlegs 0.5 × length of last tarsal segment. Prosternal process usually not distinctly narrowed posteriorly (Pl. 159:5). Male: tip of penis twisted but always parallel when seen from different angles (Pl. 159:8) .... most specimens of *Haliplus* (*Liaphlus*) *tumidus*
40 Elytral apical margin weakly serrate. Pronotum with anterior and often with vague basal mark, which may be connected to a central mark over total length of pronotum. Elytral punctures outside maculation generally weak and not clearly darkened (Pl. 146:1). Sutural apex of coxal plates rounded or square but not clearly dentate. Claws of midlegs 0.7 × length of last tarsal segment. Male: tip of penis twisted and very sharp when seen from a certain angle (Pl. 146:7, 9) .... *Haliplus* (*Liaphlus*) *signatus*
– Elytral apical margin distinctly serrate. Pronotum anteriorly with distinct blotch. Shoulders not raspy (Pl. 92:1). Male: tip of penis not sharp and not twisted (Pl. 92:8) .... sometimes *Haliplus* (*Liaphlus*) *gravidoides*
41 Length less than 3.0 mm. Pronotum without anterior blotch 42
– Length more than 4.0 mm. Pronotum with anterior round mark (Pl. 181:1) .... some specimens of *Haliplus* (*Paraliaphlus*) *grandis*
42 Prosternal process strongly impressed in anterior part, anterior margin seen from the front strongly curved (Pl. 78:5, 6). Elytral punctures in maculation not darkened (Pl. 78:1, 3) .... *Haliplus* (*Liaphlus*) *elsaltous*
– Prosternal process weakly impressed in anterior part, anterior margin seen from the front weakly curved (Pl. 110:5, 6). Elytral punctures in maculation darkened 43
43 Length less than 2.4 mm. Elytral maculation not mainly restricted to puncture rows but with continuous outline (Pl. 110:1). Sternites weakly punctured, 7th sternite very weakly punctured apically .... *Haliplus* (*Liaphlus*) *lewissii*
– Length more than 2.4 mm. Elytral maculation mainly restricted to puncture rows, often with vague connecting darkening (Pl. 130:1). Sternites moderately strongly punctured, 7th sternite apically densely punctured .... *Haliplus* (*Liaphlus*) *oklahomensis*

44 Elytral apical margin not serrate 45
– Elytral apical margin serrate 51
45 Secondary punctures in basal part of second elytral interval in a single row 46
– Secondary punctures in basal part of second elytral interval very dense and in more than one irregular row 49
46 Sutural darkening not exceeding first secondary puncture row (Pl. 129:1) 47
– Sutural darkening reaching first primary puncture row (Pl. 66:1) 48
47 Pronotum laterally margined completely (Pl. 129:1) .... part of *Haliplus* (*Liaphlus*) *novateutonianus*
– Pronotum laterally not margined in anterior half (Pl. 120:1) .... *Haliplus* (*Liaphlus*) *minimus*
48 Length more than 2.7 mm. Shoulders serrate (Pl. 66:1). Grooves on metaventral process long (Pl. 66:5). Pronotum usually anteriorly and posteriorly maculated (Pl. 66:1) .... *Haliplus* (*Liaphlus*) *confluentus*
– Length less than 2.6 mm. Shoulders usually not or weakly serrate (Pl. 52:1). Grooves on metaventral process short (Pl. 52:5). Pronotum not maculated (Pl. 52:1) .... *Haliplus* (*Liaphlus*) *annulatus*
49 Body oval to subparallel. Elytral puncture rows 5–8 not in a common strong impression 50
– Body strongly tapering, shoulders well pronounced. Elytral puncture rows 5–8 in anterior half strongly developed and in a strong common impression (Pl. 126:1) .... *Haliplus* (*Liaphlus*) *nieseri*
50 Prosternal process about parallel, anteriorly not strongly widened (Pl. 58:5) .... *Haliplus* (*Liaphlus*) *triplehorni*
– Prosternal process anteriorly clearly widened (Pl. 140:5) .... *Haliplus* (*Liaphlus*) *regili*
51 Elytral punctures outside maculation not darkened 52
– Elytral punctures outside maculation darkened 57
52 Elytral maculation consisting of isolated marks except the sutural one 53
– Elytral marks strongly confluent 54
53 Length less than 4.0 mm (Pl. 179:1) .... sometimes *Haliplus* (*Paraliaphlus*) *deceptus*
– Length more than 4.0 mm (Pl. 181:1) .... *Haliplus* (*Paraliaphlus*) *grandis*
54 Prosternal process gradually widening anteriorly 55
– Prosternal process narrowed near coxae, anteriorly wider than posteriorly (Pl. 131:5) .... most specimens of *Haliplus* (*Liaphlus*) *ornatipennis*

55 Metaventral process with sharp plica on each side running to anterior edge (Pl. 176:5) .... *Haliplus* (*Paraliaphlus*) *alberti*
– Metaventral not with sharp plica on each side 56
56 Body oval. Elytral sutural darkening covering second interval (Pl. 76:1). Prosternal process short, strongly punctured (Pl. 76:5) .... *Haliplus* (*Liaphlus*) *drechseli*
– Body tapering. Elytral sutural darkening restricted to half of first interval (Pl. 59:1). Prosternal process more slender, weakly punctured (Pl. 59:5) .... *Haliplus* (*Liaphlus*) *brasiliensis*
57 Prosternal process narrowed in the middle, anteriorly not wider than posteriorly 58
– Prosternal process not clearly narrowed in the middle, usually widest anteriorly 59
58 Secondary elytral punctures in basal part of all intervals in more than one irregular row (Pl. 108:1) .... *Haliplus* (*Liaphlus*) *langleyi*
– Secondary elytral punctures only partly double in basal part of second interval (Pl. 129:1) .... part of *Haliplus* (*Liaphlus*) *novateutonianus*
59 Second secondary puncture row single 61
– Second secondary puncture row irregular double 60
60 Prosternal process hardly or not grooved (Pl. 135:5) .... *Haliplus* (*Liaphlus*) *peruanus*
– Prosternal process distinctly grooved on each side (Pl. 93:5) .... *Haliplus* (*Liaphlus*) *gravidus*
61 Elytral maculation weak and on row 1 and 2 usually only dark lines (Pl. 101:1) .... sometimes *Haliplus* (*Liaphlus*) *indistinctus*
– Elytral maculation distinct and connected to darkened suture 62
62 Punctures outside maculation not darkened (Pl. 131: 1, 3) .... sometimes *Haliplus* (*Liaphlus*) *ornatipennis*
– Punctures outside maculation darkened 63
63 Pronotal sides not margined. Sutural elytral darkening to first primary puncture row. Shoulders raspy (Pl. 67:1) *Haliplus* (*Liaphlus*) *costaricanus*
– Pronotal sides margined. Sutural elytral darkening to first secondary puncture row. Shoulders not raspy (Pl. 92:1) .... *Haliplus* (*Liaphlus*) *gravidoide*

Species Treatment

For each species elementary details, pictures, photographs, information on biology and distribution are presented. Genera, subgenera and species are presented in alphabetical order. Full synonymy is presented in the species list.

1 Genus *Brychius* Thomson

Description/Diagnosis: Yellow to yellow-red. Body relatively small to large, pronotum about square, elytra widest in the middle, more or less strongly tapering apically.
Head moderately wide, eyes not close to each other, apical segments of maxillary and labial palpi shorter than the penultimate segments, one genal line behind eyes.
Pronotum with long basal plicae, surpassing half length.
Elytra with or without longitudinal ridges, margins partly serrate, strong primary puncture-rows, often with distinct micropunctuation, dark stripes on the puncture-rows occasionally with connecting blotches.
Ventral side yellow with darkened areas. Prosternal apophysis apically wide, laterally margined. Metasternal apophysis apically rounded. Hind coxal plates not reaching beyond the fifth sternite. Apical sternite partly grooved in the middle. Legs rather long, male with widened first two or three tarsal segment on the fore- and midlegs, claws well developed.
Fully described by Vondel (1997) and Mousseau (2004).
Biology: The species in this genus live almost exclusively in running water
Distribution: Nearctic and Palaearctic Region.

2 *Brychius elevatus* (Panzer)

Diagnosis: length 3.5–4.4 mm, width 1.8–2.1 mm. Irregular outline, pronotum about square, elytra oval, widest in the middle with longitudinal ridges. In Europe it looks like *Brychius glabratus* but the latter lacks the longitudinal ribs on the elytra (Plates 1 and 234:1).
Full description by Vondel (1997).
Biology: In oxygen-rich brooks, streams and rivers and occasionally in lakes (Holmen, 1987). The larvae live in the beds of streams and rivers among stones

© 2026 THE AUTHOR(S) | HTTPS://DOI.ORG/10.1163/9789004759671-015

covered with calcareous material, in which it excavates galleries and feeds on the algae growing on the stones (Balfour-Browne, 1940). The adults are probably omnivorous. The adults hibernate in the water (Holmen, 1987).
Immature stages: The 1st, 2nd and 3rd instar larvae are known. The actual knowledge is reviewed by Vondel (2025).
Distribution: Albania, Austria, Belgium, Bosnia Herzegovina, Bulgaria, Byelorussia, Croatia, Czech Republic, Denmark, Estonia, Finland, France, Germany, Great Britain, Iran, Iraq, Ireland, Italy, Kyrgyzstan, Latvia, Liechtenstein, Lithuania, Luxemburg, Montenegro, Netherlands, Norway, Poland, Romania, Russia (East Siberia, Central European Territory, North European Territory, South European Territory, West Siberia), Slovakia, Slovenia, Spain, Sweden, Switzerland, Ukraine, Yugoslavia (Plate 268:10).

3 *Brychius glabratus* (Villa and Villa)

Diagnosis: Length 3.8–4.3 mm, width 1.6–2.0 mm. Can be distinguished from *B. elevatus* by the lack of elytral longitudinal ridges (Plates 2 and 234:2).
Full description by Vondel (1997).
Biology: In canals with rich vegetation of filamentous algae, Water-lilies and Charaeans.
Immature stages: Unknown.
Distribution: Italy (Lombardia, Lazio, Sardegna) (Plate 268:9).

4 *Brychius hornii* Crotch

Diagnosis: length 3.0 mm, width 1.5–2.0 mm, Irregular outline, pronotum about square, elytra oval, widest in the middle, no longitudinal ridges. May be distinguished from the other Nearctic *Brychius* species by the combination of the smooth not serrate elytral humerus and the long sharp apical point of the last abdominal segment (Plates 3 and 234:3).
Full description by Mousseau (2004), Mousseau & Roughley (2007), Vondel (2021).
Biology: The habitat and life cycle of *Brychius* species was treated by Mousseau (2004). Generally found in clean running water, rarely in standing water.
Immature stages: The 2nd and 3rd instar larvae are known. The actual knowledge is reviewed by Vondel (2025).

Distribution: Canada (Alberta, British Columbia, Manitoba, Quebec? [questioned by Mousseau 2004], Saskatchewan), USA (Arizona? [questioned by Mousseau 2004], California, Colorado, Idaho, Montana, Nevada, Oregon, Utah, Washington, Wyoming) (Plate 260:1).

5 *Brychius hungerfordi* Spangler

Diagnosis: Length 3.7–4.4 mm, width 1.9–2.25 mm. Irregular outline, pronotum about square, elytra oval, widest just behind the bent humerus. May be distinguished from the other Nearctic *Brychius* species by the combination of the hooked elytral margin behind the serrate humerus and the long sharp apical point of the last abdominal segment (Plates 4 and 234:4).
Full description by Mousseau (2004), Mousseau & Roughley (2007), Vondel (2021).
Biology: In streams with *Chara.*
Immature stages: The 3rd instar larvae is known. The actual knowledge is reviewed by Vondel (2025).
Distribution: Canada (Ontario), USA (Michigan) (Plate 260:2).

6 *Brychius pacificus* Carr

Diagnosis: Length 3.3–3.7 mm, width 1.7–1.9 mm. Irregular outline, pronotum about square, elytra oval, widest behind the humerus. Can be distinguished from the other Nearctic *Brychius* species by the combination of the smooth not serrate elytral humerus and the short apical point of the last abdominal segment (Plates 5 and 234:5).
Full description by Mousseau (2004), Mousseau & Roughley (2007), Vondel (2021).
Biology: Found in streams.
Immature stages: Unknown.
Distribution: Canada (British Columbia), USA (California) (Plate 260:3).

7 Genus *Haliplus* Latreille

Diagnosis: Length 2.0–4.6 mm. Yellow to rust–coloured. Body small to large, body outline usually approximately continuously oval. Head rather narrow to

wide, Interocular distance varying from 0.8 to 2.8× the width of one eye, apical segments of maxillary and labial palpi shorter than the penultimate segments. Behind the eyes sharp grooves: the genal lines. Pronotum widest at the base, with or without basal plicae. Elytra without distinct ridges, at most the intervals a little elevated or puncture rows locally strongly impressed, with or without maculation, with or without dark stripes on the about ten primary puncture rows, with or without micropunctation. Ventral side yellow to rust-coloured, often with darkened parts. Prosternal process flat, grooved or margined, usually narrowed near the coxae. Metaventral process flat, grooved or pitted in the middle or with one or two pits or groove on each side of the middle. Hind coxal plates rounded posteriorly, not reaching beyond the fifth sternite. Apical sternite not grooved in the middle. Legs yellow to rust-coloured, often darkened towards the coxae. Hind tibia with or without a setiferous striole on the dorsal face. In the males the first one, two or three tarsal segments of fore- and midlegs widened and ventrally with a tuft of adhesive setae.
Full description by Vondel (1997).
Distribution : The genus is distributed throughout the world.
Biology: Members of the genus occur in a large variety of water-types, but the majority of species is found in stagnant water.

8 *Haliplus* Subgenus *Algophilus* Zimmermann

Diagnosis: Characterized by the wide head and posteriorly narrowed pronotum.
Full description by Vondel (2010).
Biology: Found in ponds.
Distribution: Afrotropical region.

9 *Haliplus* (*Algophilus*) *lathridioides* Zimmermann

Diagnosis: length 2.2 mm, width 1.0 mm. Body with irregular outline, widest in the middle. May be distinguished from other Afrotropical Haliplidae by the small size, the wide head and posteriorly narrowed pronotum (Plates 6 and 234:6).
Full description by Vondel (2010).
Biology: Found in ponds.
Immature stages: Unknown.
Distribution: South Africa (Plate 278:5).

10 Subgenus *Apteraliplus* Chandler

Diagnosis: Length 1.7–2.3 mm, width 0.9–1.1 mm. Body oval, narrowed at shoulders, widest behind the middle, tapering to apex. Head weakly punctured, interocular distance 2.6–3.0× width of one eye, last segment of maxillary palpus about 3/5 as long as penultimate segment, last segment of labial palpus a little shorter than penultimate segment, two genallines behind eyes. Pronotum with strong basal plicae opposite fourth elytral puncture row, strongly impressed between basal plicae, lateral sides distinctly convex, about parallel in posterior part, clearly margined. Elytra with weak primary punctures. Secondary punctures very weak and sparse, even along suture, lateral margins smooth from shoulder to apex. Prosternal process narrow, not laterally margined in anterior part, narrowed before coxae and gradually narrowed posteriorly, strongly punctured. Metaventral process narrow, impressed posteriorly with some strong punctures, weakly punctured anteriorly. Legs relatively long, without setiferous striole on dorsal face of hind tibia.
Full description by Vondel (2021).
Biology: Living in small pools.
Distribution: Nearctic region.

11 *Haliplus* (*Apteraliplus*) *parvulus* (Roberts)

Diagnosis: Same as subgenus. One of the smallest Nearctic haliplids. Body size smaller than 2.3 mm. Easily distinguished from the also small *H. annulatus* and *H. lewisii* by the strong pronotal basal plicae (Plates 7 and 234:7).
Full description by Vondel (2021).
Immature stages: The 3rd instar larva is figured by Vondel (2025)
Biology: Found in shallow permanent or temporary ponds or pools.
Distribution: USA (California, Oregon, Washington). The records for Panama and Mexico are questionable and should be checked (Plate 260:4).

12 Subgenus *Australiplus* Vondel

Diagnosis: Length 1.7–2.2 mm, width 1.1–1.3 mm. Body oval, strongly tapering posteriorly, widest in front of the middle.
Full description by Vondel (2019).
Biology: Found in creeks.
Distribution: Australia

13 *Haliplus (Australiplus) sindus* Watts

Diagnosis: Same as subgenus (Plates 8 and 234:8).
Full description by Watts (1988) and Vondel (1995a).
Biology: Found in creeks.
Immature stages: Unknown.
Distribution: Australia (Northern Territory, Queensland, Western Australia) (Plate 276:10).

14 Subgenus *Borealiplus* Vondel

Diagnosis: Length 2.5–2.9 mm. Width 1.3–1.5 mm. Body oval. Head: yellow-brown, weakly to moderately punctured, maxillary and labial palpi yellow, maxillary palpi about parallel, last segment 2/3× length of penultimate segment, labial palpi with penultimate segment apically widened, last segment narrow and about 2/3× length of penultimate segment.
Full description by Vondel (2019).
Biology: Found in lakes, ponds and creeks.
Distribution: Nearctic and Neotropical Region.

15 *Haliplus (Borealiplus) borealis* LeConte

Diagnosis: Length 2.6–2.8 mm, width 1.3–1.5 mm. Body oval, elytral shoulders well pronounced. The combination of the triangular labial palpi and the absence of basal pronotal plicae will separate this species from other Nearctic *Haliplus* species (Plates 9 and 234:9).
Full description by Vondel (2021).
Biology: Found in lakes, ponds, creeks.
Immature stages: Unknown.
Distribution: Canada (Alberta, Manitoba, New Brunswick, Nova Scotia, Ontario, Quebec, Saskatchewan), USA (California?, Colorado, Connecticut, Illinois, Indiana, Iowa, Kansas, Kentucky, Massachusetts, Michigan, Minnesota, Missouri, Montana, Nebraska, New York, North Dakota, Ohio, Pennsylvania, South Dakota, Wisconsin) (Plate 260:5).

16 *Haliplus* (*Borealiplus*) *oblongus* Zimmermann

Diagnosis: Length 2.5–2.7 mm, width 1.4–1.5 mm. Body oval to subparallel. Vague marks on elytral even intervals. Characteristic is the combination of the anteriorly strongly narrowed prosternal process and the metaventral process with an oval shallow impression on each side (Plates 10 and 235:1).
Full description by Vondel & Spangler (2008).
Biology: Found in ponds and ditches.
Immature stages: Unknown.
Distribution: Argentina, Uruguay (Plate 268:7).

17 Subgenus *Haliaplus* Ádám

Diagnosis: Length 2.5–3.6 mm, width 1.5–1.9 mm. Body with subparallel sides in central part. Head yellow-brown with brown collar, with dense and on the collar strong and darkened punctures. Maxillary palpi about parallel, last segment about 2/3× length of penultimate segment, penultimate segments of labial palpi widened apically, about 1.5× length of last segment.
Full description by Vondel (2019).
Biology: Found in slowly running waters.
Distribution: Palaearctic Region.

18 *Haliplus* (*Haliaplus*) *laminatus*

Diagnosis: Length 3.7–4.4 mm, width 1.9–2.2 mm. Body subparallel. May be distinguished from related species by the large distance between the eyes (more than 2.3× eye-width) and the metasternal process not being grooved or pitted in the middle (Plates 11 and 235:2).
Full description by Vondel (1997).
Biology: Usually in slowly running water of rivers, canals and brooks, but also found in silty ponds. The adults feed on filamentous algae and invertebrate like larvae of Chronomidae and Oligochaeta (Seeger 1971b) and freshly killed *Asellus aquatic* (Vondel 1986b).
Immature stages: The 2nd and 3rd instar larvae are known. The actual knowledge is reviewed by Vondel (2025).
Distribution: Albania, Austria, Belgium, Byelorussia, Bosnia Herzegovina, Bulgaria, Croatia, Czech Republic, Denmark, France, Germany, Great Britain,

Greece, Hungary, Iraq, Italy, Liechtenstein, Lithuania, Luxembourg, Macedonia, Montenegro, Netherlands, Poland, Romania, Russia (Central European Territory), Serbia, Slovakia, Slovenia, Spain, Sweden, Switzerland, Turkey, Ukraine, Yugoslavia (Plate 268:11).

19 Subgenus *Haliplidius* Guignot

Diagnosis: Small or medium-sized species. Most of the body covered with a clearly visible micropunctation. Elytra with weak primary puncture-rows, epipleura without strong punctures. Penultimate segments of labial palpi widened or hardly widened distally. Pronotal basal plicae weak or absent. Hind tibia without s setiferous striole on the dorsal face.
Full description by Vondel (1997).
Biology: In fresh water with Characeans.
Distribution: Palaearctic and Neotropical Region.

20 *Haliplus* (*Haliplidius*) *confinis*. Stephens

Diagnosis: Length 3.0–3.8 mm, width 1.6–2.0 mm. Body oval, sometimes subparallel, widest in the middle. Whole body covered with distinct micropunctation. It can be distinguished from the other species in the subgenus by the laterally margined prosternal process and the coloration of the pronotum (Plates 12 and 235:3).
Full description by Vondel (1997).
Biology: Living in fresh water among especially Characeans in lakes, ponds, stagnant parts of slowly running brooks, streamlets, but even brackish water. Often found in depths of more than 1 m.
Immature stages: The 1st, 2nd and 3rd as well as the pupa are known. The actual knowledge is reviewed by Vondel (2025).
Distribution: Austria, Belgium, Byelorussia, China (Xinjiang), Croatia, Czech Republic, Denmark, Estonia, Finland, France, Germany, Great Britain, Greece, Hungary, Iran, Iraq, Ireland, Italy, Kazakhstan, Latvia, Lithuania, Luxemburg, Montenegro, Netherlands, Norway, Poland, Russia (Central European Territory, East Siberia, North European Territory, South European Territory, West Siberia), Slovakia, Slovenia, Spain, Sweden, Switzerland, Transcaucasia, Turkey, Ukraine, Yugoslavia (Plate 269:2).

21 *Haliplus* (*Haliplidius*) *obliquus* (Fabricius)

Diagnosis: Length 2.9–3.8 mm, width 1.5–1.8 mm. Body oval to subparallel, shoulders usually pronounced, widest in the middle, completely covered by a strong and dense micropunctation. The knees of this species are not darkened and the pronotum is at most weakly darkened along anterior and posterior margin, which characters distinguishes it from *H. varius* (Plates 13 and 235:4).
Full description by Vondel (1997).
Biology: Living among Characeans in stagnant fresh water in ditches, lakes and gravel pits. Also recorded from brackish water. Up to an altitude of 2300 m.
Distribution: Albania, Austria, Belgium, Bosnia Herzegovina, Bulgaria, Byelorussia, Croatia, Czech Republic, Denmark, Estonia, Finland, France, Georgia, Germany, Great Britain, Greece, Hungary, Iran, Iraq, Ireland, Italy, Latvia, Liechtenstein, Lithuania, Luxembourg, Morocco, Netherlands, Norway, Poland, Romania, Russia (Central European Territory, South European Territory, Siberia), Serbia, Slovakia, Slovenia, Spain, Sweden, Switzerland, Transcaucasia, Turkey, Ukraine, Yugoslavia (Plate 269:3).

22 *Haliplus* (*Haliplidius*) *rejseki* Šťastný and Boukal

Diagnosis: Length 3.2–3.4 mm, width 2.8–3.0 mm. Body relatively long and subparallel, Elytra with black continuous stripes. May be distinguished from the other species in the subgenus with continuous black stripes *H. confinis* by the not margined prosternal process (Plates 14 and 235:5).
Full description by Šťastný & Boukal (2003), also presenting an identification key to the Palaearctic *Haliplidius* species.
Biology: Found in shallow muddy pools with sparse macrovegetation, situated on a wet meadow along a stream on 3500 m a.s.l.
Immature stages: Unknown.
Distribution: China (Qinghai, Sichuan) (Plate 269:4).

23 *Haliplus* (*Haliplidius*) *subseriatus* Zimmermann

Diagnosis: Length 3.0–3.6 mm, width 1.5–1.8 mm. Body long, parallel, widest in the middle, almost completely covered with a distinct micropunctation. Elytra not with dark stripes, but at most with vague marks. In Southern South America it can't be confused with other species (Plates 15 and 235:6).
Full description by Vondel & Spangler (2008).

Biology: Found in ponds.
Immature stages: The 1st, 2nd and 3rd instar larvae are known. The actual knowledge is reviewed by Vondel (2025).
Distribution: Argentina, Chile, Falkland Islands (Plate 268:8).

24 ***Haliplus* (*Haliplidius*) *varius* Nicolai**

Diagnosis: Length 2.5–3.2 mm, width 1.4–1.7 mm. Body oval, widest in or just before the middle. Whole body covered with a strong and dense micropunctation. This species can be distinguished from *H. obliquus* by the very dark anterior and posterior margin of the pronotum and the darkened knees (Plates 16 and 235:7).
Full description by Vondel (1997).
Biology: Living between filamentous algae and other vegetation like *Menyanthes trifoliata* L., *Hottonia palustris* L. in ditches, and ponds in peat-bog areas, in dune-lakes and -ponds, in slowly running waters with sandy bottom, in back-waters of rivers, even recorded from brackish water.
Immature stages: The 2nd and 3rd instar larvae are known. The actual knowledge is reviewed by Vondel (2025).
Distribution: Belgium, Byelorussia, Czech Republic, Estonia, Finland, France, Germany, Great Britain, Italy, Lithuania, Netherlands, Poland, Russia (Central European Territory, North European Territory, South European Territory), Slovakia, Ukraine (Plate 269:1).

25 **Subgenus *Haliplus* Latreille**

Diagnosis: Yellow to yellow-red. Body generally small, body-outline ovally. Head relatively small, penultimate segments of labial palpi very wide, with a distal angle or tooth on the inner edge. Pronotal plicae usually clearly visible, rarely hard to see or even absent. Males without visible (at 50× magnification) micropunctation on pronotum or elytra, females in some species with micropunctation on the elytra. Hind tibia without setiferous striole on the dorsal face.
Full description by Vondel (1997), Vondel et al. 2006.
Biology: Found in a large variety of water bodies.
Distribution: Present in the Nearctic and Palaearctic and most northern parts of the Neotropical and Oriental Regions.

26 *Haliplus* (*Haliplus*) *aliae* Vondel

Diagnosis: Length 2.7–3.0 mm, width 1.6–1.8 mm. Body oval, widest just before the middle. This species is closely related to *H. steppensis*. The latter has more or less interrupted dark lines on the elytral puncture rows, while in this species at most vague traces of lines can be detected. The impressions of the metasternal process of *H. steppensis* are weak, while in this species these are strong. The males differ in the shape of the aedeagus. The female of this species has elytra without micropunctures, while the elytra of female *H. steppensis* are clearly completely micropunctate (Plates 17 and 235:8).
Full description by Vondel (2003a), Vondel et al. (2006).
Biology: Unknown.
Immature stages: Unknown.
Distribution: China (Tianjin) (Plate 269:6).

27 *Haliplus* (*Haliplus*) *apicalis* Thomson

Diagnosis: Length 2.5–3.0 mm, width 1.5–1.7 mm. Body oval to subparallel, widest in the middle. Ventral side of the body often with very fine micropunctation. This species can be distinguished from the related *H. stagninus* by the smaller size, in the males by the different penis and in the females by the strong micropunctation (Plates 18 and 235:9).
Full description by Vondel (1997, 2021), Vondel et al. (2006).
Biology: Living among filamentous algae in stagnant and slowly running, brackish (sometimes in coastal areas), but also fresh water in pools, ponds, ditches and streamlets.
Immature stages: 1st, 2nd and 3rd instar larvae are known. The actual knowledge is reviewed by Vondel (2025).
Distribution: Canada (Alberta, British Columbia, Manitoba, Northwest Territories, Saskatchewan, Yukon), USA (Alaska, Colorado, Minnesota, Montana, New Mexico, North Dakota, South Dakota, Utah, Washington, Wyoming), Belgium, Denmark, Germany, Great Britain, Hungary, Ireland, Norway, Poland, Russia (North European Territory, West Siberia), Sweden, The Netherlands (Plate 269:5).

28 *Haliplus* (*Haliplus*) *blanchardi* Roberts

Diagnosis: Length 2.7–2.75 mm, width 1.6–1.65 mm. Body oval, widest behind shoulders, elytra strongly incised in apical part. Easily recognised from related species by the apically strongly incised elytra (Plates 19 and 236:1).
Full description by Vondel (2021).
Biology: In ponds in wood, along stream margins, in slowly running water, in lakes.
Immature stages: Unknown.
Distribution: Canada (Ontario, Quebec), USA (Connecticut, Georgia, Illinois, Louisiana, Massachusetts, Michigan, Minnesota, Missouri, New Jersey, New York, Ohio, Rhode Island, Wisconsin) (Plate 260:6).

29 *Haliplus* (*Haliplus*) *distinctus* Wallis

Diagnosis: Length 3.1 mm, width: 1.65 mm. Body oval, subparallel in the middle. Can be distinguished from the related *H. dorsomaculatus* by the metaventral process having a longitudinal impression on each side, in the female having a dense micropunctation on the elytra (Plates 20 and 236:2).
Full description by Vondel (2021).
Biology: In lakes and rivers, up to an altitude of 1500 m.
Immature stages: Unknown.
Distribution: Canada (British Columbia), USA (California, Colorado, Idaho, Montana, Oregon, Utah, Washington, Wyoming) (Plate 260:7).

30 *Haliplus* (*Haliplus*) *dorsomaculatus* Zimmermann

Diagnosis: Length 3.0–3.3 mm, width 1.6–1.7 mm. Body oval, widest in the middle. Can be distinguished from the related *H. distinctus* by the metaventral process without a longitudinal impression on each side, in the females by the lack of the strong micropunctation on the elytra (Plates 21 and 236:3).
Full description by Vondel (2021).
Biology: Generally found in lotic habitats like creeks, but also in lentic habitats like ponds or stagnant water bodies along rivers.
Immature stages: Unknown.
Distribution: Canada (British Columbia), USA (California, Idaho, Montana, Oregon, Utah, Washington, Wyoming) (Plate 260:8).

31 *Haliplus (Haliplus) falli* Mank

Diagnosis: Length 2.75–3.2 mm, width 1.4–1.5 mm. Body oval, subparallel. Can be distinguished from the related *H. longulus* by the more parallel shape of the pronotum and elytra and in the females by the dense micropunctation on the elytra (Plates 22 and 236:4).
Full description by Vondel (2021).
Biology: Found in rivers.
Immature stages: Unknown.
Distribution: Canada (Manitoba, Nunavat, Northwest Territories, Quebec, Yukon Territory), USA (Alaska) (Plate 260:9).

32 *Haliplus (Haliplus) fluviatilis* Aubé

Diagnosis: Length 2.5–3.2 mm, width 1.3–1.6 mm. Body oval, widest in the middle. Often confused with other species of the "*ruficollis*"-group. In the males the aedaegus and the subequal claws of the forelegs are reliable characters. Females are sometimes difficult to identify; usually they have a fairly strong micropunctuation (Plates 23 and 236:5).
Full description by Vondel (1997), Vondel et al. (2006).
Biology: Among vegetation in streams, rivers, but also in lakes with sandy or stony bottom. Attracted to light.
Immature stages: The 2nd and 3rd instar larvae are known. The actual knowledge is reviewed by Vondel (2025).
Distribution: Afghanistan, Austria, Azerbaijan, Belgium, Bosnia Herzegovina, Bulgaria, Byelorussia, Croatia, Czech Republic, Denmark, Estonia, Finland, France, Georgia, Germany, Great Britain, Greece, Hungary, Ireland, Italy, Kazakhstan, Kyrgyzstan, Latvia, Lithuania, Luxembourg, Macedonia, Netherlands, Poland, Romania, Russia (Central European Territory, East Siberia, North European Territory, South European Territory, West Siberia), Serbia, Slovakia, Slovenia, Spain, Sweden, Switzerland, Transcaucasia, Turkey, Ukraine, Yugoslavia (Plate 269:7).

33 *Haliplus (Haliplus) fulvicollis* Erichson

Diagnosis: Length 2.6–3.1 mm, width 1.4–1.7 mm. Body oval to subparallel, widest in the middle. This species can be distinguished from the other species in the '*fulvicollis*'-group by the vague elytral maculation (Plates 24 and 236:6).

Full description by Vondel (1997), Vondel et al.(2006)
Biology: Living in marshes, peat bogs and pools. In fresh, usually acid, temporary waters.
Immature stages: The 3rd instar larva is known. The actual knowledge is reviewed by Vondel (2025).
Distribution: Belgium, Bosnia Herzegovina, Bulgaria, Byelorussia, Caucasus, Croatia, Czech Republic, Denmark, Estonia, Finland, France, Germany, Greece, Hungary, Italy, Kazakhstan, Latvia, Netherlands, Norway, Poland, Russia (Central European Territory, North European Territory, South European Territory, West Siberia), Slovakia, Slovenia, Sweden, Transcaucasia, Ukraine, Yugoslavia (Plate 270:1).

34 ***Haliplus* (*Haliplus*) *furcatus* Seidlitz**

Diagnosis: Length 2.4–3.0 mm, width 1.3–1.5 mm. Body oval to subparallel, widest in the middle. Distinguished from the other species of the '*fulvicollis*'-group by the prosternal process, being grooved on each side and the dark elytral lines, being clearly interrupted (Plates 25 and 236:7).
Full description by Vondel (1997), Vondel et al. (2006).
Biology: Living in undisturbed shallow, often temporary, pools and marshes with a rich vegetation. In fresh and also in brackish water in coastal areas.
Immature stages: The third instar larva is known. The actual knowledge is reviewed by Vondel (2025).
Distribution: Armenia, Austria, Belgium, Byelorussia, China (Heilongjiang, Inner Mongolia), Croatia, Denmark, Estonia, Finland, France, Georgia, Germany, Great Britain, Hungary, Italy, Kazakhstan, Latvia, Lithuania, Mongolia, Netherlands, Poland, Romania, Russia (South European Territory, West Siberia), Slovakia, Sweden, Ukraine (Plate 270:4).

35 ***Haliplus* (*Haliplus*) *fuscicornis* Holmen, Vondel and Petrov**

Diagnosis: Length 2.7–2.9 mm, width 1.5 mm. Body subparallel, widest in the middle. Males may be distinguished from related species by the ventral ridge on the first tarsal segment of the forelegs and the ventrally extended first tarsal segment of the midlegs, while the females have completely micropunctured elytra (Plates 26 and 236:8).
Full description by Vondel et al. (2006).
Biology: No details are known.

Immature stages: Unknown.
Distribution: Russia (East Siberia), Kazakhstan (Plate 270:2).

36 ***Haliplus* (*Haliplus*) *harminae* Vondel**

Diagnosis: Length 2.6 mm, width 1.5 mm. Body oval, tapering backwards, widest just before the middle. The absence of pronotal basal plicae and the penultimate segment of the labial palpi not being extended apically separates it from related species (Plates 27 and 236:9).
Full description by Vondel (1990), Vondel et al. (2006).
Biology: No details known.
Distribution: China (Hubei, Hunan, Shaanxi, Sichuan) (Plate 270:3).

37 ***Haliplus* (*Haliplus*) *heydeni* Wehncke**

Diagnosis: Length 2.2–2.8 mm, width 1.2–1.6 mm. Body shortly oval, usually widest in front of the middle and tapering strongly behind the middle. This species is often confused with *H. ruficollis*, with which it is often living in the same locality. Males have a different aedaegus and equal claws on forelegs. Females are usually not micropunctured. In most cases the enlarged basal elytral punctures are useful to distinguish *H. heydeni* from *H. ruficollis* and other related species (Plates 28 and 237:1).
Full description by Vondel (1997), Vondel et al. (2006).
Biology: Living among filamentous algae in lakes, pools, ponds, marshes, stagnant and slow parts of rivers, brooks and streamlets. In fresh and even brackish or acid water.
Immature stages: The 1st, 2nd and 3rd instar larvae are known. The actual knowledge is reviewed by Vondel (2025).
Distribution: Armenia, Austria, Belgium, Bosnia Herzegovina, Bulgaria, Byelorussia, Caucasus, Croatia, Czech Republic, Denmark, Estonia, Finland, France, Georgia, Germany, Great Britain, Greece, Hungary, Iran, Italy, Latvia, Liechtenstein, Kazakhstan, Lithuania, Luxemburg, Macedonia, Montenegro, Netherlands, Norway, Poland, Romania, Russia (Central European Territory, East Siberia, North European Territory, South European Territory, West Siberia), Slovakia, Slovenia, Spain, Sweden, Switzerland, Transcaucasia, Turkestan, Turkey, Ukraine (Plate 270:5).

38 *Haliplus (Haliplus) immaculatus* Gerhardt

Diagnosis: Length 2.4–3.1 mm, width 1.3–1.7 mm. Body oval, sometimes subparallel, widest in the middle Similar species: Males can be distinguished from related species by the ventral ridge on the first tarsal segment of the foreleg. Females can be distinguished from related species with continuous dark elytral lines by the complete lack of elytral micropunctation (Plates 29 and 237:2).
Full description by Vondel (1997), Vondel et al. (2006).
Biology: Living among filamentous algae in fresh water in lakes, pools, ponds, marshes, slowly running brooks and streamlets. Also recorded from brackish water. Observed flying.
Immature stages: The egg, 1st, 2nd and 3rd instar larvae and the pupa are known. The actual knowledge is reviewed by Vondel (2025).
Distribution: Austria, Belgium, Byelorussia, Croatia, Czech Republic, Denmark, Estonia, Finland, France, Georgia, Germany, Great Britain, Hungary, Ireland, Latvia, Lithuania, Luxembourg, Netherlands, Norway, Poland, Romania, Russia (Central European Territory, East Siberia, Far East, North European Territory, South European Territory, West Siberia), Slovakia, Slovenia, Spain, Sweden, Switzerland, Ukraine (Plate 270:8).

39 *Haliplus (Haliplus) immaculicollis* Harris

Diagnosis: Length 2.5–2.9 mm, width 1.4–1.6 mm. Body short oval, widest behind shoulders This very common species is very variable in elytral maculation. The very short pronotal basal plicae are characteristic (Plates 30 and 237:3).
Full description by Vondel (2021).
Biology: Generally found in standing water bodies like ponds, ditches, lakes, swamps, but also in (the littoral of) creeks and rivers.
Immature stages: The egg, 1st, 2nd and 3rd instar larvae and the pupa are known. The actual knowledge is reviewed by Vondel (2025).
Distribution: Canada (Alberta, British Columbia, Labrador, Manitoba, New Brunswick, Newfoundland, Nova Scotia, Northwest Territories, Ontario, Quebec, Prince Edward Island, Saskatchewan, Yukon Territory), St. Pierre & Miquelon, USA (Alaska, Arizona, California, Colorado, Connecticut, District of Columbia, Idaho, Illinois, Indiana, Iowa, Maine, Maryland, Massachusetts, Michigan, Minnesota, Montana, Nebraska, New Hampshire, New Jersey,

New York, North Dakota, Ohio, Oregon, Pennsylvania, Rhode Island, South Carolina, South Dakota, Texas, Utah, Vermont, Virginia, Washington, West Virginia, Wisconsin, Wyoming), Mexico (Plate 260:10).

40 *Haliplus* (*Haliplus*) *interjectus* Lindberg

Diagnosis: Length 2.4–2.8 mm, width 1.4–1.6 mm. Body oval, widest in the middle. This species is often confused with related species like *H. lineolatus* or *H. sibiricus*. Males can be distinguished by the row of strong spines on ventral side of first tarsal segment of forelegs (Plates 31 and 237:4).
Full description by Vondel (1997), Vondel et al. (2006).
Biology: Living in lakes, streams and even found in a hot spring.
Immature stages: Unknown.
Distribution: Finland, Latvia, Russia (East Siberia, Far East, North European Territory, West Siberia) (Plate 271:1).

41 *Haliplus* (*Haliplus*) *kamiyai* Nakane

Diagnosis: Length 3.1 mm, width 1.6 mm. Body subparrallel, widest in the middle (Plates 32 and 237:5).
Full description by Vondel et al. (2006).
Biology: Found in paddy fields.
Immature stages: The 1st, 2nd and 3rd instar larva and the pupa are known. The actual knowledge is reviewed by Vondel (2025).
Distribution: Japan (Honshu) (Plate 270:6).

42 *Haliplus* (*Haliplus*) *kirgisiensis* Holmen and Vondel

Diagnosis: Length 3.0 mm, width 1.5 mm. Body oblong, widest in the middle (Plate 33 and 237:6).
Full description by Vondel et al. (2006).
Biology: No details known.
Immature stages: Unknown.
Distribution: Kyrgyzstan (Plate 270:7).

43 *Haliplus (Haliplus) lineolatus* Mannerheim

Diagnosis: Length 2.5–2.8 mm, width 1.4–1.5 mm. Body short oval, widest in the middle. Males can easily be distinguished from related species by the ventrally strongly incised first tarsal segment of midlegs. Characteristic females have a short wide body, narrow dark lines on elytra and a strong micropunctation on the entire elytra. Less characteristic females are often hard to distinguish from related species (Plates 34 and 237:7).
Full description by Vondel (1997), Vondel et al. (2006).
Biology: Living among filamentous algae and in Elodea–beds in clear oxygen rich fresh, but also brackish waters of lakes, slowly running waters, ditches. The adults feed on Hydrozoans. Up to altitudes of 2200 m. Observed flying.
Immature stages: The 1st, 2nd and 3rd instar larva and the pupa are known. The actual knowledge is reviewed by Vondel (2025).
Distribution: Austria, Belgium, Bulgaria, Byelorussia, Croatia, Denmark, Estonia, Finland, France, Georgia, Germany, Great Britain,? Iraq, Ireland, Italy, Kazakhstan, Latvia, Macedonia, Mongolia, Netherlands, Norway, Poland, Russia (Central European Territory, East Siberia, North European Territory, South European Territory, West Siberia), Sweden, Switzerland, Transcaucasia, Ukraine (Plate 271:4).

44 *Haliplus (Haliplus) longulus* LeConte

Diagnosis: Length 2.8–3.2 mm, width 1.5–1.7 mm. Body oval, widest in the middle. Related to *H. falli*, but less parallel in the pronotum and elytra. Female without micropunctation (Plates 35 and 237:8).
Full description by Vondel (2021).
Biology: Found in ponds, ditches, bogs, lakes, marshes and rivers.
Immature stages: Unknown.
Distribution: Canada (Alberta, British Columbia, Manitoba, New Brunswick, Northwest Territories, Nova Scotia, Ontario, Prince Edward Island, Quebec, Saskatchewan), USA (California, Colorado, Idaho, Illinois, Indiana, Iowa, Maine, Massachusetts, Michigan, Minnesota, Montana, Nevada, New Hampshire, New Jersey, New York, North Carolina, North Dakota, Ohio, Oregon, Rhode Island, South Dakota, Washington, Wisconsin) (Plate 260:11).

45 *Haliplus* (*Haliplus*) *morii* Hayashi, Iwata and Yoshotomi

Diagnosis: 3.0–3.3 mm, width 1.5–1.6 mm. Body oval, tapering backwards, widest before the middle. This species may be confused with *H.* (*Nipponiplus*) *japonicus*, but the latter has plicae on the pronotum (Plates 36 and 237:9)
Full description by Hayashi et al. (2023).
Biology: Found in rice fields.
Immature stages: Unknown.
Distribution: Japan (Honshu) (Plate 271:2).

46 *Haliplus* (*Haliplus*) *ruficollis* (De Geer)

Diagnosis: Length 2.5–3.0 mm, width 1.3–1.6 mm. Body oval, widest in or just before the middle. This species is very variable in shape and maculation and thus it can easily be confused with other species in the '*ruficollis*'-group (Plates 37 and 238:1).
Full description by Vondel (1997), Vondel et al. (2006).
Biology: Living among filamentous algae in lakes, pools, ponds, rivers, marshes, brooks, streamlets. Preferably in stagnant water. Observed flying. Attracted to light.
Immature stages: The 1st, 2nd and 3rd instar larva and the pupa are known. The actual knowledge is reviewed by Vondel (2025).
Distribution: Afghanistan, Armenia, Austria, Azerbaijan, Belgium, Bosnia Herzegovina, Bulgaria, Byelorussia, Caucasus, China (Xinjiang), Croatia, Czech Republic, Denmark, Estonia, Finland, France, Georgia, Great Britain, Germany, Greece, Hungary, Iran, Iraq, Ireland, Italy, Kazakhstan, Kyrgyzstan, Latvia, Lthuania, Luxembourg, Macedonia, Mongolia, Netherlands, Norway, Poland, Portugal, Romania, Russia (East Siberia, North European Territory, South European Territory, West Siberia), Serbia, Slovakia, Slovenia, Spain, Sweden, Switzerland, Tadzhikistan, Transcaucasia, Turkestan, Turkey, Ukraine, Uzbekistan, Yugoslavia (Plate 271:5).

47 *Haliplus* (*Haliplus*) *samojedorum* J. Sahlberg

Diagnosis: Length 2.9–3.0 mm, width 1.4–1.5 mm. Body subparallel to parallel, widest in the middle (Plates 38 and 238:2).
Full description by Vondel et al. (2006).
Biology: In rivers.

Immature stages: Unknown.
Distribution: Russia (East Siberia, Far East, West Siberia) (Plate 271:6).

48 *Haliplus* (*Haliplus*) *sibiricus* Motschulsky

Diagnosis: Length 2.8–3.1 mm, width 1.5–1.6 mm. Body oval, widest in the middle. This species may easily be confused with some other species in the *ruficollis*–group, but males can be recognised by their penis. Because only recently *H. wehnckei* is synonymised with *H. sibiricus* most West European material is known under the first name (Plates 39 and 238:3).
Full description by Vondel (1997), Vondel et al. (2006).
Biology: Living among filamentous algae in slowly running waters, brooks, streamlets, ditches, pools and lakes with clear water.
Immature stages: The 1st, 2nd and 3rd instar larva are known. The actual knowledge is reviewed by Vondel (2025).
Distribution: Andorra, Austria, Belgium, Bulgaria, Byelorussia, China (Heilongjiang, Nei Mongol, Qinghai, Xinjiang), Czech Republic, Denmark, Estonia, Finland, France, Germany, Great Britain, Hungary, Ireland, Italy, Kazakhstan, Kyrgyzstan, Latvia, Liechtenstein, Lithuania, Luxemburg, Macedonia, Mongolia, Netherlands, Norway, Poland, Romania, Russia (East Siberia, Far East, Central European Territory, North European Territory, South European Territory, West Siberia), Spain, Sweden, Switzerland, Transcaucasia, Turkey, Uzbekistan (Plate 271:8).

49 *Haliplus* (*Haliplus*) *simplex* Clark

Diagnosis: Length 2.4–3.1 mm, width 1.3–1.6 mm. Body oval, widest in the middle. The very short or even absent pronotal plicae should distinguish this species from most others, while *H. zacharenkoi* has the elytral lines more continuous in apical ⅔ (Plates 40 and 238:4).
Full description by Vondel et al. (2006).
Biology: Unknown.
Immature stages: Unknown.
Distribution: China (Anhui, Beijing, Fujian, Guangdong, Hainan, Heilongjiang, Hongkong, Jiangsu, Jilin, Liaoning, Nei Mongol, Shaanxi, Shandong, Shanxi, Taiwan, Xinjiang, Zhejiang), Japan (Hokkaido, Honshu), Mongolia, North Korea, Russia (Far East) (Plate 271:3).

50 *Haliplus (Haliplus) stagninus* Leech

Diagnosis: Length 2.95–3.55 mm, width 1.5–1.85 mm. Body oval to subparallel in the middle. Generally longer than the related *H. apicalis*, but in males with strongly bent penis and in females without elytral micropunctation (Plates 41 and 238:5).
Full description by Vondel (2021).
Biology: In beaver ponds, ephemeral wetlands, stock ponds with high turbidity, often collected from shallow margins among emergent vegetation along with *H. apicalis* and *H. immaculicollis*.
Immature stages: Unknown.
Distribution: Canada (Alberta, British Columbia, Manitoba, Northwest Territories, Saskatchewan, Yukon Territory), USA (Colorado, North Dakota, South Dakota, Washington) (Plate 260:12).

51 *Haliplus (Haliplus) steppensis* Guignot

Diagnosis: Length 2.7–3.0 mm, width 1.5–1.6 mm. Body oval, widest in the middle. The combination of the weak elytral lines without forming spots and the weak metasternal impressions, the shape of the penis in the males and the presence of elytral micropunctation in the females should distinguish this species from related species (Plates 42 and 238:6).
Full description by Vondel et al. (2006).
Biology: No details are known.
Immature stages: Unknown.
Distribution: China (Gansu, Heilongjiang, Inner Mongolia, Qinghai, Sichuan, Xizang), Mongolia, Russia (West Siberia) (Plate 271:7).

52 *Haliplus (Haliplidius) turkmenicus* Vondel

Diagnosis: Length 2.7–2.9 mm, width 1.5–1.6 mm. Body wide, widest just before the middle. Differs from *H. ruficollis* in the very reduced elytral lines and the different penis in the males (Plates 43 and 238:7).
Full description by Vondel et al. (2006).
Biology: No details known.
Immature stages: Unknown.
Distribution: Turkmenia (Plate 272:1).

53 *Haliplus* (*Haliplus*) *zacharenkoi* Gramma and Prysnyi

Diagnosis: Length 2.5–2.9 mm, width 1.4–1.6 mm. Body oval, widest in or just before the middle, shoulders weakly pronounced. In Europe this is the only species in the subgenus *Haliplus* s. str. without basal pronotal plicae. Although this species is characterized by the lack of basal pronatal plicae some specimens have one or two punctures suggesting basal plicae (Plates 44 and 238:8).
Biology: In salt lakes.
Full description by Vondel et al. (2006).
Immature stages: The 3rd instar larva is known. The actual knowledge is reviewed by Vondel (2025).
Distribution: Russia (South European Territory), Ukraine (Plate 272:2)

54 Subgenus *Liaphlus* Guignot

Diagnosis: Yellow to rust-coloured. Body small to large, body-outline usually approximately continuously oval. Head rather narrow to wide, distance between the eyes varying from 0.8 to 2.8× the width of one eye, apical segments of maxillary and labial palpi shorter than the penultimate segments. Pronotum widest at the base, with or without basal plicae. Elytra without distinct ridges, at most the intervals a little elevated, with or without maculation, with or without dark stripes on the about ten primary puncture-rows, with or without micropunctation. Ventral side yellow to rust-coloured, often with darkened parts. Prosternal apophysis flat, grooved or margined, usually narrowed near the coxae. Metasternal apophysis flat, grooved or pitted in the middle or with a pit or groove on each side of the middle. Hind coxal plates rounded posteriorly, not reaching beyond the fifth sternite. Apical sternite not grooved in the middle. Legs yellow to rust-coloured, often darkened towards the coxae. Hind tibia with or without a setiferous striole on the dorsal face, in the males the first one, two or three segments widened and ventrally provided with sucker-hairs.
Full description by Vondel (1997, 2021).
Biology: Members of the genus occur in a large variety of water-types, but the majority of species is found in stagnant water.
Distribution: Found in the Palaearctic, Oriental, Australasian, Afrotropical, Nearctic and Neotropical Regions.

55 *Haliplus* (*Liaphlus*) *abbreviatus* Wehncke

Diagnosis: Length 3.0 mm, width 1.7 mm. Body oval, widest in the middle (Plates 45 and 238:9).
Full description by Vondel (1991), Vondel & Litovkin (2017).
Biology: In rivers. Attracted to light.
Immature stages: Unknown.
Distribution: Afghanistan, China (Shanxi, Xinjiang), Cyprus, Egypt, Iran, Iraq, Israel, Kazakhstan, Lebanon, Syria, Turkey (Plate 272:4).

56 *Haliplus* (*Liaphlus*) *africanus* Aubé

Diagnosis: Length 3.1–3.5 mm, width 1.4–1.7 mm. Body subparallel, widest in the middle. This species closely resembles *H. coxaspinipes*, but can be distinguished by the absence of a sharp posterior spine on the metacoxal lobes (Plates 46 and 239:1).
Full description by Vondel (2010).
Biology: In pools.
Immature stages: Unknown.
Distribution: South Africa (Plate 278:6).

57 *Haliplus* (*Liaphlus*) *alastairi* Watts

Diagnosis: Length 3.0–3.6 mm, width 1.7–2.0 mm. Body oval, widest behind shoulders, clearly tapering in apical 1/3 part, This species can be distinguished from the related *H. timmsi* by the flat metasternal process of *H. alastairi* opposed to the metasternal process pitted on both sides of *H. timmsi* (Plates 47 and 239:2).
Full description by Watts (1988), Vondel (1995a).
Biology: Specimens were attracted to UV light placed in open forest.
Immature stages: Unknown.
Distribution: Australia (Northern Territory, Queensland, Western Australia) (Plate 276:11).

58 *Haliplus* (*Liaphlus*) *alluaudi* Régimbart

Diagnosis: Length 3.6–3.9 mm, width 2.0–2.2 mm. Body oval with regular outline, widest in the middle. The specific character of the metaventral process

having a longitudinal impression along the sides will distinguish this species from obviously related species (Plates 48 and 239:3).
Full description by Vondel (2010), Vondel & Bergsten (2012).
Biology: In temporary pools of river and attracted to light. Up to an altitude of 1620 m.
Immature stages: Unknown.
Distribution: Madagascar, Mauritius (Plate 278:7).

59 *Haliplus* (*Liaphlus*) *andalusicus* Wehncke

Diagnosis: Length 2.5–3.0 mm, width 1.4–1.7 mm. Body oval, widest in the middle. This species is most easily confused with *H. variegatus* from which it can be distinguished by the interrupted lateral plicae of the prosternal process and the apical spur of the hind tibia being as long as or longer than the first tarsal segment (Plates 49 and 239:4).
Fully described by Vondel (1991, 1997).
Biology: In fresh and especially brackish stagnant water in pools near the coast and more rarely inland.
Immature stages: Unknown.
Distribution: Algeria, France, Italy, Morocco, Portugal, Spain, Tunisia (Plate 272:3).

60 *Haliplus* (*Liaphlus*) *angusi* Vondel

Diagnosis: Length 3.1 mm, width 1.7 mm. Body oblong, subparallel, widest in the middle. This species can be distinguished from related ones by the very narrow prosternal process (Plates 50 and 239:5).
Full description by Vondel (1991).
Biology: Found in ponds. Lakes and streams.
Immature stages: Unknown.
Distribution: Russia (South European Territory, West Siberia) (Plate 272:5).

61 *Haliplus* (*Liaphlus*) *angustifrons* Régimbart

Diagnosis: Length 2.9–3.8 mm, width 1.6–2.0 mm. Body oval, widest in the middle. This species can be distinguished from the other species with a posteriorly well impressed prosternal process by the characters mentioned in the key (Plates 51 and 239:6).

Full description by Vondel (1992a), Sheth et al. (2016).
Biology: In rivers. In mixed primary forest. In watertank. Up to an altitude of 1500 m.
Immature stages: The 3rd instar larva is known. The actual knowledge is reviewed by Vondel (2025).
Distribution: Bangladesh, Burma, India, Laos, Sri Lanka, Thailand, Vietnam, Nepal, Pakistan, Japan (Nansei Islands) (Plate 275:6).

62 ***Haliplus* (*Liaphlus*) *annulatus*** **Roberts**

Diagnosis: Length 1.8–2.3 mm, width 1.0–1.3 mm. Body oval, widest in the middle. Small species with extended elytral maculation. Elytral margin smooth from shoulder to apex. Differs from *H. havaniensis* in the lateral grooves on the metaventral process instead of round impressions (Plates 52 and 239:7).
Full description by Vondel (2021).
Biology: Found in ponds and brooks.
Immature stages: Unknown.
Distribution: USA (Alabama, Arkansas, Delaware, Florida, Georgia, Illinois, Indiana, Massachusetts, Minnesota, Missouri, North Carolina, Ohio, South Carolina, Tennessee, Texas, Wisconsin), Mexico (Chiapas, Veracruz), Nicaragua (Plate 261:1).

63 ***Haliplus* (*Liaphlus*) *arrowi*** **Guignot**

Diagnosis: Length 2.8–3.1 mm, width 1.5–1.6 mm. Body oval, widest in the middle. This species is very closely related to *H. angustifrons*, but has sparser elytral puncture-rows, the eyes less close to each other, the posterior corners of the pronotum less protruding and some other slight differences, mentioned in the key (Plates 53 and 239:8).
Full description by Vondel (1993), Sheth et al. (2016).
Biology: Unknown.
Immature stages: Unknown.
Distribution: Bangladesh, Burma, India, Laos, Nepal, Pakistan (Plate 275:7).

64 *Haliplus* (*Liaphlus*) *astrakhanus* Vondel

Diagnosis: Length 3.4 mm, width 1.9 mm. Body oval, widest in the middle. This species is closely related to *H. abbreviatus* but in the males the top of the penis is narrower and more pointed (Plates 54 and 239:9).
Full description by Vondel (1991).
Biology: Unknown.
Immature stages: Unknown.
Distribution: Russia (South European Territory) (Plate 272:6).

65 *Haliplus* (*Liaphlus*) *australis* Clark

Diagnosis: Length 3.7–4.1 mm, width 2.1–2.3 mm. Body wide oval, widest just in front of middle. Males of *H. australis* have a solid digitus on the apex of the left paramere, while *H.testudo* lacks such a digitus. The penis of *H. testudo* is more curved towards the tip. The elytra of *H. australis* usually have no dark stripes on puncture rows, while *H. testudo* have well developed dark stripes on elytral puncture rows (Plates 55 and 240:1).
Full description by Vondel (1995a).
Biology: Specimens occur in rivers and ponds and have been found in trout stomachs. Also found in a well shaded woodland pond that dries out completely in summer.
Immature stages: Unknown.
Distribution: Australia (New South Wales, Queensland, South Australia, Victoria) (Plate 276:12).

66 *Haliplus* (*Liaphlus*) *bachmanni* Vidal Sarmiento and Grosso

Diagnosis: Length 3.0–3.4 mm, width: 1.8–2.1 mm. Body oval, widest in the middle. Pronotum and elytra with basal impression near fifth elytral puncture row. Metaventral process with four round impressions (Plates 56 and 240:2).
Full description by Vondel & Spangler (2008).
Biology: Unknown.
Immature stages: Unknown.
Distribution: Argentina, Brazil, Paraguay (Plate 266:5).

67 *Haliplus* (*Liaphlus*) *basinotatus* Zimmermann

Diagnosis: Length 3.1–4.0 mm, width 1.9–2.1 mm. Body widest just behind the well pronounced shoulders. This species can be distinguished from related species by the dark basal band on the elytra and its large size (Plates 57 and 240:3).
Full description by Vondel (1991).
Biology: Found in ponds, attracted to light.
Immature stages: The 3rd instar larva is known. The actual knowledge is reviewed by Vondel (2025).
Distribution: China (Heilongjiang, Liaoning, Jilin, Nei Mongol), Japan (Honshu, Kyushu, Shikoku), North Korea, Russia (Far East), Mongolia, South Korea (Plate 272:7).

68 *Haliplus* (*Liaphlus*) *bonariensis* Steinheil

Diagnosis: Length 3.8–4.2 mm, width 2.1–2.3 mm. Body subparallel. Elytra not maculate. Metaventral process with long groove on each side, which character distinguishes this species from *H. fuscipennis* (Plates 58 and 240:4).
Full description by Vondel & Spangler (2008).
Biology: Collected in a river.
Immature stages: Unknown.
Distribution: Argentina, Uruguay. (Plate 266:6).

69 *Haliplus* (*Liaphlus*) *brasiliensis* Zimmermann

Diagnosis: Length 3.1 mm, width 1.9 mm. Shoulders pronounced, body tapering backwards. Prosternal process only weakly impressed in posterior part. Setiferous striole on dorsal face of hind tibia very long (Plates 59 and 240:5).
Full description by Vondel & Spangler (2008).
Biology: Unknown.
Immature stages: The 3rd instar larva is probably known. The actual knowledge is reviewed by Vondel (2025).
Distribution: Brazil (Plate 266:7).

70 *Haliplus* (*Liaphlus*) *camposi* Guignot

Diagnosis: Length 3.2–3.6 mm, width 2.0–2.2 mm. Body oval, widest in the middle. Pronotum and elytra with basal impression. Elytra strongly maculate. Very similar to *H. tocumenus*, which has the pronotum not margined (Plates 60 and 240:6).
Full description by Vondel & Spangler (2008).
Biology: Collected in weedy roadside pools, in a roadside drainage ditch and collected at light.
Immature stages: Unknown.
Distribution: Colombia, Ecuador, Panama, Trinidad, Venezuela (Plate 266:8).

71 *Haliplus* (*Liaphlus*) *carinatus* Guignot

Diagnosis: Length 2.7–2.8 mm, width 1.7 mm. Body oval, but before apex with clear bend, widest in the middle. Characterized by the carinate impressed preapical part of the eighth elytral puncture row. The seventh row is apically also carinate, which character distinguishes it from the similar *H. unicarinatus* (Plates 61 and 240:7).
Full description by Vondel & Spangler (2008).
Biology: Unknown.
Immature stages: Unknown.
Distribution: Cuba (Plate 265:2).

72 *Haliplus* (*Liaphlus*) *chinensis* Falkenström

Diagnosis: Length 3.9–4.2 mm, width 2.1–2.2 mm. Body oval, widest in the middle. This species is closely related to *H.ovalis*. The division between the two is mainly based on the smoothly curved penis in H. chinensis while H. ovalis has the penis abruptly bent in the apical part. Probably *H. ovalis* and *H. chinensis* are conspecific (Nakajima et al. 2020, Hayashi et al. 2023) (Plates 62 and 240:8).
Full description by Vondel (1991).
Biology: Found in ponds and pools to an altitude of 1800 m.
Immature stages: Unknown.
Distribution: China (Beijing, Fujian, Guizhou, Hebei, Hubei, Jiangsu, Jiangxi, Nei Mongol, Qinghai, Shaanxi, Shandong, Shanghai, Shanxi, Sichuan, Xinjiang, Yunnan, Zhejiang), Korea (Plate 272:8).

73 *Haliplus* (*Liaphlus*) *colombiensis* Vondel and Spangler

Diagnosis: Length 2.9 mm, width 1.7 mm. Body oval to subparallel, widest in the middle. Base of fifth elytral puncture row impressed. Metaventral process with characteristic pattern of strong and deep pits (Plates 63 and 240:9).
Full description by Vondel & Spangler (2008).
Biology: Unknown.
Immature stages: Unknown.
Distribution: Colombia (Plate 266:9).

74 *Haliplus* (*Liaphlus*) *columbiensis* Wallis

Diagnosis: Length 3.5–4.0 mm, width 1.9–2.15 mm. Body oval to subparallel. Large species, more than 3.5 mm long. The combination of the metaventral process not being impressed posteriorly, the maculated elytra and the not darkened pronotal basal punctures will distinguish this species from related species like *H. fulvus*, *H. leechi*, *H. cylindricus* and *H. vancouverensis* (Plates 64 and 241:1).
Full description by Vondel (2021).
Biology: Found in rivers.
Immature stages: Unknown.
Distribution: Canada (Alberta, British Columbia, Northwest Territories), USA (Colorado, Utah) (Plate 261:2).

75 *Haliplus* (*Liaphlus*) *concolor* LeConte

Diagnosis: Length 2.9 mm, width 1.7 mm. Body oval, widest behind the middle. A small species, not longer than about 3.0 mm, without elytral maculation and with serrate elytral apex (Plates 65 and 241:2).
Full description by Vondel (2021).
Biology: Found in lakes, ponds, marshes and swamps.
Immature stages: Unknown.
Distribution: USA (Arizona, California, Colorado, Oregon, Pennsylvania?, Texas, Utah, Washington), Mexico (Baja California, Nayarit) (Plate 261:3).

76 *Haliplus* (*Liaphlus*) *confluentus* Roberts

Diagnosis: Length 2.7–2.8 mm, width 1.5–1.6 mm. Body oval, widest before the middle. Closely resembling *H. annulatus*, but a little longer and with serrate shoulders (Plates 66 and 241:3).
Full description by Vondel (2021).
Biology: Found on beaches, in pools, ponds, lakes, creeks, marshes.
Immature stages: Unknown.
Distribution: USA (District of Columbia, Florida, Georgia, Louisiana, Maryland, Missouri, North Carolina, South Carolina, Virginia), Mexico (VeraCruz), Bahamas (Plate 261:4).

77 *Haliplus* (*Liaphlus*) *costaricanus* Vondel and Spangler

Diagnosis: Length 3.0 mm, width 1.9 mm. Body short oval, strongly tapering backwards, widest just behind the shoulders. Resembles *H. signatus* or *H. tumidus*, but has the metaventral process with a diverging row of strong punctures in a common impression on each side (Plates 67 and 241:4).
Full description by Vondel & Spangler (2008).
Biology: Unknown.
Immature stages: Unknown.
Distribution: Costa Rica (Plate 265:3).

78 *Haliplus* (*Liaphlus*) *coxaspinipes* Vondel

Diagnosis: Length 3.1–3.7 mm, width 1.6–1.9 mm. Body subparallel, widest in the middle. This species closely resembles *H. africanus*, but can be distinguished by the presence of a sharp posterior spine on the metacoxal lobes (Plates 68 and 241:5).
Full description by Vondel (2010).
Biology: In pools and (probably stagnant parts of) rivers.
Immature stages: Unknown.
Distribution: South Africa (Plate 278:8)

79 *Haliplus* (*Liaphlus*) *crassus* Chapin

Diagnosis: Length 3.7–4.2 mm, width 2.1–2.5 mm. Body wide oval, widest before the middle. Large species with distinct maculation. Characterized by the very wide prosternal and metaventral process (Plates 69 and 241:6).
Full description by Vondel & Spangler (2008).
Biology: Collected in clear pond overgrown with aquatic vegetation. The following aquatic Coleoptera were found in association with *H. crassus* in Panama, Penonome: Haliplidae: *Haliplus gravidus*, Dytiscidae: *Brachyvatus*, Bidessini, *Derovatellus*, *Hydrovatus*, *Laccophilus*, *Megadytes*, *Pachydrus*, *Thermonectus*, Hydrochidae: *Hydrochus*, Hydrophilidae: *Berosus*, *Derallus*, *Helochares*, *Hydrobiomorpha*, *Hydrophilus*, *Tropisternus*, Noteridae: *Hydrocanthus*, *Suphis*, *Suphisellus*.
Immature stages: Unknown.
Distribution: Brazil, Panama, Paraguay, Peru, Venezuela (266:10).

80 *Haliplus* (*Liaphlus*) *cribrarius* LeConte

Diagnosis: Length 4.3–4.7 mm, width 2.3–2.6 mm. Body oval, widest just before the middle. Strongly resembling *H. nitens*, but longer than 4.3 mm, while the maximum length of *H. nitens* is 4.0 mm. The elytral maculation is generally coarser than in *H. nitens*. In males the width of the parallel tip of the penis is about 1.8× the length, while in *H. nitens* this is about 2.2× (Plates 70 and 241:7).
Full description by Vondel (2021).
Biology: Found in creeks, ponds, ditches, lakes and swamps. Attracted to light. Occurs up to an altitude of 3100 m.
Immature stages: The 2nd and 3rd instar larva are known. The actual knowledge is reviewed by Vondel (2025).
Distribution: Canada (Alberta, British Columbia, Manitoba, New Brunswick, Newfoundland & Labrador, Northwest Territories, Nova Scotia, Ontario, Prince Edward Island, Quebec, Saskatchewan, Yukon), St. Pierre & Miquelon, USA (Alaska, Colorado, Connecticut, Dakota [N. or S.?], Illinois, Indiana, Maine, Massachusetts, Michigan, Minnesota, Nebraska, New Hampshire, New Jersey, New York, Pennsylvania, Vermont, Wisconsin) (Plate 261:5).

81 *Haliplus* (*Liaphlus*) *cubensis* Chapin

Diagnosis: Length 2.6 mm, width 1.6 mm. Body wide oval, widest in the middle. Relatively small. Shoulders at most weakly raspy. Resembles *H. youngi* very much, but the latter is more than 2.9 mm long and has the shoulders distinctly raspy (Plates 71 and 241:8).
Full description by Vondel & Spangler (2008).
Biology: Collected in a river and at light.
Immature stages: Unknown.
Distribution: Bahamas, Cayman Islands, Cuba, Mexico, USA (Virgin Islands) (Plate 265:4).

82 *Haliplus* (*Liaphlus*) *cylindricus* Roberts

Diagnosis: Length 4.0–4.2 mm, width 2.1–2.3 mm. Body oblong, parallel in the middle. Related to *H. vancouverensis*, but with darkened elytral punctures (Plates 72 and 241:9).
Full description by Vondel (2021).
Biology: Found in lakes.
Immature stages: Unknown.
Distribution: Canada (British Columbia), USA (Arizona, California) (Plate 261:6).

83 *Haliplus* (*Liaphlus*) *dalmatinus* G. Müller

Diagnosis: Length 3.2–3.9 mm, width 1.8–2.2 mm. Body oval, sides somewhat parallel, widest in the middle. This species is closely related to *H. fulvus*, but the latter has shorter metatarsi, a more narrowed prosternal process, in the males a longer digitus on the right paramere and generally is of a larger size (Plates 73 and 242:1).
Full description by Vondel (1991, 1997).
Biology: Unknown.
Immature stages: Unknown.
Distribution: Albania, Azerbaijan, Bosnia Herzegovina, Croatia, Georgia, Greece, Montenegro, ? Transcaucasia, Turkey (Plate 272:9).

84 *Haliplus* (*Liaphlus*) *diopus* Guignot

Diagnosis: Length 2.8–3.1 mm, width 1.6–1.7 mm. Body oval, widest in or just before the middle. This species can be distinguished from *H. pulchellus* by the wider distance between the eyes and from *H. indicus* by the clearly separated pits on the metasternal process (Plates 74 and 242:2).
Full description by Vondel (1993).
Biology: Unknown.
Immature stages: Unknown.
Distribution: Vietnam (Plate 275:8).

85 *Haliplus* (*Liaphlus*) *discessus* Guignot

Diagnosis: Length 3.5–3.8 mm, width 2.1–2.2 mm. Body with regular outline, widest in the middle. The strongly concave lateral sides of the pronotum and the raspy shoulders generally may distinguish this species from related species (Plates 75 and 242:3).
Full description by Vondel (2010).
Biology: In ponds, brooklets, rivers, swamps. Attracted to light.
Immature stages: Unknown.
Distribution: Benin, Burkina Faso, Cameroun, Gambia, Ghana, Guinea, Ivory Coast, Mali, Nigeria, Somalia, Senegal, Sudan, Tchad, Togo, Zaire (Plate 278:9).

86 *Haliplus* (*Liaphlus*) *drechseli* Vondel and Spangler

Diagnosis: Length 2.2–2.6 mm, width 1.3–1.5 mm. Body short oval, widest in the middle. Small species with vague elytral maculation. Can be distinguished from *H.ornatipennis* by the prosternal process not being narrowed near the coxae (Plates 76 and 242:4).
Full description by Vondel & Spangler (2008).
Biology: Unknown.
Immature stages: Unknown.
Distribution: Argentina, Paraguay (Plate 266:11).

87 *Haliplus* (*Liaphlus*) *ebolovensis* Guignot

Diagnosis: Length 3.6–3.7 mm, width 2.1–2.2 mm. Body with regular outline, widest in the middle. This species can be distinguished from related species such as *H. lamottei* by the parallel sided prosternal process (Plates 77 and 242:5).
Full description by Vondel (2010).
Biology: Unknown.
Immature stages: Unknown.
Distribution: Cameroun, Zaire (Plate 278:10).

88 *Haliplus* (*Liaphlus*) *elsaltous* Vondel and Spangler

Diagnosis: Length 2.1–2.7 mm, width 1.1–1.4 mm. Body narrow, subparallel, widest in the middle. Small subparallel to parallel species. Resembles *H. lewisii*, but can be distinguished by the narrow prosternal process, being strongly concave when seen from the front (Plates 78 and 242:6).
Full description by Vondel & Spangler (2008).
Biology: Collected in a temporary grassy meadow pool and pond with grassy margins. The following aquatic Coleoptera were found in association with *H. elsaltous* in Mexico, El Salto: Dytiscidae: *Deronectes*, Haliplidae: *Haliplus grandis*, *Peltodytes mexicanus*, Hydrophilidae: *Helochares* and in Mexico, La Ciudad: Dytiscidae: *Agabus*, *Bidessus*, *Dytiscus*, *Hydroporus*, *Hygrotus*, Hydraenidae: *Ochthebius*, Hydrophilidae: *Berosus*, *Helophorus*.
Immature stages: Unknown.
Distribution: Mexico (Plate 265:5).

89 *Haliplus* (*Liaphlus*) *excoffieri* Vondel

Diagnosis: Length 3.7–3.8 mm, width 1.9–2.0 mm. Body oblong oval, tapering apically, widest in or just before the middle. This species is closely related to *H. chinensis* and *H. ovalis*, but the distance between the eyes is usually less than 1.4× eye-width, the darkened suture is only reaching to the first secondary puncture-row and in the males the penis is clearly widened before the apex (Plates 79 and 242:7).
Full description by Vondel (1991).
Biology: Found in ponds.
Immature stages: Unknown.
Distribution: China (Guizhou, Sichuan, Yunnan) (Plate 272:10).

90 *Haliplus* (*Liaphlus*) *eximius* Clark

Diagnosis: Length 3.4–3.9 mm, width 1.9–2.1 mm. Body oval, widest in or just before the middle. This species can be distinguished from related species in the East Palaearctic and Oriental region by the total lack of maculation between the elytral puncture-rows. It resembles the West Palaearctic *H. flavicollis*, which however lacks the lateral plicae in the anterior part of the prosternal process (Plates 80 and 242:8).
Full description by Vondel (1991, erroneously as *H. eximis*).
Biology: Found in lakes up to an altitude of 2400 m.
Immature stages: The 3rd instar larva and pupa are known. The actual knowledge is reviewed by Vondel (2025).
Distribution: Indonesia (Sumatra), Vietnam, China (Beijing, Fujian, Guangdong, Guizhou, Hebei, Heilongjiang, Hunan, Jiangsu, Jiangxi, Liaoning, Shanghai, Sichuan, Xinjiang, Yunnan, Zhejiang), Japan (Honshu), Korea (Plate 272:11).

91 *Haliplus* (*Liaphlus*) *exsecratus* Guignot

Diagnosis: Length 3.2–3.6 mm, width 1.8–1.9 mm. Body rounded, widest in the middle. This species can be distinguished from the related *H. rufescens* by the prosternal process being only slightly wider posteriorly than anteriorly (Plates 81 and 242:9).
Full description by Vondel (2010).
Biology: In ponds, pools, rivers, rain pools.
Immature stages: The 3rd instar larva is probably known. The actual knowledge is reviewed by Vondel (2025).
Distribution: South Africa, Zimbabwe (Plate 278:11).

92 *Haliplus* (*Liaphlus*) *ferruginipes* Régimbart

Diagnosis: Length 3.3–3.6 mm, width 1.8–1.9 mm. Body oval, widest in middle. This species can be distinguished from related species by the widely darkened elytral base in combination with puncture-row 5 not impressed basally (Plates 82 and 243:1).
Full description by Vondel (1995).
Biology: Unknown.
Immature stages: Unknown.
Distribution: Australia (Northern Territory, Queensland), Indonesia (West New Guinea), Papua New Guinea (Plate 277:1).

93 *Haliplus (Liaphlus) figuratus* J. Sahlberg

Diagnosis: Length 2.6 mm, width 1.4 mm. Body subparallel. The immaculate pronotum and the metaventral process with two impressions distinguish this species from related species without serrate elytral apex (Plates 83 and 243:2).
Full description by Vondel (2010).
Biology: Unknown.
Immature stages: Uknown.
Distribution: Egypt (Plate 278:12).

94 *Haliplus (Liaphlus) fiorentini* Benetti and Hamada

Diagnosis: Length 4.0 mm, width 2.37 mm. Body oval, tapering backwards, widest just behind the shoulders, with narrow pronotum and shoulders slightly pronounced. This species can be distinguished from other neotropical species of *Haliplus* by the following combination of characteristics: elytra yellow-brown with extensive and well-defined maculation consisting of black patches, prosternal process parallel, straight, wider and short, about 1.15 times longer than wide at base, metaventral process with a long diverging groove on each side and not impressed in the middle, and the median lobe of aedeagus in lateral view curved, subparallel, gradually narrowed, distally elongate and with rounded apex (Plates 84 and 243:3).
Full description by Benetti & Hamada (2017).
Biology: Collected in a shallow pond with macrophytes near the shores.
Immature stages: Unknown.
Distribution: Brazil (Plate 266:12).

95 *Haliplus (Liaphlus) flavicollis* Sturm

Diagnosis: Length 3.6–4.1 mm, width 1.8–2.1 mm. Body fairly slender, sometimes subparallel, widest before the middle. This species is related to *H. rubidus*. In *H. rubidus* the basal elytral punctures are larger than the distance between the punctures of the first and second row in the basal part, while in *H. flavicollis* these punctures are clearly smaller than that distance. In specimens from Morocco, however, that difference is less distinct. The males of *H. flavicollis* have the left paramere narrower at the base than in *H. rubidus*. Specimens with vague blotches on the elytra can be distinguished from *H. fulvus* by the lack of the lateral plicae in the anterior part of the prosternal process (Plates 85 and 243:4).

Full description by Vondel (1991, 1997).
Biology: In clean and clear stagnant or very slowly running water of lakes, ditches, canals and pools with rich vegetation, especial Characeans (Seeger, 1971c), Najadaceans (Burmeister, 1939) or *Cabomba*-beds in Kazakhstan (Konev, 1976). Also recorded from brackish water. The adults feed on eggs and larvae of Chironomidae (Seeger, 1971c, but his observation probably concerns *H. fulvus*, as he gave a picture of the mandible of *fulvus* under the name of *flavicollis*), aquatic macro-invertebrates as *Asellus* (only when already or almost dead), *Sialis* larvae (probably predated), dead water beetles, even of their own kind and probably Characeans or filamentous algae. Seeger (1971a, b, c) provided much information on the life-cycle. This species is capable to fly (Jackson, 1973, taken at light).
Immature stages: The 2nd and 3rd instar larva and the pupa are known. The actual knowledge is reviewed by Vondel (2025).
Distribution: Austria, Azerbaijan, Belgium, Bosnia Herzegovina, Bulgaria, Byelorussia, Croatia, Czech Republic, Denmark, Estonia, Finland, France, Georgia, Germany, Great Britain, Greece, Hungary, Iran, Iraq, Ireland, Italy, Kazakhstan, Latvia, Lithuania, Luxembourg, Macedonia, Mongolia, Montenegro, Netherlands, Norway, Poland, Romania, Russia (Central European Territory, North Euopean Territory, South European Territory, West Siberia), Serbia, Slovakia, Slovenia, Spain, Sweden, Switzerland, Transcaucasia, Turkestan, Turkey, Ukraine, Uzbekistan, Yugoslavia (Plate 273:1).

96 *Haliplus (Liaphlus) fulvus* (Fabricius)

Diagnosis: Length 3.6–4.5 mm, width 2.0–2.4 mm. Body elongate oval, widest in the middle. This species is very variable in dimension, maculation and coloration. Of all Haliplidae it has the largest distribution, covering most of the palaearctic and the north of the nearctic region. Mainly because of the variation in elytral maculation a number of varieties, subspecies and new species as *H. lapponum* and *H. pyraeneus* have been described, but they are all considered to be conspecific with *H. fulvus*. In North America closely related to *H. leechi*, but basal pronotal punctures not stronger than primary elytral discal punctures on second and third row (Plate 86 and 243:5).
Full description by Vondel (1991, 1997, 2021).
Biology: In clean clear or peaty stagnant or slowly running water of lakes, pools, canals, rivers and ditches with rich vegetation, especially Characeans. Even recorded from brackish water. The larvae probably feed on Characeans

such as *Nitella*, while the adults are omnivorous with a diet of algae, pollen and invertebrates like Oligochaeta (Seeger, 1971b). The beetle was seen flying and is caught at light.
Immature stages: The 3rd instar larva and the pupa are known. The actual knowledge is reviewed by Vondel (2025).
Distribution: Canada (Alberta, British Columbia, Manitoba, New Brunswick, Newfoundland & Labrador, Northwest Territories, Nova Scotia, Ontario, Quebec, Saskatchewan, Yukon), USA (Arizona, California, Colorado, Idaho, Illinois, Indiana, Maine, Massachusetts, Michigan, Minnesota, Montana, Nebraska, Nevada, New Mexico, New York, North Dakota, Oregon, South Dakota, Utah, Vermont, Washington, Wisconsin, Wyoming), Albania, Algeria, Andorra, Austria, Azerbaijan, Belgium, Bosnia Herzegovina, Bulgaria, Byelorussia, Caucasus, Croatia, Czech Republic, Denmark, Estonia, Faeroe Islands, Finland, France, Germany, Great Britain, Greece, Hungary, Iceland, Iraq, Ireland, Italy, Kazakhstan, Latvia, Lithuania, Luxembourg, Macedonia, Mongolia, Morocco, Netherlands, Norway, Poland, Portugal, Romania, Russia (Central European Territory, East Siberia, North European Territory, South European Territory, West Siberia), Serbia, Slovakia, Slovenia, Spain, Sweden, Switzerland, Transcaspia, Transcaucasia, Turkestan, Turkey, Ukraine, Yugoslavia (Plate 273:3).

97 *Haliplus* (*Liaphlus*) *fuscipennis* Germain

Diagnosis: Length 4.0 mm, width 2.0 mm. Body long oval, subparallel, widest in the middle, shoulders weakly pronounced. Body parallel to subparallel. Elytra not maculate. Metaventral process with short groove on each side. Very similar to *H. valdiviensis*, which is larger and has the elytra vaguely maculate (Plates 87 and 243:6).
Full description by Vondel & Spangler (2008).
Biology: Unknown.
Immature stages: Unknown.
Distribution: Chile (Plate 267:1)

98 *Haliplus* (*Liaphlus*) *gamma* Vondel and Bergsten

Diagnosis: Length 2.7 mm, width 1.7 mm. Body oval. The very peculiar elytral marks (resembling Greek characters) distinguish it from related species (Plates 88 and 243:7).

Full description by Vondel & Bergsten (2012).
Biology: Unknown.
Immature stages: Unknown.
Distribution: Madagascar (Plate 279:1).

99 *Haliplus (Liaphlus) garambanus* Guignot

Diagnosis: Length 2.9–3.5 mm, width 1.7–1.9 mm. Body subparallel. This species can be distinguished from related species by the metaventral process having an anteriorly angulate impression on each side (Plates 89 and 243:8).
Full description by Vondel (2010).
Biology: Unknown.
Immature stages: Unknown.
Distribution: Zaire (Plate 279:2).

100 *Haliplus (Liaphlus) genieri* Vondel

Diagnosis: Length 3.4 mm, width 2.0 mm. Body oval to subparallel. The anteriorly very wide prosternal process allows to distinguish it from related species (Plates 90 and 243:9).
Full description by Vondel (2010).
Biology: Attracted to light.
Immature stages: Unknown.
Distribution: Senegal (Plate 279:3).

101 *Haliplus (Liaphlus) gracilis* Roberts

Diagnosis: Length 3.6–3.8 mm, width 1.8–1.9 mm. Related to *H. cribrarius* and *H. nitens*, but without a dark mark on anterior margin of pronotum (Plates 91 and 244:1).
Full description by Vondel (2021).
Biology: Found in ponds.
Immature stages: Unknown.
Distribution: USA (California, Nevada, Oregon, Washington) (Plate 261:7).

102 *Haliplus* (*Liaphlus*) *gravidoides* Vondel and Spangler

Diagnosis: Length 3.1–3.3 mm, width 1.8–1.9 mm. Body oval, slightly tapering backwards, widest in the middle. Resembles *H. gravidus* very much, but can be distinguished by the confluent elytral maculation instead of the striped appearance of *H. gravidus* (Plates 92 and 244:2).
Full description by Vondel & Spangler (2008).
Biology: Collected in ponds and in a weedy roadside ditch. The following aquatic and semi-aquatic Coleoptera were found in association with *H. gravidoides* in Mexico, Campeche: Dytiscidae: *Bidessus, Hydrovatus, Laccophilus, Neoclypeodytes, Thermonectus*, Haliplidae: *Haliplus signatus*, Hydrophilidae: *Berosus, Hydrophilus, Tropisternus*, Noteridae: *Hydrocanthus, Suphis.*
Immature stages: Unknown.
Distribution: Belize, Cuba, Guadeloupe, Mexico (Plate 265:6).

103 *Haliplus* (*Liaphlus*) *gravidus* Aubé

Diagnosis: Length 2.8–3.5 mm, width 1.7–2.2 mm. Body wide, strongly pronounced shoulders, widest before the middle. Elytra with striped maculation, shoulders pronounced. Metaventral process on each side with long groove, in which two impressions may be recognized. *H. gravidoides* may be confused with this species, but has a transverse extended elytral maculation instead of stripes on puncture rows. This species is very variable in shape. Body from subparallel to distinctly tapering apically. Anterior margin of pronotum usually, but not always elevated (Plates 93 and 244:3).
Full description by Vondel & Spangler (2008).
Biology: Collected in ponds and pools (with vegetation like *Chara* or grass), in drainage ditches, in rivers and at light. The following aquatic Coleoptera were found in association with *H. gravidus* in Puerto Rico, in a weedy roadside ditch near Los Torres: Dytiscidae: *Copelatus, Laccophilus*, Hydrophilidae: *Enochrus*, Hydrobiomorpha. In Panama, Los Algarrobos, it was found in association with: Dytiscidae: *Thermonectus*, Hydrophilidae: *Berosus, Hydrophilus, Tropisternus*, Noteridae: *Hydrocanth*us. In Panama, Penonome, it was found in association with: Dytiscidae: *Brachyvatus*, Bidessini, *Derovatellus, Hydrovatus, Laccophilus, Megadytes, Pachydrus, Thermonectus*, Haliplidae: *Haliplus crassus*, Hydrochidae: *Hydrochus*, Hydrophilidae: *Berosus, Derallus, Helochares*, Hydrobiomorpha, *Hydrophilus, Tropisternus*, Noteridae: *Hydrocanthus, Suphis, Suphisellus*. In the Dominican Republic, Guerra, it was found in association with Dytiscidae:

Anodocheilus, Derovatellus, Laccophilus, Thermonectus, Hydrophilidae: *Enochrus, Tropisternus lateralis, T. collaris*, Hydrochidae: *Hydrochus*, Noteridae: *Hydrocanthus, Notomicrus sharpi.*
Immature stages: Unknown.
Distribution: Antigua, Argentina, Barbados, Bolivia, Bonaire, Brazil, Colombia, Costa Rica, Curaçao, Dominican Republic, Ecuador, Galapagos Islands, Guadeloupe, Guatemala, Guyana, Haiti, Margarita, Maria Galante, Mexico, Nicaragua, Panama, Paraguay, Puerto Rico, Saint Lucia, Saint Martin, Surinam, Trinidad, Venezuela (Plate 266:4).

104 ***Haliplus* (*Liaphlus*) *guignoti* Vondel**

Diagnosis: Length 2.9–3.2 mm, width 1.6–1.7 mm. Body subparallel, clear angle between elytral and pronotal margin. The parallel shape and the almost equal primary and secondary elytral punctures clearly separates it from related species (Plates 94 and 244:4).
Full description by Vondel (2010).
Biology: Attracted to light.
Immature stages: Unknown.
Distribution: D.R. Congo, Tanzania (Plate 279:4).

105 ***Haliplus* (*Liaphlus*) *guttatus* Aubé**

Diagnosis: Length 3.6–4.1 mm, width 1.7–2.1 mm. Body parallel to subparallel. This species is very variable in shape, from subparallel to strongly parallel, maculation not always clearly visible. The large basal punctures on the pronotum, the narrow prosternal process and the subparallel to parallel shape of the body distinguishes it from related species (Plates 95 and 244:5).
Full description by Vondel (1991, 1997).
Biology: In stagnant water, fresh or brackish, even in brooks.
Immature stages: Unknown.
Distribution: Algeria, Croatia, France, Greece, Hungary, Italy, Malta, Morocco, Spain, Tunisia (Plate 273:2).

106 *Haliplus* (*Liaphlus*) *halsei* Watts and McRae

Diagnosis: Length 3.5–4.1 mm. Width 2.2–2.5 mm. Body broad, somewhat spindle-shaped, broadest at shoulders, subparallel in middle (Plates 96 and 244:6).
Full description by Watts & McRae (2010).
Biology: Found in a large fresh, turbid, seasonal, open water claypan with sand and clay sediment.
Immature stages: The 1st and 3rd instar larva are known. The actual knowledge is reviewed by Vondel (2025).
Distribution: Australia (Western Australia) (Plate 277:2).

107 *Haliplus* (*Liaphlus*) *havaniensis* Wehncke

Diagnosis: Length 2.8–3.1 mm, width 1.7–1.9 mm. Body oval and fairly short (length more than 1.7× width). widest before the middle. Pronotum and elytra maculate. Metaventral process with round pit on each side, sometimes followed by a row of punctures or even a weak groove. May be confused with *H. cubensis*, which is shorter (length less than 1.5× width). Comparison of the male genitalia, especially the tip of the aedeagus, are sometimes necessary for a reliable determination (Plates 97 and 244:7).
Full description by Vondel & Spangler (2008).
Biology: Collected in a cattle pool with Characeans, in brackish pools and at light.
Immature stages: Unknown.
Distribution: USA (Florida, Texas), Mexico (Chiapas), Bahama Islands, Cuba, Dominican Republic (Plate 261:8).

108 *Haliplus* (*Liaphlus*) *heppneri* Vondel and Spangler

Diagnosis: Length 2.4–2.5 mm, width 1.4–1.5 mm. Body short oval, slightly tapering, widest just before the middle. Elytra with isolated marks, suture not darkened. Metaventral process with four round pits (Plates 98 and 244:8).
Full description by Vondel & Spangler (2008).
Biology: Collected in subtropical moist forest.
Immature stages: Unknown.
Distribution: Peru (Plate 267:2).

109 *Haliplus* (*Liaphlus*) *incrassatus* Régimbart

Diagnosis: Length 4.0 mm, width 2.4 mm. Body oval to subparallel in the middle. The immaculate pronotum and elytra, together with the serrate shoulder and elytral apex can distinguish it from related species (Plates 99 and 244:9).
Full description by Vondel (2010), Vondel & Bergsten (2012).
Biology: Unknown.
Immature stages: Unknown.
Distribution: Madagascar (Plate 279:5).

110 *Haliplus* (*Liaphlus*) *indicus* Régimbart

Diagnosis: Length 2.8–3.1 mm, width 1.6–1.7 mm. Body oval, widest in or just before the middle. This species, belonging to the group with double grooved prosternal process, can be distinguished from *H. pulchellus* by the wider distance between the eyes (Plates 100 and 245:1).
Full description by Vondel (1993).
Biology: Unknown.
Immature stages: Unknown.
Distribution: Bangladesh, India, Nepal (Plate 275:9).

111 *Haliplus* (*Liaphlus*) *indistinctus* Zimmermann

Diagnosis: Length 2.8–3.6 mm, width 1.5–2.0 mm. Body oval, elytral maculation fairly weak and not connected to sutural darkening (Plates 101 and 245:2).
Full description by Vondel & Spangler (2008).
Biology: Collected in temporary and permanent pools and at light.
Immature stages: The 1st and 2nd instar larva are known. The actual knowledge is reviewed by Vondel (2025).
Distribution: Argentina, Bolivia, Brazil, Chile, Colombia, Paraguay, Trinidad, Venezuela (Plate 267:3).

112 *Haliplus* (*Liaphlus*) *insularis* Guignot

Diagnosis: Length 3.5 mm. Width 2.0 mm. Body oval to subparallel in the middle. The marks on the even intervals of the posterior half of the elytra should distinguish it from related species (Plates 102 and 245:3).
Full description by Vondel & Bergsten (2012).

Biology: Unknown.
Immature stages: Unknown.
Distribution: Madagascar (Plate 279:6).

113 *Haliplus* (*Liaphlus*) *javanicus* Vondel

Diagnosis: Length 2.8–3.0 mm, width 1.5–1.7 mm. Body oval, widest in the middle. This species can be distinguished from other species with a double grooved prosternal process by the combination of the central pronotal blotch, the smooth elytral margin near shoulders and the elytral basal darkening reaching sixth puncture-row (Plates 103 and 245:4).
Full description by Vondel (1993).
Biology: Unknown.
Immature stages: Unknown.
Distribution: Indonesia (Java) (Plate 275:10).

114 *Haliplus* (*Liaphlus*) *kantneri* Vondel

Diagnosis: Length 2.6–3.3 mm, width 1.5–1.8 mm. Body with regular outline, widest in the middle. The flat prosternal process allows to distinguish it from related species; specimens with a weak impression on each side in posterior half of the prosternal process are also assigned to this species (Plates 104 and 245:5).
Full description by Vondel (2010).
Biology: Attracted to light.
Immature stages: Unknown.
Distribution: Angola, Burkina Faso, Gambia, Kenya, South Africa, Sudan, Tanzania (Plate 279:7).

115 *Haliplus* (*Liaphlus*) *kapuri* Vazirani

Diagnosis: Length: 3.2 mm, width 1.7 mm, body oval, widest in the middle.
This species resembles *H. indicus* in many aspects, but the elytral maculation is not confluent with the sutural darkening. From *H. perroti* it can be istinguished by the smaller width between the eyes (Plate 105 and 245:6).
Full description by Vondel (1993), Vondel & Ghosh (2018).
Biology: Unknown.

Immature stages: Unknown.
Distribution: India (Plate 275:11).

116 *Haliplus* (*Liaphlus*) *kotoshonis* Kano and Kamiya

Diagnosis: Length 2.8–3.6 mm, width 1.6–2.0 mm. Body oval, widest in the middle. A very variable species. Specimens with and without black blotches or bands on the base of the elytra and specimens with weak or with very strong and close punctures on the head are considered to belong to this species (Plates 106 and 245:7).
Full description by Vondel (1991, as *H.diruptus*)
Biology: Found in ponds and ditches. Up to an altitude of 4000 m. Attracted to light.
Immature stages: Unknown.
Distribution: Burma, India, Vietnam, China (Anhui, Beijing, Fujian, Guangdong, Guizhou, Hainan, Hebei, Heilongjiang, Hong Kong, Hubei, Hunan, Jiangsu, Liaoning, Shaanxi, Shandong, Shanghai, Shanxi, Sichuan, Tianjin, Xinjiang, Yunnan), Japan (Okinawa, Ryukyu Islands), Korea, Laos, Russia (Far East), Taiwan, Vietnam (Plate 273:4).

117 *Haliplus* (*Liaphlus*) *lamottei* Legros

Diagnosis: Length 3.5 mm, width 1.9 mm. Body with regular outline, widest in the middle. The anteriorly widened prosternal process and the posterioly widened pronotal dark mark distinguish it from the related *H. ebolovensis* (Plates 107 and 245:8).
Full description by Vondel (2010).
Biology: Unknown.
Immature stages: Unknown.
Distribution: Guinea (Plate 279:8).

118 *Haliplus* (*Liaphlus*) *langleyi* Vondel and Spangler

Diagnosis: Length 3.3–3.6 mm, width 1.9–2.1 mm. Body oval. Elytral secondary punctures very dense in more than one row. Grooves on prosternal process posteriorly ending in a pit (Plates 108 and 245:9).
Full description by Vondel & Spangler (2008).

Biology: Collected at blacklight.
Immature stages: Unknown.
Distribution: Brazil, Ecuador (Plate 267:4).

119 *Haliplus* (*Liaphlus*) *leechi* Wallis

Diagnosis: Length 3.85–4.6 mm, width 2.1–2.4 mm. Body elongate oval, widest about in the middle. Related to *H. fulvus*, but discal elytral punctures of second and third row stronger than basal pronotal punctures opposite these rows and with dark mark on anterior edge of pronotum (Plates 109 and 246:1).
Full description by Vondel (2021).
Biology: This species is found in lakes, beaver ponds, ephemeral ponds and marshy areas along the headwaters of streams. Occurs up to an altitude of 3200 m.
Immature stages: Unknown.
Distribution: Canada (Alberta, British Columbia, Northwest Territories, Saskatchewan, Yukon), USA (Alaska, California, Colorado, Idaho, Montana, New Mexico, Nevada, North Dakota, Oregon, South Dakota, Texas, Utah, Washington, Wyoming), Mexico (Tamaulipas) (Plate 261:9).

120 *Haliplus* (*Liaphlus*) *lewisii* Crotch

Diagnosis: Length 2.2–2.4 mm, width 1.2–1.4 mm. Body oval, widest in the middle. Relative small species with variable elytral maculation. Prosternal process parallel, anterior margin weakly concave when seen from the front. Resembles *H. elsaltous*, which, however, has a strongly concave anterior margin of the prosternal process (Plates 110 and 246:2).
Full description by Vondel (2021).
Biology: Collected in ponds and swamps.
Immature stages: Unknown.
Distribution: USA (Indiana, Oklahoma, Texas, Wisconsin), Mexico (Tamaulipas) (Plate 261:10).

121 *Haliplus* (*Liaphlus*) *maculatus* Motschulsky

Diagnosis: Length 3.6–4.0 mm, width 1.9–2.2 mm. Body oval, widest in the middle. This species closely resembles small specimens of *H. fulvus*. The male can

be distinguished from *H. fulvus* by the narrow and somewhat pointed top of the penis (Plates 111 and 246:3).
Full description by Vondel (1991, 1997).
Biology: In temporary pools, in brooks.
Immature stages: The 3rd instar and probably the 1st and 2nd instar larva are known. The actual knowledge is reviewed by Vondel (2025).
Distribution: Afghanistan, Austria, Azerbaijan, Bosnia and Herzegovina, Czech Republic, Greece, Hungary, Israel, Iran, Iraq, ? Italy, Kazakhstan, Kyrgyzstan, Lebanon, ? Poland, Romania, Russia (South European Territory), Slovakia, Syria, Turkmenistan, Turkey, Ukraine, Yugoslavia (Plate 273:5).

122 ***Haliplus* (*Liaphlus*) *maculicollis* Zimmermann**

Diagnosis: Length 3.2–3.5 mm, width 1.9–2.1 mm. Body oval, widest in the middle, shoulders slightly pronounced. Characterized by the basally grooved third to seventh elytral puncture rows. Most parts of elytral puncture rows with dark stripe. Metaventral process very wide (Plates 112 and 246:4).
Full description by Vondel & Spangler (2008).
Biology: Collected in a river.
Immature stages: Unknown.
Distribution: Argentina, Uruguay (Plate 267:5).

123 ***Haliplus* (*Liaphlus*) *maculipennis* Schaum**

Diagnosis: Length 3.0–4.2 mm, width 1.6–2.3 mm. Body wide with subparallel sides. This species is very variable with respect to maculation and punctation. The presence of small dark marks on the ninth elytral puncture row is characteristic of this species (Plates 113 and 246:5).
Full description by Vondel (1991).
Biology: In pools and brooklets. Attracted to light.
Immature stages: Unknown.
Distribution: Benin, Egypt, Ethiopia, Ivory Coast, Mali, Mozambique, South Africa, Sudan, Tchad, Zaire (Plate 279:9).

124 *Haliplus* (*Liaphlus*) *madagascariensis* Vondel and Bergsten

Diagnosis: Length 2.3–2.7 mm, width 1.5 mm. Body oval to subparallel. The discal elytral dark mark and the absence of other markings distinguish it from related species (Plates 114 and 246:6).
Full description by Vondel & Bergsten (2012), Vondel & Matsumoto (2021).
Biology: Unknown.
Immature stages: Unknown.
Distribution: Madagascar (Plate 279:10).

125 *Haliplus* (*Liaphlus*) *megapunctatus* Vondel and Spangler

Diagnosis: Length 3.7 mm, width 2.4 mm. Body very wide oval, strongly tapering backwards, widest before the middle. Elytral puncture row in anterior part extremely strong. Metaventral process very wide (Plates 115 and 246:7).
Full description by Vondel & Spangler (2008).
Biology: Unknown.
Immature stages: Unknown.
Distribution: Brazil (Plate 267:6).

126 *Haliplus* (*Liaphlus*) *mesoamericanus* Vondel and Spangler

Diagnosis: Length 3.3–4.0 mm, width 2.0–2.3 mm. Body oval, widest just behind the shoulders. Prosternum and elytra with basal impression. Pronotum with central dark mark from anterior to posterior edge (Plates 116 and 246:8).
Full description by Vondel & Spangler (2008).
Biology: Collected in a roadside pond.
Immature stages: Unknown.
Distribution: Costa Rica, El Salvador, Guatemala, Honduras, Mexico, Nicaragua, Panama, Trinidad (Plate 265:7).

127 *Haliplus* (*Liaphlus*) *methneri* Zimmermann

Diagnosis: Length 2.7–3.7 mm, width 1.5–2.0 mm. Body oval, widest in the middle. This species closely resembles *H. mimulus*. The shape of the penis is slightly different. It is smoothly rounded on outer margin of *H. mimulus*, while

the outer margin is hooked before the top in *H. methneri*. Probably it is not more than a variation and are both species conspecific (Plates 117 and 246:9).
Full description by Vondel (2010).
Biology: In rice fields, water reservoirs. Attracted to light.
Immature stages: Unknown.
Distribution: Guinea-Bissau, Ivory Coast, Kenya, Mozambique, Namibia, Nigeria, Sudan, Tanzania, Tchad, Zimbabwe (Plate 279:11).

128 ***Haliplus* (*Liaphlus*) *mexicanus* Vondel and Spangler**

Diagnosis: Length 4.0–4.1 mm, width 2.3–2.4 mm. Body large oval, widest in the middle. Pronotum with large central mark from anterior to posterior edge. Prosternal and metaventral process very wide (Plates 118 and 247:1).
Full description by Vondel & Spangler (2008).
Biology: Unknown.
Immature stages: Unknown.
Distribution: Mexico (Plate 265:8).

129 ***Haliplus* (*Liaphlus*) *mimulus* Guignot**

Diagnosis: Length 2.9–3.9 mm, width 1.6–2.0 mm. Body oval, widest in the middle. This species closely resembles *H.methneri*. The shape of the penis is slightly different. In *H. mimulus* it is smoothly rounded on outer margin, while in *H. methneri* the outer margin is hooked before the top. Probably it is not more than a variation and are both species conspecific (Plates 119 and 247:2).
Full description by Vondel (2010).
Biology: In pools. Attracted to light.
Immature stages: Unknown.
Distribution: Botswana, Ghana, Guinea Bissau, Kenya, Mozambique, Senegal, Tanzania (Plate 279:12).

130 ***Haliplus* (*Liaphlus*) *minimus* Vondel and Spangler**

Diagnosis: Length 2.2–2.5 mm, width 1.3–1.5 mm. Small species, body oval, widest in the middle, side of pronotum making an angle to the shoulder of the elytra. Elytral punctures strong in anterior half. Metaventral process with long diverging row of strong punctures on each side (Plates 120 and 247:3).
Full description by Vondel & Spangler (2008).

Biology: Collected in roadside drainage ditch.
Immature stages: Unknown.
Distribution: Ecuador (Plate 267:7).

131 *Haliplus* (*Liaphlus*) *mucronatus* Stephens

Diagnosis: Length 4.0–4.4 mm, width 1.9–2.2 mm. This species can be distinguished from related species by the large distance between the eyes (more than 2.3× eye-width) and the metasternal process not being grooved or pitted in the middle (Plates 121 and 247:4).
Full description by Vondel (1991, 1997).
Biology: In fresh or brackish water, particularly in clay and gravel pits with Characeans.
Immature stages: The 1st, 2nd and 3rd instar larva are known. The actual knowledge is reviewed by Vondel (2025).
Distribution: Albania, Algeria, Austria, Belgium, Bulgaria, Croatia, France, Great Britain, Germany, Greece, Iran, Iraq, Italy, Kyrgyzstan, Libya, Macedonia, Montenegro, Morocco, Netherlands, Portugal, Russia (North European Territory), Slovenia, Spain, Switzerland, Tunisia, Turkey, Ukraine, Yugoslavia (Plate 273:7).

132 *Haliplus* (*Liaphlus*) *nanus* Guignot

Diagnosis: Length 2.2–2.4 mm, width 1.3–1.4 mm. Body very small, short oval, widest in the middle (Plates 122 and 247:5).
Full description by Vondel & Spangler (2008).
Biology: Unknown.
Immature stages: Unknown.
Distribution: Cuba (Plate 265:9).

133 *Haliplus* (*Liaphlus*) *napolovi* Vondel

Diagnosis: Length 2.4 mm, width 1.4 mm. Body wide oval, widest in the middle, strongly tapering in apical third. This species closely resembles *H. philippinus*, but has one impression on the metasternal process instead of two and the anterior edge of the prosternal process is strongly bent upwards, while in *H. philippinus* this is not the case (Plates 123 and 247:6).

Full description by Vondel (2003).
Biology: Unknown.
Immature stages: Unknown.
Distribution: Vietnam (Plate 275:12).

134 *Haliplus* (*Liaphlus*) *natalensis* Wehncke

Diagnosis: Length 3.3–3.7 mm, width 1.7–2.0 mm. Body with regular outline, widest in the middle. This species can be distinguished from related species with the same pronotal and elytral maculation by the metaventral process having one pit in the middle, in which two closely set finer pits may be present. In the other species the separate two pits are more evident and less close to each other (Plates 124 and 247:7).
Full description by Vondel (2010).
Biology: In ponds, ditches, rain pools, rivers. Attracted to light. Up to an altitude of 2400 m.
Immature stages: Unknown.
Distribution: Angola, Benin, Botswana, Burkina Faso, Ethiopia, Gabon, Ivory Coast, Kenya, Malawi, Mozambique, Namibia, Nigeria, Senegal, Sierra Leone, Somalia, South Africa, Tanzania, Togo, Uganda, Zaire, Zimbabwe (Plate 280:1).

135 *Haliplus* (*Liaphlus*) *nedungaduensis* Vondel

Diagnosis: Length 2.7 mm, width 1.5 mm. Body oval, widest in the middle. This species can be distinguished from the other species with double grooved prosternal process by the absence of elytral markings (Plates 125 and 247:8).
Full description by Vondel (1993).
Biology: Unknown.
Immature stages: Unknown.
Distribution: India (Plate 276:1).

136 *Haliplus* (*Liaphlus*) *nieseri* Vondel and Spangler

Diagnosis: Length 3.5–3.7 mm, width 2.2–2.3 mm. Body oval, slightly tapering backwards. Elytral punctures very strong in basal half of fourth to eighth row. Metaventral process with two confluent impressions on each side. Elytral margin apically not serrate (Plates 126 and 247:9).
Full description by Vondel & Spangler (2008).

Biology: Collected at light and in a sand pool.
Immature stages: Unknown.
Distribution: Brazil (Plate 267:8).

137 *Haliplus* (*Liaphlus*) *nigerianus* Vondel

Diagnosis: Length 2.9–3.0 mm, width 1.6–1.7 mm. Body with regular outline, widest in the middle. This species can be distinguished from related species by the immaculate pronotum (Plates 127 and 248:1).
Full description by Vondel (2010).
Biology: In rice fields.
Immature stages: Unknown.
Distribution: Nigeria (Plate 280:2).

138 *Haliplus* (*Liaphlus*) *nitens* LeConte

Diagnosis: Length 3.75–4.0 mm, width 2.05–2.4 mm. Body oblong oval, widest in the middle. Strongly resembling *H. cribrarius*, but generally not longer than 4.0 mm, while the minimum length of *H. cribarius* is 4.3 mm. The elytral maculation is generally less coarse than in *H. cribrarius*. In males the width of the parallel tip of the penis is about 2.2× the length, while in *H. cribarius* this is about 1.8× (Plates 128 and 248:2).
Full description by Vondel (2021).
Biology: Found in ponds and lakes and along the edge of rivers.
Immature stages: Unknown.
Distribution: Canada (Alberta, British Columbia, Manitoba, New Brunswick, Newfoundland & Labrador, Northwest Territories, Nova Scotia, Ontario, Prince Edward Island, Quebec, Saskatchewan, Yukon), USA (Alaska, Colorado, Maine, Massachusetts, Michigan, Minnesota, Montana, New York, North Dakota, South Dakota, Texas, Vermont, Wisconsin) (Plate 261:11).

139 *Haliplus* (*Liaphlus*) *novateutonianus* Vondel

Diagnosis: Length: 3.1–3.6 mm, width 1.7–2.1 mm. Body oval, slightly tapering apically. This species resembles *H. brasiliensis* in dorsal view, but the prosternal process is strongly grooved on each side, the elytral apex is at most weakly serrate, the last sternite is completely densely punctured and the aedeagus is different (Plates 129 and 248:3).

Full description by Vondel (2016a).
Biology: Unknown.
Immature stages: Unknown.
Distribution: Brasil (Plate 267:9).

140 ***Haliplus* (*Liaphlus*) *oklahomensis* Wallis**

Diagnosis: Length 2.4–2.6 mm, width 1.35–1.45 mm. Body oval, widest in the middle. A little larger than *H. lewisii*, which has an apically weaker punctured seventh (last) sternite (Plates 130 and 248:4).
Full description by Vondel (2021).
Biology: Found in a ditch.
Immature stages: Unknown.
Distribution: USA (Oklahoma, Texas), Mexico (San Luis Potosi, Tamaulipas) (Plate 261:12).

141 ***Haliplus* (*Liaphlus*) *ornatipennis* Zimmermann**

Diagnosis: Length 2.4–2.8 mm, width 1.4–1.6 mm. Body oval and relatively short, widest in the middle, shoulders pronounced. Elytral maculation transverse. Resembles *H. drechseli*, but the latter has the prosternal process gradually narrowed anteriorly instead of narrowed near coxae (Plates 131 and 248:5).
Full description by Vondel & Spangler (2008).
Biology: Collected in ponds and pools and at light.
Immature stages: Unknown.
Distribution: Argentina, Bolivia, Brazil, Paraguay, Peru (Plate 267:10).

142 ***Haliplus* (*Liaphlus*) *ovalis* Sharp**

Diagnosis: Length 4.1–4.3 mm, width 2.1–2.2 mm. Body oval, widest in the middle. This species usually differs from the closely related *H. chinensis* in having the distance between the eyes more than 1.4× the eye-width and in the male by the abruptly bent top of the penis. Probably *H. ovalis* and *H. chinensis* are conspecific (Nakajima et al. 2020, Hayashi et al. 2023) (Plates 132 and 248:6).
Full description by Vondel (1991).
Biology: Found in ponds. Attracted to light.

Immature stages: The 3rd instar larva is known. The actual knowledge is reviewed by Vondel (2025).
Distribution: Japan (Honshu, Kyushu, Shikoku) (Plate 273:6).

143 ***Haliplus* (*Liaphlus*) *panamanus* Chapin**

Diagnosis: Length 2.9–3.1 mm, width 1.7–1.8 mm. Body oval, widest in the middle. Elytra distinctly maculate. Metaventral process with four round impressions of which the posterior ones are more distant from each other than the anterior ones. (Plate 133 and 248:7).
Full description by Vondel & Spangler (2008).
Biology: Collected in a swampy area in pasture and at light.
Immature stages: Unknown.
Distribution: Bolivia, Colombia, Guyana, Panama, Surinam, Venezuela (Plate 267:11).

144 ***Haliplus* (*Liaphlus*) *perroti* Guignot**

Diagnosis: Length 3.2–3.4 mm, width 1.7–1.9 mm. Body oblong oval, widest in the middle. This species can be recognised by the longitudinal elytral marks, which are not connected to the darkened suture (Plates 134 and 248:8).
Full description by Vondel (1993).
Biology: Unknown.
Immature stages: Unknown.
Distribution: India, Nepal, Vietnam, Pakistan (Plate 276:2).

145 ***Haliplus* (*Liaphlus*) *peruanus* Zimmermann**

Diagnosis: Length 3.6 mm, width 2.2 mm. Body oval, widest in the middle. This species is only known by the female holotype. It resembles *H. gravidus*, with which it is probably conspecific although it has the prosternal process hardly grooved (Plates 135 and 248:9).
Full description by Vondel & Spangler (2008).
Biology: No details known, but probably collected in a river.
Immature stages: Unknown.
Distribution: Peru (Plate 267:12).

146 *Haliplus* (*Liaphlus*) *philippinus* Chapin

Diagnosis: Length 2.4–2.7 mm, width 1.3–1.5 mm. Body broad, more or less parallel in basal half of elytra. This species can be distinguished from the related *H. pulchellus* by the four punctures on the mesepimeron (Plates 136 and 249:1).
Full description by Vondel (1993).
Biology: Attracted to light. Up to an altitude of 920 m.
Immature stages: Unknown.
Distribution: Burma, China (Yunnan), Indonesia (Java, Sulawesi), Laos, Philippines, Thailand, Vietnam (Plate 276:3).

147 *Haliplus* (*Liaphlus*) *pilbaraensis* Watts and McRae

Diagnosis: Length 3.3–4.0 mm, width 1.9–2.2 mm. Body broad, somewhat spindle-shaped, broadest at shoulders, subparallel in the middle (Plates 137 and 249:2).
Full description by Watts & McRae (2010).
Biology: Found in fresh water pools, up to 4 m deep, at the base of a cliff, with dense patches of *Chara* and *Nijas*. The sediments were coarse and dominated by pebble, cobble, boulder and gravel.
Immature stages: The 3rd instar larva is known. The actual knowledge is reviewed by Vondel (2025).
Distribution: Australia (Western Australia) (Plate 277:3).

148 *Haliplus* (*Liaphlus*) *pruthii* Vazirani

Diagnosis: Length 3.0–3.1 mm, width 1.6–1.7 mm. This species resembles *H. angustifrons* and *H. manipurensis* in many aspects, but can be distinguished by the two longitudinal dark marks on the pronotum (Plates 138 and 249:3).
Full description by Vondel (1993), Vondel & Ghosh (2018).
Biology: Unknown.
Immature stages: Unknown.
Distribution: India (Plate 276:4).

149 *Haliplus* (*Liaphlus*) *pulchellus* Clark

Diagnosis: Length 2.5–3.4 mm, width 1.5–2.0 mm. Body oval, widest just before the middle. This species can be distinguished from the other species with a

double grooved prosternal process by the characters mentioned in the key (Plates 139 and 249:4).
Full description by Vondel (1993).
Biology: Collected up to an altitude of 920 m. Attracted to light.
Immature stages: Unknown.
Distribution: India, Indonesia (Java, Sulawesi, Sumatra, Sumba), Laos, Sri Lanka, Thailand, Vietnam, China (Fujian, Guangxi, Yunnan) (Plate 276:5).

150 ***Haliplus* (*Liaphlus*) *regili* Benetti and Hamada**

Diagnosis: Length 3.36 mm, width 2.13 mm. Body oval, tapering backwards, widest just behind the shoulders, with very narrow pronotum and shoulders pronounced. This species can be distinguished from other neotropical species of *Haliplus* by the following combination of characteristics: elytra brown with extensive but not clearly defined maculation consisting of dark patches; prosternal process gradually widening anteriorly, slightly narrowed in posterior part; about 1.9 times longer than wide at base; metaventral process with two well-defined impressions on both sides and median lobe of aedeagus in lateral view curved, straight in the apical third, with the apex strongly truncated. This species resembles *H. langleyi* and *H. triplehorni*, but differs from these species in the shape of prosternal process, gradually widening anteriorly (almost parallel in *H. langleyi* and *H. triplehorni*); male median lobe, shorter and less curved downward in *H. langleyi* and *H. triplehorni* and right paramere, without distal appendage (with distal appendage in *H. langleyi* and *H. triplehorni*) (Plates 140 and 249:5).
Full description by Benetti & Hamada (2017).
Biology: Collected in a shallow pond with macrophytes covering the entire surface.
Immature stages: Unknown.
Distribution: Brazil (Plate 268:1).

151 ***Haliplus* (*Liaphlus*) *rubidus* Perris**

Diagnosis: Length 3.9–4.4 mm, width 2.0–2.3 mm. Body with strong shoulders, tapering strongly behind the middle. Closely related to *H. flavicollis*, from which it differs in the larger punctures on the elytral base, the more pronounced shoulders, the wider distance between the eyes and the usually larger size (Plates 141 and 249:6).
Full description by Vondel (1991, 1997).

Biology: Stagnant water in pools, dune-pools.
Immature stages: The 3rd instar larva is known. The actual knowledge is reviewed by Vondel (2025).
Distribution: Canary Islands, France, Italy, Morocco, Portugal, Spain, Tunisia (Plate 273:8).

152 *Haliplus* (*Liaphlus*) *rufescens* Régimbart

Diagnosis: Length 3.6 mm, width 2.0 mm. Body oval, widest in the middle. This species is closely related to *H. exsecratus*, but in the latter species the prosternal process is posteriorly about as wide as anteriorly, while in *H. rufescens* the prosternal process is much wider posteriorly than anteriorly (Plates 142 and 249:7).
Full description by Vondel (2010).
Biology: Found in rivers.
Immature stages: Unknown.
Distribution: South Africa (Plate 280:3).

153 *Haliplus* (*Liaphlus*) *samosirensis* Vondel

Diagnosis: Length 3.3–3.5 mm, width 1.9–2.0 mm. Body oval, widest in the middle. This species can be distinguished from related species by the elytra, which are provided with a basal dark band reaching margins and very irregular dark marks (Plates 143 and 249:8).
Full description by Vondel (1993), Mazzoldi & Vondel (1997).
Biology: Up to an altitude of 850 m.
Immature stages: Unknown.
Distribution: Indonesia (Kalimantan, Sumatra) (Plate 276:6).

154 *Haliplus* (*Liaphlus*) *sharpi* Wehncke

Diagnosis: Length 3.6 mm, width 1.9 mm. Body oval, widest before the middle. This species could be confused with specimens of *H. kotoshonis* in which a basal dark band is present on the elytra. The prosternal process in *H. sharpi* is wide and hardly narrowed near the coxae, while *H. kotoshonis* has a narrow prosternal process, which is strongly narrowed near the coxae. In the male the penis of both species is clearly different (Plates 144 and 249:9).

Full description by Vondel (1991).
Biology: In stagnant pools. Caught at light and in a hotel room.
Immature stages: The 2nd and 3rd instar larva and the pupa are known. The actual knowledge is reviewed by Vondel (2025).
Distribution: China (Anhui, Chongqing, Fujian, Guizhou, Jiangsu, Liaoning, Shanghai, Yunnan), Japan (Hokkaido, Honshu, Kyushu, Shikoku), North Korea, Taiwan (Plate 274:1).

155 *Haliplus* (*Liaphlus*) *signatipennis* Régimbart

Diagnosis: Length 3.4–3.7 mm, width 1.9–2.1 mm. Body oval, parallel behind sligthly protruding shoulders. This species can be distinguished from other species in the region by the longitudinal dark mark on the pronotum (Plates 145 and 250:1).
Full description by Vondel (1995a).
Biology: Specimens are collected in a stock pond, in a Sago swamp and in bomb craters.
Immature stages: Unknown.
Distribution: Indonesia (Seram), Papua New Guinea (Plate 277:4).

156 *Haliplus* (*Liaphlus*) *signatus* Sharp

Diagnosis: Length 2.9–3.3 mm, width 1.7–2.0 mm. Body short oval, widest behind shoulders. This is a variable species and resembles *H. tumidus* very much. Males can bedistinguished by the aedeagus, but no characters are known to distinguish females of *H.signatus* and *H. tumidus* (Plates 146 and 250:2).
Full description by Vondel & Spangler (2008).
Biology: Collected in pools, ditches and at light. The following aquatic Coleoptera were found in association with *H. signatus* in Mexico, Campeche: Dytiscidae: *Bidessus, Hydrovatus, Laccophilus, Neoclypeodytes, Thermonectus*, Haliplidae: *Haliplus gravidoides*, Hydrophilidae: *Berosus, Hydrophilus, Tropisternus*. Noteridae: *Hydrocanthus, Suphis*. In Costa Rica (VillaColon) it was found in association with Hydrophilidae: *Derallus*.
Immature stages: Unknown.
Distribution: Costa Rica, Guatemala, Mexico, Nicaragua, Panama (Plate 265:10).

157 *Haliplus* (*Liaphlus*) *solitarius* Sharp

Diagnosis: Length 2.9–3.2 mm, width 1.7–1.9 mm. Body oval, shoulders pronounced, widest in the middle This species is characterized by the combination of a medially maculate pronotum and a metaventrite with a wide central impression (Plates 147 and 250:3).
Full description by Vondel & Spangler (2008).
Biology: Collected in a roadside temporary grassy ditch and at light. The following aquatic Coleoptera were found in association with *H. solitarius* in Mexico, Guadalajara: Dytiscidae: *Bidessus, Hydaticus, Hydrovatus, Hygrotus, Laccophilus, Megadytes, Rhantus, Stictotarsus, Thermonectus*, Hydrophilidae: *Berosus, Enochrus, Helophorus, Hydrophilus, Paracymus, Tropisternus*, Noteridae: *Suphis*.
Immature stages: Unknown.
Distribution: El Salvador, Guatemala, Honduras, Mexico, Nicaragua (Plate 265:11).

158 *Haliplus* (*Liaphlus*) *srilankanus* Vondel

Diagnosis: Length 3.2 mm, width 1.8 mm. Body oval, widest in the middle. This species is related *H. angustifrons* and *H. arrowi*. It can be distinguished from them by the characters given in the key (Plates 148 and 250:4).
Full description by Vondel (1993).
Biology: Unknown.
Immature stages: Unknown.
Distribution: Sri Lanka (Plate 276:7).

159 *Haliplus* (*Liaphlus*) *stepheni* Watts

Diagnosis: Length 2.8–3.0 mm, width 1.5–1.6 mm. Body long oval, widest in the middle This species can be distinguished from related species by the combination of a darkened elytral base and elytral puncture-row 5 with a strong transverse impression on base (Plates 149 and 250:5).
Full description by Watts (1988), Vondel (1995).
Biology: Specimens were collected in a creek and were attracted to light.
Immature stages: Unknown.
Distribution: Australia (Northern Territory, Queensland, Western Australia) (Plate 277:5).

160 *Haliplus* (*Liaphlus*) *sulawesicus* Vondel

Diagnosis: Length: 3.6–3.8 mm, width 1.7–1.9 mm. Body oval, subparallel in the middle. Metaventral process flat with weak impressions on each side. This species has much in common with *H. signatipennis*. *H. sulawesicus* can be distinguished from *H. signatipennis* by a few clear differences. The shoulders of *H. sulawesicus* are serrate and in *H. signatipennis* they are smooth. The secondary elytral punctures of *H. sulawesicus* are strong and in parts of the elytra nearly as strong as the primary punctures, while in *H. signatipennis* the secondary punctures are weak (Plates 150 and 250:6).
Full description by Vondel (2008).
Biology: Collected at light and in a lake.
Immature stages: Unknown.
Distribution: Indonesia (Seram, Sulawesi) (Plate 276:8).

161 *Haliplus* (*Liaphlus*) *tantoyucanus* Vondel and Spangler

Diagnosis: Length 3.0–3.2 mm, width 1.7–1.8 mm. Body oval, tapering backwards, widest before the middle. Basal elytral punctures in a transverse impression, preapical punctures of seventh and eighth row stronger (Plates 151 and 250:7).
Full description by Vondel & Spangler (2008).
Biology: Collected in a pond in pasture.
Immature stages: Unknown.
Distribution: Guatemala, Mexico (Plate 265:12).

162 *Haliplus* (*Liaphlus*) *tanzanianus* Vondel

Diagnosis: Length 3.7–4.0 mm, width 2.1–2.2 mm. Body subparallel in the middle. This species can be distinguished from related species by the elytral maculation, consisting of separate marks, which are not connected to the suture (Plates 152 and 250:8).
Full description by Vondel (2010).
Biology: Unknown.
Immature stages: Unknown.
Distribution: Kenya, Tanzania (Plate 280:4).

163 *Haliplus* (*Liaphlus*) *testaceus* Zimmermann

Diagnosis: Length 3.5–3.7 mm, width 2.1 mm. Body with narrow pronotum, shoulders well pronounced, elytra subparallel. Body not maculate. Elytral punctures not darkened (Plates 153 and 250:9).
Full description by Vondel & Spangler (2008).
Biology: Unknown.
Immature stages: Unknown.
Distribution: Argentina, Brazil, Uruguay (Plate 268:2).

164 *Haliplus* (*Liaphlus*) *testudo* Clark

Diagnosis: Length 3.5–4.0 mm, width 1.9–2.2 mm. Body oval, widest just in front of the middle. This species can be distinguished from *H. australis* by the presence of dark stripes on the basal area of the elytra, the prostemal process usually not being wider anteriorly than posteriorly, the coloured pronotal anterior punctures, the clearly serrate apical elytral margin and in the male the penis being gradually curved to the top and the left paramere not having a solid digitus. It can be distinguished from *H. wattsi* by the proepisternum being hardly punctured (Plates 154 and 251:1).
Full description by Vondel (1995).
Biology: This species is found in rivers and in lentic freshwater and is attracted to light.
Immature stages: The 3rd instar larva is known. The actual knowledge is reviewed by Vondel (2025).
Distribution: Australia (New South Wales, Northern Territory, Queensland, South Australia, Victoria) (Plate 277:6).

165 *Haliplus* (*Liaphlus*) *thoracicus* Zimmermann

Diagnosis: Length 3.7–3.8 mm, width 2.3–2.4 mm. Body oval, subparallel, widest in the middle, shoulders slightly pronounced. This species is characterized by the combination of the medially strongly maculate pronotum and the anteriorly distinctly narrowed prosternal process (Plates 155 and 251:2).
Full description by Vondel & Spangler (2008).
Biology: Unknown.
Immature stages: Unknown.
Distribution: Brazil (Plate 268:3).

166 *Haliplus* (*Liaphlus*) *timmsi* Vondel

Diagnosis: Length 3.0–3.6 mm, width 1.7–2.0 mm. Body oval, widest behind shoulders, clearly tapering in apical 1/3. This species can be distinguished from the related *H. alastairi* by the metasternal process being impressed on both sides (Plates 156 and 251:3).
Full description by Vondel (1995).
Biology: Specimens were attracted to light.
Immature stages: The 3rd instar larva is known. The actual knowledge is reviewed by Vondel (2025).
Distribution: Australia (Northern Territory, Queensland) (Plate 277:7).

167 *Haliplus* (*Liaphlus*) *tocumenus* Vondel and Spangler

Diagnosis: Length 2.9–3.1 mm, width 1.6–1.8 mm. Body oval, widest in the middle. Pronotum and elytra with basal impression. Resembles *H. camposi* very much but can be distinguished by the pronotum not being margined (Plates 157 and 251:4).
Full description by Vondel & Spangler (2008).
Biology: Unknown.
Immature stages: Unknown.
Distribution: Panama, Trinidad, Venezuela (Plate 268:4).

168 *Haliplus* (*Liaphlus*) *triplehorni* Vondel and Spangler

Diagnosis: Length 3.2–3.3 mm, width 1.9 mm. Body oval to subparallel, widest in the middle. Elytra basally with dense and in more than one row standing secondary punctures. Prosternal process also impressed between the two longitudinal grooves (Plates 158 and 251:5).
Full description by Vondel & Spangler (2008).
Biology: Collected at light.
Immature stages: Unknown.
Distribution: Brazil (Plate 268:5).

169 *Haliplus* (*Liaphlus*) *tumidus* LeConte

Diagnosis: Length 3.0–3.2 mm, width 1.9–2.0 mm. Body oval, widest just behind shoulders. Elytra maculate, shoulders raspy. Pronotum usually not distinctly maculate. Elytral maculation varying from only along the suture to covering most of the elytra. Pronotum usually immaculate, but sometimes with central mark which may be present from anterior to posterior margin. This species can easily be confused with *H. signatus* from which the males can be separated by comparison of the tip of the penis. No characters are known to distinguish the females of *H. signatus* and *H. tumidus* (Plates 159 and 251:6).
Full description by Vondel & Spangler (2008).
Biology: Collected in rivers, a mangrove swamp and at light. The following aquatic and semiaquatic Coleoptera were found in association with *H. tumidus* in Mexico, Oaxaca: Dryopidae: *Helichus*, Dytiscidae: *Bidessus*, *Deronectes*, *Thermonectus marmoratus*, Hydrophilidae: *Tropisternus ellipticus*, *T. lateralis*. In the mangrove swamp in Yucatan it was found in association with Hydraenidae: *Ochthebius*, Hydrophilidae: *Enochrus*, *Paracymus*, *Tropisternus quadristriatus*.
Immature stages: Unknown.
Distribution: USA (Arizona, California, Oklahoma, Texas), Cuba, Guatemala, Mexico (VeraCruz, Yucatan) (Plate 262:1).

170 *Haliplus* (*Liaphlus*) *unicarinatus* Vondel and Spangler

Diagnosis: Length 2.9 mm, width 1.7 mm. Body oval, widest just before the middle. Resembles *H. carinatus* very much, but has the seventh elytral puncture row apically not impressed (Plates 160 and 251:7).
Full description by Vondel & Spangler (2008).
Biology: Unknown.
Immature stages: Unknown.
Distribution: Mexico, Guatemala (Plate 266:1).

171 *Haliplus* (*Liaphlus*) *uniformis* Zimmermann

Diagnosis: Length 3.0–3.1 mm, width 1.5–1.6 mm. Body narrow oval with regular outline, widest in the middle. This species can be distinguished from related species without maculation and with serrate apical margin of elytra by the metaventral process having a central pit (Plates 161 and 251:8).

Full description by Vondel (2010).
Biology: Unknown.
Immature stages: Unknown.
Distribution: South Africa (Plate 280:5).

172 *Haliplus* (*Liaphlus*) *valdiviensis* Moroni

Diagnosis: Length 4.6 mm, width 2.4 mm. Body long, subparallel, widest behind shoulders. Elytra with weak maculation. Looks like *H.fuscipennis*, which is smaller and has no maculate elytra (Plates 162 and 251:9).
Full description by Vondel & Spangler (2008).
Biology: Collected in a roadside pond.
Immature stages: The 3rd instar larva is known. The actual knowledge is reviewed by Vondel (2025).
Distribution: Chile (Plate 268:6).

173 *Haliplus* (*Liaphlus*) *vancouverensis* Matheson

Diagnosis: Length 4.35 mm, width 3.05 mm. Body oval, widest in the middle. Large species without distinct elytral maculation or darkened row of punctures (Plates 163 and 252:1).
Full description by Vondel (2021).
Biology: Unknown.
Immature stages: Unknown.
Distribution: Canada (Alberta, British Columbia) (Plate 262:2).

174 *Haliplus* (*Liaphlus*) *varicator* Guignot

Diagnosis: Length 2.9–3.1 mm, width 1.7–1.8 mm. Body oval, widest in the middle. This species can be distinguished from related species with unmaculated pronotum by the metaventral process having a central impression (Plates 164 and 252:2).
Full description by Vondel (2010).
Biology: Unknown.
Immature stages: Unknown.
Distribution: D.R. Congo, Mozambique (Plate 280:6).

175 *Haliplus* (*Liaphlus*) *variegatus* Sturm

Diagnosis: Length 2.8–3.1 mm, width 1.5–1.7 mm. Body oval, widest before the middle. This species is very variable in size, colour and maculation. The distance between the eyes, 1.7–2.0× eye-width, usually distinguishes it from related species of the same size (Plates 165 and 252:3).
Full description by Vondel (1991, 1997).
Biology: In stagnant waterbodies with fresh water, often with a vegetation of Characeans, *Sphagnum* or *Hypnum*, in temporary dune-pools, in peat bogs.
Immature stages: The 1st, 2nd and 3rd instar larva and the pupa are known. The actual knowledge is reviewed by Vondel (2025).
Distribution: Algeria, Austria, Azerbaijan, Belgium, Bosnia Herzegovina, Bulgaria, Byelorussia, Croatia, Cyprus, Czech Republic, Denmark, Estonia, Finland, France, Georgia, Great Britain, Germany, Greece, Hungary, Ireland, Israel, Italy, Kyrgyzstan, Latvia, Lithuania, Luxembourg, Macedonia, Montenegro, Netherlands, Norway, Poland, Romania, Russia (Central European Territory, South European Territory), Serbia, Slovakia, Slovenia, Spain, Sweden, Switzerland, Syria, Transcaucasia, Turkey, Ukraine, Yugoslavia (Plate 274:2).

176 *Haliplus* (*Liaphlus*) *venustus* Régimbart

Diagnosis: Length 3.2–3.6 mm, width 1.8–2.1 mm. Body with regular outline, subparallel in the middle. This species looks very much like *H. natalensis*, but the latter species has the metaventral process with one deep pit, while in *H. venustus* two clear pits are present (Plates 166 and 252:4).
Full description by Vondel (2010).
Biology: In rice fields. Attracted to light.
Immature stages: Unknown.
Distribution: Guinea-Bisau, Ivory Coast, Mocambique, Namibia, Nigeria, South Africa, Tanzania, Tchad, Zimbabwe (Plate 280:7).

177 *Haliplus* (*Liaphlus*) *volpei* Vondel

Diagnosis: Length 3.1–3.5 mm, width 1.7–1.9 mm. Body oval, widest in the middle. Closely related to *H. angustifrons*, but can be distinguished by the characters mentioned in the identification key (Plates 167 and 252:5).
Full description by Vondel (2009), Vondel & Volpe (2015).
Biology: Found in muddy pool.

Immature stages: Unknown.
Distribution: Myanmar (Plate 276:9).

178 *Haliplus* (*Liaphlus*) *watsoni* Vondel

Diagnosis: Length 2.6–2.8 mm, width 1.4–1.6 mm. Body with regular outline, widest in the middle. This species can be distinguished from related species by the small mark on pronotal anterior margin and in the males by the very long extension on the right paramere (Plates 168 and 252:6).
Full description by Vondel (2010).
Biology: In rice fields.
Immature stages: Unknown.
Distribution: D.R. Congo, Mocambique, Nigeria (Plate 280:8).

179 *Haliplus* (*Liaphlus*) *wattsi* Vondel

Diagnosis: Length 3.4–3.6, width 1.9–2.1 mm. Body oval, widest behind shoulders, strongly tapering behind the middle This species can easily be distinguished from *H. australis* and *H. testudo* by the proepistemum being strongly and densely or even coarsely punctured. It can also be distinguished from *H. testudo* by the absence of dark stripes on the basal area of the elytra, the prosternal process usually being wider anteriorly than posteriorly, the uncoloured pronotal anterior punctures, the weakly serrate or nearly smooth apical elytral margin and in the male the penis not being gradually curved to the top and the left paramere having a solid, sometimes very small, digitus (Plates 169 and 252:7).
Full description by Vondel (1995).
Biology: Specimens have been found in rivers and are attracted to light.
Immature stages: Unknown.
Distribution: Australia (Northern Territory, Queensland, Western Australia), Papua New Guinea (Plate 277:8).

180 *Haliplus* (*Liaphlus*) *youngi* Vondel and Spangler

Diagnosis: Length 2.8–3.3 mm, width 1.8–2.1 mm. Body short oval, tapering backwards, widest before the middle. Resembles *H. cubensis*, but is larger and has the shoulders distinctly serrate (Plates 170 and 252:8).

Full description by Vondel & Spangler (2008).
Biology: Collected in pools and rivers. The following aquatic Coleoptera were found in association with *H. youngi* in Puerto Rico, Cueva del Indio: Dytiscidae: *Laccophilus*, *Pachydrus*, *Rhantus calidus*, *Thermonectus*, Hydraenidae: *Hydraena*, Hydrochidae: *Hydrochus*, Hydrophilidae: *Enochrus*, *Ydrobiomorpha*.
Immature stages: Unknown.
Distribution: Bahamas, Cuba, Dominican Republic, Haïti, Jamaica, Puerto Rico, St. Croix, St. Thomas (Plate 266:2).

181 Subgenus *Neohaliplus* Netolitzky

Diagnosis: Very small to medium-sized species. Body subparallel, pronotum with long strongly curved plicae. Pronotum and plicae without distinct micropunctation. Penultimate segments of labial palpi very wide. Hind legs without setiferous striole on dorsal side of tibae.
Full description by Vondel (1997).
Biology: Found in all kinds of water bodies.
Distribution: West Palaearctic region.

182 *Haliplus* (*Neohaliplus*) *lineatocollis* (Marsham)

Diagnosis: Length 2.2–3.5 mm, width 1.1–1.8 mm. Body oval, subparallel and widest in the middle. Small specimens of this species can be distinguished from *H. ruficeps* by the grooved punctures at elytral base. Investigation of DNA of specimens from all over Europe and North Africa by Dr Johannes Bergsten (pers. comm.) suggests that the *H. lineatocollis* complex contains at least three different species, but so far no reliable characters have been found to separate them (Plates 171 and 252:9)
Full description by Vondel (1997).
Biology: Living in all kinds of fresh and also in brackish water bodies, stagnant or slowly running. The adults have been seen flying relatively often (Jackson, 1956a, b). Recorded from altitudes up to 2200 m.
Immature stages: The 1st, 2nd and 3rd instar larva and the pupa are known. The actual knowledge is reviewed by Vondel (2025).
Distribution: Albania, Algeria, Armenia, Austria, Azerbaijan, Belgium, Bosnia Herzegovina, Bulgaria, Byelorussia, Canary Islands, Caucasus, Croatia, Czech Republic, Denmark, Egypt, Eritrea, Ethiopia, France, Georgia, Germany, Great Britain, Greece, Hungary, Iran, Ireland, Israel, Italy, Lebanon, Liechtenstein,

Libya, Lithuania, Luxembourg, Macedonia, Malta, Montenegro, Morocco, Netherlands, Poland, Portugal, Romania, Russia (Central European Territory, South European Territory), Saudi Arabia, Slovakia, Slovenia, Spain, Sudan, Sweden, Switzerland, Syria, Tchad, Transcaucasia, Tunisia, Turkey, Ukraine, Yemen, Yugoslavia (Plate 274:3).

183 *Haliplus* (*Neohaliplus*) *ruficeps*, Chevrolat

Diagnosis: Length 1.9–2.2 mm, width 0.9–1.3 mm. Body oval, subparallel, widest in the middle. In North Africa *H. ruficeps* and *H. lineatocollis* can be found together (Plates 172 and 253:1).
Full description by Vondel (1997).
Biology: Unknown.
Immature stages: Unknown.
Distribution: Algeria, Morocco (Plate 274:4).

184 Subgenus *Nipponiplus* Vondel

Diagnosis: Species with length 2.4–3.5 mm. Body widest behind shoulders, tapering to apex. Head with two genal lines, penultimate segment of labial palpi strongly widened apically. Pronotum with basal plicae. Elytra with strong primary puncture rows, dark interrupted lines on puncture rows and small marks at intervals. Prosternal process narrowed near coxae, laterally grooved in posterior half. Metacoxal plates with distinctive spines on posterior edge, the depression underneath only surrounded by a sharp ridge on the fourth sternite, sternites 3 and 4 not fused (only visible when coxal plate is removed). Metatibia without setiferous striole on dorsal side. Males with a more or less rectangular left paramere.
Full description by Vondel (2019).
Distribution: East Palaearctic Region.

185 *Haliplus* (*Nipponiplus*) *japonicus* Sharp

Diagnosis: Length 2.6–3.5 mm, width 1.5–1.8 mm. Body oval, tapering backwards, widest before the middle. This species may be confused with *H. regimbarti*, but the latter has an impression on each side of the metasternal process (Plates 173 and 253:2).

Full description by Vondel et al. (2006).
Biology: In rivers, streams, ponds and lakes. Up to altitudes of 2500 m.
Immature stages: The 3rd instar larva is known. The actual knowledge is reviewed by Vondel (2025).
Distribution: China (Beijing, Chongqing, Gansu, Guizhou, Hunan, Jiangsu, Sichuan, Shanghai, Yunnan, Zhejiang), Japan (Hokkaido, Honshu, Kyushu), Russia (Far East) (Plate 274:5).

186 *Haliplus (Nipponiplus) latreillei* Jia and Vondel

Diagnosis: Length 2.9–3.0 mm, width 1.6–1.7 mm. Body oval, tapering backwards, widest before the middle. This species is close to *H. japonicus* and *H. regimbarti* in body size and shape, arrangement of elytral dark spots and black lines, punctuation and the row of setae on posterior edge of the metacoxal plates. However, the new species lacks pronotal basal plicae, its pronotum lacks the transverse basal rim and it differs from the above species in the shape of the median lobe and parameres of the aedaeagus (Plates 174 and 253:3).
Full description by Jia & Vondel (2011).
Biology: Unknown.
Immature stages: Unknown.
Distribution: China (Guizhou) (Plate 274:6).

187 *Haliplus (Nipponiplus) regimbarti* Zaitzev

Diagnosis: Length 2.4–3.0 mm, width 1.4–1.7 mm. Body oval, strongly tapering backwards, widest before the middle. Similar species: This species may be confused with *H. japonicus*, but the latter has no impressions on each side of the metaternal process (Plates 175 and 253:4).
Full description by Vondel et al. (2006).
Biology: In streams and rivers.
Immature stages: Unknown.
Distribution: China (Anhui, Fujian, Guangdong, Guangxi, Guizhou, Henan, Hubei, Hunan, Jiangsu, Jiangxi, Shaanxi, Shandong, Yunnan, Zhejiang), Taiwan, Japan (Yonaguni Island) (Plate 274:7).

188 Subgenus *Paraliaphlus* Guignot

Diagnosis: Yellow to yellow-brown or yellow-red. Body medium to large-sized, body outline oval, length 3.0–4.5 mm, width 1.7–2.5 mm. Head yellow to yellow-red or yellow-brown, weakly to strongly punctured, antennae yellow to yellow-brown, maxillary palpi parallel, labial palpi parallel, interocular distance 1.0–1.8× the width of one eye, two genal lines behind the eyes. Pronotum widened posteriorly, lateral sides margined, base without plicae, anterior margin sometimes beaded at sides, base often with larger darkened punctures. Elytra yellow to yellow-brown or yellow-red, with or without all kinds of maculation. Ventral side yellow to yellow-red or brown, prosternal process with a groove on each side, metaventral process with a sharp plica on each side from anterior to posterior end. Carina on the metepisternum not clearly darkened except in *Haliplus pantherinus*. Metacoxal plates reaching to fourth abdominal sternite. Legs yellow to brown, hind tibia with or without a setiferous striole on the dorsal face.
Full description by Vondel (2021).
Biology: Found in all kinds of fresh stagnant and running water.
Distribution: Nearctic and Northern part of Neotropical Region.

189 *Haliplus* (*Paraliaphlus*) *alberti* Megna, Hendrich and Balke

Diagnosis: Length 3.2–3.4 mm, width 1.9–2.0 mm. Body oval, tapering backwards, widest just behind the shoulders. With narrow pronotum and shoulders slightly pronounced. This species can be readily distinguished from other species by its size, elytra testaceus with well-defined maculation consisting of black patches; prosternal process widened anteriorly; metaventral process with a long groove on each side; elytral apical margin weakly serrate; elytral punctures outside maculation not darkened and median lob of aedaeagus in lateral view curved, subparallel, distally elongate and with rounded apex. It can be distinguished from *H. deceptus* and *H. mutchleri* by its testaceus pronotum without dark spot. It can be distinguished from *H. fasciatus* and *H. pseudofasciatus* by the shape of the prosternal process and the hind tibia without setiferous striole on dorsal face (Plates 176 and 253:5).
Full description by Megna et al. (2018).
Biology: Collected in a permanent lagoon with abundant aquatic vegetation, muddy bottom and macrophytes near the shores at 100 m a.s.l.
Immature stages: Unknown.
Distribution: Cuba (Plate 266:3).

190 *Haliplus* (*Paraliaphlus*) *apostolicus* Wallis

Diagnosis: Length 4.05–4.4 mm, width 2.15–2.3 mm. Body oval, widest in the middle. Large species. The combination of the absence of a beaded anterior pronotal margin and a distinctly narrowed anterior half of the prosternal process will separate this species from all other species in this sub-genus except *H. lucigelinae*, but the latter is much smaller (Plates 177 and 253:6).
Full description by Vondel (2021).
Biology: Found in ponds.
Immature stages: Unknown.
Distribution: Canada (Nova Scotia, Ontario, Quebec), USA (Minnesota, Nebraska, New Hampshire, Wisconsin) (Plate 262:3).

191 *Haliplus* (*Paraliaphlus*) *connexus* Matheson

Diagnosis: Length 3.6–3.7 mm, width 1.85–2.1 mm. Body slender oval, widest just behind shoulders, tapering backwards. This species is easily recognisable by the very narrow elytral sutural darkening and the elytral apex clearly serrate up to the posterior tip (Plates 178 and 253:7).
Full description by Vondel (2021).
Biology: Found in ponds, swamps, on margins of brooks and streams.
Immature stages: Unknown.
Distribution: Canada (Manitoba, New Brunswick, Newfoundland & Labrador, Nova Scotia, Ontario, Prince Edward Island, Quebec), USA (Connecticut, District of Columbia, Illinois, Iowa, Maine, Massachusetts, Michigan, Minnesota, Nebraska, New Hampshire, New York, North Carolina, North Dakota, Ohio, Pennsylvania, Vermont, Wisconsin) (Plate 262:4).

192 *Haliplus* (*Paraliaphlus*) *deceptus* Matheson

Diagnosis: Length 3.1–3.6 mm, width 1.8–2.0 mm. Body oval, widest just before the middle. Smaller than the related *H. grandis* (Plates 179 and 253:8).
Full description by Vondel (2021, 2023).
Biology: In small ponds, margins of lakes, reservoirs, streams and irrigation ditches. Found in strip-mine ponds with *Nitella* sp. or *Chara* sp., *Potamogeton* sp., *Veronica* sp. and the filamentous alga *Oedegonium* sp.. On some specimens of this species the ascomycete fungus (Laboulbeniales) *Hydraeomycetes halipli* was found.

Immature stages: The 3rd instar larva is known. The actual knowledge is reviewed by Vondel (2025).
Distribution: USA (Colorado, Illinois, Kansas, Missouri, Montana, New Mexico, Ohio, Oklahoma, South Dakota, Texas, Wyoming), Mexico (Coahuila, Nuevo Lion) (Plate 262:5).

193 *Haliplus* (*Paraliaphlus*) *fasciatus* Aubé

Diagnosis: Length 4.05–4.1 mm, width 2.25–2.4 mm. Body oval, shoulders slightly pronounced. Anterior margin of pronotum beaded, elytral apex not or very weakly serrate, dorsal side of metatibia with distinct setiferous striole (Plates 180 and 253:9).
Full description by Vondel (2021).
Biology: Occurs in lakes, ponds, creeks and cypress swamps. The species is known as a predator of chironomid eggs.
Immature stages: Unknown.
Distribution: Canada (New Brunswick, Nova Scotia, Ontario, Quebec), USA (Alabama, Arkansas, California, Connecticut, District of Columbia, Florida, Georgia, Illinois, Indiana, Iowa, Kansas, Kentucky, Louisiana, Maine, Maryland, Massachusetts, Michigan, Mississippi, Missouri, Nebraska, New Hampshire, New Jersey, New York, North Carolina, Ohio, Oklahoma, Pennsylvania, Rhode Island, South Carolina, Tennessee, Texas, Virginia, West Virginia, Wisconsin) (Plate 262:6).

194 *Haliplus* (*Paraliaphlus*) *grandis* Vondel and Spangler

Diagnosis: Length 4.2–4.5 mm, width 2.2–2.4 mm. Body oblong oval to nearly parallel in the middle, widest in the middle. Large subparallel species. Metaventral process with short impression on each side as part of the fine plica running to anterior end (Plates 181 and 254:1).
Full description by Vondel & Spangler (2008), Vondel (2021).
Biology: Collected in a ditch/pond, in a temporary grassy meadow pond and in a residual pool of a stream bed. Occurs up to an altitude of 2400 m. The following aquatic Coleoptera were found in association with *H. grandis* in Mexico, El Salto: Dytiscidae: *Deronectes* sp., Haliplidae: *Haliplus elsaltous*, Hydrophilidae: *Helochares* sp.
Immature stages: Unknown.

Distribution: USA (New Mexico), Mexico (Aguascalientes, Durango, Jalisco, Ciudad de Mexico, Michoacán, Puebla, Zacatecas) (Plate 262:7).

195 *Haliplus* (*Paraliaphlus*) *kenneri* Vondel

Diagnosis: Length 3.4–3.8 mm, width 1.9–2.0 mm. Body oval. General colour ferrugineous, elytra slightly darker than head and pronotum. This species can be distinguished from *H. mimeticus* by the strongly serrate elytral apical margin, while in *H. mimeticus* this apical margin is undulate. The elytra of *H. kenneri* are unicoloured, while in *H. mimeticus*, the elytra have lighter spots on the sides (Plates 182 and 254:2).
Full description by Vondel (2014, 2021).
Biology: This species has been collected in clear, warm water in shallow pools slowly flowing over limestone bedrock containing abundant emergent and submerged macrophytes as well as filamentous algae and *Chara* spp. It is also known from marshy pools and a pool in a saline meadow.
Immature stages: Unknown.
Distribution: Mexico (Baja California) (Plate 262:8).

196 *Haliplus* (*Paraliaphlus*) *leopardus* Roberts

Diagnosis: Length 4.05–4.35 mm, width 2.2–2.45 mm. Body oval, widest behind the shoulders. Distinguished from related species by the strong punctures on posterior side of mesotrochanter (Plates 183 and 254:3).
Full description by Vondel (2021).
Biology: Found in lakes, ponds and streams. Also attracted to UV light
Immature stages: Unknown.
Distribution: Canada (British Columbia, New Brunswick, Nova Scotia, Ontario, Quebec), USA (Alabama, Connecticut, District of Columbia, Georgia, Illinois, Indiana, Louisiana, Maine, Maryland, Massachusetts, Michigan, Minnesota, Missouri, New Jersey, New York, North Carolina, Ohio, Pennsylvania, Rhode Island, South Carolina, Texas, Virginia, Wisconsin) (Plate 262:9).

197 *Haliplus* (*Paraliaphlus*) *lucigelinae* Vondel and Alarie

Diagnosis: Length 3.0–3.1 mm, width 1.7–1.8 mm. Body V-shaped, widest before the middle, strongly tapering posteriorly. This species can be distinguished from any other Nearctic species of *Haliplus* by the following combination of characters: penultimate segment of labial palpus not triangular but about parallel; pronotum without basal plicae, anterior margin not beaded at sides, frequently with anterior median spot; metaventral process on each side with a sharp carina running to anterior edge; prosternal process distinctly margined at sides, anteriorly clearly narrower than posteriorly; metatibia without setiferous striole on dorsal face; elytra with distinct maculation, elytral margin smooth in humeral and apical part; total length less than 3.4 mm. Owing to its size, *H. lucigelinae* resembles *H. confluentus* and *H. annulatus* from which it differs by the shape of the prosternal and metaventral process. In these latter two species the prosternal process is about parallel (narrowed in anterior part in *H. lucigelinae*), and the metaventral process lacks a continuous sharp carina (present in *H. lucigelinae*). Additionally the smooth humeral and apical elytral margin will usually distinguish this species from other members of the subgenus *Paraliaphlus* (Plates 184 and 254:4).
Full description by Vondel & Alarie (2016), Vondel (2021)
Biology: All specimens were collected in shallow streams with rocky bottom and moderately to fast-flowing clear water. Collections were made along the littoral of these habitats. At the locality in Quebec, *H. lucigelinae* was associated with three other haliplid species, *H. fasciatus*, *H. pantherinus*, and *Peltodytes edentulus* as well as one Dytiscidae, *Neoporus spurius* (LeConte). At the Ontario locality this species was collected with two additional haliplid species, *Haliplus immaculicollis* and *Peltodytes duodecimpunctatus*, as well as the following adephagan species: *Laccophilus maculosus* Say, *Neoporus undulatus* (Say), *Uvarus granarius* (Aubé) and *Dineutus nigrior* Roberts.
Immature stages: Unknown.
Distribution: Canada (Ontario, Quebec) (Plate 262:10).

198 *Haliplus* (*Paraliaphlus*) *mimeticus* Matheson

Diagnosis: Length 3.4–3.8 mm, width 1.9–2.3 mm. Body oval. General colour ferrugineous, elytra slightly darker than head and pronotum. This species can be distinguished from *H. kenneri* by the undulate elytral apical margin; in *H. kenneri* this apical margin is strongly and sharply serrate. The elytra of

H. mimeticus have paler spots, while in *H. kenneri* the elytra are unicoloured (Plates 185 and 254:5).
Full description by Vondel (2021).
Biology: This species has been collected in clear, warm water in shallow pools and in clear slow water from springs slowly running over limestone bedrock containing abundant emergent and submerged macrophytes as well as filamentous algae and *Chara* spp. It is known also from cool pools near streams.
Immature stages: Unknown.
Distribution: USA (Arizona, California, Nevada, New Mexico, Utah), Mexico (Baja California) (Plate 262:11).

199 *Haliplus (Paraliaphlus) mutchleri* Wallis

Diagnosis: Length 3.0–3.8 mm, width 1.8–2.0 mm. Body oval, widest just behind the middle. Related to *H. punctatus*, but the prosternal process is not strongly widened anteriorly (Plates 186 and 254:6).
Full description by Vondel (2021).
Biology: Unknown.
Immature stages: Unknown.
Distribution: USA (Florida, Louisiana, North Carolina) (Plate 262:12).

200 *Haliplus (Paraliaphlus) pantherinus* Aubé

Diagnosis: Length 3.35–3.8 mm, width 1.85–2.2 mm. Body oval, widest in the middle. May be distinguished from the related *H. apostolicus, H. grandis, H. deceptus* by the darkened carina on the metepisternum (Plates 187 and 254:7).
Full description by Vondel (2021).
Biology: Found in lakes, ponds, ditches and creeks.
Immature stages: Unknown.
Distribution: Canada (New Brunswick, Nova Scotia, Ontario, Quebec), USA (Alabama, Alaska?, District of Columbia, Florida, Georgia, Illinois, Indiana, Kansas, Kentucky, Louisiana, Maine, Maryland, Massachusetts, Michigan, Minnesota, Montana, New Hampshire, New Jersey, New York, North Carolina, Ohio, Pennsylvania, Rhode Island, South Carolina, South Dakota, Virginia, West Virginia, Wisconsin) (Plate 263:1).

201 *Haliplus (Paraliaphlus) pseudofasciatus* Wallis

Diagnosis: Length 3.55 mm, width 2.0 mm. Body oval, widest behind the shoulders. Closely related to *H. fasciatus*, but with very short or absent setiferous striole on dorsal side of hind tibia (Plates 188 and 254:8).
Full description by Vondel (2021).
Biology: Attracted to light.
Immature stages: Unknown.
Distribution: USA (District of Columbia, Florida, Georgia, Kansas, Kentucky, Massachusetts, North Carolina, Oklahoma, South Carolina, Texas, Virginia) (Plate 263:2).

202 *Haliplus (Paraliaphlus) punctatus* Aubé

Diagnosis: Length 3.2–4.0 mm, width 1.8–2.2 mm. Body oval, widest before the middle. Closely related to *H. mutchleri*, but the prosternal process is more widened anteriorly (Plates 189 and 254:9).
Full description by Vondel (2021).
Biology: Found in lakes, ponds and streams.
Immature stages: Unknown.
Distribution: USA (Alabama, Arizona, Arkansas, Colorado, District of Columbia, Florida, Georgia, Illinois, Indiana, Louisiana, Maryland, New Jersey, North Carolina, Oklahoma, Oregon, Pennsylvania, South Carolina, Tennessee, Texas, Vermont, Virginia) (Plate 263:3).

203 *Haliplus (Paraliaphlus) tortilipenis* Brigham and Sanderson

Diagnosis: Length 3.6–4.0 mm, width 2.0–2.2 mm. Body oval, widest behind the shoulders. Related to *H. connexus*, but the serration on elytral apex is getting weaker to the posterior tip. In males the penis is very characteristic (Plates 190 and 255:1).
Full description by Vondel (2021).
Biology: Found in a cemetery pond of about 0.4 ha, with a maximum depth of 1 m, with dominant vegetation of *Scirpa* sp. and *Typha* sp. and dense growth of filamentous algae. The pond occasionally dries up. Also found in a strip-mine pond and a gravel pit.

Immature stages: Unknown.
Distribution: USA (Colorado, Illinois, Iowa, Kansas, Minnesota, Missouri, Oklahoma, South Dakota, Texas, Wisconsin) (Plate 263:4).

204 *Haliplus* (*Paraliaphlus*) *triopsis* Say

Diagnosis: Length 3.2–3.7 mm, width 1.8–2.15 mm. Body oval. Prosternal process not narrowed in anterior half. Sutural darkening reaching to first elytral primary row of punctures. Elytral ground color yellowish brown. Male with a characteristic penis with a strongly bent narrowed tip (Plates 191 and 255:2).
Full description by Vondel (2021).
Biology: Found in lakes, ponds and rivers. Attracted to UV light.
Immature stages: The 2nd and 3rd instar larva and the pupa are known. The actual knowledge is reviewed by Vondel (2025).
Distribution: Canada (Alberta, Manitoba, New Brunswick, Nova Scotia, Ontario, Quebec), USA (Alabama, Arizona, Arkansas, Colorado, Connecticut, District of Columbia, Florida, Georgia, Illinois, Indiana, Iowa, Kansas, Kentucky, Louisiana, Maine, Maryland, Massachusetts, Michigan, Minnesota, Mississippi, Missouri, Nebraska, "New England", New Hampshire, New Jersey, New Mexico, New York, North Carolina, Ohio, Oklahoma, Pennsylvania, Rhode Island, South Carolina, South Dakota, Tennessee, Texas, Virginia, West Virginia, Wisconsin) (Plate 263:5).

205 Genus *Peltodytes* Régimbart

Diagnosis: Yellow to brown, usually with some markings on dorsal side. Body widely oval with subparallel or even parallel sides, pronotum tapering anteriorly, elytra widest in the middle. Head with or without two marks between eyes, eyes relatively close to each other, apical segments of maxillary and labial palpi longer than penultimate segments. Pronotum with strong basal punctures and usually basal marks opposite third elytral puncture-row. Elytra with strong, occasionally interrupted, primary puncture-rows, usually some markings on intervals, sutural stria on at least posterior half, sutural interval darkened. Ventral side yellow with darkened areas. Prosternal process widened posteriorly. Metasternal process with lateral posteriorly diverging striae. Hind coxal plates with or without a posterior tooth, reaching beyond the sixth or seventh sternite. Legs rather short, male with widened first two tarsal segment on the fore- and midlegs, claws well developed.

Full description by Vondel (1997)
Biology: The species in this genus live in stagnant and slowly running water.
Distribution: Afrotropical, Nearctic, Neotropical, Oriental and Palaearctic Region.

206 Subgenus *Neopeltodytes* Satô

Diagnosis: Metasternal process without a central impression.
Full description and discussion about the status of this subgenus by Vondel (2021)
Biology: In all kinds of fresh stagnant and running water.
Distribution: Nearctic and Northern part of Neotropical region.

207 *Peltodytes* (*Neopeltodytes*) *bradleyi* Young

Diagnosis: Length 3.3–3.6 mm, width 2.0–2.2 mm. Body short oval. Very similar to *P. pedunculatus*, but can in the male easily been separated by the different aedaeagus (Plates 192 and 255:3).
Full description by Vondel (2021).
Biology: Occurs in rivers, swamps and in calcareous springs. Attracted to light.
Immature stages: Unknown.
Distribution: USA (Florida, Georgia, North Carolina, South Carolina, Virginia) (Plate 263:6).

208 *Peltodytes* (*Neopeltodytes*) *callosus* (LeConte)

Diagnosis: Length 3.1–3.8 mm, width 1.9–2.2 mm. Body short oval, widest in the middle. Together with the *P. roughleyi*, this is the only *Peltodytes* sp, with elytral tubercles. In *P. callosus* the tubercle is smooth, while in *P. roughleyi* the tubercle is sharp (Plates 193 and 255:4).
Full description by Vondel (2021).
Biology: The species occurs in creeks, brooks, streams, lakes, ponds (even brackish) and ditches up to an elevation of ca 2750 m.
Immature stages: Unknown.
Distribution: Canada (Alberta, British Columbia, Saskatchewan), USA (Arizona, California, Colorado, Idaho, Kansas, Minnesota, Montana, Nevada, New

Mexico, North Dakota, Oregon, South Carolina?, South Dakota, Utah, Washington, Wyoming), Mexico (Baja California, Chihuahua, Durango) (Plate 263:7).

209 *Peltodytes* (*Neopeltodytes*) *darlingtoni* Young

Diagnosis: Length 3.7–3.8 mm, width 2.1–2.3 mm. Body oval, widest before the middle. Similar to *P. dietrichi*, but first elytral interval is totally darkened in anterior part (Plates 194 and 255:5).
Full description by Vondel (2021).
Biology: Unknown.
Immature stages: Unknown.
Distribution: Cuba (Plate 263:8).

210 *Peltodytes* (*Neopeltodytes*) *dietrichi* Young

Diagnosis: Length 3.7–4.0 mm, width 2.2–2.25 mm. Body short oval, widest in the middle. Differs from *P. tortulosus* by the smaller size and from *P.darlingtoni* by the narrow sutural darkening, not reaching to first primary row of punctures (Plates 195 and 255:6).
Full description by Vondel (2021).
Biology: Found in ditches, fresh water and brackish ponds.
Immature stages: Unknown.
Distribution: Canada (Quebec), USA (Delaware, District of Columbia, Florida, Georgia, Louisiana, Maryland, Mississippi, New Jersey, New York, North Carolina, South Carolina, Virginia) (Plate 263:9).

211 *Peltodytes* (*Neopeltodytes*) *dunavani* Young

Diagnosis: Length 3.1–3.6 mm, width 1.8–2.1 mm. Body short oval, widest in the middle. Similar to *P. floridensis*, but males of *P. dunavani* have bold parameres, without fringes of hair, while males of *P. floridensis* have fringes of hair on top of the parameres (Plates 196 and 255:7).
Full description by Vondel (2021).
Biology: In lakes, pools, rivers, streams, ditches and swamps. Attracted to UV light.
Immature stages: Unknown.

Distribution: Canada (Alberta?), USA (Alabama, Arkansas, Florida, Georgia, Illinois, Indiana, Kentucky, Louisiana, Mississippi, Missouri, North Carolina, Ohio, Oklahoma, South Carolina, Tennessee, Texas, Virginia) (Plate 263:10).

212 *Peltodytes* (*Neopeltodytes*) *duodecimpunctatus* (Say)

Diagnosis: Length 3.2–3.8 mm, width 1.8–2.3 mm. Body short oval, widest in the middle. The yellow band on the hind femur distinguishes this species from other *Peltodytes* spp. except *P. edentulus*, but the latter has a black collar (Plates 197 and 255:8).
Full description by Vondel (2021).
Biology: In ponds, lakes, creeks and rivers. Predators of chironomid eggs. At UV light.
Immature stages: The 1st, 2nd and 3rd instar larva and the pupa are known. The actual knowledge is reviewed by Vondel (2025).
Distribution: Canada (New Brunswick, Nova Scotia, Ontario, Quebec), USA (Alabama, Arizona, Arkansas, California, Connecticut, Delaware, District of Columbia, Florida, Georgia, Illinois, Indiana, Iowa, Kansas, Kentucky, Louisiana, Maine, Maryland, Massachusetts, Michigan, Minnesota, Missouri, Montana, Nebraska, New Hampshire, New Jersey, New York, North Carolina, Ohio, Oklahoma, Oregon, Pennsylvania, Rhode Island, South Carolina, South Dakota, Tennessee, Texas, Vermont, Virginia, West Virginia, Washington?, Wisconsin) (Plate 263:11).

213 *Peltodytes* (*Neopeltodytes*) *edentulus* (LeConte)

Diagnosis: Length 3.3–4.05 mm, width 1.85–2.3 mm. The black collar on the head distinguishes this species from the other species with a yellow-ringed femur: *P. duodecimpunctatus* (Plates 198 and 255:9).
Full description by Vondel (2021).
Biology: In rivers, streams, ponds, lakes and ditches. The species is known as a predator of chironomid eggs.
Immature stages: The eggs, the 1st, 2nd and 3rd instar larva and the pupa are known. The actual knowledge is reviewed by Vondel (2025).
Distribution: Canada (Alberta, British Columbia, Manitoba, New Brunswick, Nova Scotia, Ontario, Prince Edward Island, Quebec, Saskatchewan), USA (California, Colorado, Connecticut, Delaware, District of Columbia, Illinois,

Indiana, Iowa, Kansas, Kentucky, Louisiana, Maine, Maryland, Massachusetts, Michigan, Minnesota, Missouri, Mississippi, Montana, Nebraska, Nevada, New Hampshire, New Jersey, New York, North Dakota, Ohio, Oklahoma, Pennsylvania, Rhode Island, South Dakota, Texas, Utah, Vermont, Washington, Wisconsin, Wyoming) (Plate 263:12).

214 *Peltodytes* (*Neopeltodytes*) *festivus* (Wehncke)

Diagnosis: Length 2.9–3.0 mm, width 1.8 mm. Body oval, widest in the middle. This species is characterised by the small size and the pronotal basic plicae (Plates 199 and 256:1).
Full description by Vondel (2021).
Biology: In streams, creeks, ponds.
Immature stages: Unknown.
Distribution: USA (Arkansas, Missouri, Texas) (Plate 264:1).

215 *Peltodytes* (*Neopeltodytes*) *floridensis* Matheson

Diagnosis: Length 3.05–3.8 mm, width 1.8–2.25 mm. Body oval (Similar to *P. dunavani*, but males of *P. dunavani* have bold parameres, without fringes of hair, while males of *P. floridensis* have fringes of hair on top of the parameres) (Plates 200 and 256:2).
Full description by Vondel (2021).
Biology: In lakes, ponds, cypress sloughs, and open marshy areas, in beach drift, at UV light.
Immature stages: Unknown.
Distribution: USA (Alabama, Florida, Georgia, Indiana, Louisiana, Maryland, Missouri, North Carolina, South Carolina, Virginia) (Plate 264:2).

216 *Peltodytes* (*Neopeltodytes*) *litoralis* Matheson

Diagnosis: Length 3.3–3.9 mm, width 2.0–2.3 mm. Body oval. The yellow hind femur distinguishes this species from related species (Plates 201 and 256:3).
Full description by Vondel (2021).
Biology: In rtivers, creeks, ponds, attracted to light.
Immature stages: The 3rd instar larva is known. The actual knowledge is reviewed by Vondel (2025).

Distribution: Canada (New Brunswick), USA (Arizona, Arkansas, Colorado, District of Columbia, Illinois, Indiana, Iowa, Kansas, Louisiana, Mississippi, Missouri, Nebraska, New Hampshire, New York, Oklahoma, Pennsylvania, South Carolina?, Texas) (Plate 264:3).

217 ***Peltodytes* (*Neopeltodytes*) *muticus*** **(LeConte)**

Diagnosis: Length 3.25–3.6 mm, width of body 2.0–2.3 mm. Body short oval, widest in the middle. The combination of the absence of an elytral humeral spot, the absence of an elytral tubercle and the presence of the elytral sutural dark mark being connected to the suture will separate this species from other *Peltodytes* spp. (Plates 202 and 256:4)
Full description by Vondel (2021).
Biology: This species occurs in lakes, rivers, streams, ponds, canals and is attracted to UV light.
Immature stages: The eggs, the 1st, 2nd and 3rd instar larva and the pupa are known. The actual knowledge is reviewed by Vondel (2025).
Distribution: Canada (Manitoba, Ontario, Quebec), USA (Alabama, Arizona, Arkansas, California, Connecticut, Delaware, District of Columbia, Florida, Georgia, Illinois, Indiana, Kansas, Kentucky, Louisiana, Maryland, Massachusetts, Michigan, Mississippi, Missouri, Nevada, New Jersey, New York, North Carolina, Ohio, Oklahoma, Pennsylvania, Rhode Island, South Carolina, Tennessee, Texas, Utah, Vermont, Virginia, West Virginia), Mexico (Michoacán?, Nuevo Léon) (Plate 264:4).

218 ***Peltodytes* (*Neopeltodytes*) *oppositus*** **Roberts**

Diagnosis: Length 3.6–3.7 mm, width 2.2–2.3 mm. Body oval. Similar to *P. shermani*, but in the males easily separated by the shape of the penis and the mesotarsus (Plates 203 and 256:5).
Full description by Vondel (2021).
Biology: Occurs in lakes, rivers, swamps, muddy ponds, streams and cypress swamps. Attracted to UV light.
Immature stages: Unknown.
Distribution: USA (Alabama, Florida, Georgia, Louisiana, Massachusetts, Minnesota, Mississippi, New York, North Carolina, South Carolina, Texas, Virginia) (Plate 264:5).

219 *Peltodytes (Neopeltodytes) ovalis* Zimmermann

Diagnosis: Length 3.1–3.4 mm, width 1.9–2.0 mm. Body oval, widest before the middle, maculation not strongly developed, elytral punctures weak. The angulate hind coxal plate in combination with the metaventral process not having lateral plicae in anterior part distinguishes this species from the other Mexican species (Plates 204 and 256:6).
Full description by Vondel (2021).
Biology: Collected in pools of a stream bed.
Immature stages: Unknown.
Distribution: Mexico (Guanajuato, Colima, Guerrero, Oaxaca, Puebla, Querétaro, Veracruz, Zacatecas) (Plate 264:6).

220 *Peltodytes (Neopeltodytes) pedunculatus* (Blatchley)

Diagnosis: Length 3.0–3.5 mm, width 2.0 mm. Body oval. Very similar to *P. bradleyi*, but males can easily been distinguished by the different aedaeagus (Plates 205 and 256:7).
Full description by Vondel (2021).
Biology: Found in marshy ponds. Attracted to UV light.
Immature stages: Unknown.
Distribution: USA (Florida?, Georgia, Illinois, Indiana, Kansas, Kentucky, Louisiana, Maine, Michigan, Minnesota, Missouri, New York, North Carolina, Ohio, Oklahoma, South Carolina, Texas) (Plate 264:7).

221 *Peltodytes (Neopeltodytes) roughleyi* Vondel

Diagnosis: Length 3.2–3.7 mm, width 1.9–2.2 mm. Body short oval, widest in the middle. Together with *P. callosus,* this is the only species with a tubercle (or callosity) at the centre of each elytron. In *P. callosus*, the tubercle is smooth, while in *P. roughleyi* the tubercle is sharp (Plates 206 and 256:8).
Full description by Vondel (2021).
Biology: In pools in creek bed, temporary ponds, up to an altitude of 2114 m (6937 ft).
Immature stages: Unknown.
Distribution: USA (Arizona, New Mexico) (Plate 264:8).

222 *Peltodytes* (*Neopeltodytes*) *sexmaculatus* Roberts

Diagnosis: Length 3.65–3.75 mm, width 2.1–2.2 mm. Body oval. Usually the absence of an elytral humeral spot and the free medial spots on the elytra will separate this species from other *Peltodytes* spp. with dark hind femora. Specimens with a humeral spot may be distinguished from *P. festivus* by the absence of basal pronotal plicae (Plates 207 and 256:9).
Full description by Vondel (2021).
Biology: Occurs in lakes, rivers, slow streams, ditches, pools and on mud flats. Attracted to UV light. The species is known as a predator of chironomid eggs.
Immature stages: The eggs, the 1st, 2nd and 3rd instar larva and the pupa are known. The actual knowledge is reviewed by Vondel (2025).
Distribution: Canada (Ontario, Quebec), USA (Alabama, Arizona, Arkansas, Delaware, Florida, Georgia, Illinois, Indiana, Iowa, Kansas, Kentucky, Louisiana, Maine, Maryland, Massachusetts, Michigan, Minnesota, Mississippi, Missouri, Montana, New Jersey, North Carolina, Ohio, Oklahoma, Pennsylvania, South Carolina, Tennessee, Texas, Virginia, West Virginia, Wisconsin) (Plate 264:9).

223 *Peltodytes* (*Neopeltodytes*) *shermani* Roberts

Diagnosis: Length 3.55–3.7 mm, width, 2.0–2.2 mm. Body oval. Similar to *P. oppositus*, but males can easily be distinguished by the shape of the penis and the mesotarsus (Plates 208 and 257:1).
Full description by Vondel (2021).
Biology: Occurs in lakes, rivers, streams, pools and swamps. Attracted to UV light. The species is known as a predator of chironomid eggs.
Immature stages: Unknown.
Distribution: USA (Alabama, Connecticut, District of Columbia, Florida, Georgia, Louisiana, Maine, Maryland, Massachusetts, Mississippi, New Hampshire, New Jersey, New York, North Carolina, Ohio, Pennsylvania, Rhode Island, South Carolina, Tennessee, Texas, Virginia) (Plate 264:10).

224 *Peltodytes* (*Neopeltodytes*) *simplex* (LeConte)

Diagnosis: Length 3.5–3.9 mm, width 2.0–2.3 mm. Oval to subparallel, widest in the middle. Base of elytra not or weakly darkened and without humeral spot.

These characters will generally separate this species from other *Peltodytes* spp. (Plates 209 and 257:2).
Full description by Vondel (2021).
Biology: Collected in a stream and in a roadside ditch with algae, *Jussica* sp. and *Ranunculus* sp. The following aquatic Coleoptera were found in association with *P. simplex* in Mexico, Patzcuaro: Dytiscidae: *Hydroporus* sp., *Thermonectus* sp., Haliplidae: *Haliplus elsaltous*, Hydrochidae: *Hydrochus* sp., Noteridae: *Suphisellus* sp.
Immature stages: Unknown.
Distribution: Canada (British Columbia, Saskatchewan), USA (Arizona, Arkansas, California, Colorado, District of Columbia?, Louisiana, Minnesota, Missouri, Nevada, New Mexico, Pennsylvania?, Tennessee, Texas, Utah), Mexico (Baja California, Baja California Sur, Distrito Federal, Durango, Guanajuato, Hidalgo, Jalisco, Michoacán, Oaxaca, Puebla, Sinaloa, Sonora, Veracruz) (Plate 264:11).

225 *Peltodytes* (*Neopeltodytes*) *tamaulipensis* Young

Diagnosis: Length 3.3–3.4 mm, width 1.9–2.0 mm. Body oval, widest just behind the shoulders. Metaventral process with strong wide impression on each side, which character should distinguish this species from other Mexican species (Plates 210 and 257:3).
Full description by Vondel (2021).
Biology: Collected in a stream in gallery forest and in pools in stream bed.
Immature stages: Unknown.
Distribution: Mexico (Tamaulipas) (Plate 264:12).

226 *Peltodytes* (*Neopeltodytes*) *tortulosus* Roberts

Diagnosis: Length 4.0–4.2 mm, width 2.4–2.6 mm. Body oval, slightly subparallel. Large species with pale-coloured hind femur (Plates 211 and 257:4).
Full description by Vondel (2021).
Biology: Occurs in pools, swamps and creeks.
Immature stages: Unknown.
Distribution: Canada (Alberta, British Columbia, Manitoba, NewBrunswick, Nova Scotia, Ontario, Prince Edward Island, Quebec, Saskatchewan), USA (Connecticut, Iowa, Indiana, Maine, Massachusetts, Michigan, Minnesota, Missouri, Montana, Nebraska, New Hampshire, New York, North Dakota, Ohio, South Dakota, Wisconsin) (Plate 265:1).

227 Subgenus *Peltodytes* Régimbart

Diagnosis: Metasternal process with a central impression.
Full description and discussion about the status of this subgenus by Vondel (2021)
Biology: In all kinds of fresh stagnant and running water.
Distribution: Palaearctic, Oriental and Afrotropical Region.

228 *Peltodytes* (*Peltodytes*) *caesus* (Duftschmid)

Diagnosis: Length 3.3–3.8 mm, width 1.7–2.0 mm. Body subparallel to parallel, widest in the middle. This species can be distinguished from *P. rotundatus* by the blunt denticle on the metacoxal plate (Plates 212 and 257:5).
Full description by Vondel (1992a, 1997).
Biology: In fresh or occasionally brackish stagnant water of ponds or ditches and slowly running streams. Among vegetation of *Batrachium*, *Callitriche*, *Elodes* and *Chara*. Feeding on filamentous algae and probably Characeans. The eggs are laid on the surface of waterplants. The adults are attracted to light (Konev, 1976 and pers. obs.).
Immature stages: The 1st, 2nd and 3rd instar larva and the pupa are known. The actual knowledge is reviewed by Vondel (2025).
Distribution: Afghanistan, Albania, Algeria, Armenia, Austria, Azerbaijan, Belgium, Bosnia Herzegovina, Bulgaria, Byelorussia, China (Xinjiang), Croatia, Czech Republic, Denmark, France, Georgia, Germany, Great Britain, Greece, Hungary, Iran, Iraq, Israel, Italy, Kazakhstan, Kyrgyzstan, Latvia, Lebanon, Liechtenstein, Lithuania, Luxembourg, Macedonia, Montenegro, Morocco, Netherlands, Poland, Portugal, Romania, Russia (Central European Territory, South European Territory), Serbia, Slovakia, Slovenia, Spain, Sweden, Switzerland, Syria, Tadzhikistan, Transcaucasia, Tunisia, Turkmenistan, Turkey, Ukraine, Uzbekistan, Yugoslavia (Plate 274:9).

229 *Peltodytes* (*Peltodytes*) *congoensis* Zimmermann

Diagnosis: Length 3.5–3.9 mm, width 2.1–2.3 mm. Body oval, widest before the middle. *P. congoensis* is closely related to the variable *P. speratus* and probably conspecific (Plates 213 and 257:6).
Full description by Vondel (2010).
Biology: Attracted to light.

Immature stages: Unknown.
Distribution: Cameroun, Congo, Gabon, Kenya, Zaire (Plate 280:9).

230 *Peltodytes* (*Peltodytes*) *coomani* Peschet

Diagnosis: Length 3.2–3.5 mm, width 1.9–2.1 mm. Body wide, subparallel in the middle. Closely resembles *P. sumatrensis*. Probably both species are conspecific (Plates 214 and 257:7).
Full description by Vondel (1992a).
Biology: Unknown.
Immature stages: Unknown.
Distribution: China (Guangdong, Guangxi, Hainan), Laos, Thailand, Vietnam (Plate 275:4).

231 *Peltodytes* (*Peltodytes*) *dauricus* Zimmermann

Diagnosis: Length 4.2 mm, width 2.4 mm. Body wide, subparallel (Plates 215 and 257:8).
Full description by Vondel (1992a).
Biology: Unknown.
Immature stages: Unknown.
Distribution: China (Heilongjiang, Liaoning, Nei Mongol), Russia (Far East) (Plate 274:8).

232 *Peltodytes* (*Peltodytes*) *ghanaensis* Vondel

Diagnosis: Length 3.4–3.7 mm, width 1.9–2.2 mm. Body subparallel. This species can be distinguished from related species by the absence of a row of punctures in pronotal basal impression (Plates 216 and 257:9).
Full description by Vondel (2010).
Biology: Attracted to light.
Immature stages: Unknown.
Distribution: Ghana (Plate 280:10).

233 *Peltodytes (Peltodytes) intermedius* (Sharp)

Diagnosis: Length 4.2 mm, width 2.4 mm. Body wide, subparallel (Plates 217 and 258:1).
Full description by Vondel (1992a).
Biology: Found in ponds and in a river.
Immature stages: Unknown.
Distribution: China (Beijing, Guangdong, Hubei, Jiangsu, Jiangxi, Liaoning, Shanghai, Sichuan, Zhejiang), Japan (Hokkaido, Honshu, Kyushu, Shikoku, Tsushima), Russia (Far East), South Korea, Taiwan (Plate 274:10).

234 *Peltodytes (Peltodytes) nodieri* Guignot

Diagnosis: Length 3.7–3.8 mm, width 2.2–2.3 mm. Body oval to slightly subparallel. This species is characterized by the strong impression around scutellum and the rough large punctures on basal fifth and sixth elytral puncture row (Plates 218 and 258:2).
Full description by Vondel (2010).
Biology: In ponds. Attracted to light.
Immature stages: Unknown.
Distribution: Benin, Bourkina Faso, D.R. Congo, Guinea, Ivory Coast, Mali, Nigeria, Senegal, Tchad (Plate 280:11).

235 *Peltodytes (Peltodytes) pekinensis* Vondel

Diagnosis: Length 3.6–3.9 mm, width 2.1–2.2 mm. Body widest just behind shoulders, in the middle subparallel (Plates 219 and 258:3).
Full description by Vondel (1992a).
Biology: Unknown.
Immature stages: Unknown.
Distribution: China (Beijing, Fujian, Guangdong, Guangxi, Hebei, Heilongjiang, Liaoning, Shaanxi, Shandong, Shanghai, Tianjin), Russia (Far East) (Plate 275:1).

236 *Peltodytes (Peltodytes) quadratus* Régimbart

Diagnosis: Length 3.7–4.1 mm, width 2.2–2.4 mm. Body oval, widest before the middle. This species is different from the other *Peltodytes* species in lacking the lateral margin on the metacoxal lobes (Plates 220 and 258:4).

Full description by Vondel (2010).
Biology: In ditches with filamentous algae and rivers. Attracted to light.
Immature stages: The 3rd instar larva is known. The actual knowledge is reviewed by Vondel (2025).
Distribution: Madagascar (Plate 280:12).

237 *Peltodytes* (*Peltodytes*) *rotundatus* (Aubé)

Diagnosis: Length 3.3–3.9 mm, width 2.0–2.2 mm. Body wide, subparallel, widest in the middle. This species can be distinguished from *P. caesus* by the sharp pointed denticle on the metacoxal plate (Plates 221 and 258:5).
Full description by Vondel (1992a, 1997).
Biology: Living between filamentous algae in lakes, pools, ponds, rivers, marshes, brooks, streamlets. Prefers running water. Among vegetation of *Utricularia* and *Nuphar*.
Immature stages: The 3rd instar larva is known. The actual knowledge is reviewed by Vondel (2025).
Distribution: Albania, Algeria, Bosnia Herzegovina, Croatia, France, Greece, Hungary, Iran, Italy, Luxembourg, Montenegro, Morocco, Portugal, Slovenia, Spain, Switzerland, Tunisia, Turkey, Yugoslavia (Plate 275:2).

238 *Peltodytes* (*Peltodytes*) *sinensis* (Hope)

Diagnosis: Length 3.4–3.8 mm, width 1.9–2.2 mm. Body oval, subparallel, widest in the middle (Plates 222 and 258:6).
Full description by Vondel (1992a).
Biology: In stagnant water, in rivers. Up to an altitude of 2100 m.
Immature stages: Unknown.
Distribution: China (Anhui, Beijing, Chongqing, Fujian, Guangdong, Guangxi, Guizhou, Hainan, Hebei, Henan, Hubei, Hunan, Jiangsu, Jiangxi, Jilin, Liaoning, Shaanxi, Shandong, Shanghai, Sichuan, Tianjin, Yunnan, Zhejiang), Japan (Ryukyu Islands, Ishigaki), Korea, Laos, Philippines, Taiwan, Vietnam (Plate 275:3).

239 *Peltodytes* (*Peltodytes*) *speratus* Régimbart

Diagnosis: Length 3.6–4.1 mm, width 2.0–2.3 mm. Body oval to subparallel, widest before the middle. The basal pronotal puncture row and the angulate coxal lobes distinguish it from related species, although the difference with *P. congoensis* is not always clear (see remarks above) (Plates 223 and 258:7).
Full description by Vondel (2010).
Biology: In lakes. Attracted to light. Up to an altitude of 1400 m.
Immature stages: Unknown.
Distribution: Kenya, Mauritius, Ruanda, Tanzania, Uganda, Zaire, Zambia (Plate 281:1).

240 *Peltodytes* (*Peltodytes*) *sumatrensis* Régimbart

Diagnosis: Length 3.4–3.7 mm, width 1.9–2.1 mm, Body subparallel, widest in the middle (Plates 224 and 258:8).
Full description by Vondel (1992a).
Biology: In stagnant water.
Immature stages: Unknown.
Distribution: China (Yunnan), Indonesia (Sulawesi, Sumatra), Laos, Myanmar, Thailand (Plate 275:5).

241 *Peltodytes* (*Peltodytes*) *ugandanus* Vondel

Diagnosis: Length 3.5–3.6 mm, width 2.0–2.1 mm. Body subparallel to oval, widest before the middle. The penis of this species will immediately distinguish it from *P. speratus* or *P. congoensis* (Plates 225 and 258:9).
Full description by Vondel (2010).
Biology: Unknown.
Immature stages: Unknown.
Distribution: D.R. Congo, Uganda (Plate 281:2).

242 Genus *Phalilus* Guignot

Diagnosis: Head without distinctive genal lines behind the eyes. Penultimate segment of labial palpi slender. Prostenal process parallel with strong marginal puncture row. Male protarsus and mesotarsus with the first segment strongly and the second segment weakly widened, both segments ventrally with adhesive setae.
Full description by Vondel (2019).
Biology: In a variety of fresh and salt water bodies.
Distribution: Australasian Region.

243 *Phalilus bistriatus* (Wehncke)

Diagnosis: Length 2.5–3.4 mm, width 1.2–1.6 mm. Body oval, parallel to subparallel in the middle. This species can be distinguished from *P. fuscatus* and *P. gibbus* by the long curved pronotal plicae, from *P. hydei* by the basally curved pronotal plicae and the metasternal process not covered with dense hairy punctures and from *P. oberthuri* and *P. storeyi* by the smooth last sternite. Most males can be distinguished from those of related species by the solid digitus on top of the left paramere (Plates 226 and 259:1).
Full description by Vondel (1995).
Biology: Specimens have been attracted to light in places in open forest.
Immature stages: Unknown.
Distribution: Australia (New South Wales, Northern Territory, Queensland, South Australia, Western Australia) (Plate 277:9).

244 *Phalilus fortescueensis* (Watts and McRae)

Diagnosis: Length 2.4–2.9 mm, width 1.4–1.6 mm. Body somewhat spindle-shaped, subparallel in the middle, widest at shoulders (Plates 227 and 259:2).
Full description by Watts & McRae (2010).
Biology: Found in a very large salt marsh at the terminus of the upper portion of the Fortescue River, fresh when first filled, it becomes more saline as it dries.
Immature stages: The 3rd instar larva is known. The actual knowledge is reviewed by Vondel (2025).
Distribution: Australia (Western Australia) (Plate 277:10).

245 *Phalilus fuscatus* (Clark)

Diagnosis: Length 2.4–3.2 mm, width 1.2–1.8 mm. Body oval, but parallel to subparallel in the middle. In females no reliable characters are known to distinguish *P. fuscatus* and *P. gibbus*. The males of those species have a distinctly different penis (Plates 228 and 259:3).
Full description by Vondel (1995).
Biology: This species occurs in swamps, temporary ponds, rivers and lakes. A specimen was found in a trout stomach. It is attracted to light.
Immature stages: Unknown.
Distribution: Australia (New South Wales (?), Queensland, South Australia, Victoria, Western Australia) (Plate 277:11).

246 *Phalilus gibbus* (Clark)

Diagnosis: Length 2.4–3.2 mm, width 1.2–1.6 mm. Body oval, but parallel to subparallel in the middle. This species closely resembles *P. fuscatus*, from which the male can be distinguished by the penis, which is broader in *P. gibbus*. No reliable characters are known to distinguish the females of *P. gibbus* and *P. fuscatus* (Plates 229 and 259:4).
Full description by Vondel (1995).
Biology: This species occurs in swamps, temporary ponds and rivers and has been collected from submerged vegetation at the margin of slow moving streams.
Immature stages: Unknown.
Distribution: Australia (New South Wales (?), South Australia, Tasmania, Victoria, Western Australia) (Plate 277:12).

247 *Phalilus hydei* (Vondel)

Diagnosis: Length 2.7–3.0 mm, width 1.4–1.6 mm. Body oval, parallel to subparallel in the middle. This species can be distinguished from *P. bistriatus*, *P. fuscatus* and *P. gibbus* by the pronotal plicae, being long and straight at base and the male with clearly unequal protarsal claws (Plates 230 and 259:5).
Full description by Vondel (1995).
Biology: Unknown.
Immature stages: Unknown.
Distribution: Australia (Queensland) (Plate 278:1).

248 *Phalilus oberthuri* (Guignot)

Diagnosis: Length 3.0–3.3 mm, width 1.6–1.9 mm. Body broad, parallel in the middle. The male of this species is easy to distinguish from other species with pronotal plicae by the strong ridge on last stemite (Plates 231 and 259:6).
Full description by Vondel (1995), Vondel et al. (2021).
Biology: Found in a marsh and in dolines.
Immature stages: Unknown.
Distribution: Australia (New South Wales, Queensland), New Caledonia (Plate 278:2).

249 *Phalilus pinderi* (Watts and McRae)

Diagnosis: Length 2.6–2.9 mm, width 1.5–1.6. Deep bodied, oval, subparallel in the middle, pronotal/elytral junction indented, widest in the middle (Plate 232 and 259:7).
Full description by Watts & McRae (2010).
Biology: Found in small shallow billabong with *Chara* and a base of sand and rock. Also collected in deep clear pools and in fresh water pools up to 4 m deep with patches of *Chara* and *Nijas*. Sediments coarse, dominated by pebble, cobble, boulder and gravel.
Immature stages: The 3rd instar larva is known. The actual knowledge is reviewed by Vondel (2025).
Distribution: Australia (Western Australia) (Plate 278:3)

250 *Phalilus storeyi* (Vondel)

Diagnosis: Length 2.4–2.5 mm, width 1.2–1.3 mm. Body broad, parallel in the middle. This species differs from other species with pronotal plicae by the ridge on last sternite and from *H. oberthuri* by the faintly impressed pronotal plicae and its smaller size (Plates 233 and 259:8).
Full description by Vondel (1995).
Biology: Unknown.
Immature stages: Unknown.
Distribution: Australia (Northern Territory) (Plate 278:4).

References

Baca, S.M., A. Alexander, G.T. Gustafson & A.E.Z. Short. 2017. Ultraconserved elements show utility in phylogenetic inference of Adephaga (Coleoptera) and suggest paraphyly in 'Hydradephaga'. – Systematic Entomology pp. 1–10. DOI:10.1111/syen.12244.

Balfour-Browne, F. 1940. British water beetles. Vol. 1. London: Ray Society, 375 pp. + 5 pls.

Balfour-Browne, F. 1953. Coleoptera Hydradephaga. – Handbooks for the Identification of British Insects, 4(3): 1–33.

Beier, M. 1929. Zur Kenntniss der Lebenweise von *Haliplus wehnckei* Gerh. – Zeitschrift für Morphologie und Öekologie der Tiere. 14: 191–223.

Benetti, C.J. & N. Hamada. 2017. Two new species of *Haliplus* Latreille , 1802 (Coleoptera, Haliplidae) from Northern Brazil. – Zootaxa 4282(3) : 584–592.

Beutel, R.G. & R.E. Roughley. 1988. On the systematic position of the family Gyrinidae (Coleoptera: Adephaga). – Z. zool. Syst. Evolut.-forsch. 26: 380–400.

Beutel, R.G. & Ruhnau S. 1990. Phylogenetic analysis of the genera of Haliplidae (Coleoptera) based on characters of adults. – Aquatic Insects 12(1): 1–17.

Beutel, R.G. 1997. Über Phylogenese und Evolution der Coleopteren (Insecta), insbesondere der Adephaga. – Abhandlungen des Naturwissenschaftlichen Vereins in Hamburg (NF) 31:1–164.

Beutel, R.G., I. Ribera & O.R.P. Bininda-Emonds. 2008. A genus-level supertree of Adephaga (Coleoptera). – Organisms, Diversity & Evolution 7: 255–269.

Brakman, P.J. 1966. Lijst van Coleoptera uit Nederland en het omliggend gebied. – Monographieen van de Nederlandsche Entomologische Vereniging 2. Amsterdam: 219 pp.

Brocher, F. 1922. Observations biologiques sur les Haliplidés (Coléoptères). – Annales de Biologie Lacustre 11(1): 7–18.

Burmeister, E.G. 1976. Der Ovipositor der Hydradephaga (Coleoptera) und seine phylogenetische Bedeutung unter besonderer Berücksichtigung der Dytiscidae. – Zoomorphologie 85(3): 165–257.

Burmeister, E.G. 1980. Funktionsmorphologie und Evolution des Ovipositor der Adephaga (Coleoptera). – Verhandlungen des naturwissenschaftlichen Vereins Hamburg (NF) 24(1): 89–184.

Everts, E. 1898–1922. Coleoptera Neerlandica. De schildvleugelige insekten van Nederland en het aangrenzend gebied. I (1898), II (1903), III (1922). – The Hague.

Falkenström, G. 1926. Beiträge zur Kenntnis der Biologie der Haliplidae und der Metamorphose von *Haliplus immaculatus* Gerh. – Entomologisk Tidskrift. 47: 1–28.

Foster, G.N. 1981. Atlas of British Water Beetles – Part 1. – The Balfour-Browne Club Newsletter 22: 1–18.

Guignot, F. 1931–33. Les Hydrocanthares de France, Hygrobiidae, Haliplidae, Dytiscidae et Gyrinidae de la France continentale avec notes sur les espèces de la Corse et de ‘l Afrique du nord française. – Toulouse. 1057 pp.

Hansen, V. 1930. Biller VIII. Vandkalve og Hvirvlere (Haliplidae, Dytiscidae & Gyrinidae). – Danmarks Fauna, 34: 1–233.

Hayashi, M., T. Iwata & H. Yoshitomi. 2023. Revision of the family Haliplidae (Insecta, Coleoptera) in Japan. – ZooKeys 1168: 267–294.

Hendrich, L. & Vondel, B.J. van. 2010. Haliplidae (Coleoptera), pp. 237–242 – In: Jäch, M.A. & Balke, M. (eds.): Water beetles of New Caledonia (part 1). – Monographs on Coleoptera 3: IV+449 pp.

Hickman, J.R. 1930. Life-Histories of Michigan Haliplidae (Coleoptera). – Studies of the Michigan Academy (1929), 12: 399–424.

Hickman, J.R. 1931a. Contributions to the biology of the Haliplidae (Coleoptera). – Annals of the Entomological Society of America. 24: 129–142.

Hickman, J.R. 1931b. Respiration of the Haliplidae (Coleoptera). – Papers of the Michigan Academy of Science, Arts and Letters. 13: 277–289.

Hickman, J.R. 1937. Family Haliplidae. In: Galtsoff, P.S. et al (eds.). Culture Methods for Invertebrate Animals. New York, 589 pp.

Hoch, K. 1960. Bestimmungstabelle der mitteleuopäischen Arten der Untergattung *Haliplinus* Guignot 1947 (*ruficollis*- Gruppe) der Gattung *Haliplus* Latr. – Entomologische Blätter 56: 49–69.

Holmen, M. 1987. The aquatic Adephaga (Coleoptera) of Fennoscandia and Denmark. I. Gyrinidae, Haliplidae, Hygrobiidae and Noteridae. – Fauna Entomologica Scandinavica. 20: 9–17, 1–168.

Jackson, D.J. 1952. Observations on the capacity for flight of water beetles. – Proceedings of the Royal Entomological Society of London (A) 27: 57–70.

Jackson, D.J. 1956a. Observations on flying and flightless waterbeetles. – Journal of the Linnean Society of London 43: 18–42.

Jackson, D.J. 1956b. The capacity for flight of certain water beetles and its bearing on their origin in the western Scottish Isles. – Proceedings of the Linnean Society of London 167: 76–96.

Jackson, D.J. 1973. The influence of flight capacity on the distribution of aquatic Coleoptera in Fife and Kinross-shire. – Entomologist’s Gazette 24: 247–293.

Jia, F., Z. Mai, Z.Jiang & Z. Liang. 2024. Illustrated Handbook of Aquatic Beetles of China (Myxophaga, Hydradephaga, Hydrophiloidea). – Straits Books Publishing House, 316 pp. [in Chinese]

Jia, F., & Vondel, B.J. van. 2011. Annotated catalogue of the Haliplidae of China with the descriptionof a new species and new records from China (Coleoptera, Adephaga). – Zookeys 133:1–17.

Kitano, T, & S. Sano. 2011. Haliplidae larvae eat algae. In: H. Karube (Ed.), Suisei Kontyu daihyakka, Kanagawa prefectural Museum (pp. 48–49). Odawara. [in Japanese].

Konev, A.A. 1976. Water beetles of the suborder Adephaga (Coleoptera) in central Kazakhstan. – Entomological Review 55(4):56–58.

Lee, D-H., Vondel, B.J. van, Lee, S.H. & Ahn, K-J. 2015. Taxonomic review of the Haliplidae (Coleoptera) in the Korean peninsula. – Entomological Research 45(2015): 123–134.

Lindroth, C.H. (ed.). 1960. Catalogus Coleopterorum Fennoscandiae et Daniae. Lund.

Löbl, I. & D. Löbl. 2017. Catalogue of Palaearctic Coleoptera, Volume 1. Archostemata-Myxophaga-Adephaga. Revised and Updated Edition. 1443 pp.

Matheson, R. 1912. The Haliplidae of North America, North of Mexico. – Journal of the New York Entomological Society. 20: 156–193.

Mazzoldi, P. & Vondel, B.J. van. 1997. *Haliplus samosirensis* Vondel recorded from Kalimantan (Coleoptera: Haliplidae). – Koleopterologische Rundschau 67:119–120.

Megna, Y.S., L. Hendrich & M. Balke. 2018. A new species of *Haliplus* Latreille, 1802 from Cuba (Coleoptera: Haliplidae). – Aquatic Insects, 2018. pp. 1–9.

Mousseau T. 2004. Taxonomy, Classification, Reconstructed Phylogeny, Biogeography, and Natural History of Nearctic Species of *Brychius* Thomson (Coleoptera: Haliplidae). xvi+144 pp., 42 figs, 5 app. – Thesis.

Mousseau, T. & Roughley, R.T. 2007. Taxonomy, Classification, reconstructed Phylogeny and Biography of Nearctic Species of *Brychius* Thomson (Coleoptera: Haliplidae). – The Coleopterist's Bulletin 61(3): 351–397.

Nakajima, J., M. Hayashi, K. Ishida, T. Kitano & H. Yoshimura. 2020. Aquatic Coleoptera and Hemiptera of Japan. 352 pp. Japan. [in Japanese].

Scholz, R. 1929. Zur Systematik der Haliplidae. – Entomologisches Nachrichtenblatt 3: 28–29.

Seeger, W. 1971a. Morphologie, Bionomie und Ethologie von Halipliden, unter besonderer Berücksichtigungfunktionsmorphologischer Gesichtspunkte (Haliplidae: Coleoptera). – Archiv für Hydrobiologie. 68: 400–435.

Seeger, W. 1971b. Autökologische Laboruntersuchungen an Halipliden mit zoogeographischen Anmerkungen (Haliplidae; Coleoptera). – Archiv für Hydrobiologie 68: 528–547.

Seeger, W. 1971c. Die Biotopwahl bei Halipliden, zugleich ein Beitrag zum Problem der syntopischen (sympatrischen s.str.) Arten (Haliplidae; Coleoptera). – Archiv für Hydrobiologie 69: 175–199.

Sheth, S.D., Ghate, H.V. & Vondel, B.J. van. 2016. Illustrated redescription of *Haliplus* (*Liaphlus*) *arrowi* Guignot, 1936 (Coleoptera: Haliplidae) from the Western Ghats, India, and notes on the closely related *H. angustifrons* Régimbart, 1892. – Zootaxa 4127 (2): 355–364.

Short, A.E.Z., Torres, P.J., &Vondel, B.J. 2010. A review of the crawling water beetles of Mongolia (Coleoptera: Haliplidae). – Proceedings of the Academy of Natural Sciences of Philadelphia 159: 205–212.

Šťastný, J. & Boukal, M. 2003. Haliplidae: III. *Haliplus* (*Haliplidius*) *rejseki* sp. n. from Sichuan (Coleoptera), pp. 295–300. In: Jäch, M.A. & Ji, L. (eds.): Water Beetles of China, Vol. III. Wien: Zoologisch-Botanishe Gesellschaft in Österreich and Wiener Coleopterologenverein, 572 pp.

Vondel, B.J. van. 1985. Verspreiding van Haliplidae in Nederland. Een tussenstand. – Nieuwsbrief European Invertebrate Survey-Nederland. 16: 31–34.

Vondel, B.J. van. 1986a. About two teratological Haliplidae (Coleoptera). – Entomologische Berichten, Amsterdam 46: 12.

Vondel, B.J. van. 1986b. Description of the second and third instar larvae of *Haliplus laminatus* (Schaller) with notes on the subgeneric status (Coleoptera: Haliplidae). Entomologische Berichten, Amsterdam 46, 128–132.

Vondel, B.J. van. 1991. Revision of the palaearctic species of *Haliplus* subgenus *Liaphlus* Guignot (Coleoptera: Haliplidae). – Tijdschrift voor Entomologie 134(1): 75–144.

Vondel, B.J. van, 1992a. Revision of the Palaearctic and Oriental species of *Peltodytes* Régimbart (Coleoptera: Haliplidae). – Tijdschrift voor Entomologie 135: 275–297.

Vondel, B.J. van. 1992b. Haliplidae – In: M.B.P. Drost, H.P.J.J. Cuppen, E.J. van Nieukerken & M. Schreijer (red.), 1992. – De Waterkevers van Nederland: 73–85. – Uitgeverij K.N.N.V., Utrecht.

Vondel, B.J. van. 1993. Revision of the *Liaphlus* species of the Oriental region, excluding China (Coleoptera: Haliplidae). – Tijdschrift voor Entomologie 136: 289–316.

Vondel, B.J. van. 1995a. Revision of the Haliplidae (Coleoptera) of the Australian Region and the Moluccas. – Records of the South Australian Museum 28 (1): 61–101.

Vondel, B.J. van, 1995b. Haliplidae: Review of the Haliplidae of China (Coleoptera), pp. 111–154. In M.A. Jäch and L. Ji, eds. – Waterbeetles of China. Zoologisch-Botanische Gesellschaft, Vienna. 410 pp.

Vondel, B.J. van. 1997. Haliplidae. pp. 1–95. In: Schwoerbel, J. & Zwick, P. (eds.) Süsswasserfauna von Mitteleuropa 20.2. Gustav Fischer, Stuttgart.

Vondel, B.J. van. 2003a. Additional notes on Chinese Haliplidae, with description of a new species (Coleoptera). pp. 289–294. In: Jäch M.A. & Ji L. (Eds.): Water beetles of China, Vol. 3. Wien: Zoologisch-Botanische Gesellschaft in Österreich and Wiener Coleopterologenverein, vi + 572 pp.

Vondel, B.J. van. 2003b. Review of the Haliplidae of Vietnam with the description of a new species (Coleoptera). – Tijdschrift voor Entomologie 146: 45–47.

Vondel, B.J. van. 2005. Haliplidae, pp. 20–86. In: Nilsson A.N. & Vondel B.J. van. Amphizoidae, Aspidytidae, Haliplidae, Noteridae and Paelobiidae (Coleoptera: Adephaga). – In: World Catalogue of Insects 7: 1–171.

Vondel, B.J. van. 2007. World Catalogue of Haliplidae – corrections and additions, 1 (Coleoptera: Haliplidae). – Koleopterologische Rundschau 77, 89–96.

Vondel, B.J. van. 2008. A new species of *Haliplus* from Sulawesi (Coleoptera: Haliplidae). – Tijdschrift voor Entomologie 151: 141–143.

Vondel, B.J. van. 2009. Review of the Haliplidae of Myanmar with description of *Haliplus volpei* (Coleoptera). – Tijdschrift voor Entomologie 152: 333–338, Figs 1–12.

Vondel, B.J. van. 2010. Revision of the Haliplidae of the Afrotropical region including North Africa (Coleoptera). – Tijdschrift voor Entomologie 153: 239–314, Figs 1–326.

Vondel, B.J. van, 2013. World Catalogue of Haliplidae – corrections & additions, 2 (Coleoptera: Haliplidae). – Koleopterologische Rundschau 83:23–34.

Vondel, B.J. van. 2014. Reconsideration of the Nearctic species *Haliplus mimeticus* Matheson and its relatives with description of a new species (Coleoptera: Haliplidae). – Tijdschrift voor Entomologie 157: 137–144.

Vondel, B.J. van. 2016a. Corrections and additions to the "Revision of the Haliplidae of the Neotropical Region including Mexico" (Vondel & Spangler 2008) with description of a new species (Coleoptera: Haliplidae). – Koleopterologische Rundschau, 86, 51–59.

Vondel, B.J. van. 2016b. Haliplidae. pp. 89–98 in: Beutel, R.G. & Leschen, R.A.B. (Eds.) Coleoptera, Beetles. Morphology and Systematics. Archostemata, Adephaga, Myxophaga, Polyphaga Partim. Vol. 1, 2nd edition. Handbook of Zoology, Beutel, R.G. & Kristensen, N.P. (Eds.). Arthropoda: Insecta I–XVII, 1–684.

Vondel, B.J. van. 2019. Features of the metacoxal air-storage space as additional characters for reconstructing the phylogeny of Haliplidae (Coleoptera). – Tijdschrift voor Entomologie 162 (2019): 13–32.

Vondel, B.J. van. 2021. Revision of the Nearctic Haliplidae (Coleoptera). – Tijdschrift voor Entomologie 163 (2–3): 101–298.

Vondel, B.J. van. 2023. World Catalogue of Haliplidae – Corrections and additions 3.– Koleopterologische Rundschau 92: 73–84.

Vondel, B.J. van. 2023. Revision of the Nearctic Haliplidae (Coleoptera). Corrections and additions I.– Tijdschrift voor Entomologie 166: 117–124.

Vondel, B.J. van. 2025. Immature stages of Haliplidae. Review of the current knowledge, additions and corrections (Coleoptera). – Zoologica 167: 1–100.

Vondel, B.J. van & Alarie, Y. 2016. A New Species of *Haliplus* Latreille, 1802 (Coleoptera: Adephaga: Haliplidae) from Canada. – The Coleopterist's Bulletin 70(4): 801–804.

Vondel, B.J. van & Bergsten, J. 2012. Review of the Haliplidae of Madagascar, with descriptions of two new species (Coleoptera). – Tijdschrift voor Entomologie 155: 57–66.

Vondel, B.J. van & Ghosh, S.K. 2018. Re-evaluation of Vazirani's *Haliplus* species from India (Coleoptera: Haliplidae). – Tijdschrift voor Entomologie 161: 3–10.

Vondel, B.J. van, M. Holmen and P.N. Petrov. 2006. Review of the Palearctic and Oriental species of the subgenus *Haliplus s.str.* (Coleoptera: Haliplidae: *Haliplus*) with descriptions of three new species. – Tijdschrift voor Entomologie 149, 227–273.

Vondel, B.J. & Litovkin, S. 2017. Five new synonymies in *Haliplus* subg. *Liaphlus* Guignot, based on the variability of the left paramere (Coleoptera: Haliplidae). – Koleopterologische Rundschau 87: 31–35.

Vondel, B.J. van & K. Matsumoto. 2021. First record and diagnosis of a male *Haliplus (Liaphlus) madagascariensis* Vondel & Bergsten and additional faunistic records of African Haliplidae (Coleoptera: Adephaga). – Tijdschrift voor Entomologie 164: 1–14.

Vondel, B.J. van, C. Pöllabauer, J-F Parpet, N. Mary, N. Charpin & M.A. Jäch. 2021. On the rediscovery of *Phalilus oberthuri* (Guignot, 1935) in New Caledonia (Coleoptera: Haliplidae). – Koleopterologische Rundschau 91 311–322.

Vondel, B.J. van & Spangler, P.J. 2008. Revision of the Haliplidae of the Neotropical Region including Mexico (Coleoptera: Haliplidae). – Koleopterologische Rundschau 78, 69–194.

Vondel, B.J. van & Volpe, V. 2015. Additional notes on *Haliplus volpei* Vondel (Coleoptera, Haliplidae). – Latissimus 36, 5–7.

Ward, R.D. 1979. Metathoracic wing structure as phylogenetic indicators in the Adephaga (Coleoptera). In: Ball et al.(ed.): Carabid Beetles: Their Evolution, Natural History and Classification. The Hague/Boston/London.

Watts, C.H.S. & McRae, J., 2010. The identity of *Haliplus* (Coleoptera: Haliplidae) from the Pilbara region of Australia, including the description of four new species. – Records of the Western Australian Museum 25: 387–398.

Watts, C.H.S. 1988. Revision of Australian Haliplidae (Coleoptera). – Records of the South Australian Museum 22(1): 21–28.

Yamasaki, S., K. Hirasawa & S. Sano. 2020. Knowledge of larval and pupal period of *Haliplus eximius*. Sayabane. New Series 40: 17–18. [in Japanese].

Young, F.N. 1961. Pseudosibling species in the genus *Peltodytes* (Coleoptera: Haliplidae). – Annals of the Entomological Society of America. 54: 214–222.

Zimmermann, A., 1920. Dytiscidae, Haliplidae, Hygrobiidae, Amphizoidae. – In: S. Schenkling (ed.). Coleopterorum Catalogus. Vol. 4, pars 71. – W. Junk, Berlin. 326 pp.

Zimmermann, A. 1924. Die Halipliden der Welt. Bestimmungs-tabellen der Haliplidae mit gleichzeitig beigefügten Detailbeschreibungen der einzelnen Arten. – Entomologische Blätter 20 (1): 1–16; 20 (2): 65–80; 20 (3): 129–144; 20 (4): 193–213.

Plates 1–233. Figures of Species in Alphabetical Order

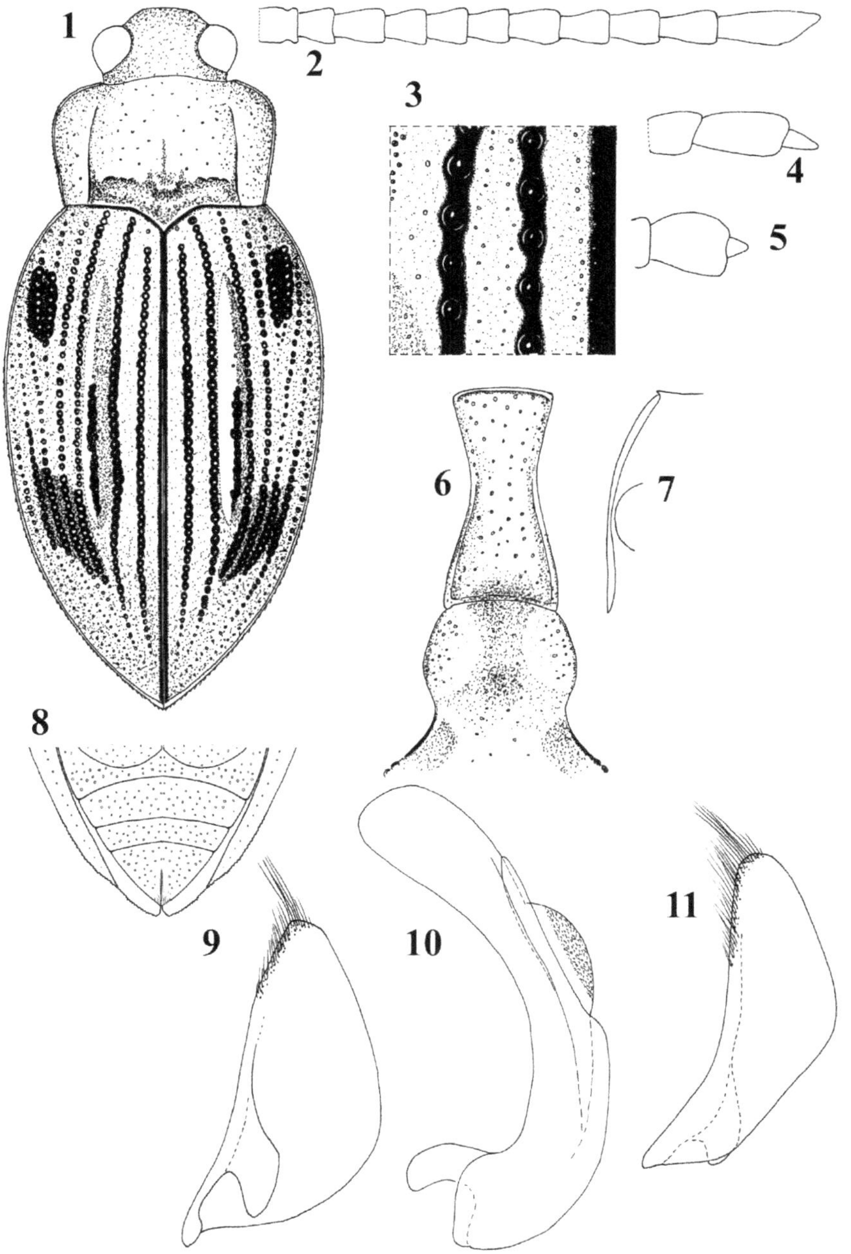

PLATE 1 *Brychius elevatus*. 1, dorsal view; 2, antenna; 3, elytral punctures near base and suture; 4, maxillary palpus; 5, labial palpus; 6, prosternal and metaventral process; 7, prosternal process in lateral view; 8, last sternite; 9, left paramere; 10, penis; 11, right paramere

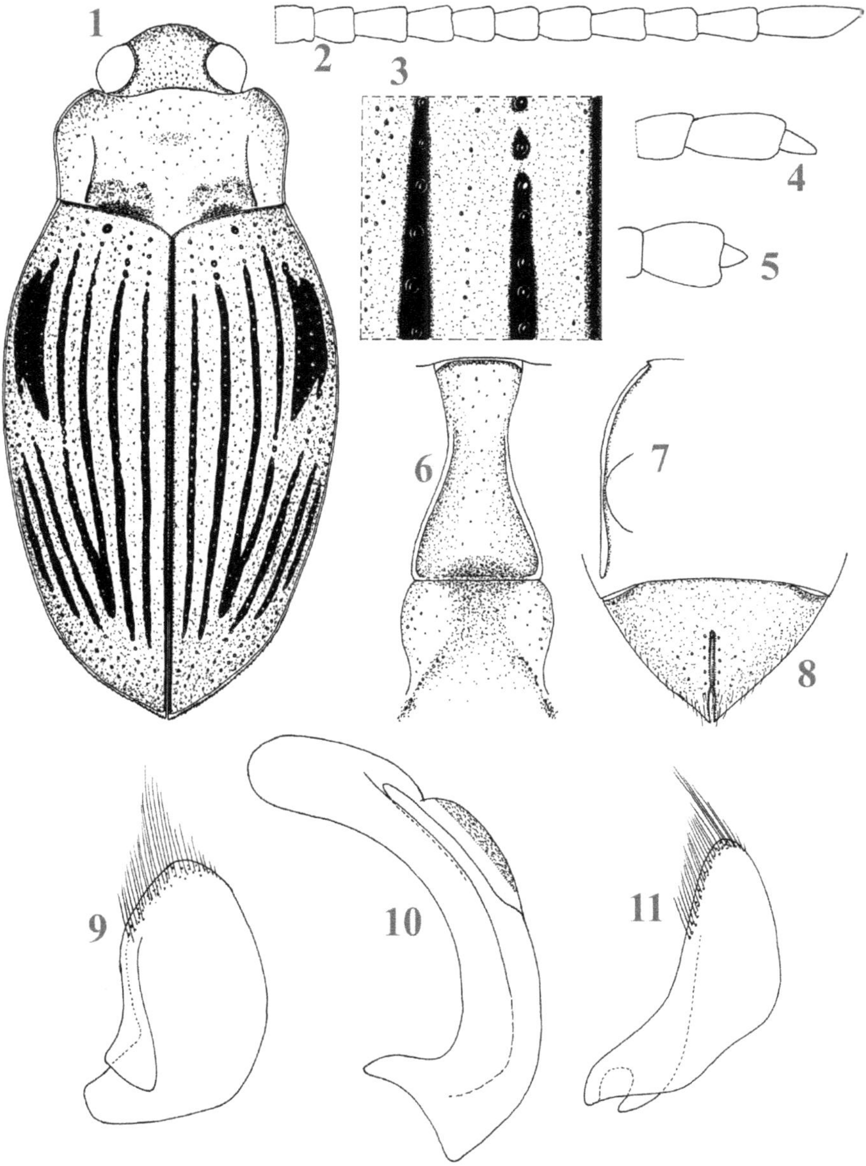

PLATE 2 *Brychius glabratus*. 1, dorsal view; 2, antenna; 3, elytral punctures near base and suture; 4, maxillary palpus; 5, labial palpus; 6, prosternal and metaventral process; 7, prosternal process in lateral view; 8, sternites and epipleuron; 9, left paramere; 10, penis; 11, right paramere

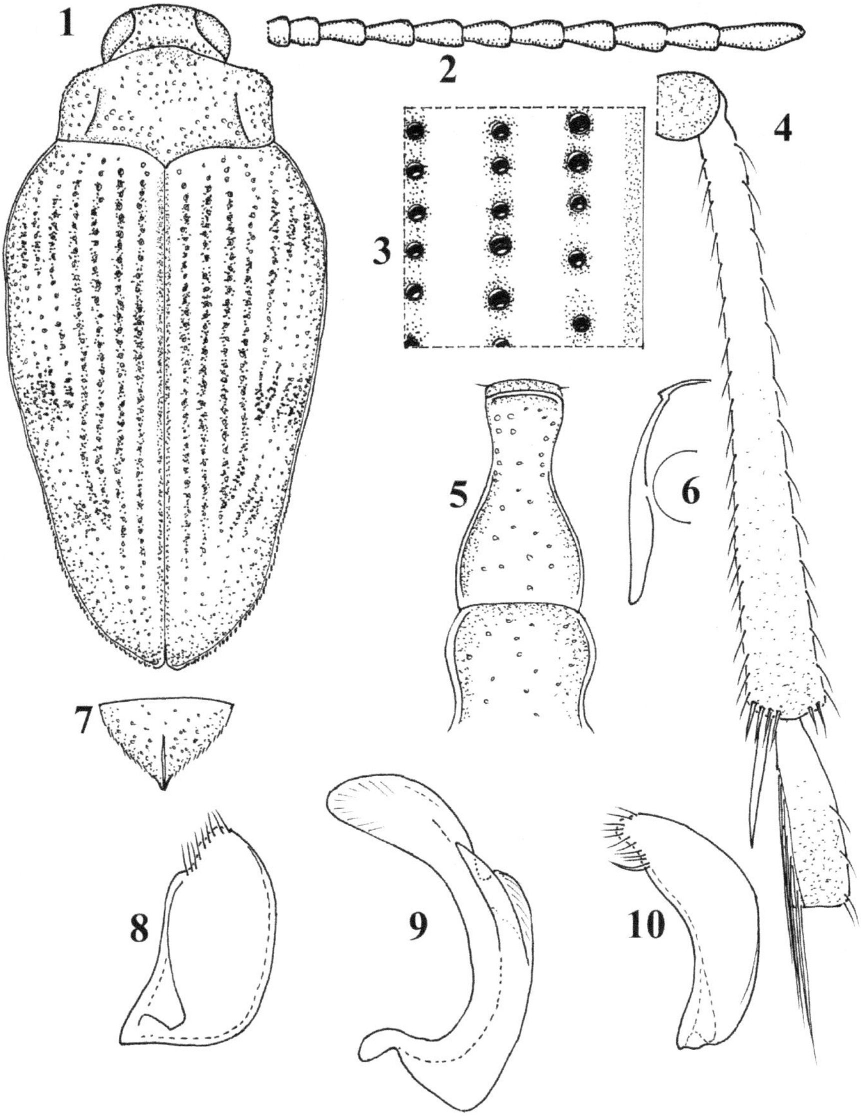

PLATE 3 *Brychius hornii*. 1, dorsal view; 2, antenna; 3, elytral punctures near base and suture; 4, hind tibia in dorsal view; 5, prosternal and metaventral process; 6, prosternal process in lateral view; 7, last sternite; 8, left paramere; 9, penis; 10, right paramere

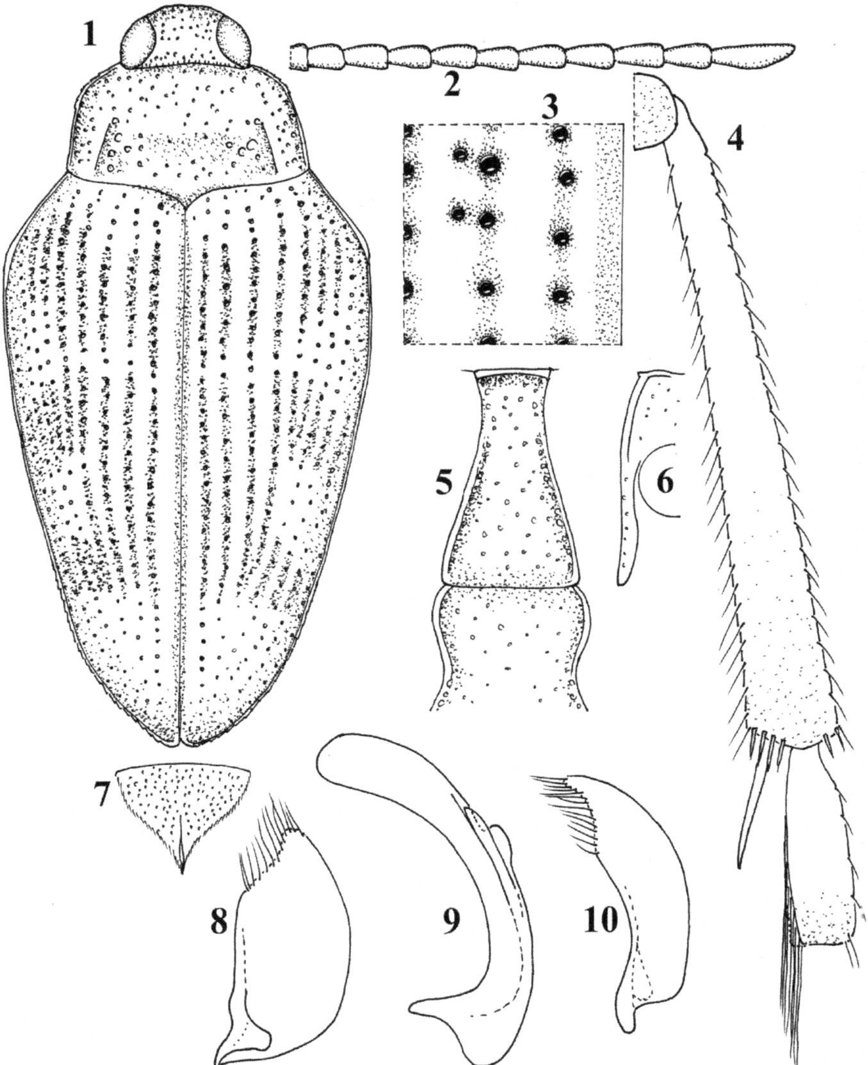

PLATE 4 *Brychius hungerfordi.* 1, dorsal view; 2, antenna; 3, elytral punctures near base and suture; 4, hind tibia in dorsal view; 5, prosternal and metaventral process; 6, prosternal process in lateral view; 7, last sternite; 8, left paramere; 9, penis; 10, right paramere

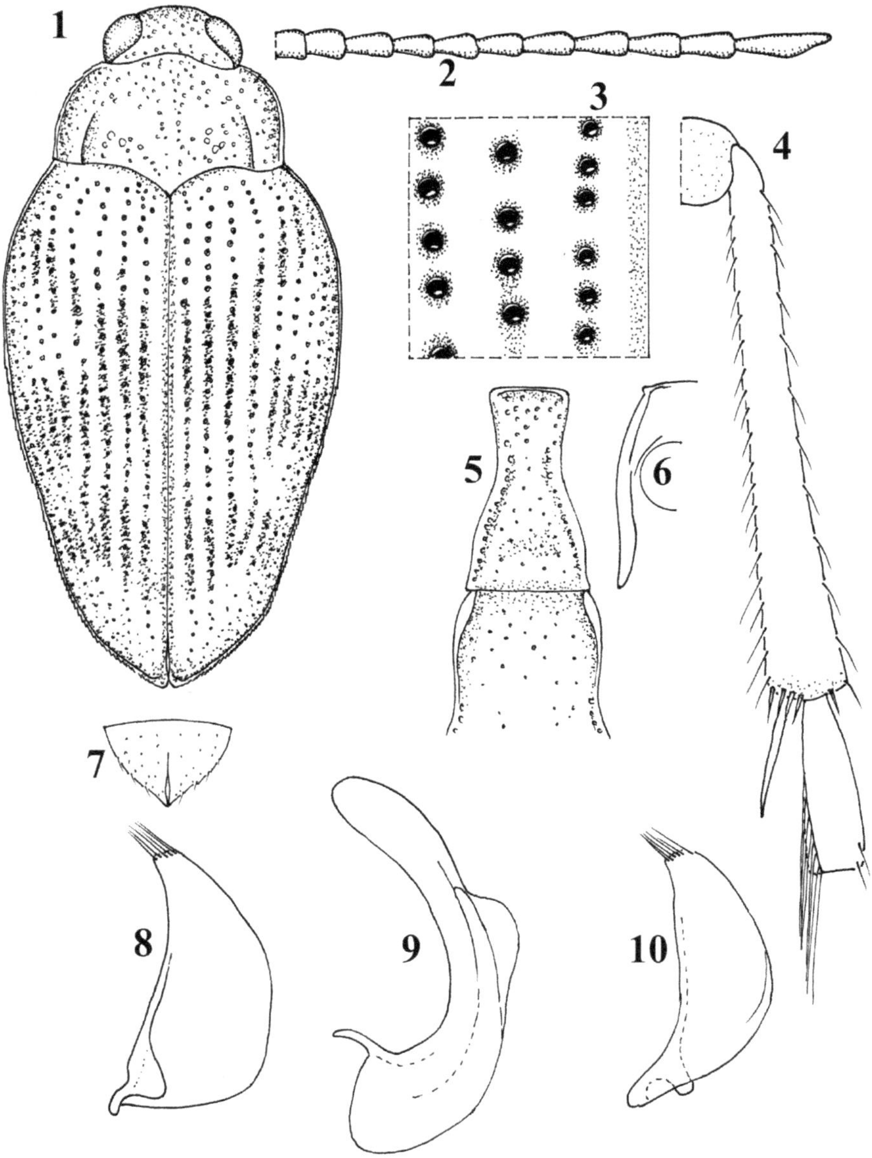

PLATE 5 *Brychius pacificus*. 1, dorsal view; 2, antenna; 3, elytral punctures near base and suture; 4, hind tibia in dorsal view; 5, prosternal and metaventral process; 6, prosternal process in lateral view; 7, last sternite; 8, left paramere; 9, penis; 10, right paramere

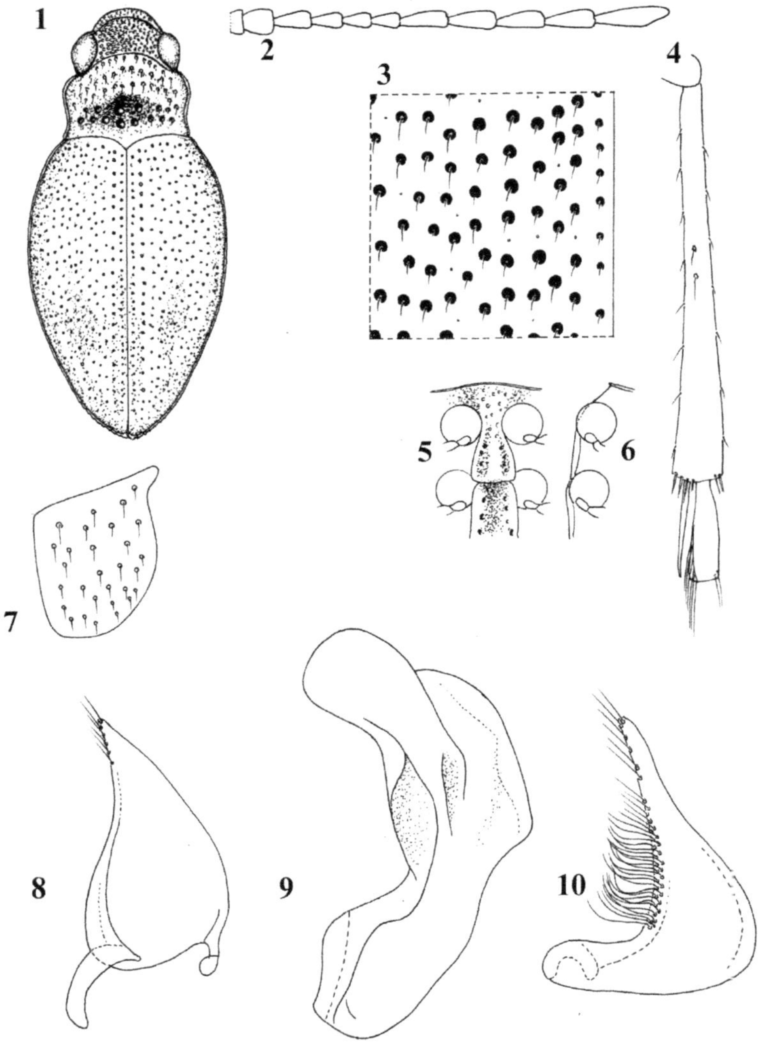

PLATE 6 *Haliplus* (*Algophilus*) *lathridioides*. 1, dorsal view; 2, antenna; 3, elytral punctures near base and suture; 4, hind tibia in dorsal view; 5, prosternal and metaventral process; 6, prosternal process in lateral view; 7, hind coxal plate; 8, left paramere; 9, penis; 10, right paramere

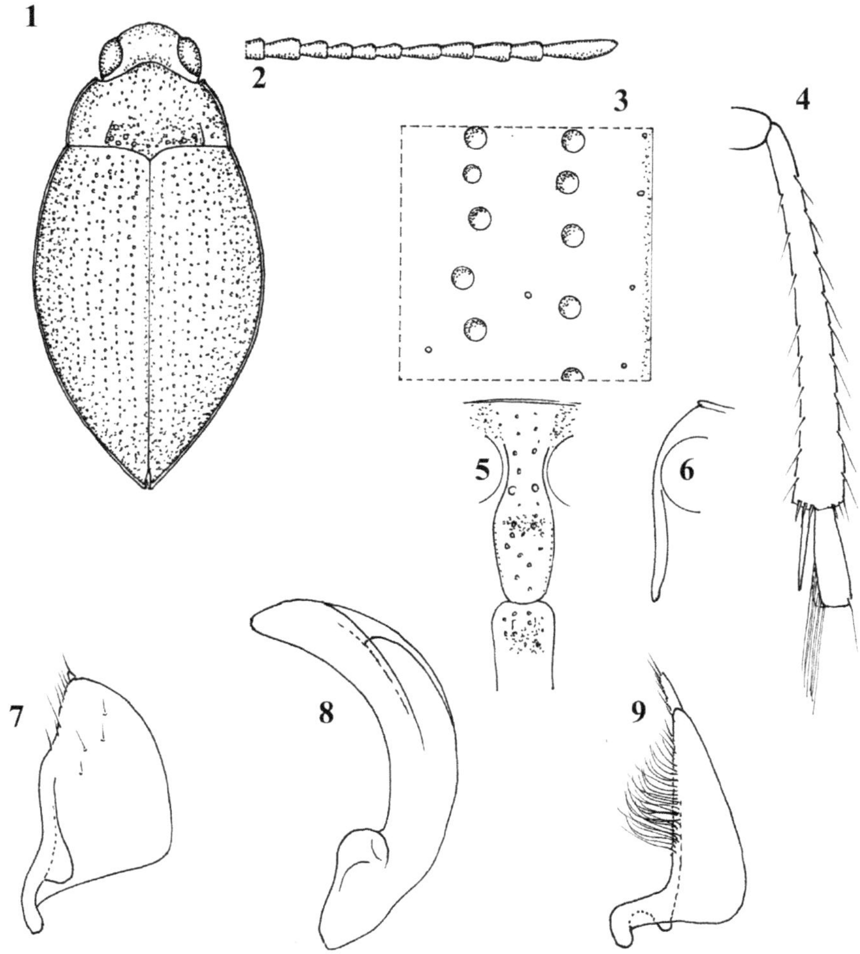

PLATE 7 *Haliplus* (*Apteraliplus*) *parvulus*. 1, dorsal view; 2, antenna; 3, elytral punctures near base and suture; 4, hind tibia in dorsal view; 5, prosternal and metaventral process; 6, prosternal process in lateral view; 7, left paramere; 8, penis; 9, right paramere

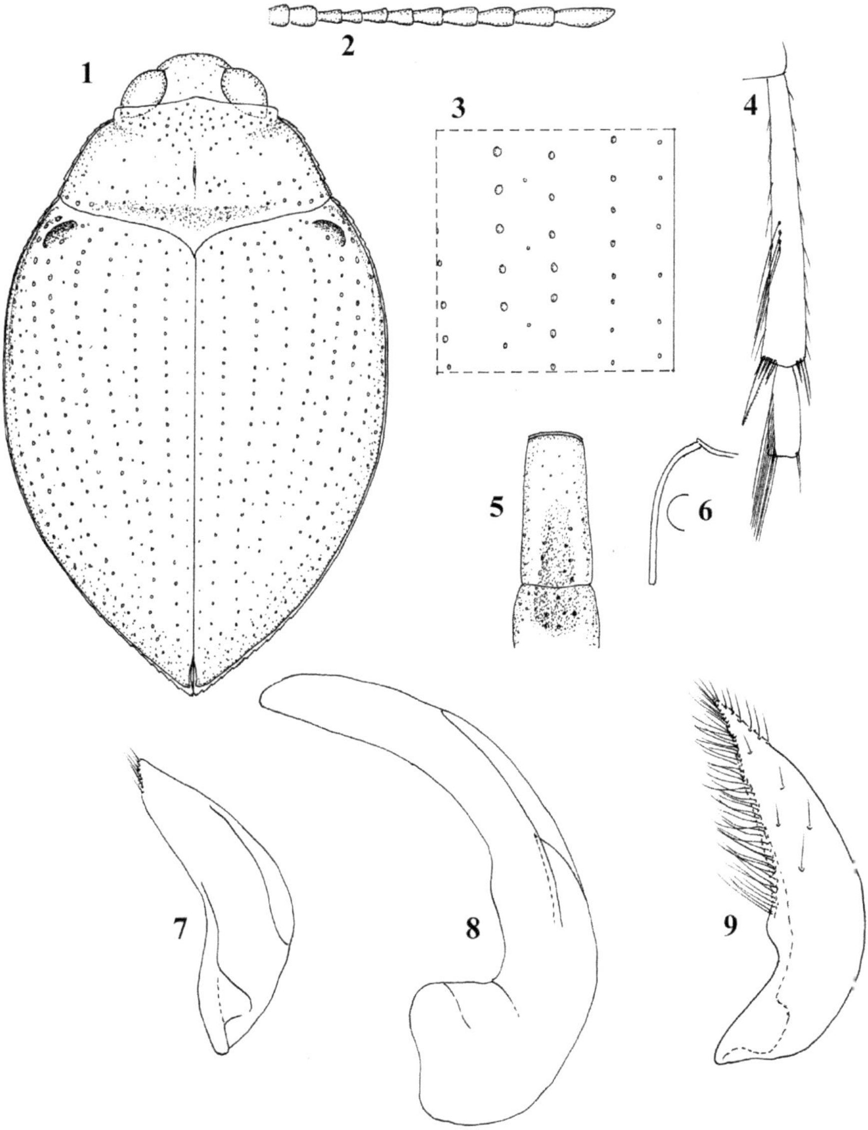

PLATE 8 *Haliplus* (*Australiplus*) *sindus*. 1, dorsal view; 2, antenna; 3, elytral punctures near base and suture; 4, hind tibia in dorsal view; 5, prosternal and metaventral process; 6, prosternal process in lateral view; 7, left paramere; 8, penis; 9, right paramere

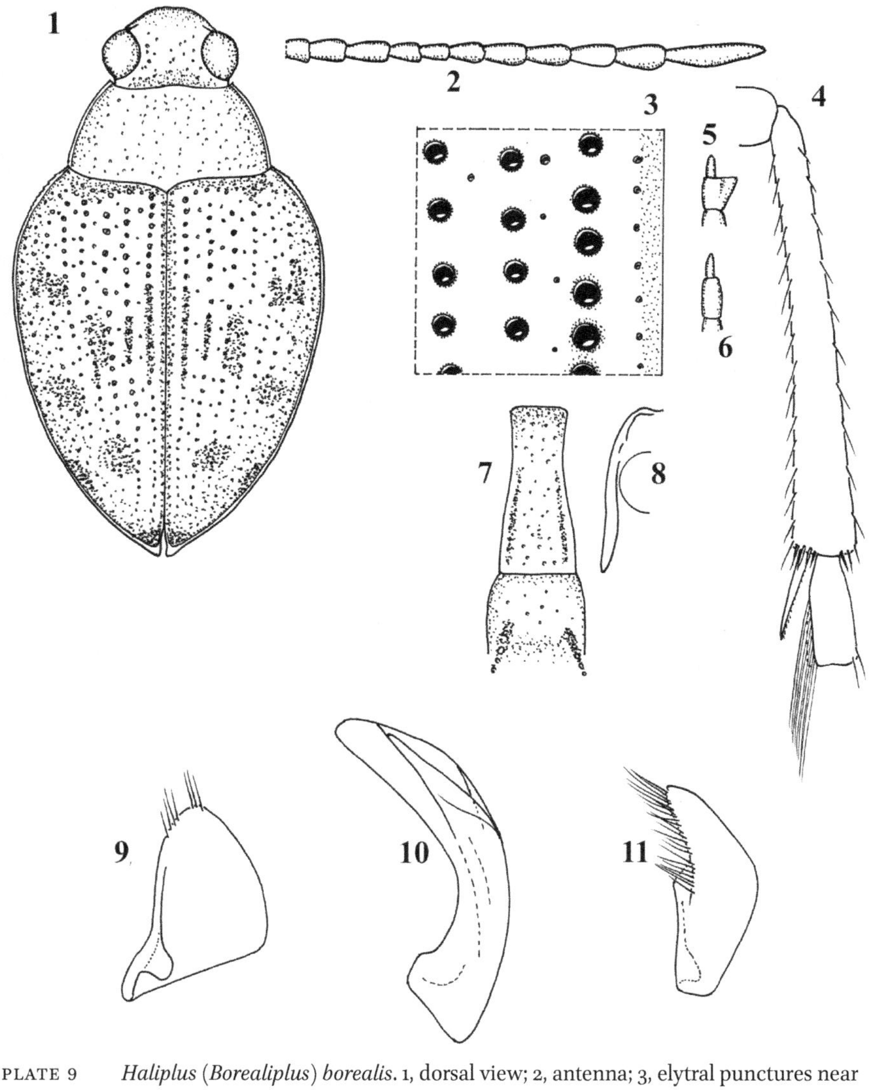

PLATE 9 *Haliplus* (*Borealiplus*) *borealis*. 1, dorsal view; 2, antenna; 3, elytral punctures near base and suture; 4, hind tibia in dorsal view; 5, labial palpus; 6, maxillary palpus; 7, prosternal and metaventral process; 8, prosternal process in lateral view; 9, left paramere; 10, penis; 11, right paramere

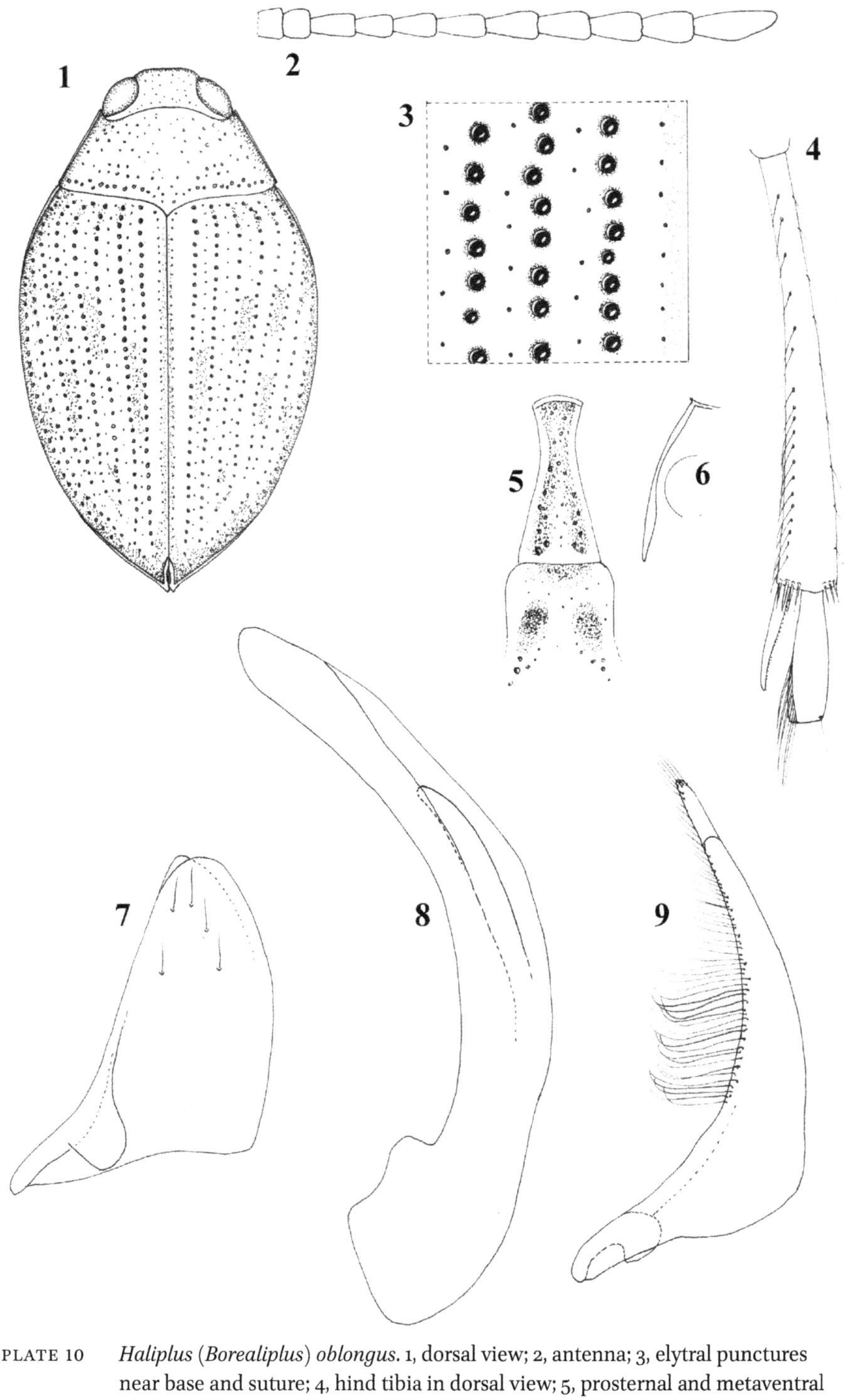

PLATE 10 *Haliplus* (*Borealiplus*) *oblongus*. 1, dorsal view; 2, antenna; 3, elytral punctures near base and suture; 4, hind tibia in dorsal view; 5, prosternal and metaventral process; 6, prosternal process in lateral view; 7, left paramere; 8, penis; 9, right paramere

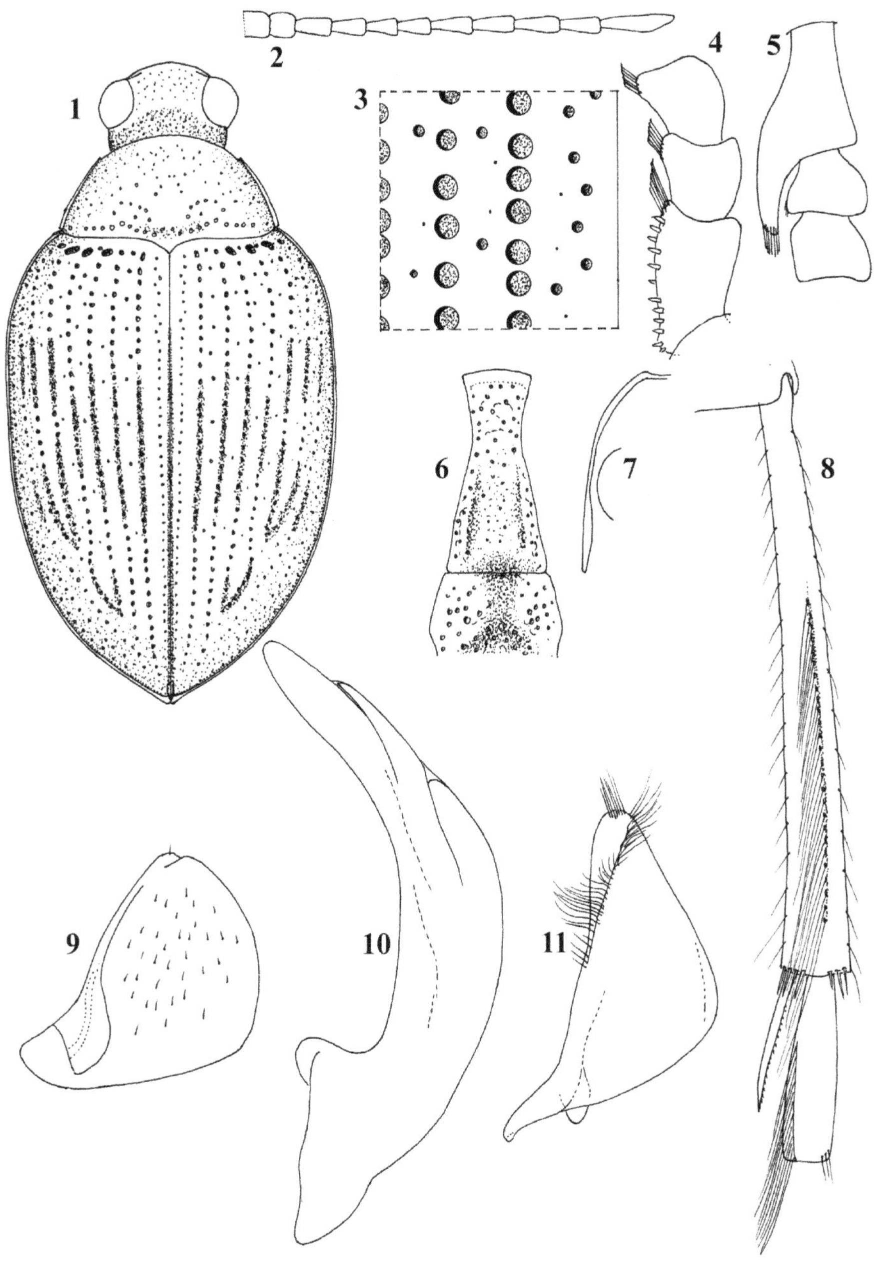

PLATE 11 *Haliplus* (*Haliaplus*) *laminatus*. 1, dorsal view; 2, antenna; 3, elytral punctures near base and suture; 4, first tarsal segments of male foreleg; 5, first tarsal segments of male midleg; 6, prosternal and metaventral process; 7, prosternal process in lateral view; 8, hind tibia in dorsal view; 9, left paramere; 10, penis; 11, right paramere

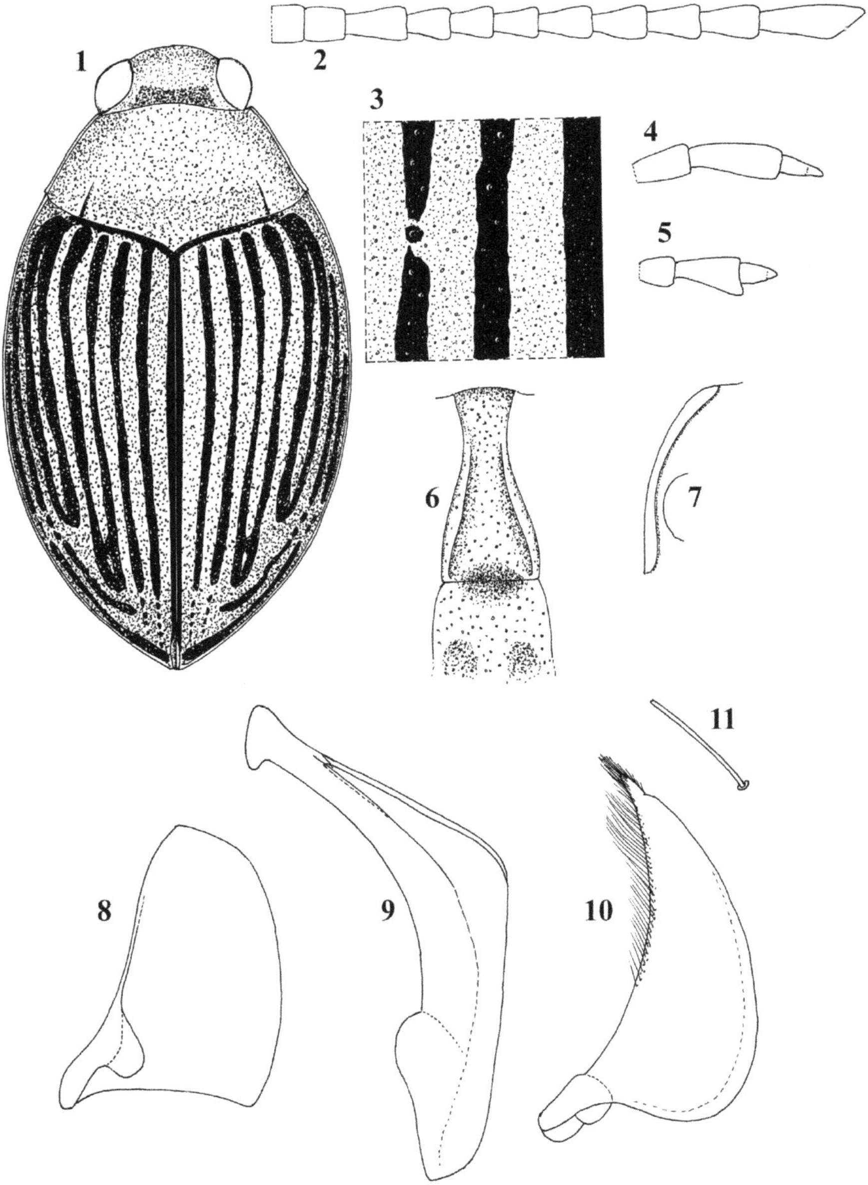

PLATE 12 *Haliplus* (*Haliplidius*) *confinis*. 1, dorsal view; 2, antenna; 3, elytral punctures near base and suture; 4, maxillary palpus; 5, labial palpus; 6, prosternal and metaventral process; 7, prosternal process in lateral view; 8, left paramere; 9, penis; 10, right paramere; 11, setae of right paramere

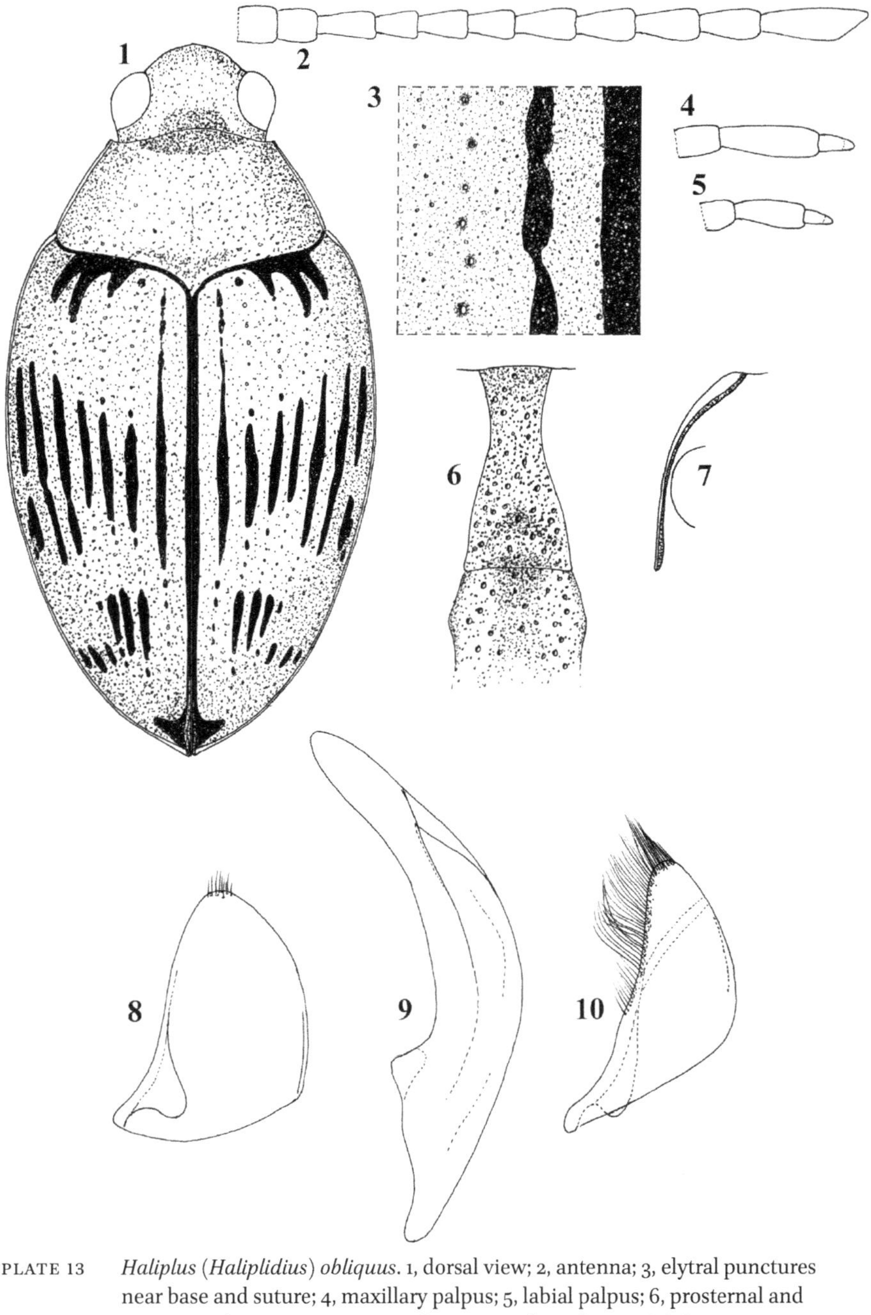

PLATE 13 *Haliplus* (*Haliplidius*) *obliquus*. 1, dorsal view; 2, antenna; 3, elytral punctures near base and suture; 4, maxillary palpus; 5, labial palpus; 6, prosternal and metaventral process; 7, prosternal process in lateral view; 8, left paramere; 9, penis; 10, right paramere

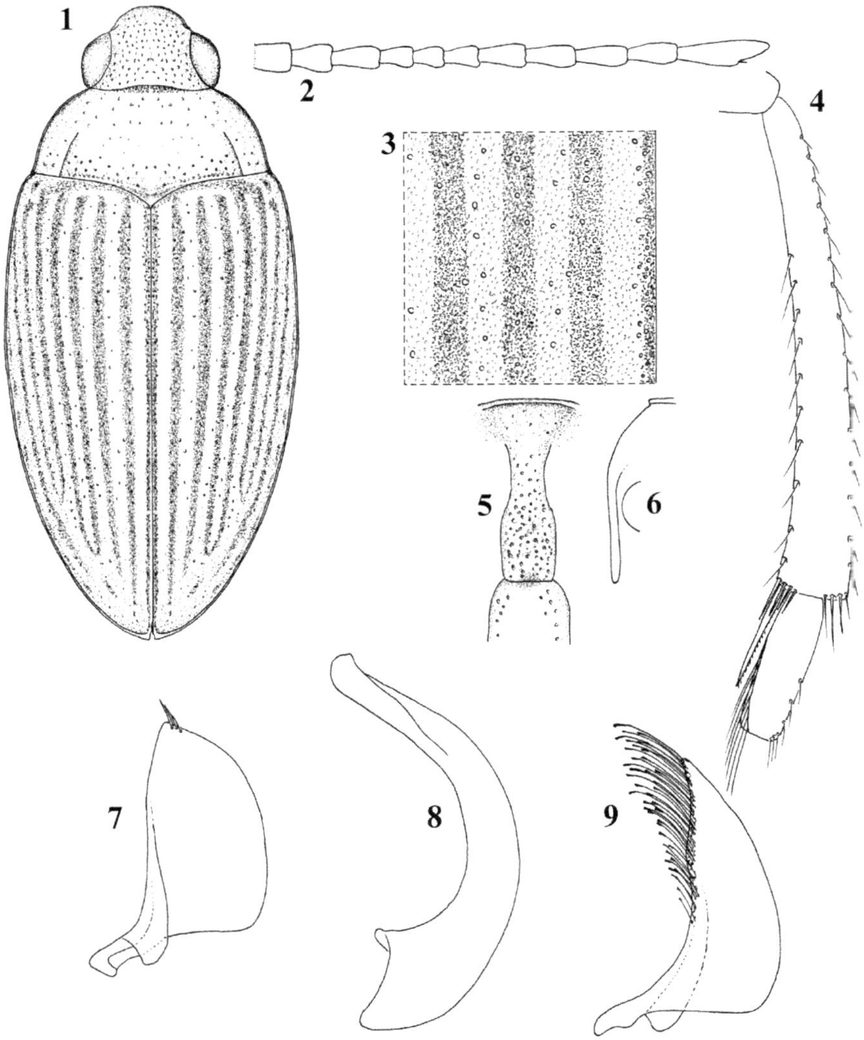

PLATE 14 *Haliplus* (*Haliplidius*) *rejseki*. 1, dorsal view; 2, antenna; 3, elytral punctures near base and suture; 4, hind tibia in dorsal view; 5, prosternal and metaventral process; 6, prosternal process in lateral view; 7, left paramere; 8, penis; 9, right paramere

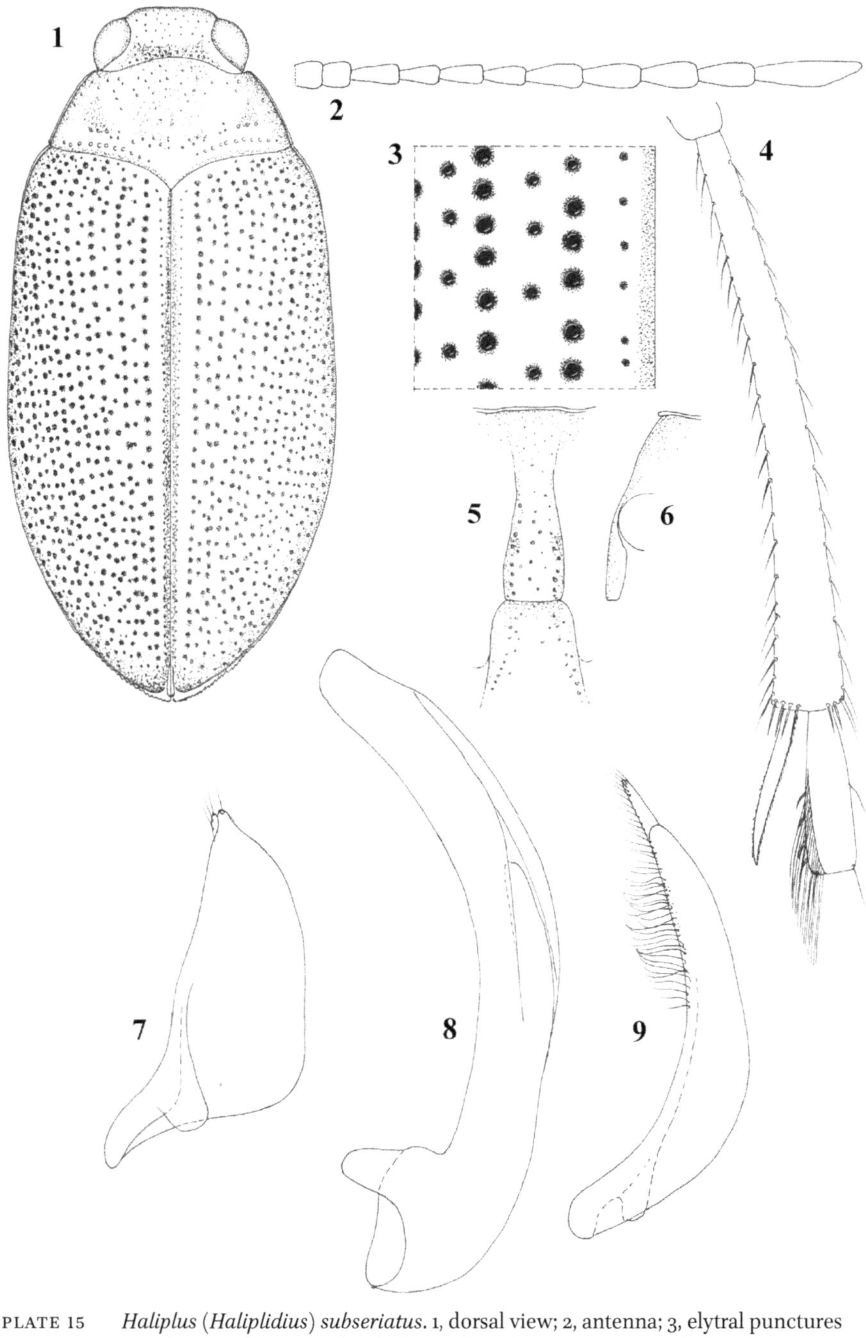

PLATE 15 *Haliplus* (*Haliplidius*) *subseriatus*. 1, dorsal view; 2, antenna; 3, elytral punctures near base and suture; 4, hind tibia in dorsal view; 5, prosternal and metaventral process; 6, prosternal process in lateral view; 7, left paramere; 8, penis; 9, right paramere

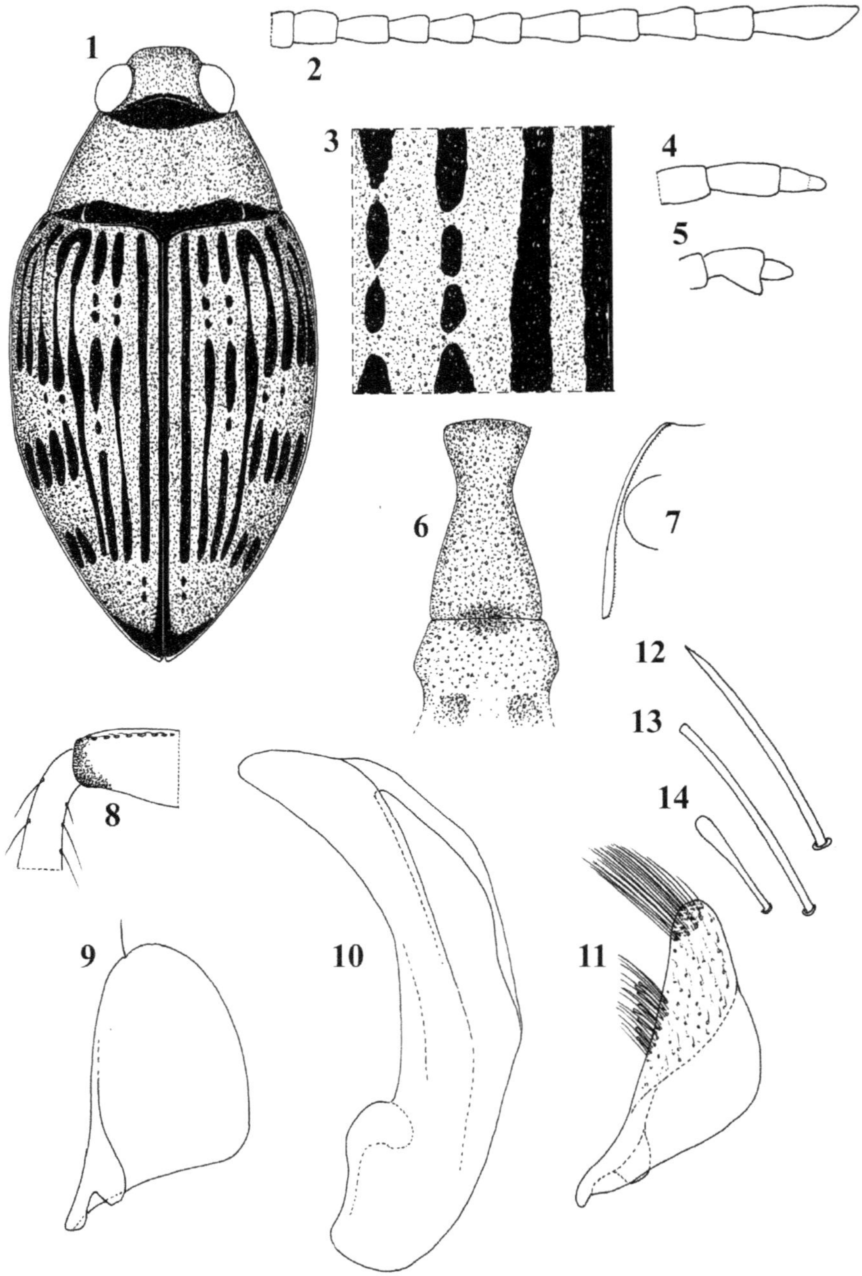

PLATE 16 *Haliplus* (*Haliplidius*) *varius*. 1, dorsal view; 2, antenna; 3, elytral punctures near base and suture; 4, maxillary palpus; 5, labial palpus; 6, prosternal and metaventral process; 7, prosternal process in lateral view; 8, apical part of femur (knee); 9, left paramere; 10, penis; 11, right paramere; 12–14, setae of right paramere

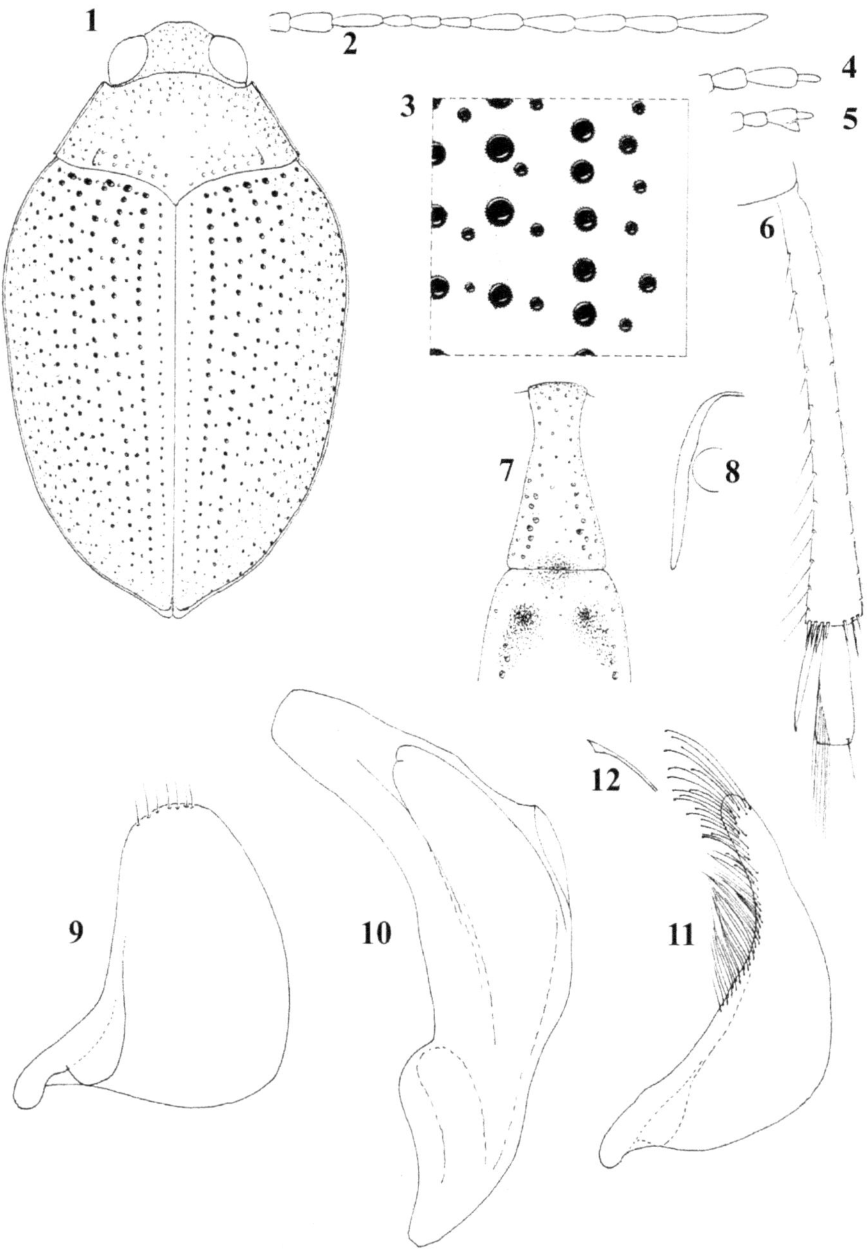

PLATE 17 *Haliplus* (*Haliplus*) *aliae*. 1, dorsal view; 2, antenna; 3, elytral punctures near base and suture; 4, maxillary palpus; 5, labial palpus; 6, hind tibia in dorsal view; 7, prosternal and metaventral process; 8, prosternal process in lateral view; 9, left paramere; 10, penis; 11, right paramere; 12, setae of right paramere

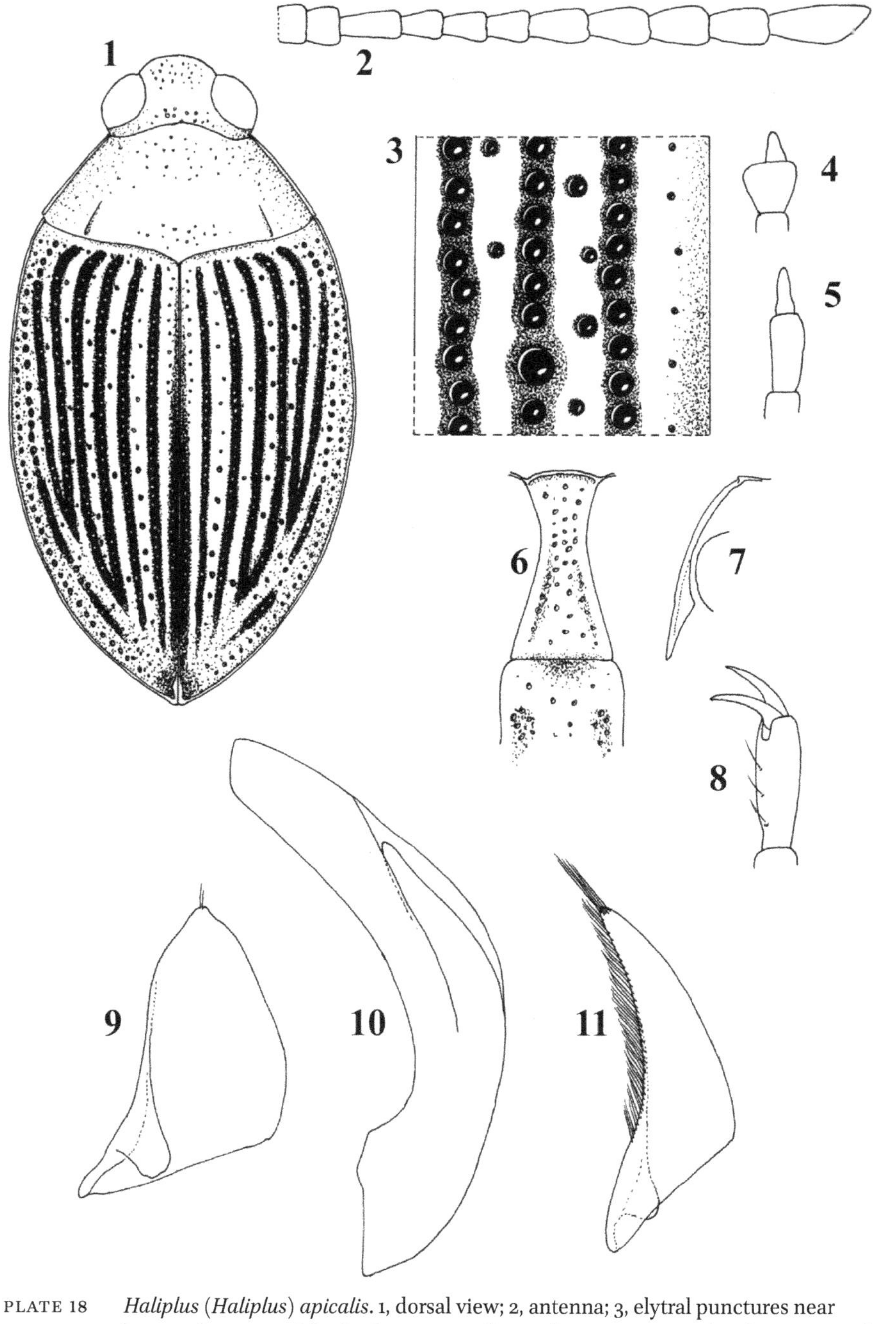

PLATE 18 *Haliplus* (*Haliplus*) *apicalis*. 1, dorsal view; 2, antenna; 3, elytral punctures near base and suture; 4, labial palpus 5, maxillary palpus; 6, prosternal and metaventral process; 7, prosternal process in lateral view; 8, claws and last tarsal segment of male foreleg; 9, left paramere; 10, penis; 11, right paramere

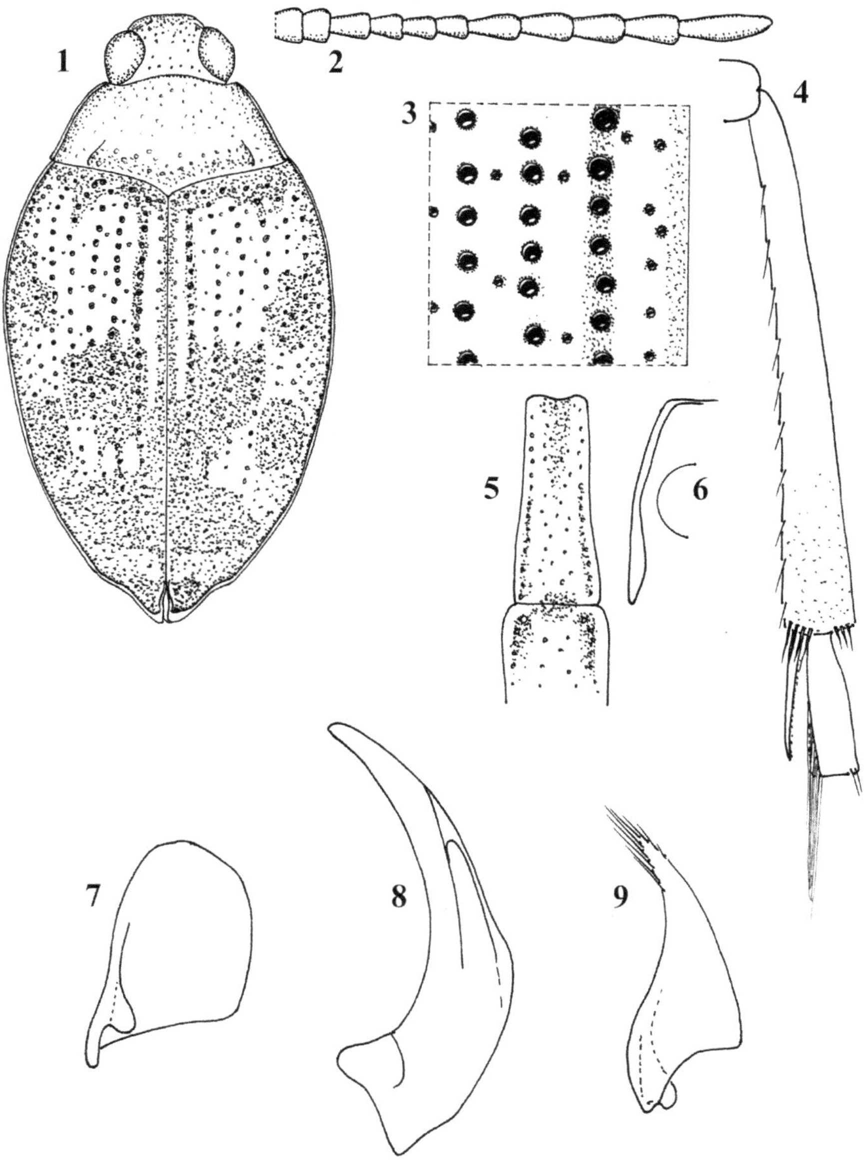

PLATE 19 *Haliplus* (*Haliplus*) *blanchardi*. 1, dorsal view; 2, antenna; 3, elytral punctures near base and suture; 4, hind tibia in dorsal view; 5, prosternal and metaventral process; 6, prosternal process in lateral view; 7, left paramere; 8, penis; 9, right paramere

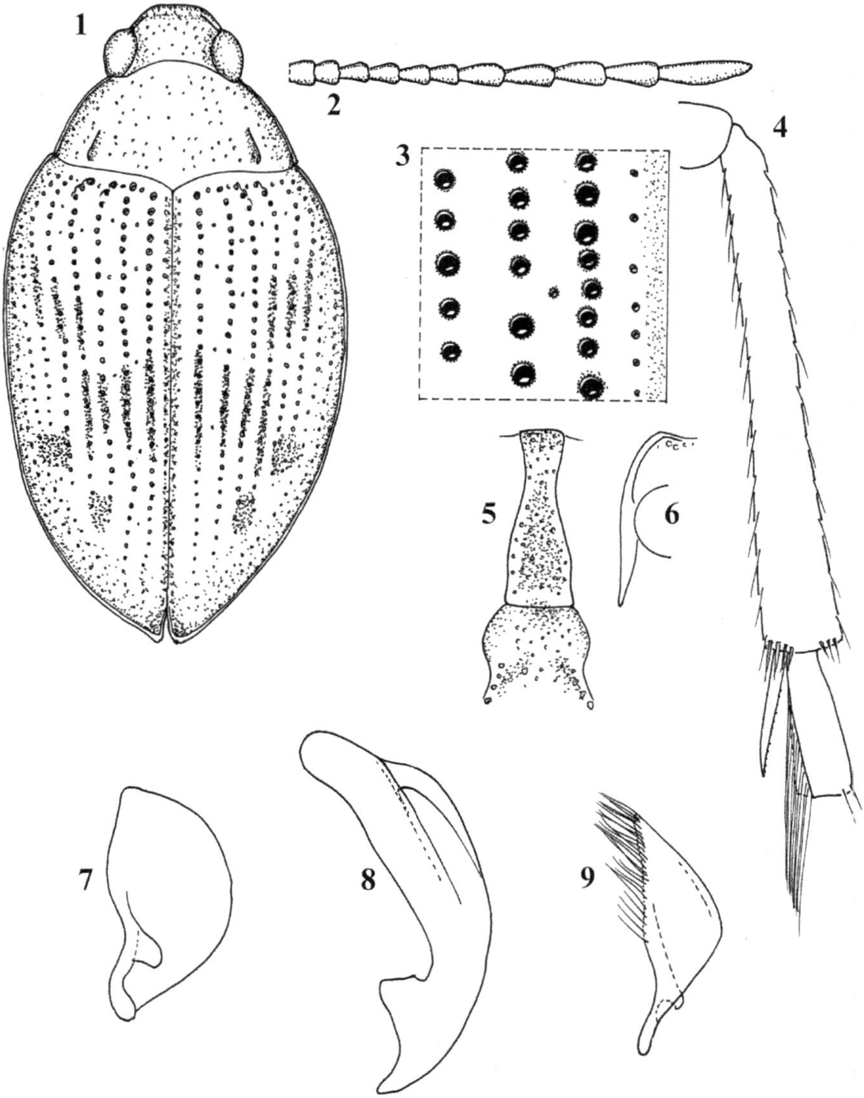

PLATE 20 *Haliplus* (*Haliplus*) *distinctus*. 1, dorsal view; 2, antenna; 3, elytral punctures near base and suture; 4 hind tibia in dorsal view; 5, prosternal and metaventral process; 6, prosternal process in lateral view; 7, left paramere; 8, penis; 9, right paramere

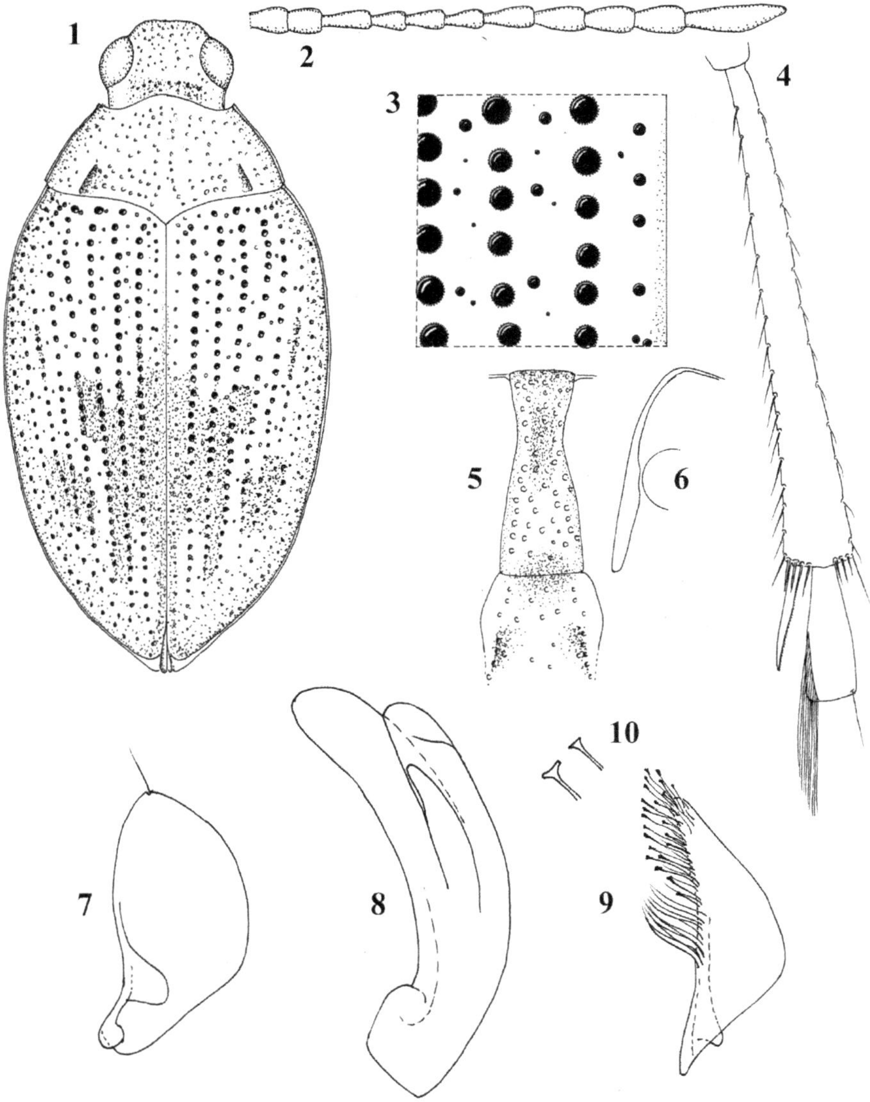

PLATE 21 *Haliplus* (*Haliplus*) *dorsomaculatus*. 1, dorsal view; 2, antenna; 3, elytral punctures near base and suture; 4, hind tibia in dorsal view; 5, prosternal and metaventral process; 6, prosternal process in lateral view; 7, left paramere; 8, penis; 9, right paramere; 10, setae of right paramere

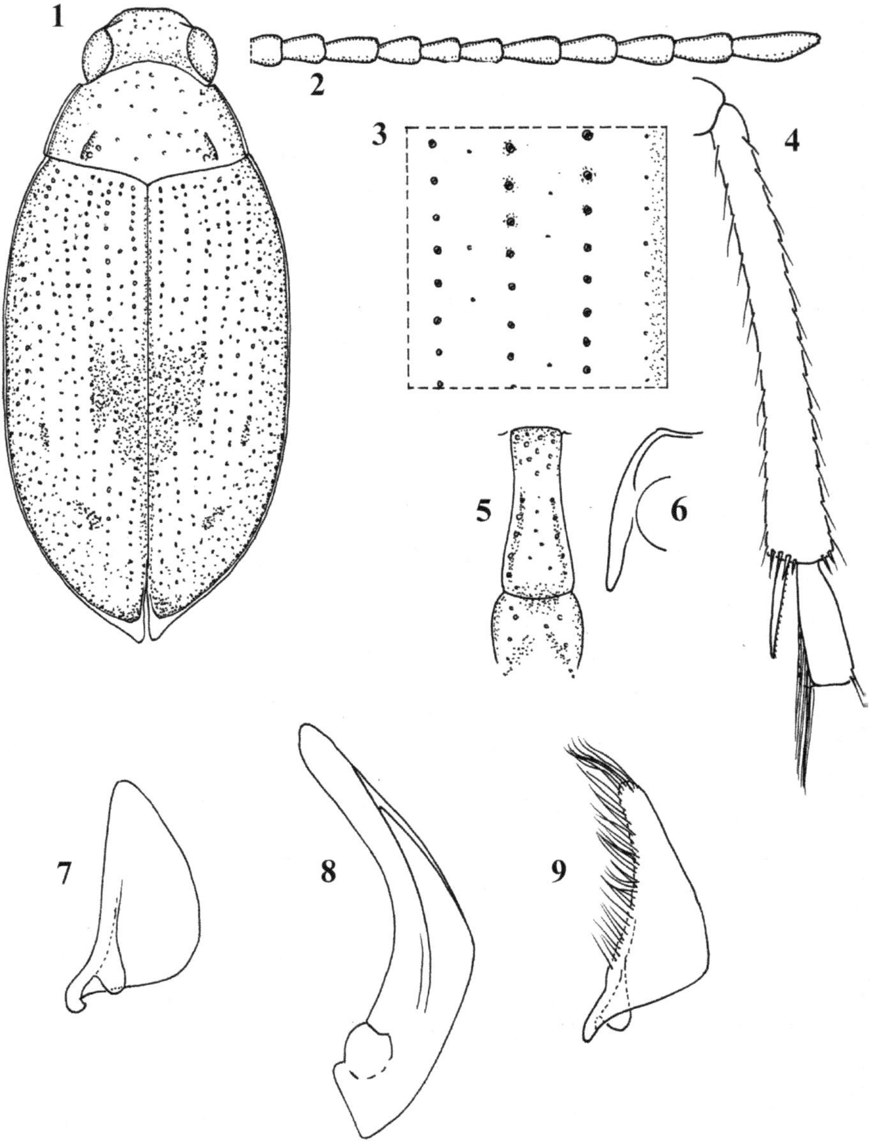

PLATE 22 *Haliplus* (*Haliplus*) *falli*. 1, dorsal view; 2, antenna; 3, elytral punctures near base and suture; 4, hind tibia in dorsal view; 5, prosternal and metaventral process; 6, prosternal process in lateral view; 7, left paramere; 8, penis; 9, right paramere

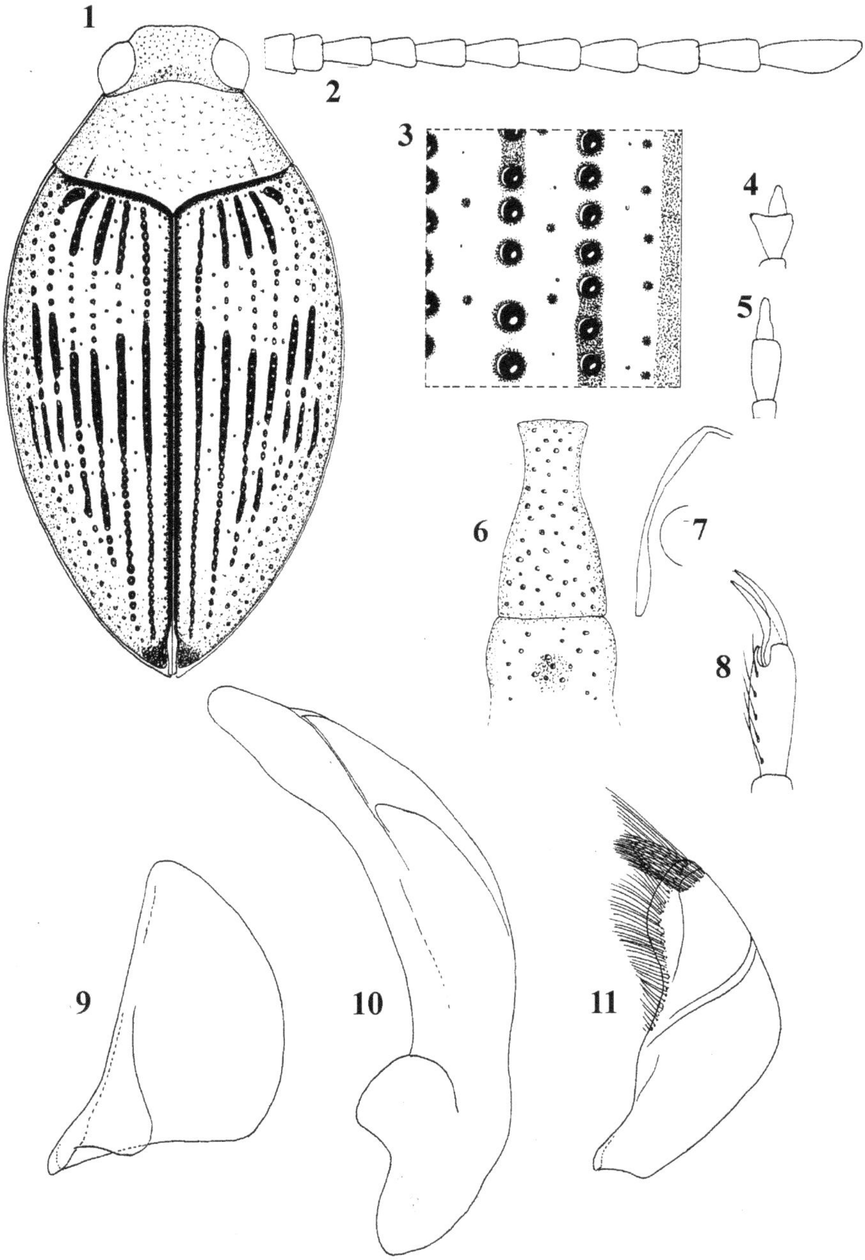

PLATE 23 *Haliplus* (*Haliplus*) *fluviatilis*. 1, dorsal view; 2, antenna; 3, elytral punctures near base and suture; 4, labial palpus 5, maxillary palpus; 6, prosternal and metaventral process; 7, prosternal process in lateral view; 8, claws and last tarsal segment of male foreleg; 9, left paramere; 10, penis; 11, right paramere

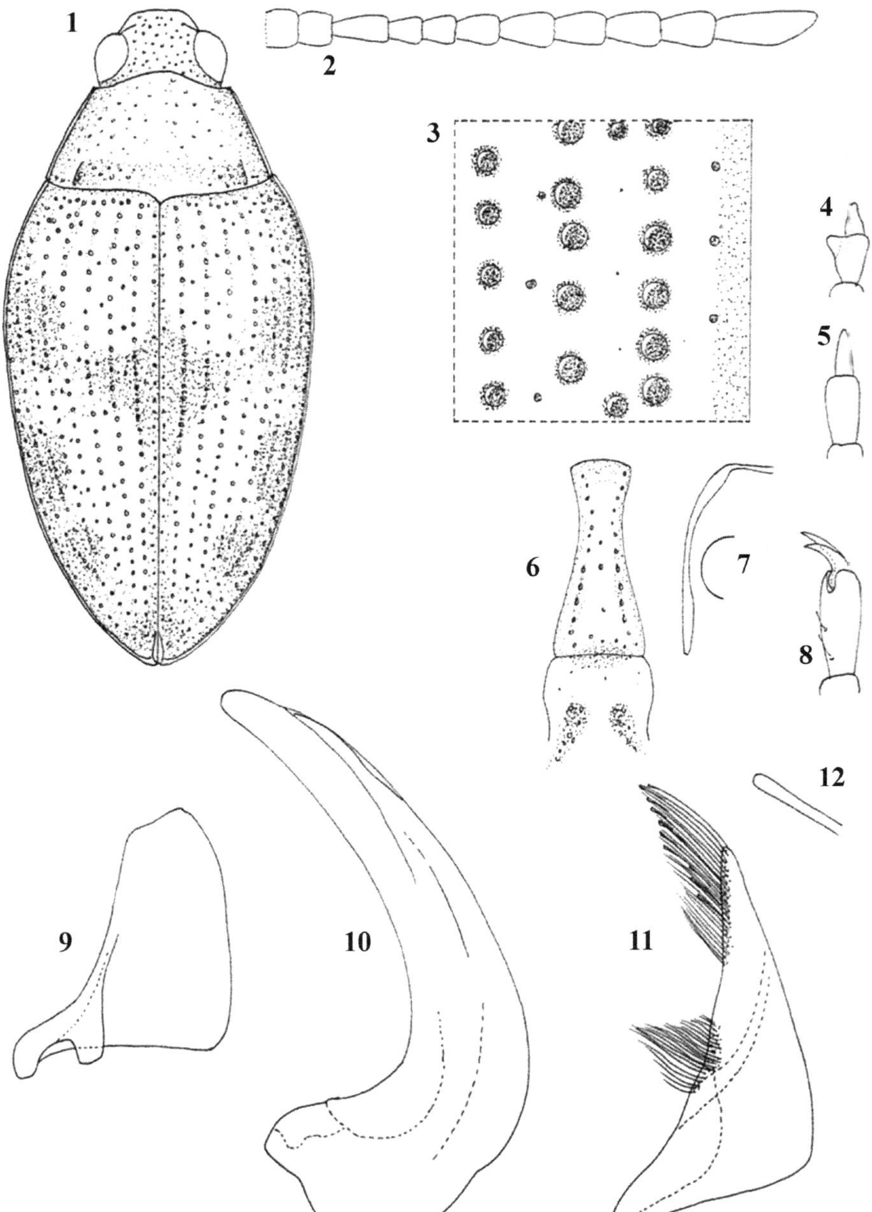

PLATE 24 *Haliplus* (*Haliplus*) *fulvicollis*. 1, dorsal view; 2, antenna; 3, elytral punctures near base and suture; 4, labial palpus 5, maxillary palpus; 6, prosternal and metaventral process; 7, prosternal process in lateral view; 8, claws and last tarsal segment of male foreleg; 9, left paramere; 10, penis; 11, right paramere; 12, setae of right paramere

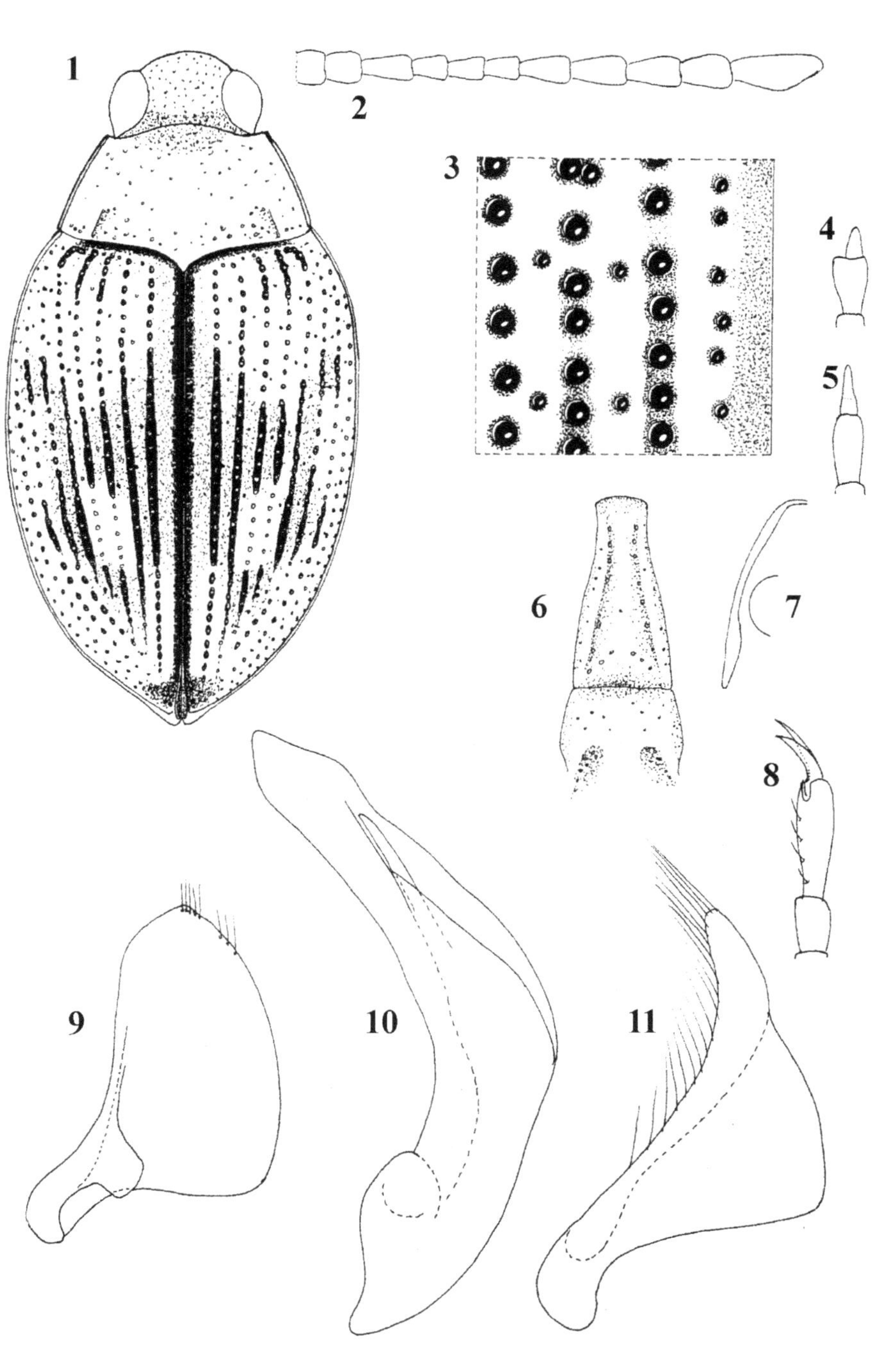

PLATE 25 *Haliplus* (*Haliplus*) *furcatus*. 1, dorsal view; 2, antenna; 3, elytral punctures near base and suture; 4, labial palpus 5, maxillary palpus; 6, prosternal and metaventral process; 7, prosternal process in lateral view; 8, claws and last tarsal segment of male foreleg; 9, left paramere; 10, penis; 11, right paramere

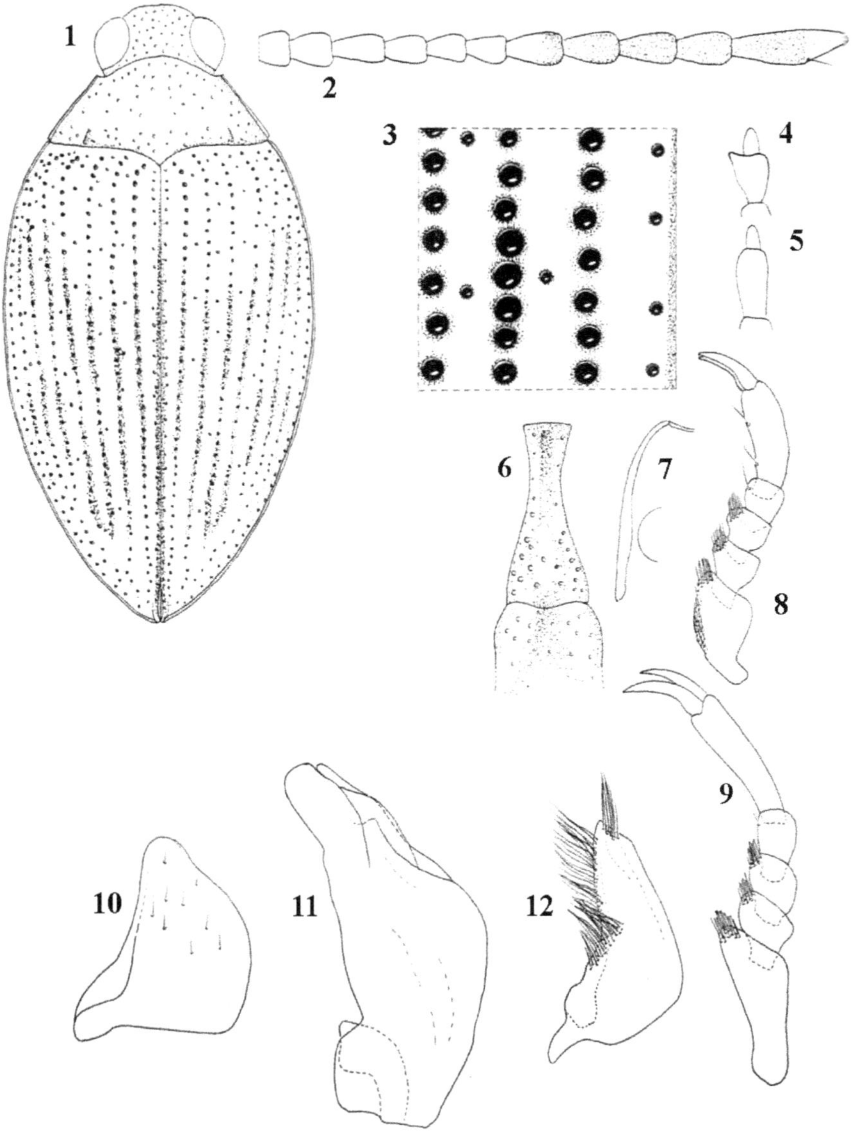

PLATE 26 *Haliplus* (*Haliplus*) *fuscicornis*. 1, dorsal view; 2, antenna; 3, elytral punctures near base and suture; 4, labial palpus 5, maxillary palpus; 6, prosternal and metaventral process; 7, prosternal process in lateral view; 8, tarsus of male foreleg; 9, tarsus of male midleg; 10, left paramere; 11, penis; 12, right paramere

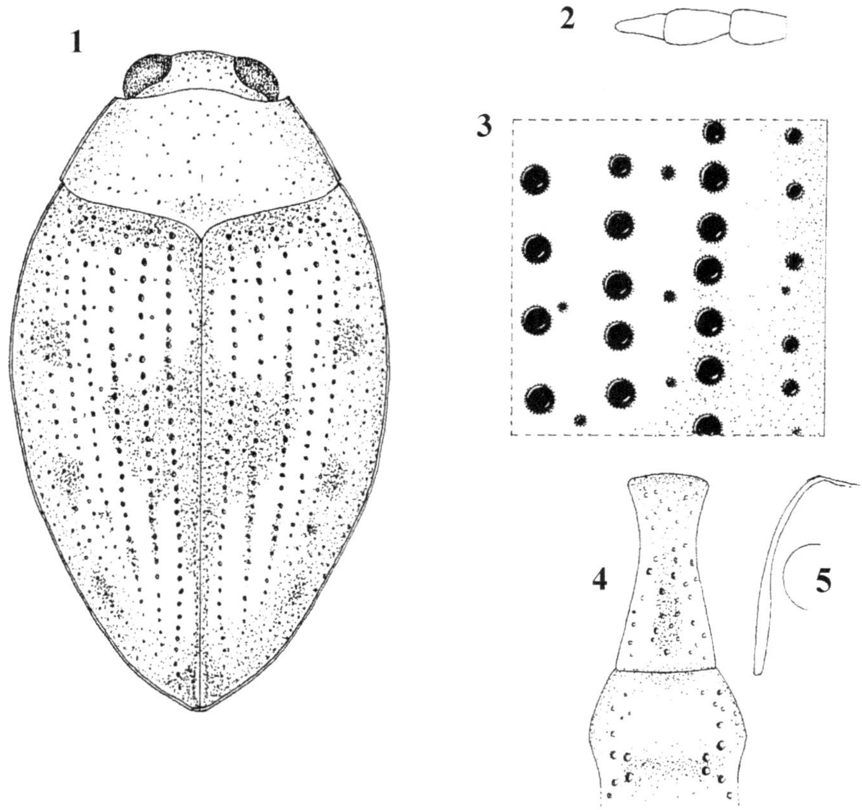

PLATE 27 *Haliplus* (*Haliplus*) *harminae*, female. 1, dorsal view; 2, labial palpus; 3, elytral punctures near base and suture; 4, prosternal and metaventral process; 5, prosternal process in lateral view

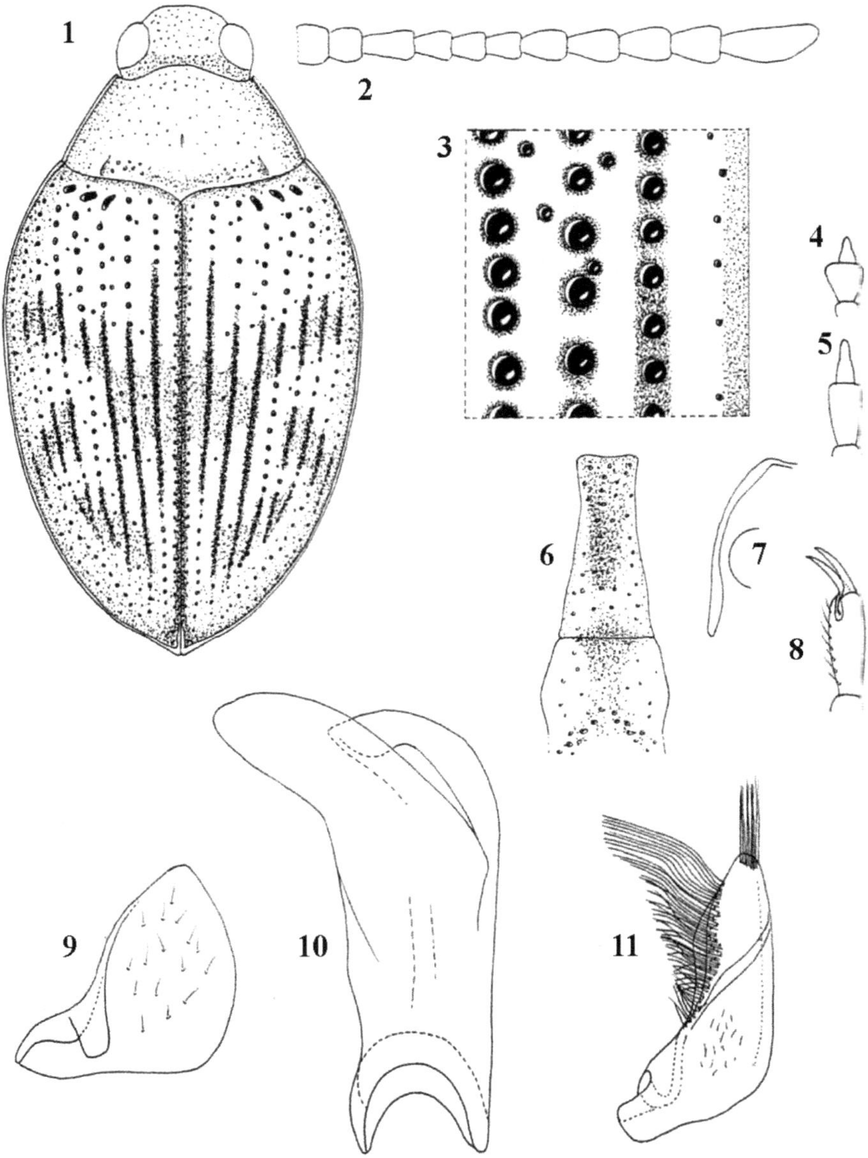

PLATE 28 *Haliplus* (*Haliplus*) *heydeni*. 1, dorsal view; 2, antenna; 3, elytral punctures near base and suture; 4, labial palpus 5, maxillary palpus; 6, prosternal and metaventral process; 7, prosternal process in lateral view; 8, claws and last tarsal segment of male foreleg; 9, left paramere; 10, penis; 11, right paramere

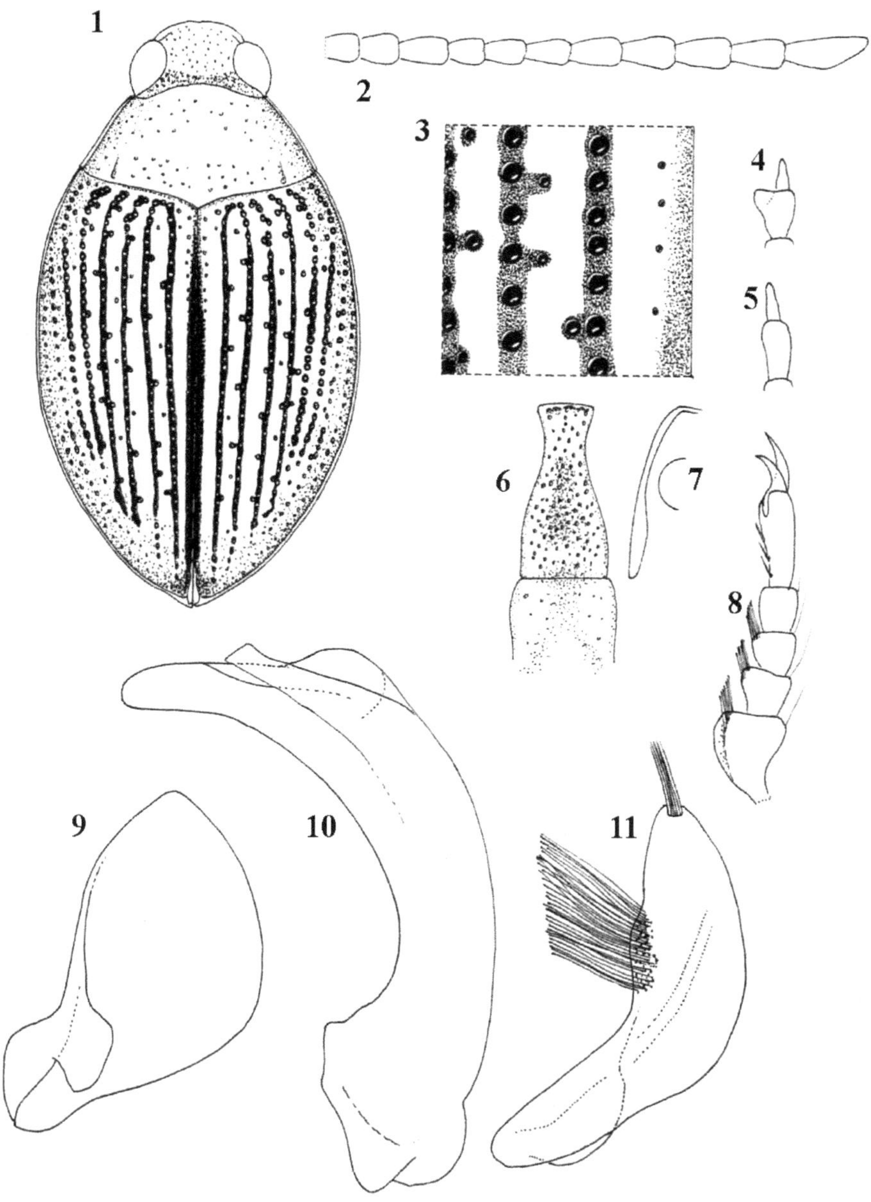

PLATE 29 *Haliplus* (*Haliplus*) *immaculatus*. 1, dorsal view; 2, antenna; 3, elytral punctures near base and suture; 4, labial palpus 5, maxillary palpus; 6, prosternal and metaventral process; 7, prosternal process in lateral view; 8, tarsus of male foreleg; 9, left paramere; 10, penis; 11, right paramere

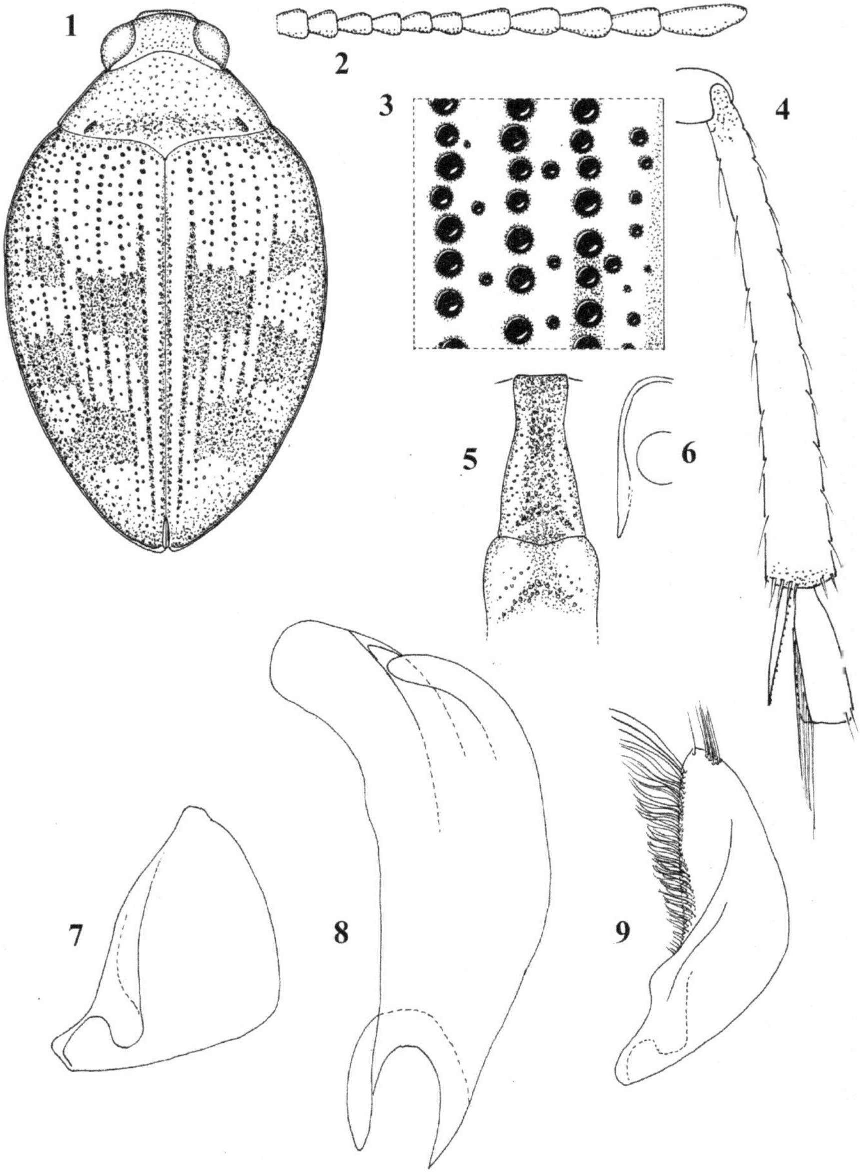

PLATE 30 *Haliplus* (*Haliplus*) *immaculicollis*. 1, dorsal view; 2, antenna; 3, elytral punctures near base and suture; 4, hind tibia in dorsal view; 5, prosternal and metaventral process; 6, prosternal process in lateral view; 7, left paramere; 8, penis; 9, right paramere

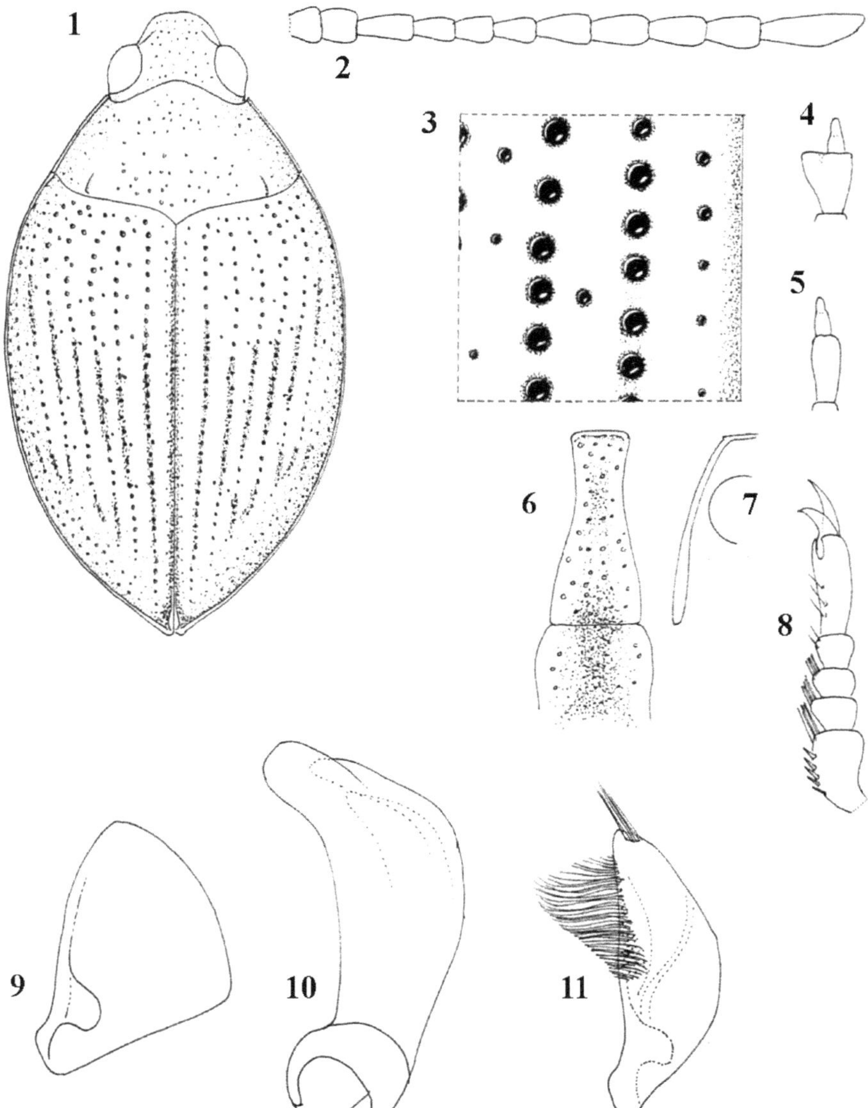

PLATE 31 *Haliplus* (*Haliplus*) *interjectus*. 1, dorsal view; 2, antenna; 3, elytral punctures near base and suture; 4, labial palpus 5, maxillary palpus; 6, prosternal and metaventral process; 7, prosternal process in lateral view; 8, tarsus of male foreleg; 9, left paramere; 10, penis; 11, right paramere

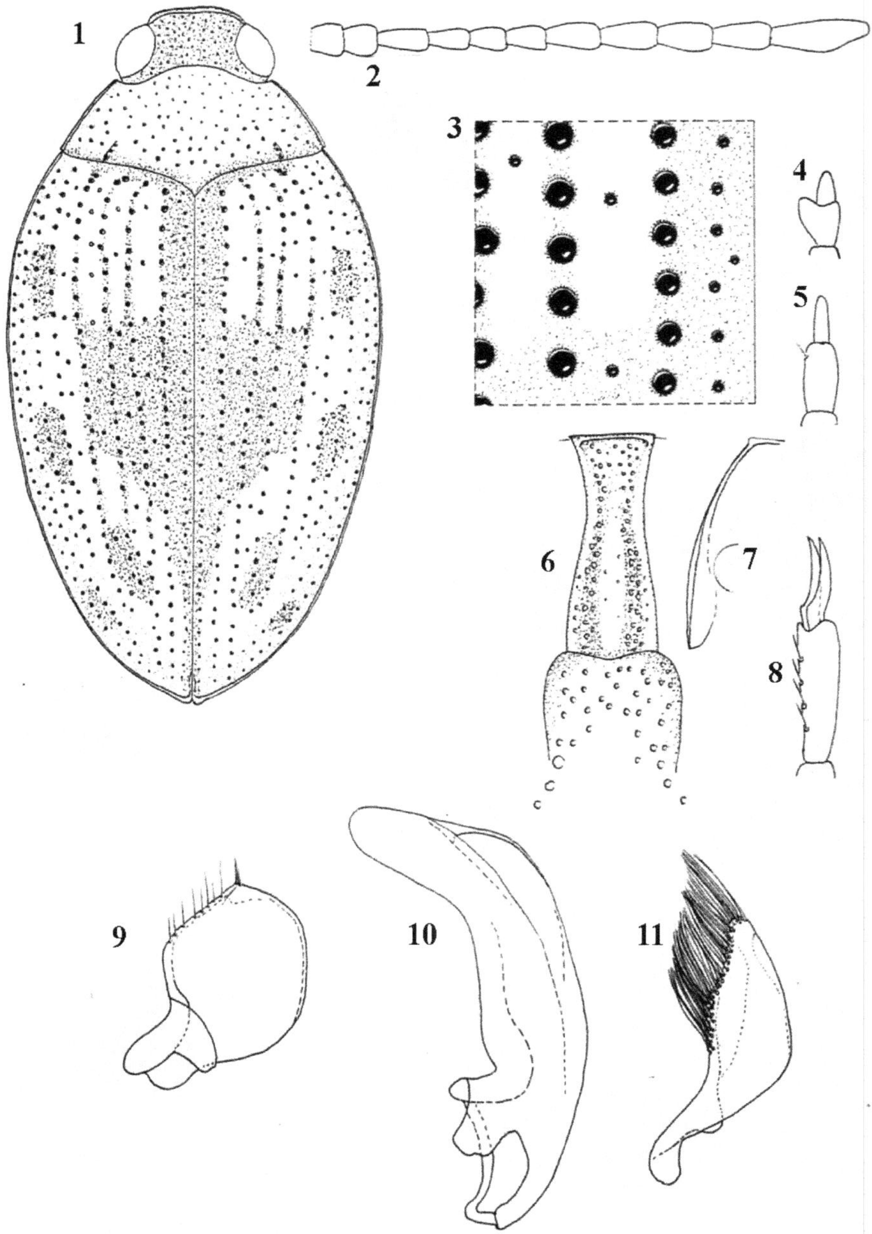

PLATE 32 *Haliplus* (*Haliplus*) *kamiyai*. 1, dorsal view; 2, antenna; 3, elytral punctures near base and suture; 4, labial palpus 5, maxillary palpus; 6, prosternal and metaventral process; 7, prosternal process in lateral view; 8, claws and last tarsal segment of male foreleg; 9, left paramere; 10, penis; 11, right paramere

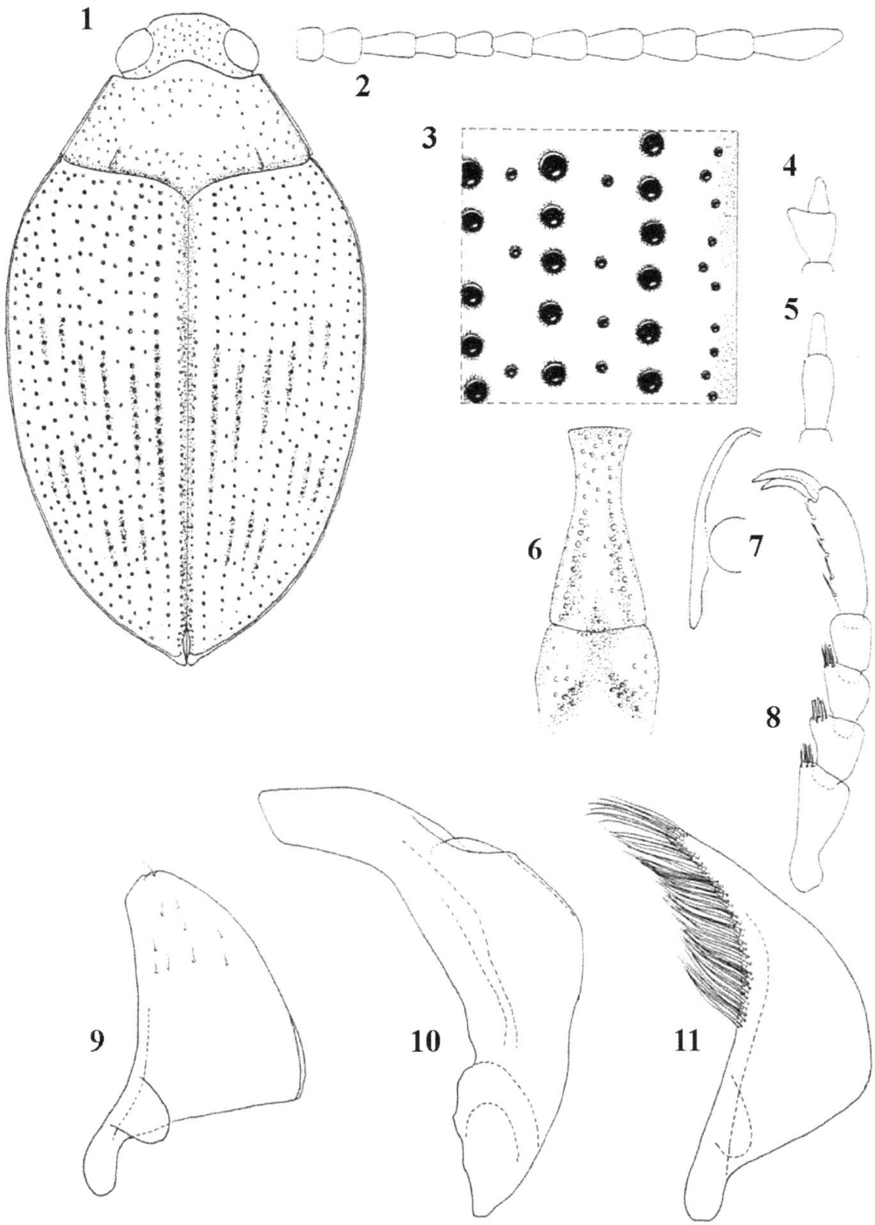

PLATE 33 *Haliplus* (*Haliplus*) *kirgisiensis*. 1, dorsal view; 2, antenna; 3, elytral punctures near base and suture; 4, labial palpus 5, maxillary palpus; 6, prosternal and metaventral process; 7, prosternal process in lateral view; 8, tarsus of male foreleg; 9, left paramere; 10, penis; 11, right paramere

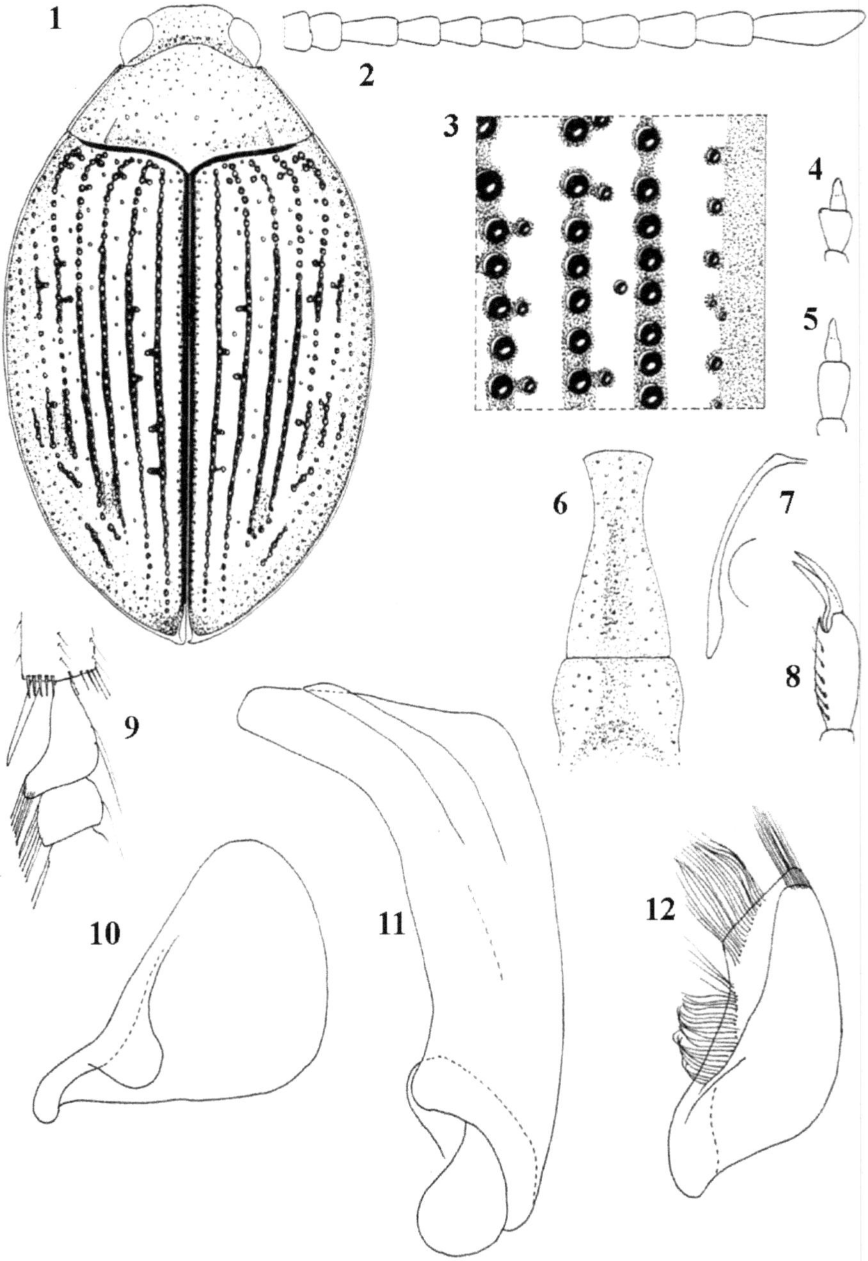

PLATE 34 *Haliplus* (*Haliplus*) *lineolatus*. 1, dorsal view; 2, antenna; 3, elytral punctures near base and suture; 4, labial palpus 5, maxillary palpus; 6, prosternal and metaventral process; 7, prosternal process in lateral view; 8, claws and last tarsal segment of male foreleg; 9, first tarsal segment of male midleg; 10, left paramere; 11, penis; 12, right paramere

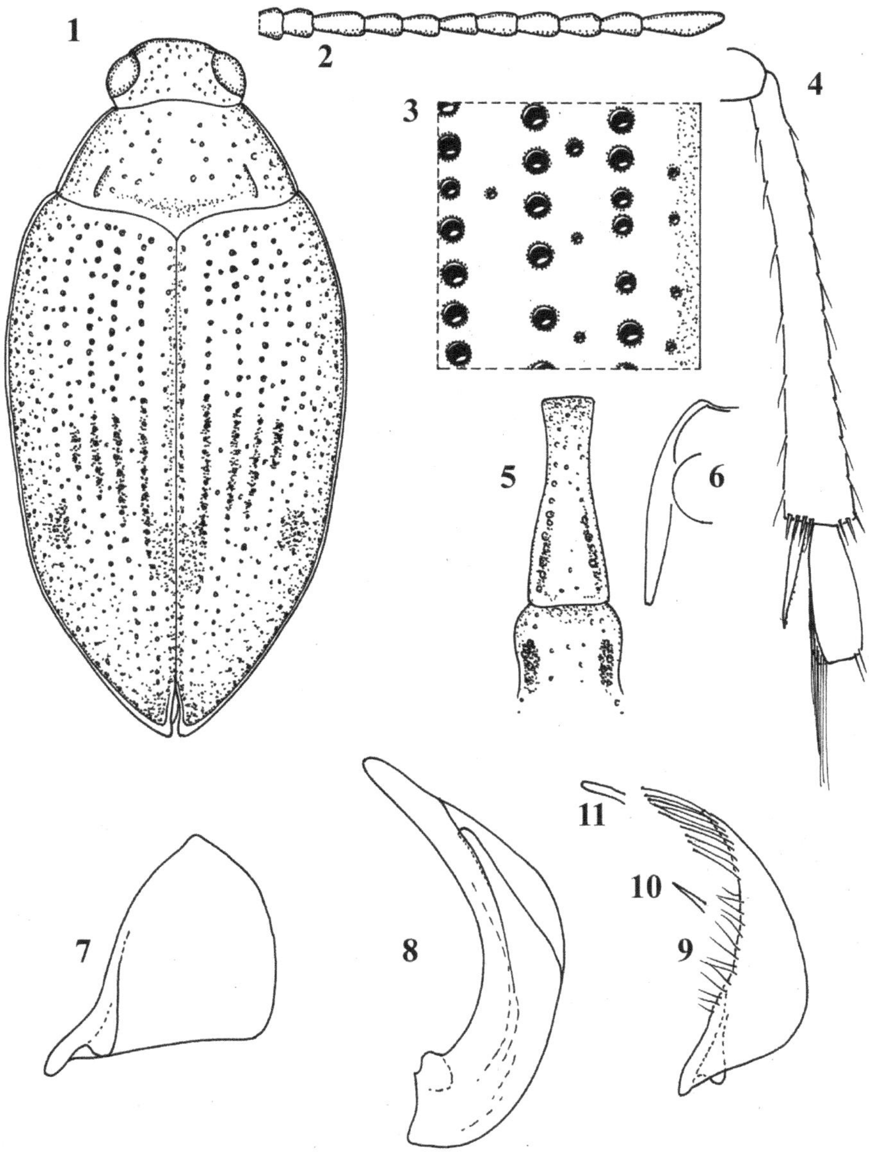

PLATE 35 *Haliplus* (*Haliplus*) *longulus*. 1, dorsal view; 2, antenna; 3, elytral punctures near base and suture; 4, hind tibia in dorsal view; 5, prosternal and metaventral process; 6, prosternal process in lateral view; 7, left paramere; 8, penis; 9, right paramere; 10, 11, setae of right paramere

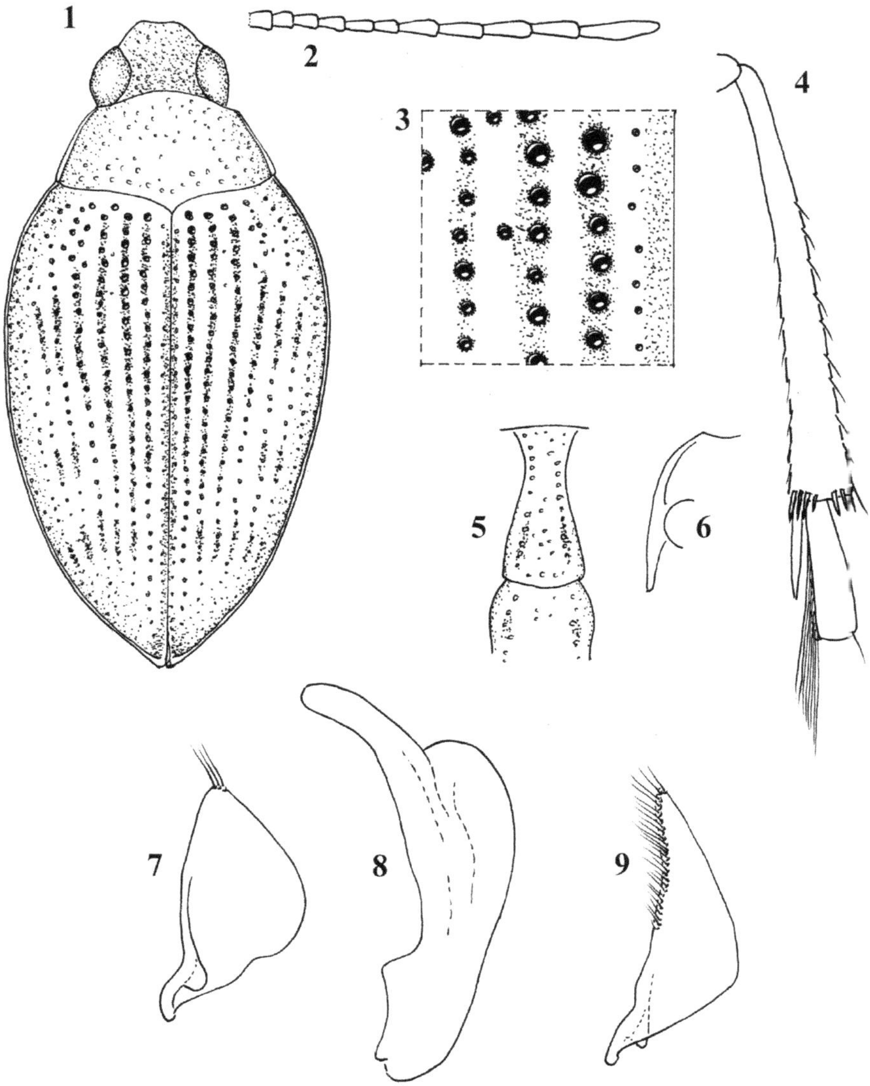

PLATE 36 *Haliplus* (*Haliplus*) *morii*. 1, dorsal view; 2, antenna; 3, elytral punctures near base and suture; 4, hind tibia in dorsal view; 5, prosternal and metaventral process; 6, prosternal process in lateral view; 7, left paramere; 8, penis; 9, right paramere

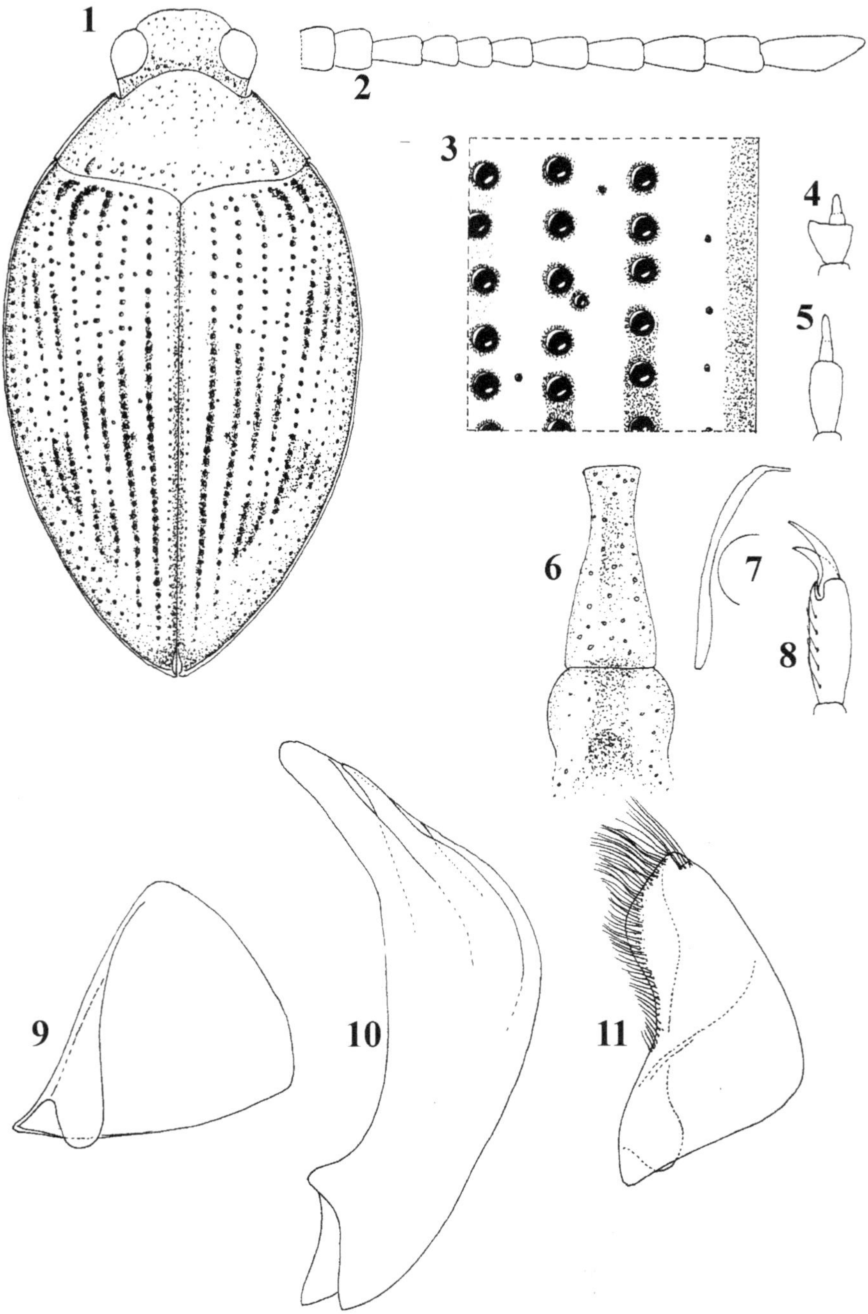

PLATE 37 *Haliplus* (*Haliplus*) *ruficollis*. 1, dorsal view; 2, antenna; 3, elytral punctures near base and suture; 4, labial palpus 5, maxillary palpus; 6, prosternal and metaventral process; 7, prosternal process in lateral view; 8, claws and last tarsal segment of male foreleg; 9, left paramere; 10, penis; 11, right paramere

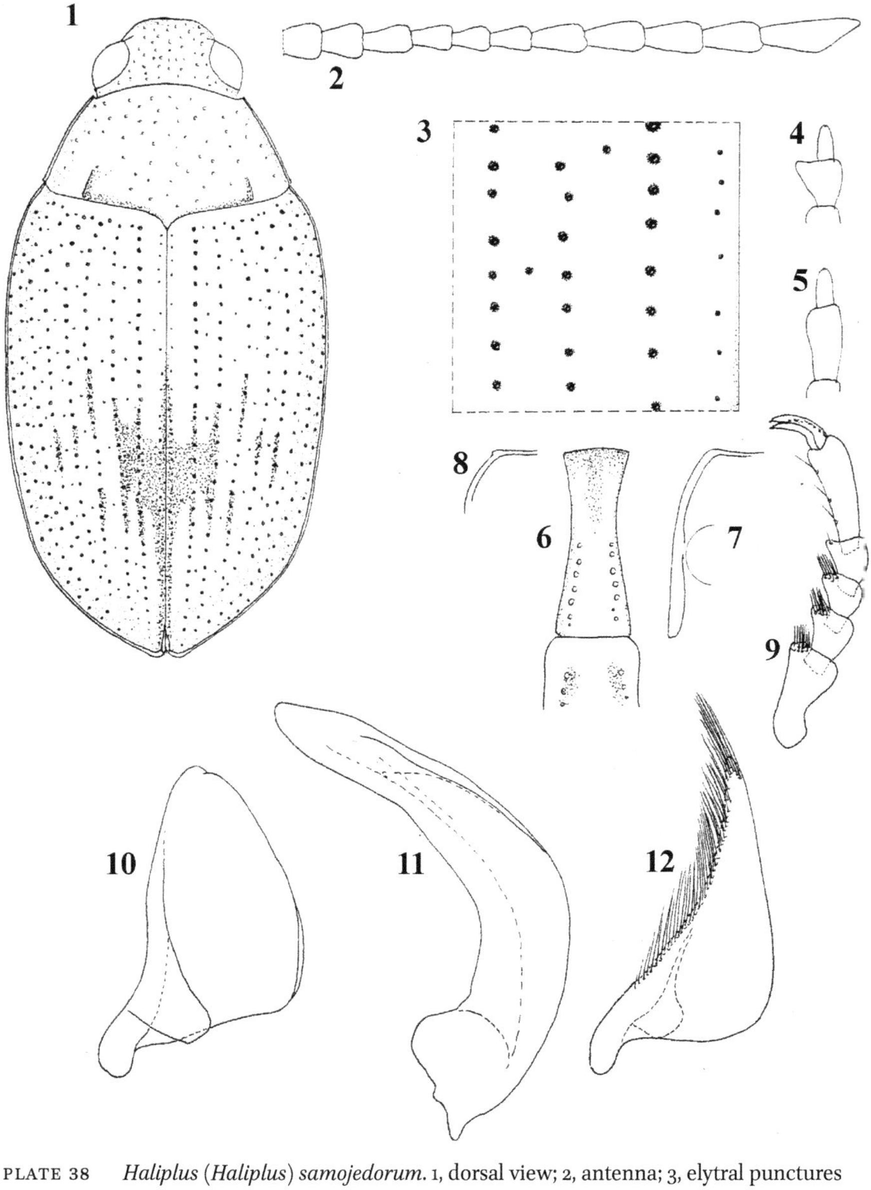

PLATE 38 *Haliplus* (*Haliplus*) *samojedorum*. 1, dorsal view; 2, antenna; 3, elytral punctures near base and suture; 4, labial palpus 5, maxillary palpus; 6, prosternal and metaventral process; 7, 8, prosternal process in lateral view; 9, tarsus of male foreleg; 10, left paramere; 11, penis; 12, right paramere

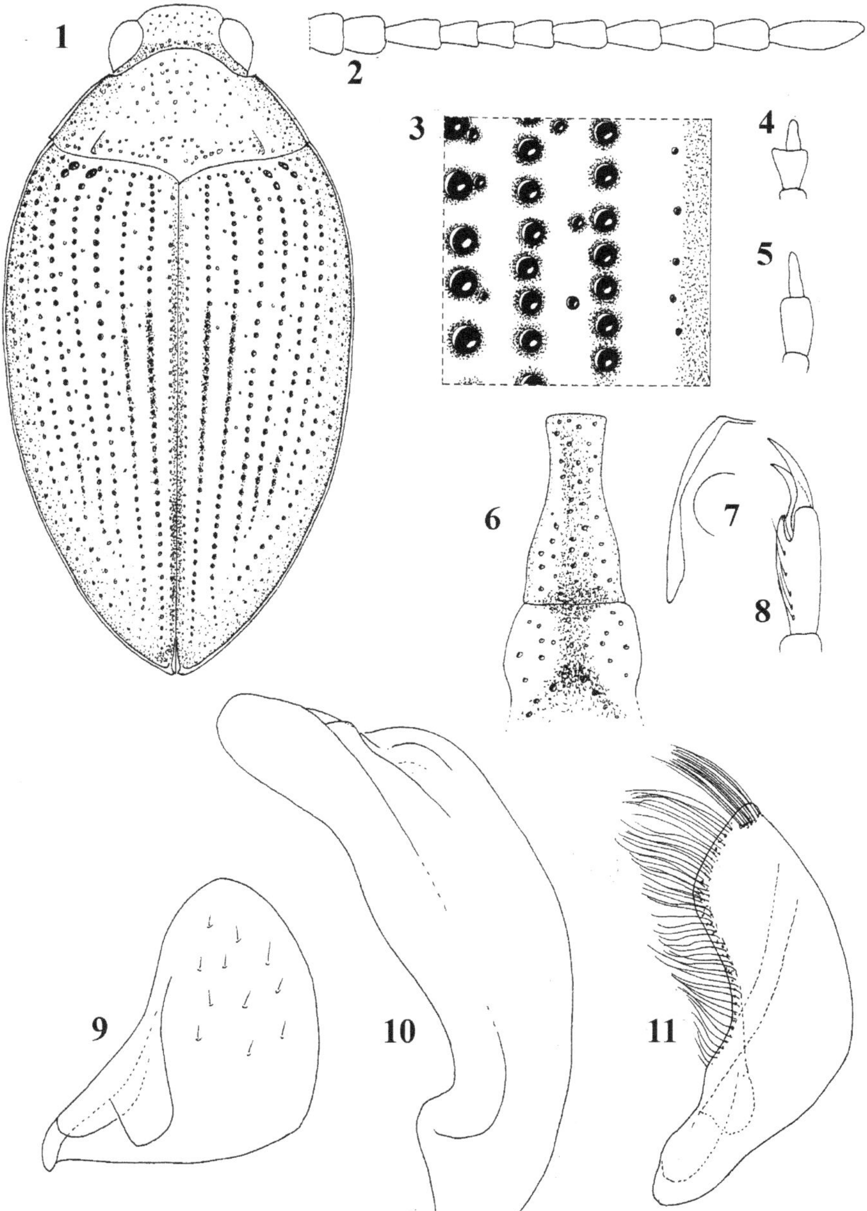

PLATE 39 *Haliplus* (*Haliplus*) *sibiricus*. 1, dorsal view; 2, antenna; 3, elytral punctures near base and suture; 4, labial palpus 5, maxillary palpus; 6, prosternal and metaventral process; 7, prosternal process in lateral view; 8, claws and last tarsal segment of male foreleg; 9, left paramere; 10, penis; 11, right paramere

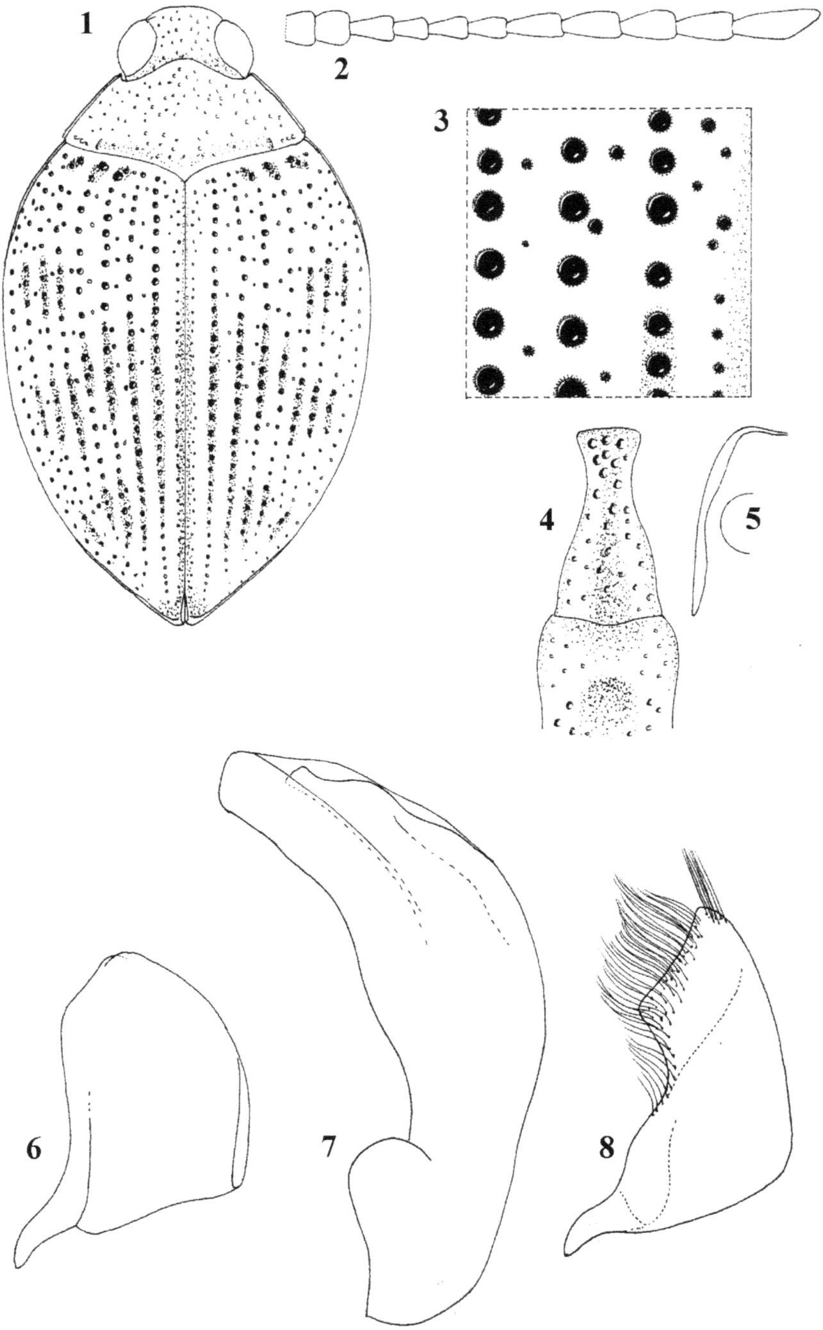

PLATE 40 *Haliplus* (*Haliplus*) *simplex*. 1, dorsal view; 2, antenna; 3, elytral punctures near base and suture; 4, prosternal and metaventral process; 5, prosternal process in lateral view; 6, left paramere; 7, penis; 8, right paramere

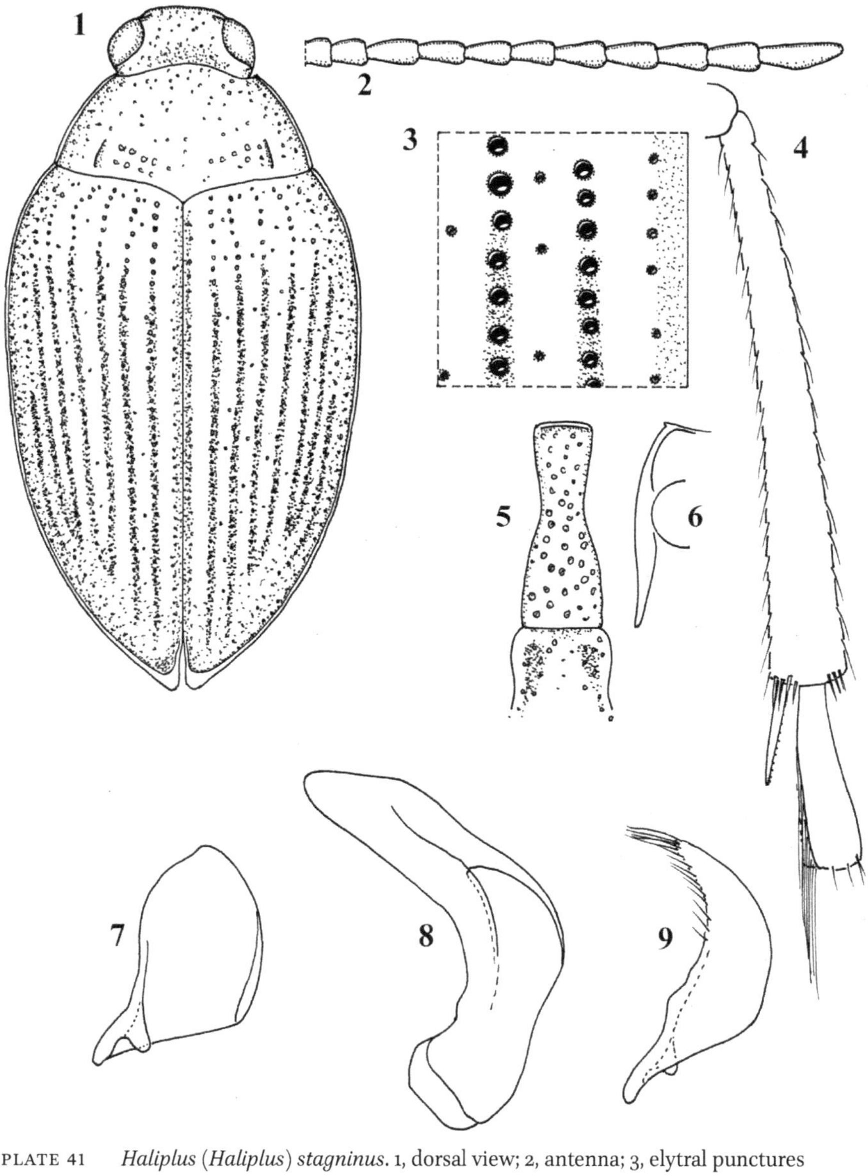

PLATE 41 *Haliplus* (*Haliplus*) *stagninus*. 1, dorsal view; 2, antenna; 3, elytral punctures near base and suture; 4, hind tibia in dorsal view; 5, prosternal and metaventral process; 6, prosternal process in lateral view; 7, left paramere; 8, penis; 9, right paramere

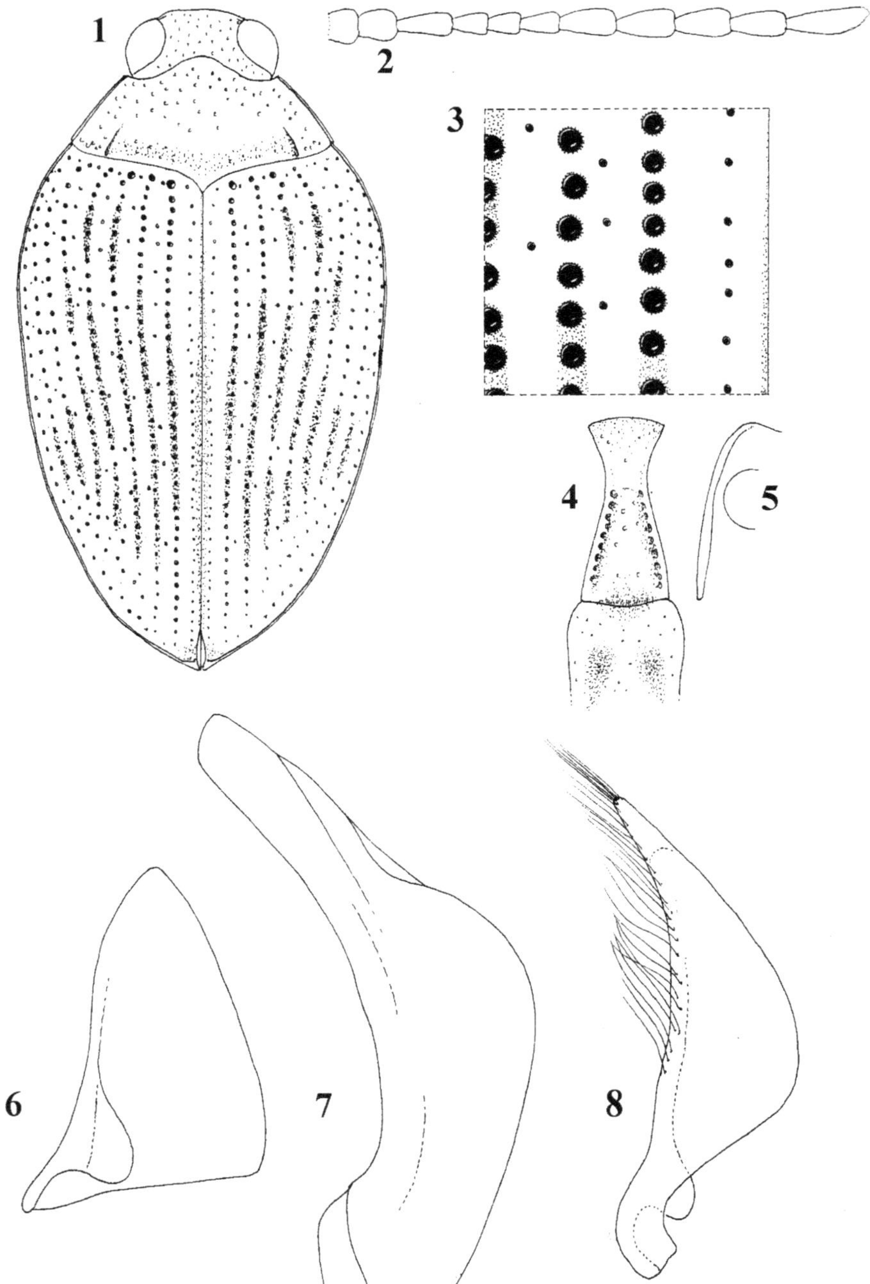

PLATE 42 *Haliplus* (*Haliplus*) *steppensis*. 1, dorsal view; 2, antenna; 3, elytral punctures near base and suture; 4, prosternal and metaventral process; 5, prosternal process in lateral view; 6, left paramere; 7, penis; 8, right paramere

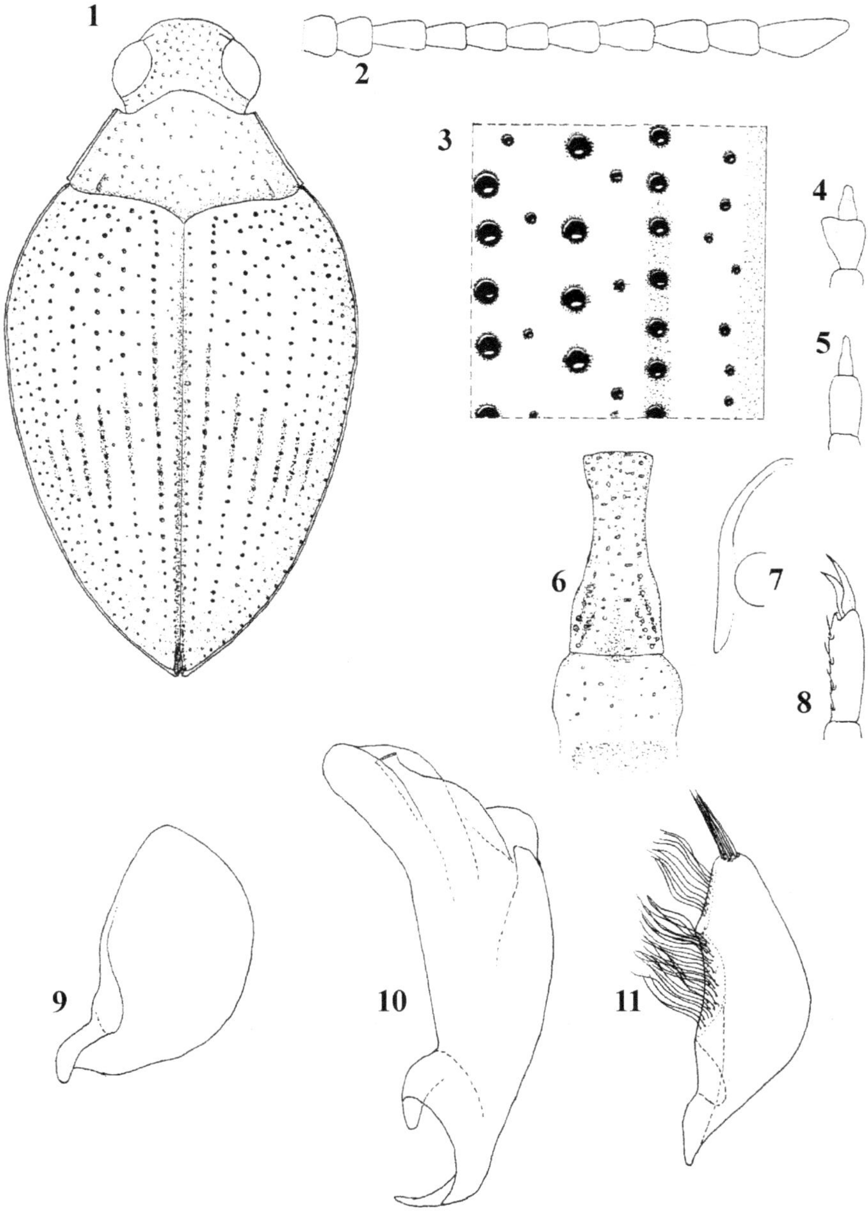

PLATE 43 *Haliplus* (*Haliplus*) *turkmenicus*. 1, dorsal view; 2, antenna; 3, elytral punctures near base and suture; 4, labial palpus 5, maxillary palpus; 6, prosternal and metaventral process; 7, prosternal process in lateral view; 8, claws and last tarsal segment of male foreleg; 9, left paramere; 10, penis; 11, right paramere

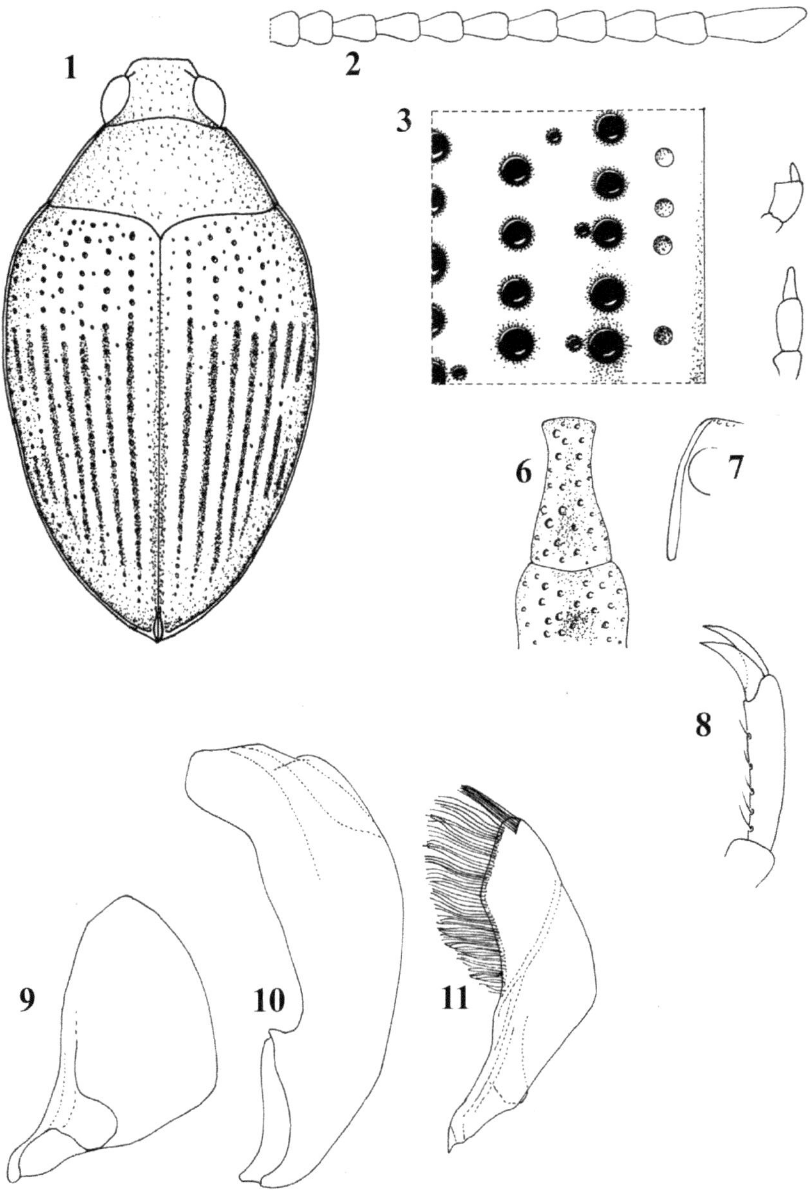

PLATE 44 *Haliplus* (*Haliplus*) *zacharenkoi*. 1, dorsal view; 2, antenna; 3, elytral punctures near base and suture; 4, labial palpus 5, maxillary palpus; 6, prosternal and metaventral process; 7, prosternal process in lateral view; 8, claws and last tarsal segment of male foreleg; 9, left paramere; 10, penis; 11, right paramere

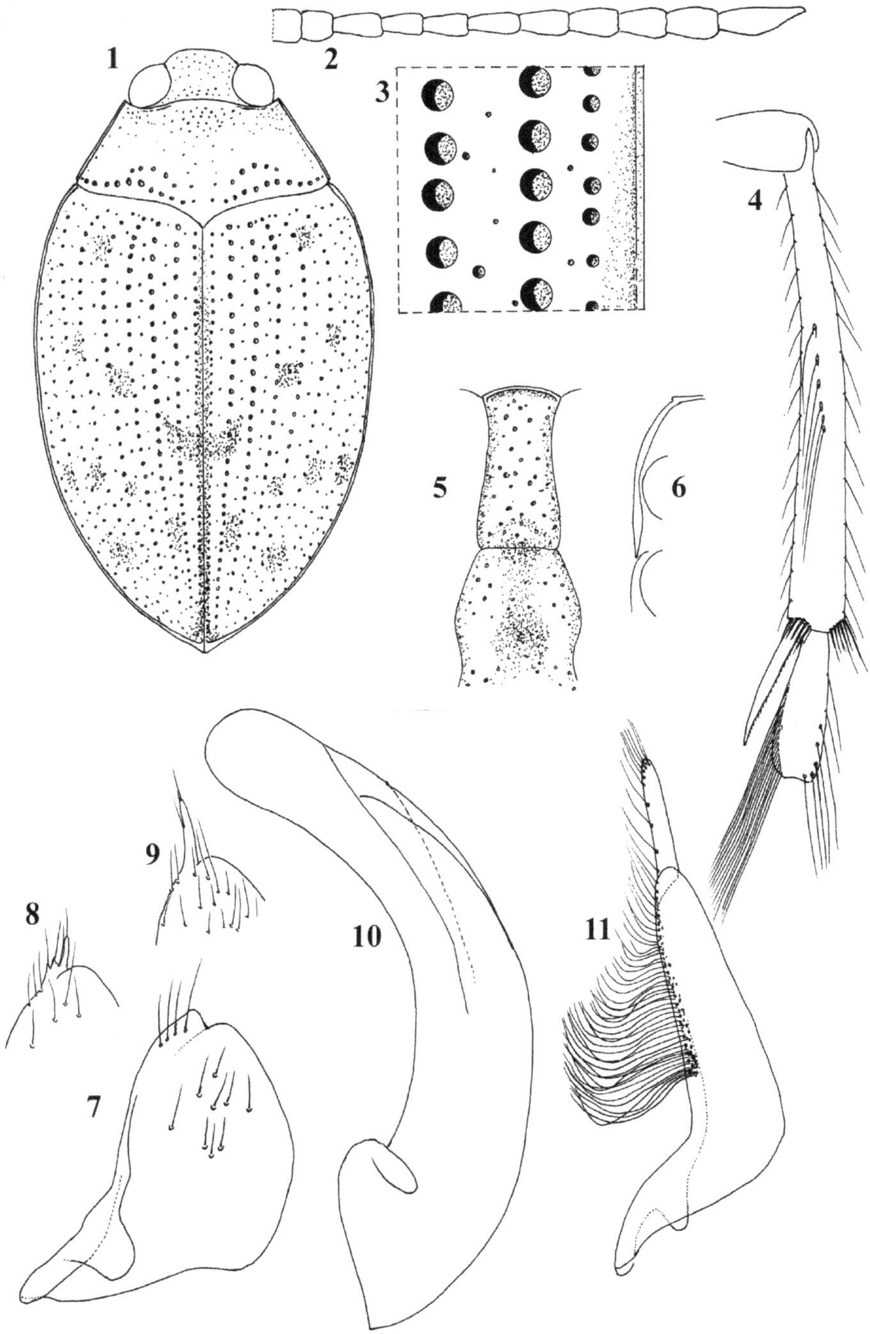

PLATE 45 *Haliplus* (*Liaphlus*) *abbreviatus*. 1, dorsal view; 2, antenna; 3, elytral punctures near base and suture; 4, hind tibia in dorsal view; 5, prosternal and metaventral process; 6, prosternal process in lateral view; 7–9, left paramere; 10, penis; 11, right paramere

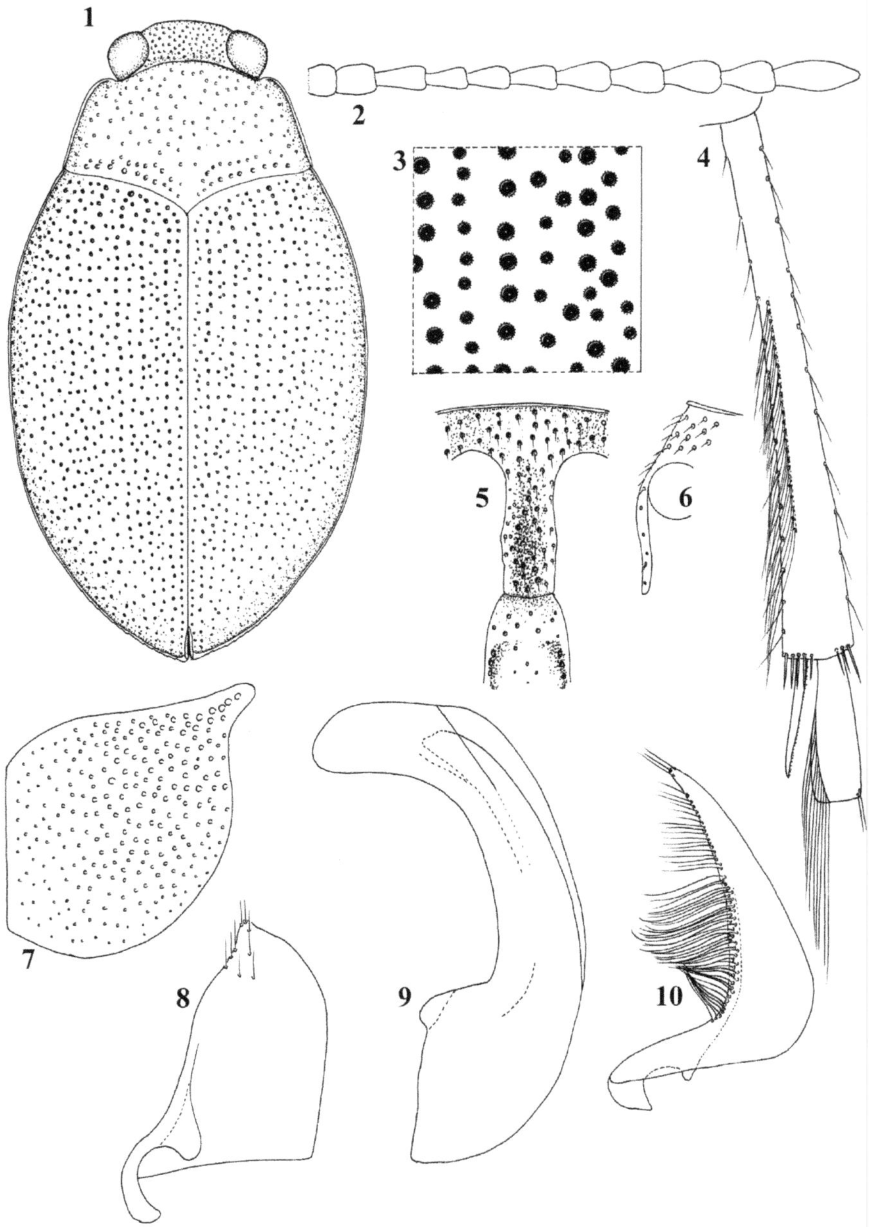

PLATE 46 *Haliplus* (*Liaphlus*) *africanus*. 1, dorsal view; 2, antenna; 3, elytral punctures near base and suture; 4, hind tibia in dorsal view; 5, prosternal and metaventral process; 6, prosternal process in lateral view; 7, hind coxal plate; 8, left paramere; 9, penis; 10, right paramere

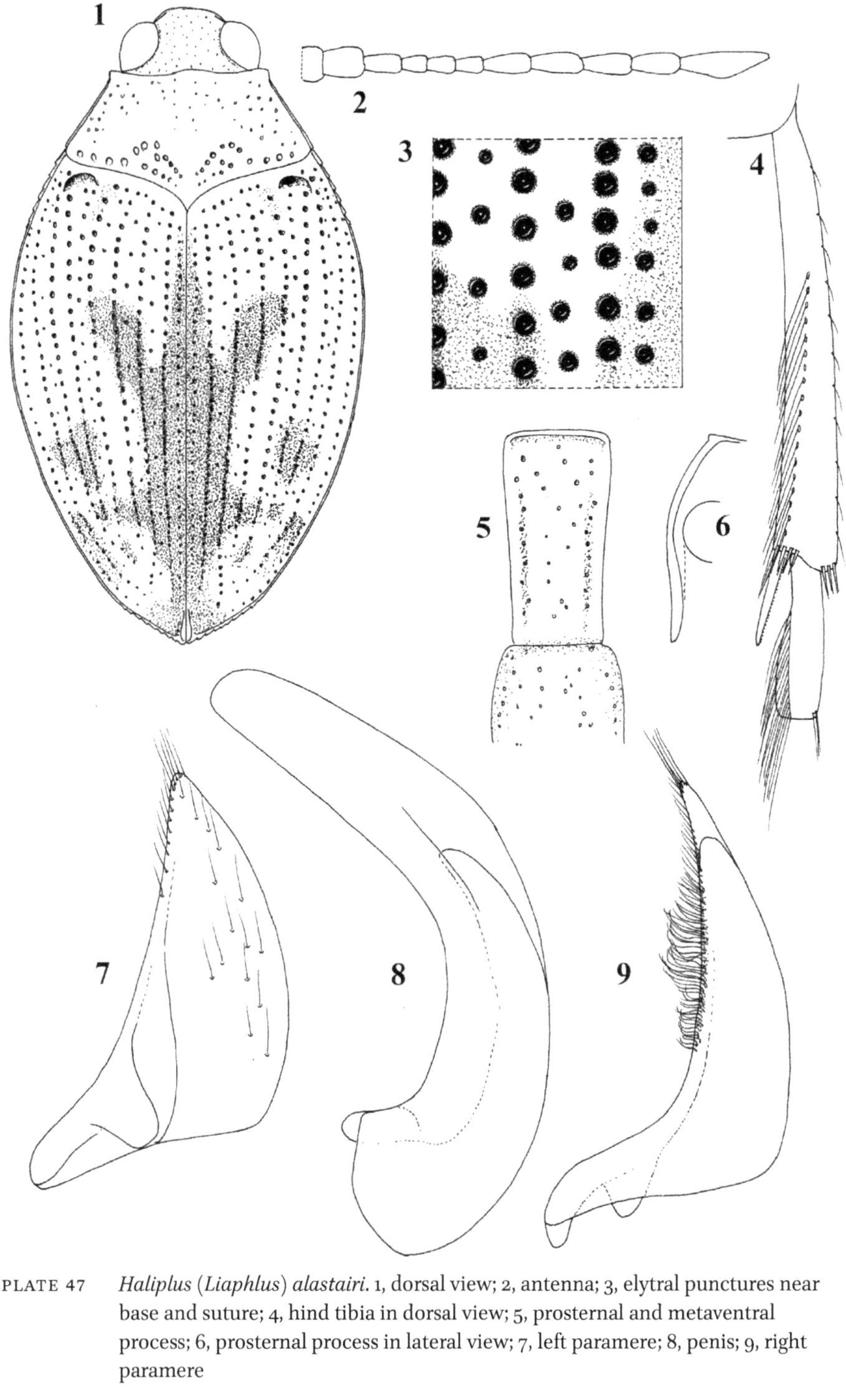

PLATE 47 *Haliplus* (*Liaphlus*) *alastairi*. 1, dorsal view; 2, antenna; 3, elytral punctures near base and suture; 4, hind tibia in dorsal view; 5, prosternal and metaventral process; 6, prosternal process in lateral view; 7, left paramere; 8, penis; 9, right paramere

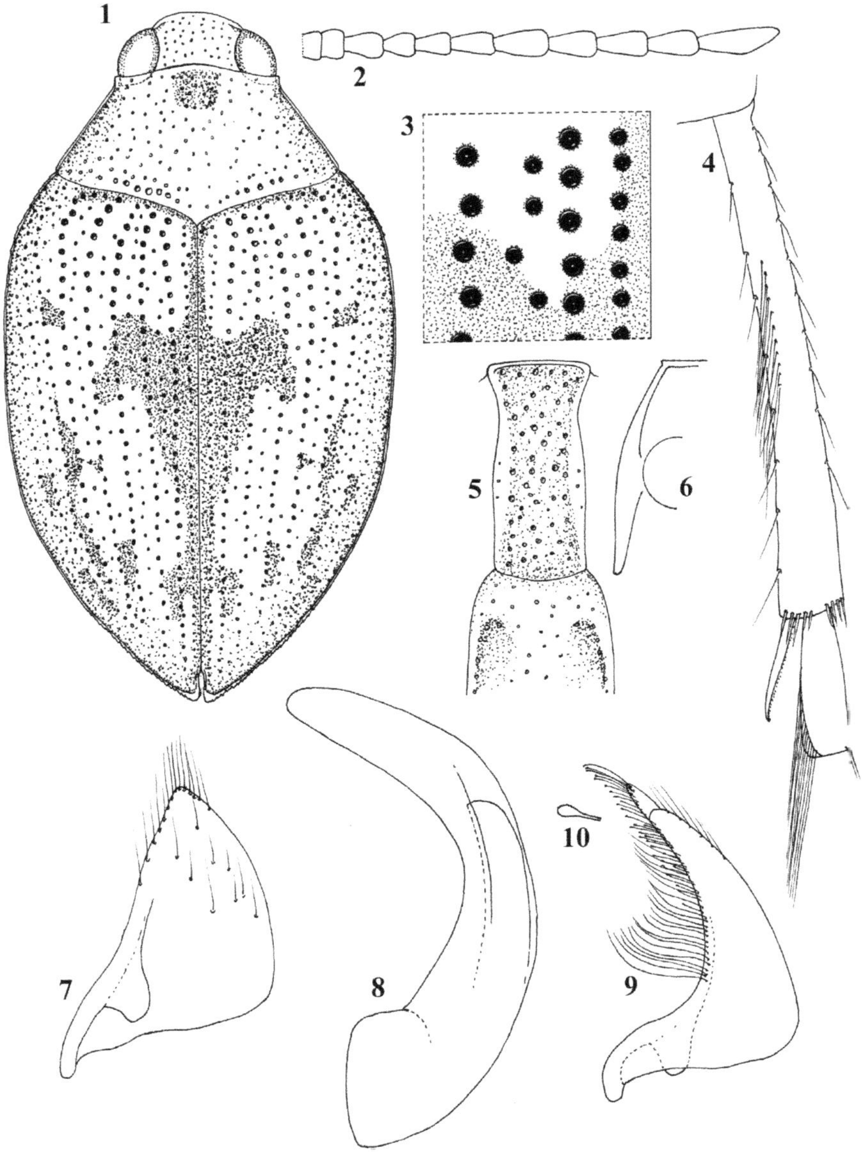

PLATE 48 *Haliplus* (*Liaphlus*) *alluaudi*. 1, dorsal view; 2, antenna; 3, elytral punctures near base and suture; 4, hind tibia in dorsal view; 5, prosternal and metaventral process; 6, prosternal process in lateral view; 7, left paramere; 8, penis; 9, right paramere; 10, setae of right paramere

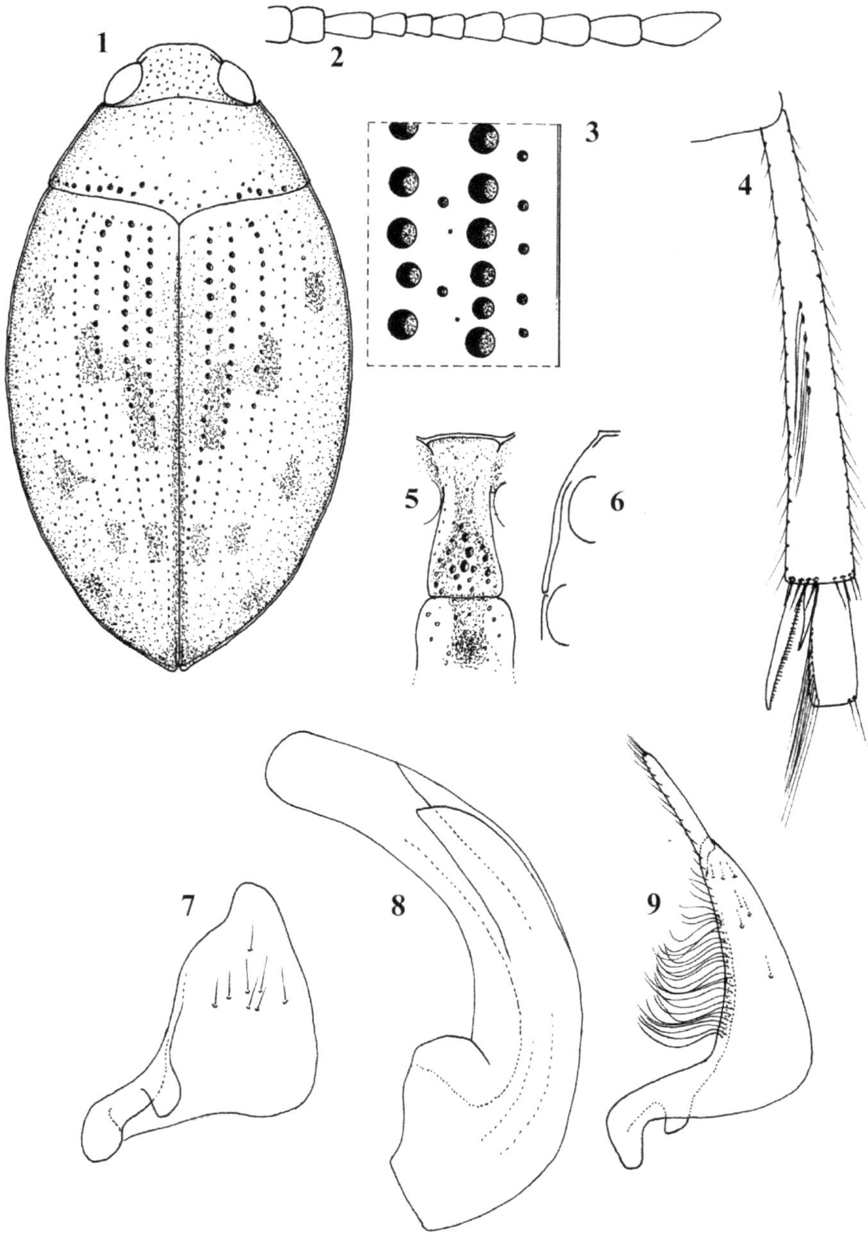

PLATE 49 *Haliplus* (*Liaphlus*) *andalusicus*. 1, dorsal view; 2, antenna; 3, elytral punctures near base and suture; 4, hind tibia in dorsal view; 5, prosternal and metaventral process; 6, prosternal process in lateral view; 7, left paramere; 8, penis; 9, right paramere

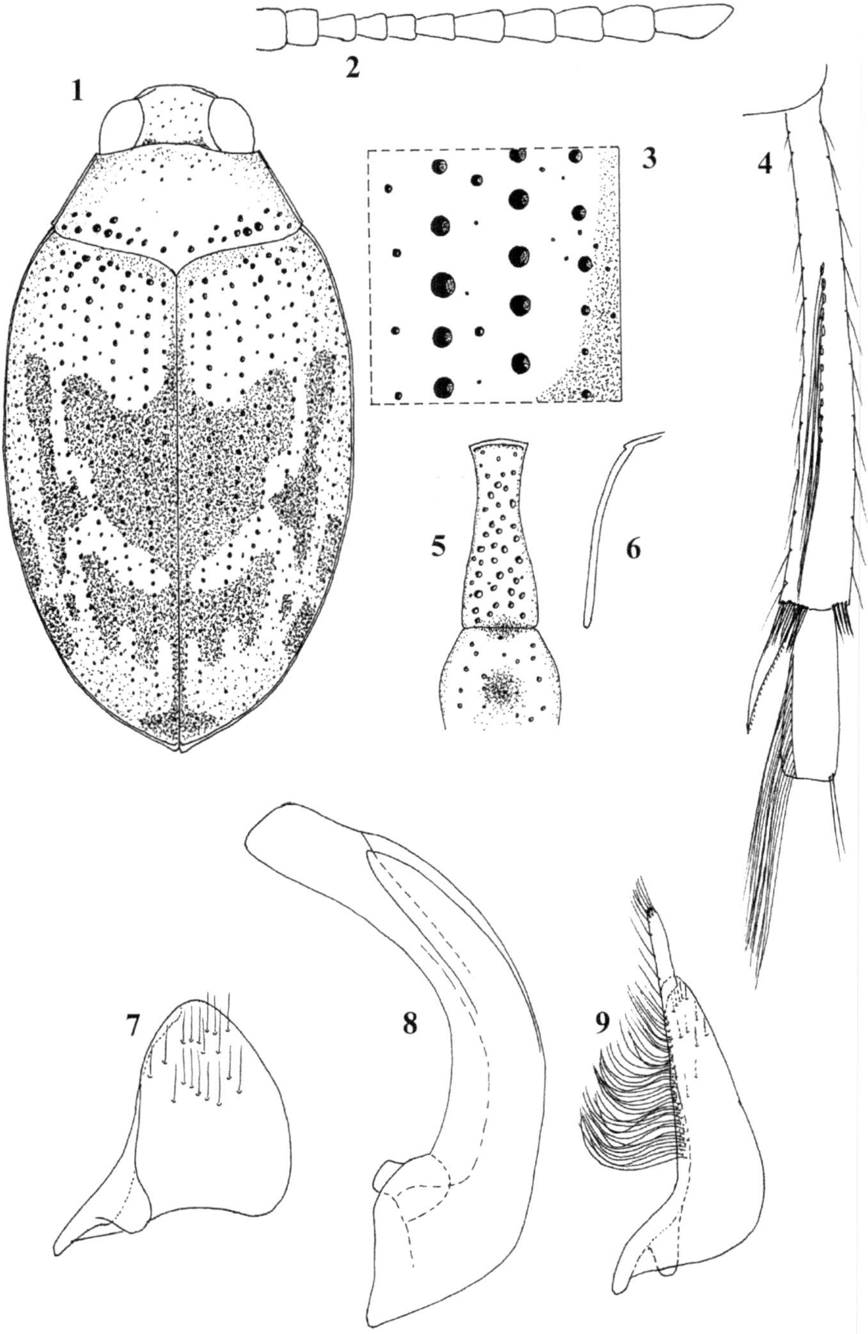

PLATE 50 *Haliplus* (*Liaphlus*) *angusi*. 1, dorsal view; 2, antenna; 3, elytral punctures near base and suture; 4, hind tibia in dorsal view; 5, prosternal and metaventral process; 6, prosternal process in lateral view; 7, left paramere; 8, penis; 9, right paramere

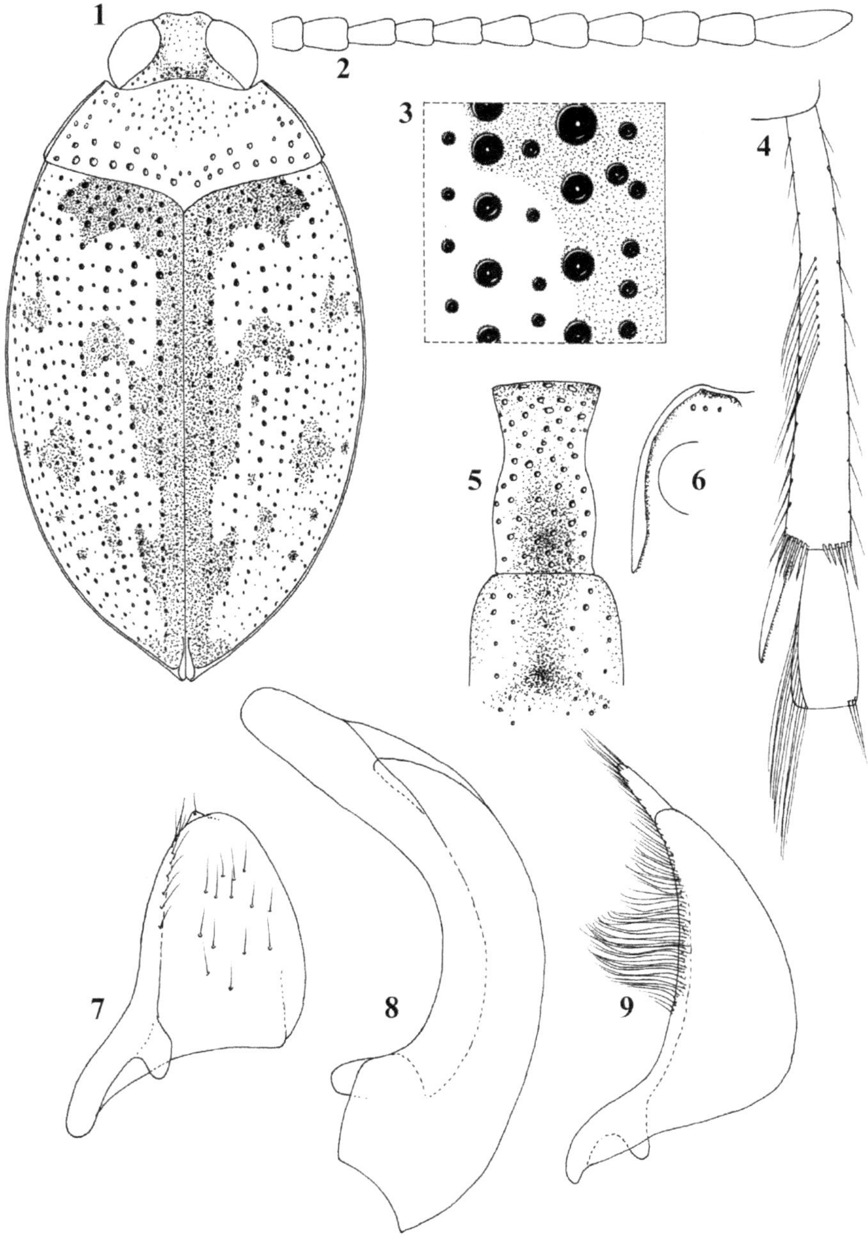

PLATE 51 *Haliplus* (*Liaphlus*) *angustifrons*. 1, dorsal view; 2, antenna; 3, elytral punctures near base and suture; 4, hind tibia in dorsal view; 5, prosternal and metaventral process; 6, prosternal process in lateral view; 7, left paramere; 8, penis; 9, right paramere

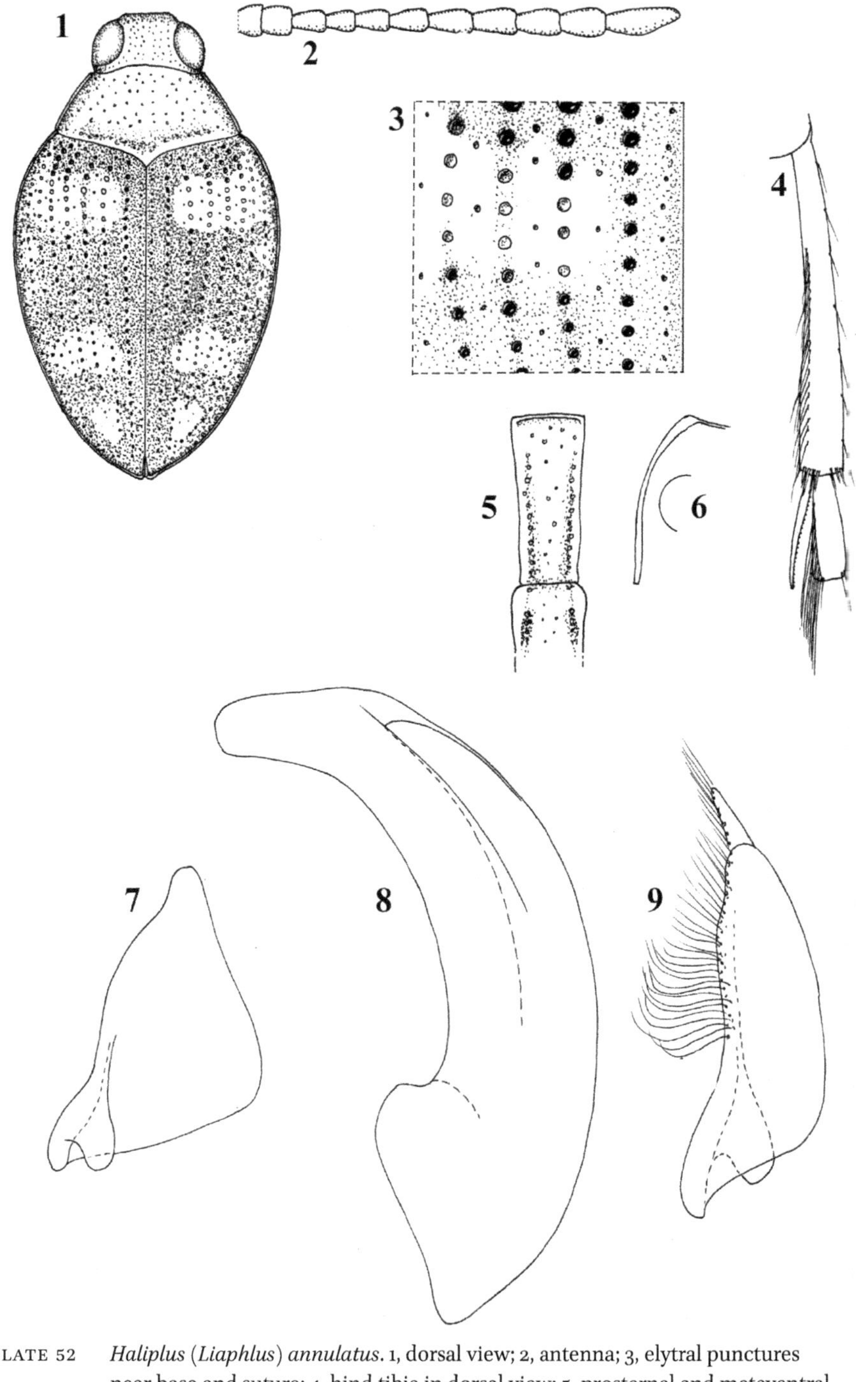

PLATE 52 *Haliplus* (*Liaphlus*) *annulatus*. 1, dorsal view; 2, antenna; 3, elytral punctures near base and suture; 4, hind tibia in dorsal view; 5, prosternal and metaventral process; 6, prosternal process in lateral view; 7, left paramere; 8, penis; 9, right paramere

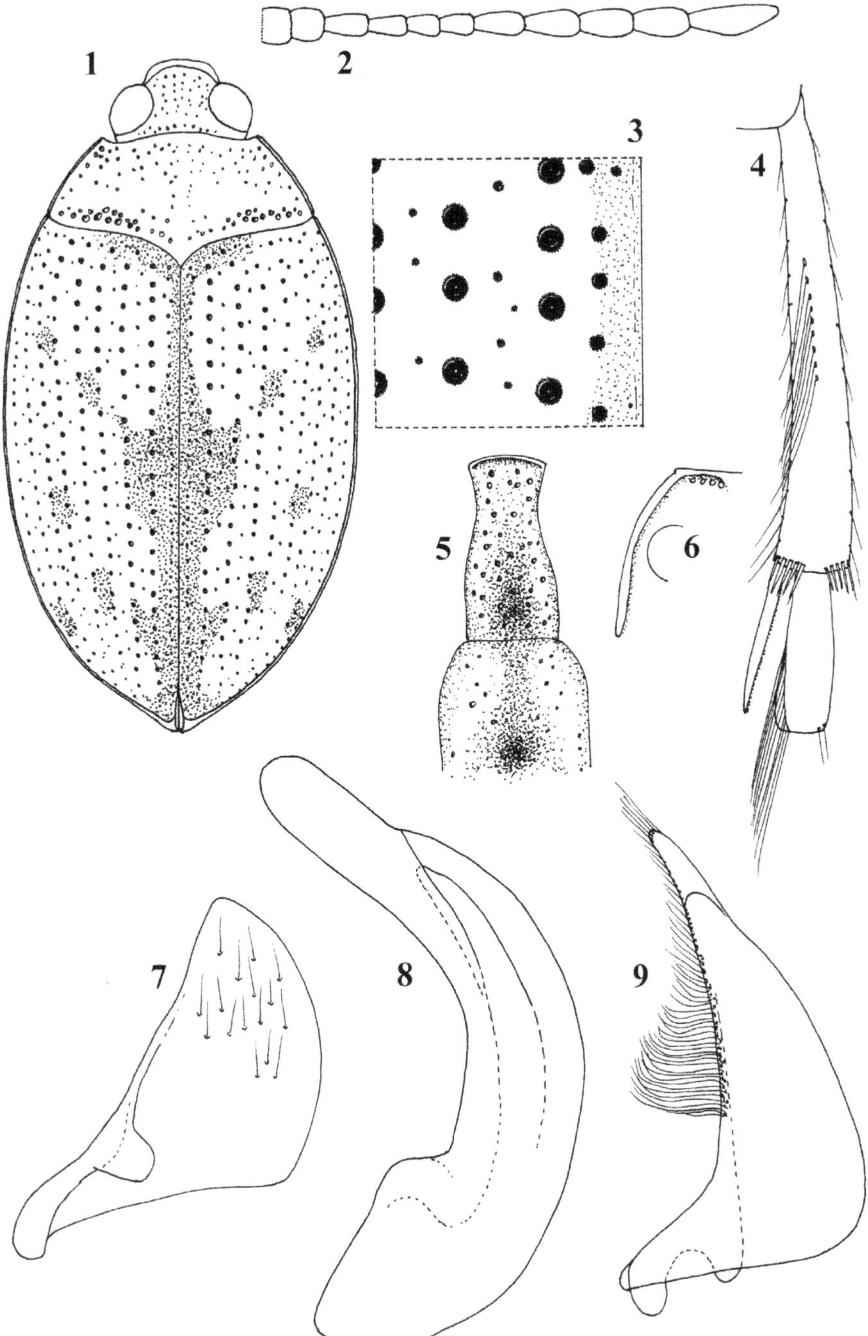

PLATE 53 *Haliplus* (*Liaphlus*) *arrowi*. 1, dorsal view; 2, antenna; 3, elytral punctures near base and suture; 4, hind tibia in dorsal view; 5, prosternal and metaventral process; 6, prosternal process in lateral view; 7, left paramere; 8, penis; 9, right paramere

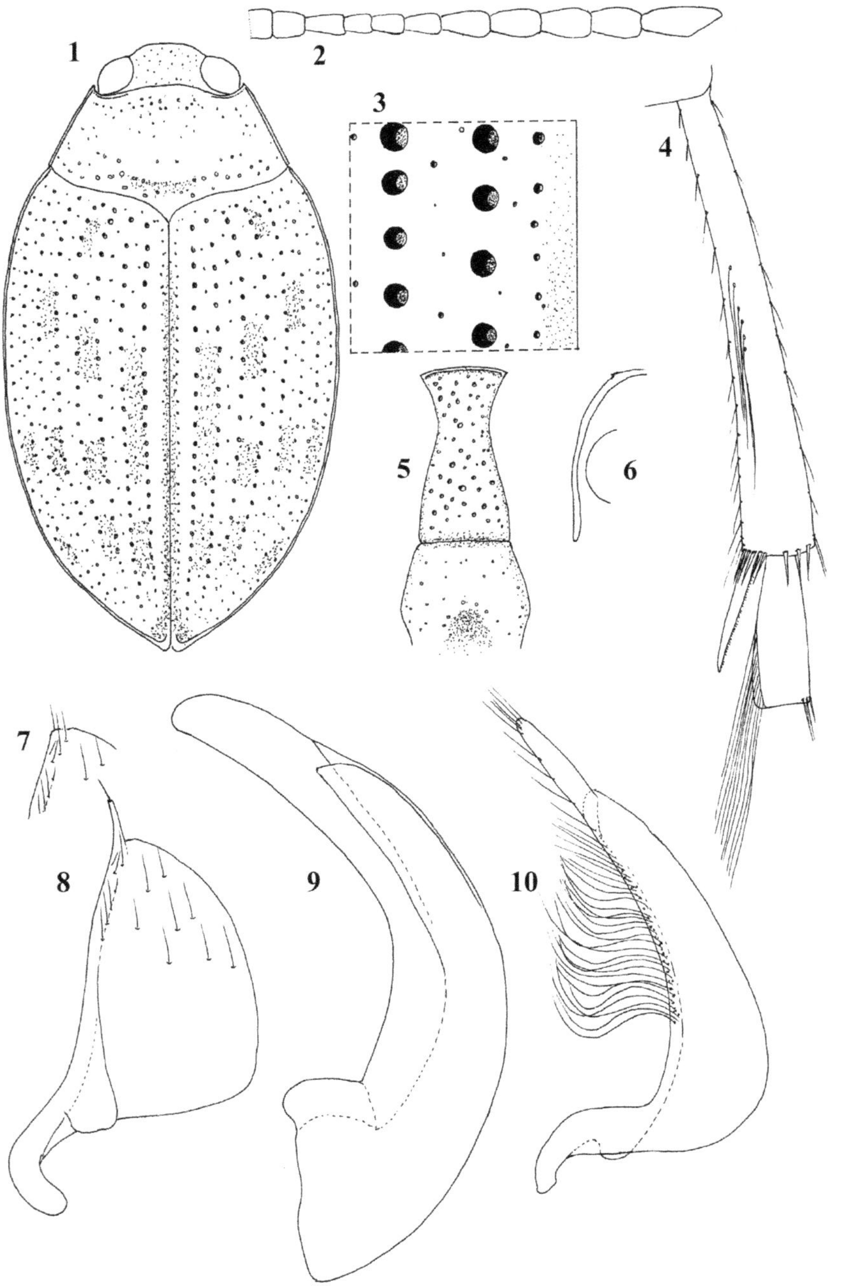

PLATE 54 *Haliplus* (*Liaphlus*) *astrakhanus*. 1, dorsal view; 2, antenna; 3, elytral punctures near base and suture; 4, hind tibia in dorsal view; 5, prosternal and metaventral process; 6, prosternal process in lateral view; 7, 8, left paramere; 9, penis; 10, right paramere

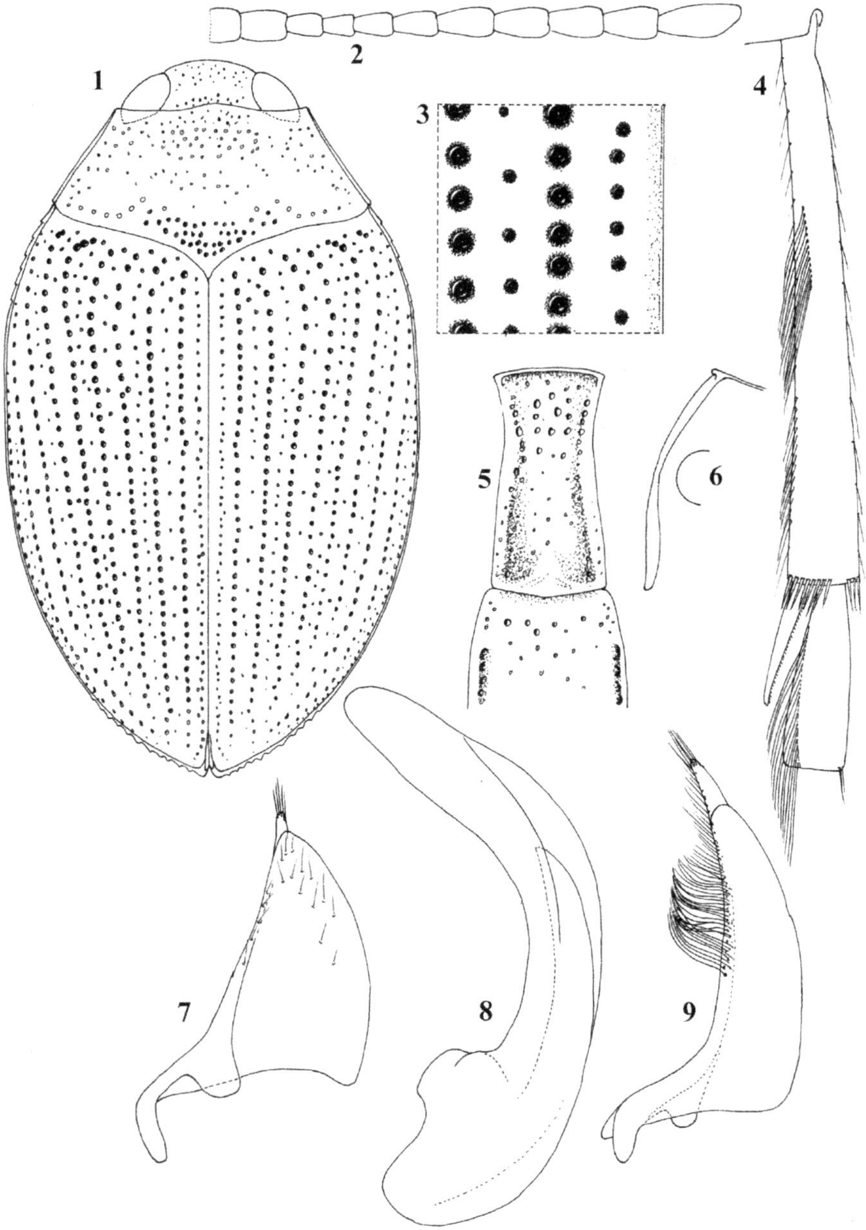

PLATE 55 *Haliplus* (*Liaphlus*) *australis*. 1, dorsal view; 2, antenna; 3, elytral punctures near base and suture; 4, hind tibia in dorsal view; 5, prosternal and metaventral process; 6, prosternal process in lateral view; 7, left paramere; 8, penis; 9, right paramere

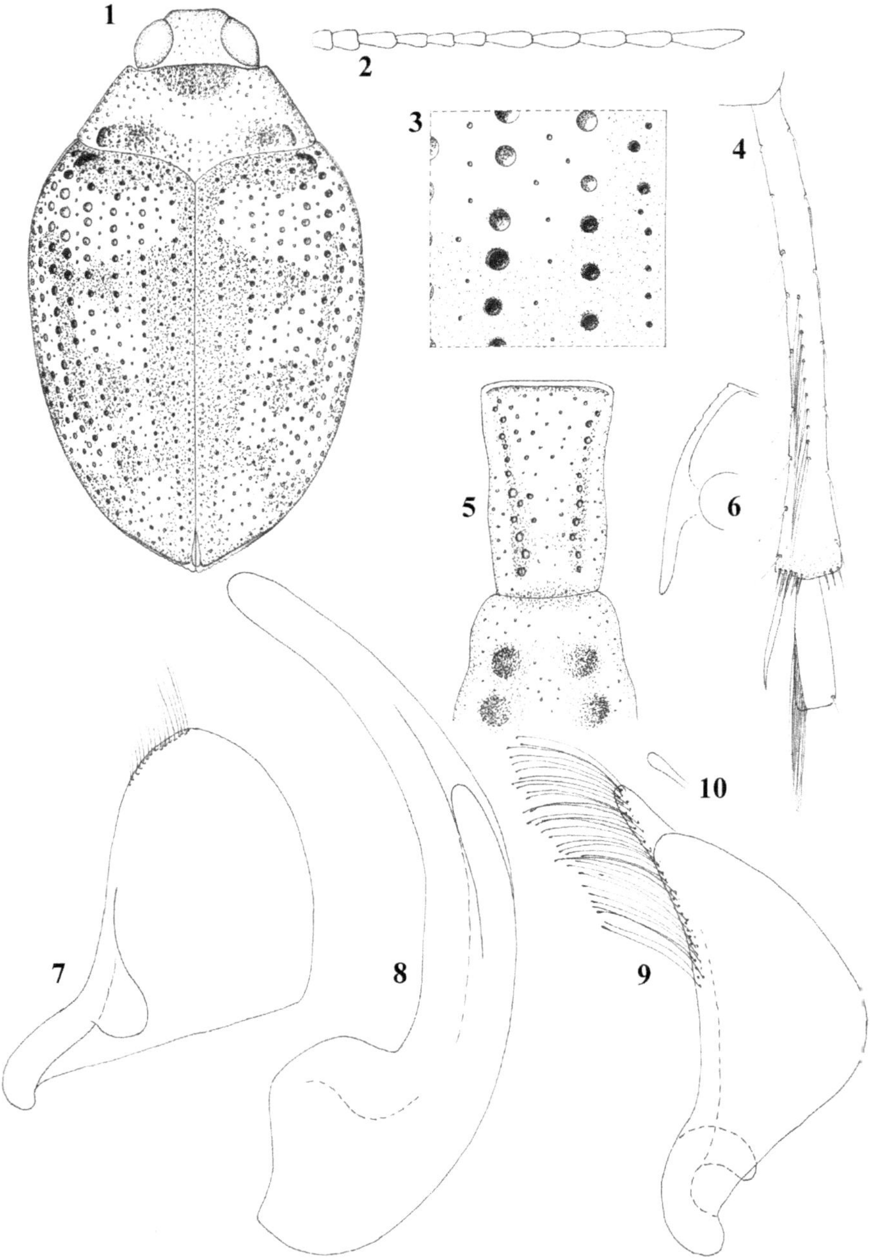

PLATE 56 *Haliplus* (*Liaphlus*) *bachmanni*. 1, dorsal view; 2, antenna; 3, elytral punctures near base and suture; 4, hind tibia in dorsal view; 5, prosternal and metaventral process; 6, prosternal process in lateral view; 7, left paramere; 8, penis; 9, right paramere; 10 setae of right paramere

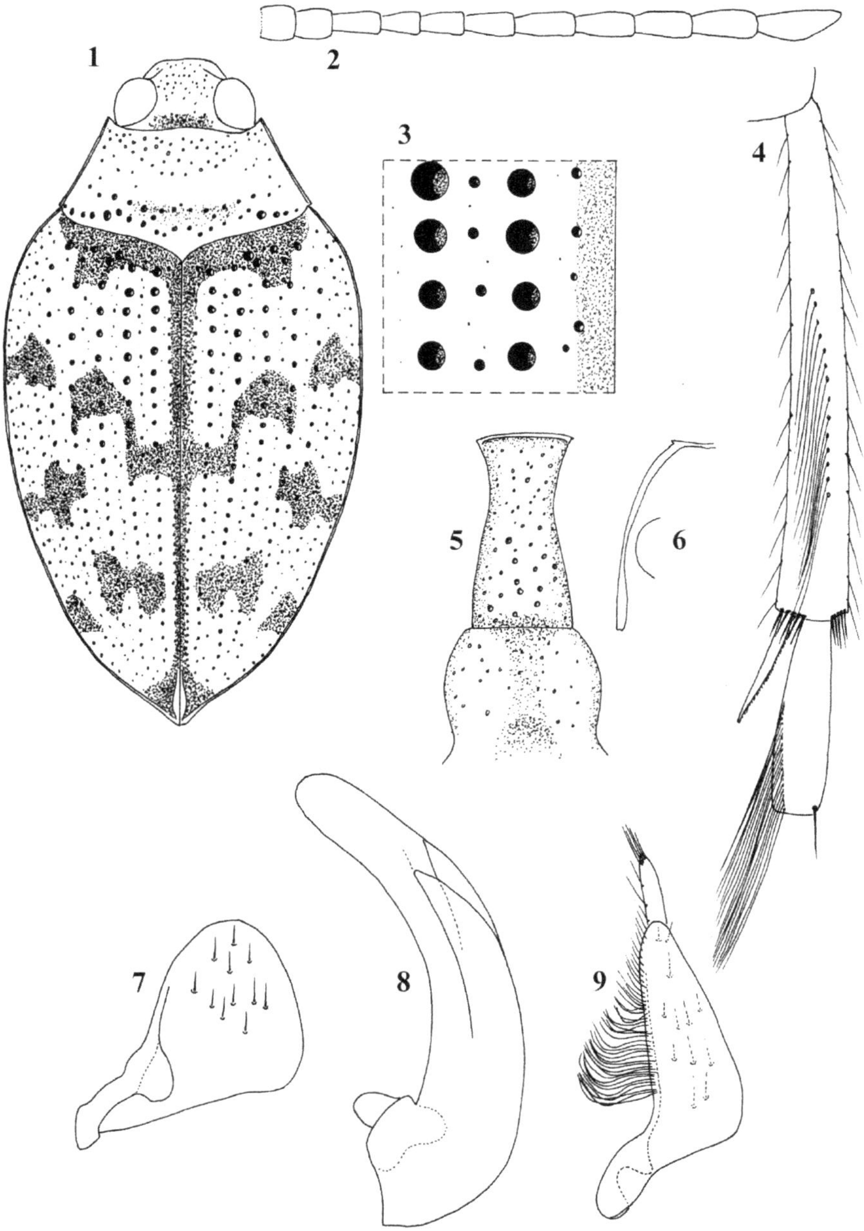

PLATE 57 *Haliplus* (*Liaphlus*) *basinotatus*. 1, dorsal view; 2, antenna; 3, elytral punctures near base and suture; 4, hind tibia in dorsal view; 5, prosternal and metaventral process; 6, prosternal process in lateral view; 7, left paramere; 8, penis; 9, right paramere

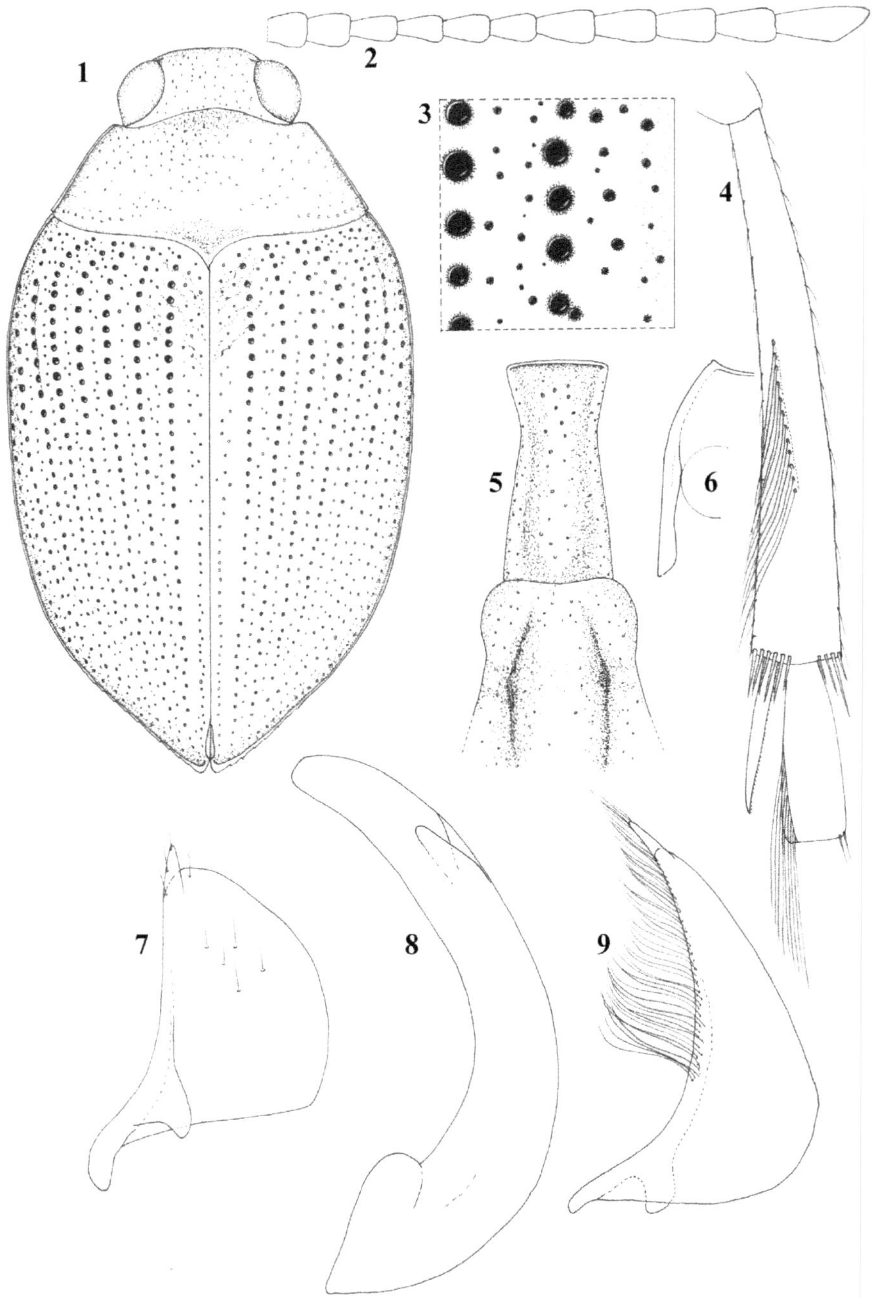

PLATE 58 *Haliplus* (*Liaphlus*) *bonariensis*. 1, dorsal view; 2, antenna; 3, elytral punctures near base and suture; 4, hind tibia in dorsal view; 5, prosternal and metaventral process; 6, prosternal process in lateral view; 7, left paramere; 8, penis; 9, right paramere

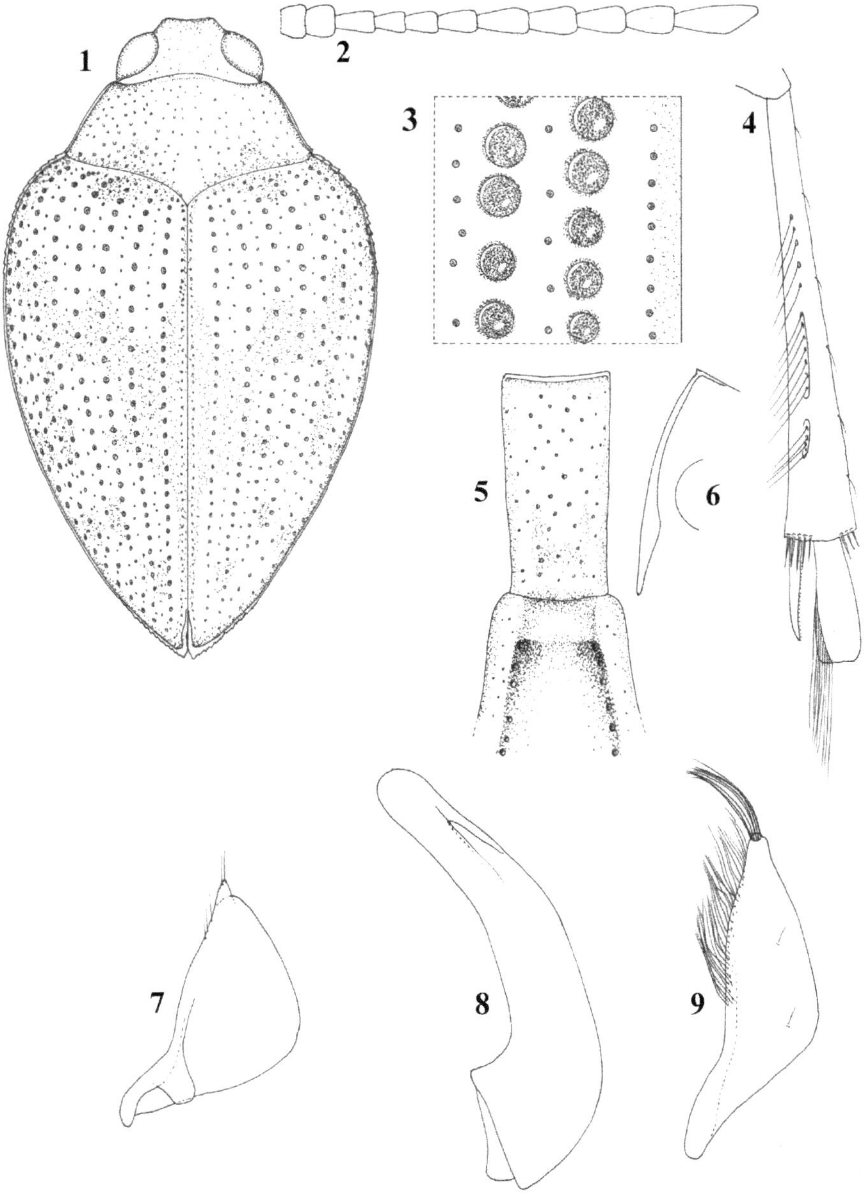

PLATE 59 *Haliplus* (*Liaphlus*) *brasiliensis*. 1, dorsal view; 2, antenna; 3, elytral punctures near base and suture; 4, hind tibia in dorsal view; 5, prosternal and metaventral process; 6, prosternal process in lateral view; 7, left paramere; 8, penis; 9, right paramere

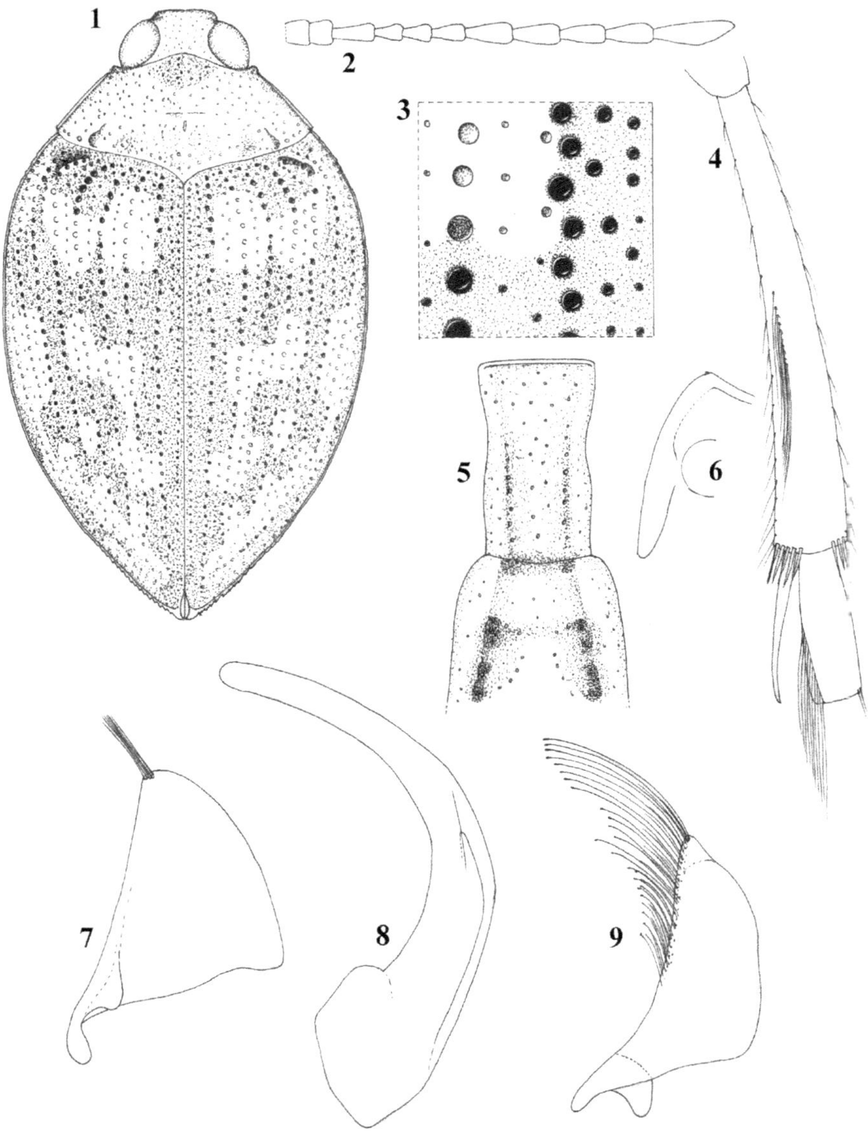

PLATE 60 *Haliplus* (*Liaphlus*) *camposi*. 1, dorsal view; 2, antenna; 3, elytral punctures near base and suture; 4, hind tibia in dorsal view; 5, prosternal and metaventral process; 6, prosternal process in lateral view; 7, left paramere; 8, penis; 9, right paramere

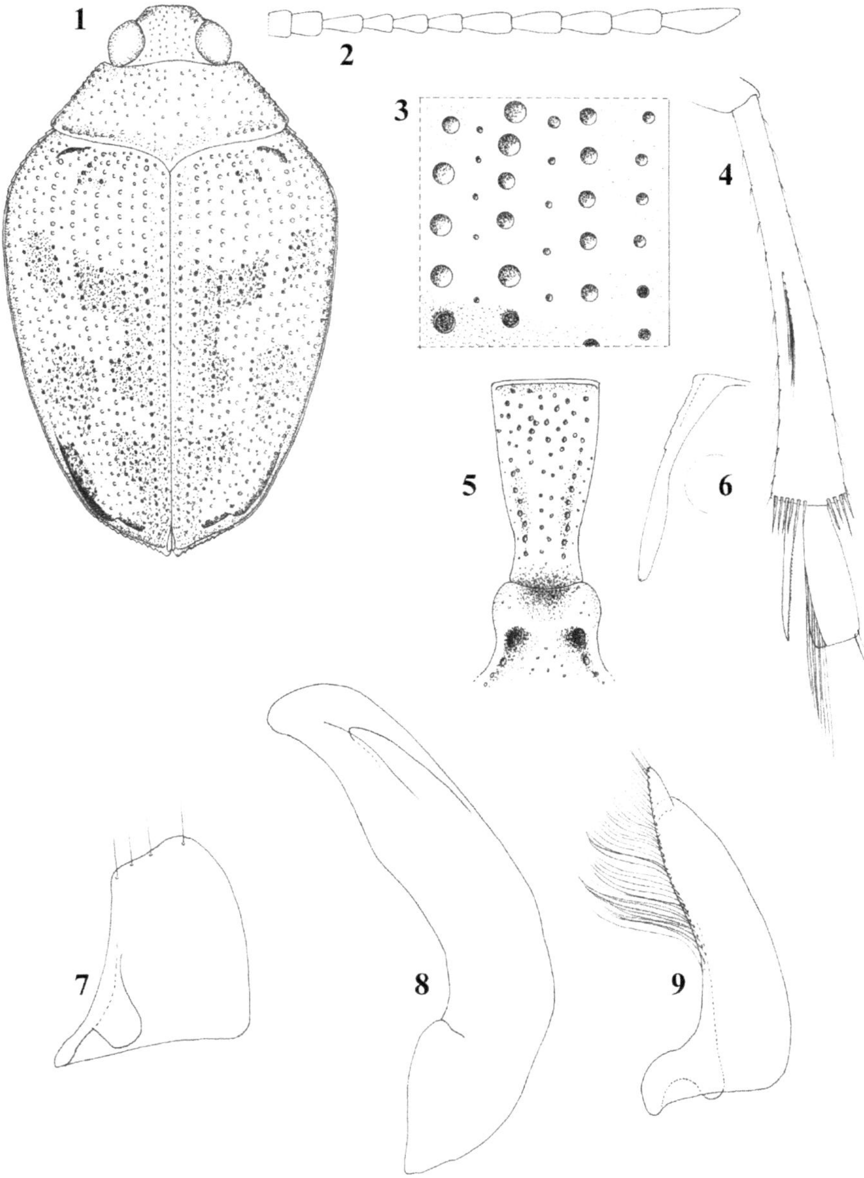

PLATE 61 *Haliplus* (*Liaphlus*) *carinatus*. 1, dorsal view; 2, antenna; 3, elytral punctures near base and suture; 4, hind tibia in dorsal view; 5, prosternal and metaventral process; 6, prosternal process in lateral view; 7, left paramere; 8, penis; 9, right paramere

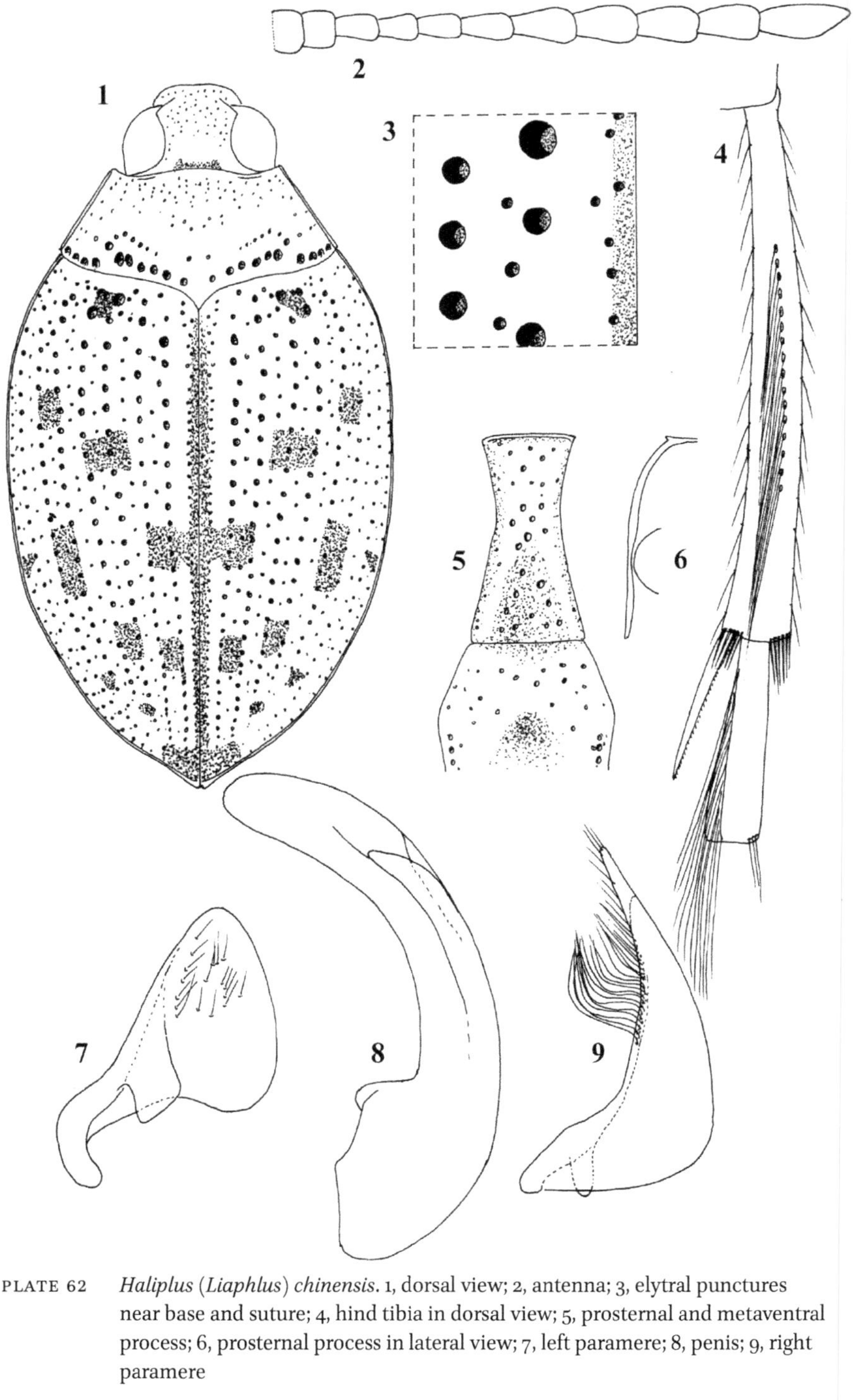

PLATE 62 *Haliplus* (*Liaphlus*) *chinensis*. 1, dorsal view; 2, antenna; 3, elytral punctures near base and suture; 4, hind tibia in dorsal view; 5, prosternal and metaventral process; 6, prosternal process in lateral view; 7, left paramere; 8, penis; 9, right paramere

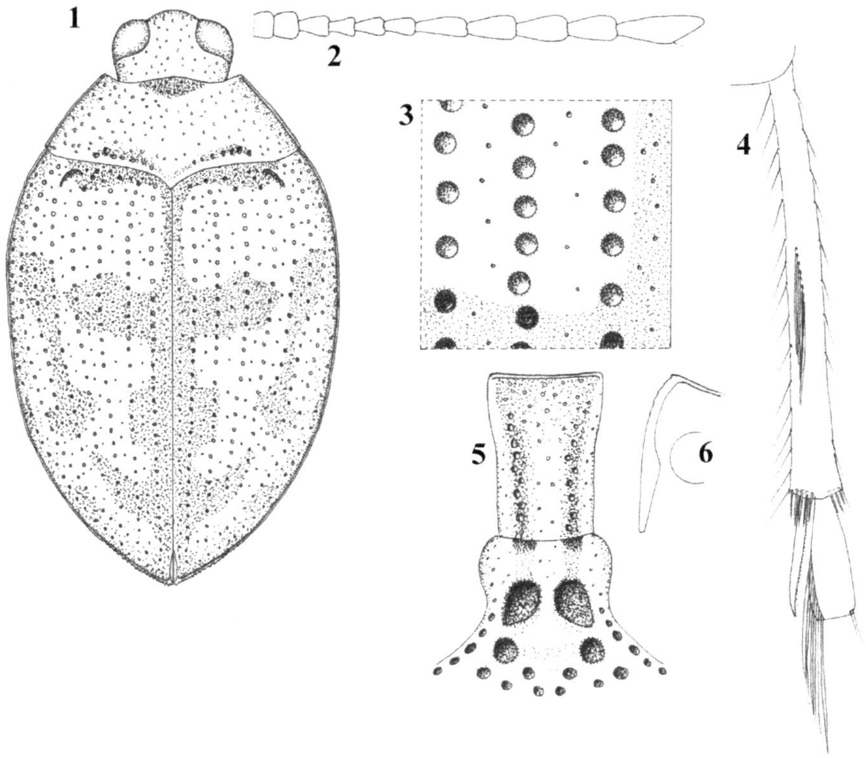

PLATE 63 *Haliplus* (*Liaphlus*) *colombiensis* female. 1, dorsal view; 2, antenna; 3, elytral punctures near base and suture; 4, hind tibia in dorsal view; 5, prosternal and metaventral process; 6, prosternal process in lateral view

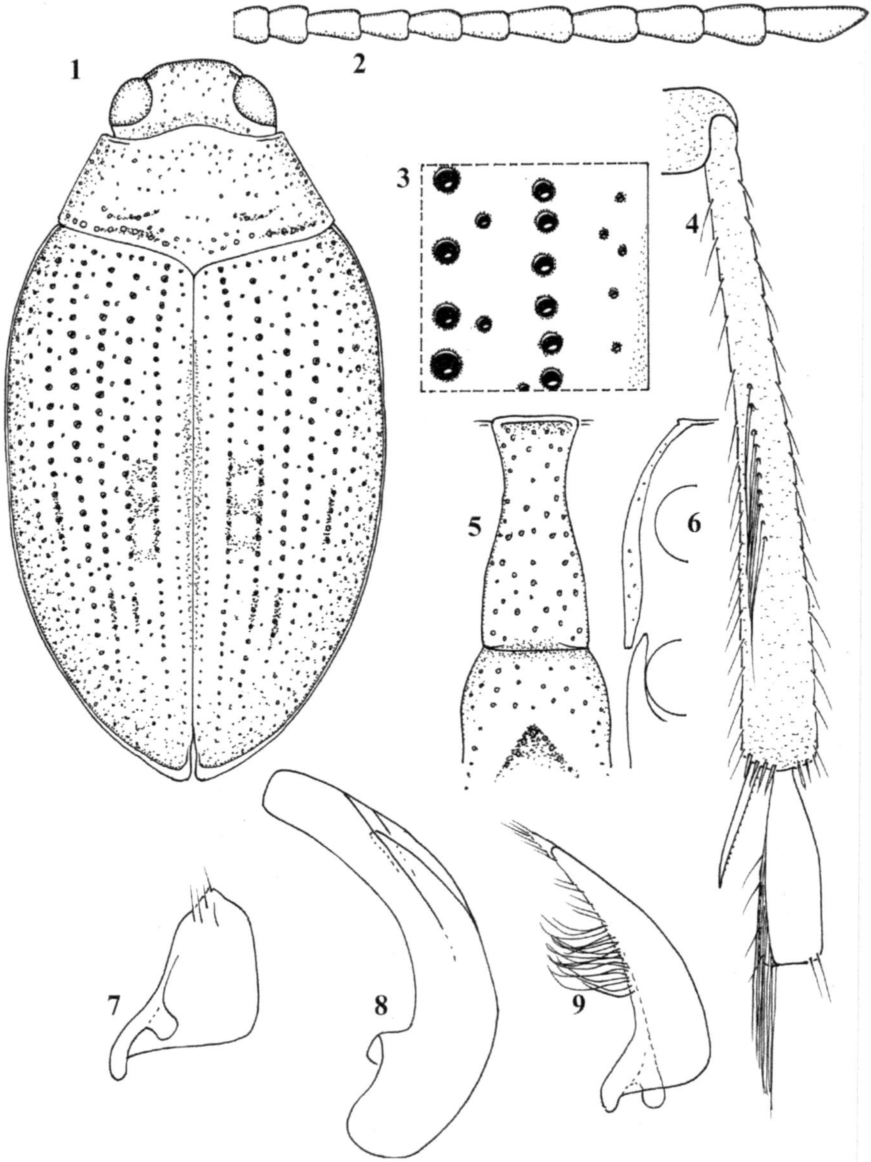

PLATE 64 *Haliplus* (*Liaphlus*) *columbiensis*. 1, dorsal view; 2, antenna; 3, elytral punctures near base and suture; 4, hind tibia in dorsal view; 5, prosternal and metaventral process; 6, prosternal process in lateral view; 7, left paramere; 8, penis; 9, right paramere

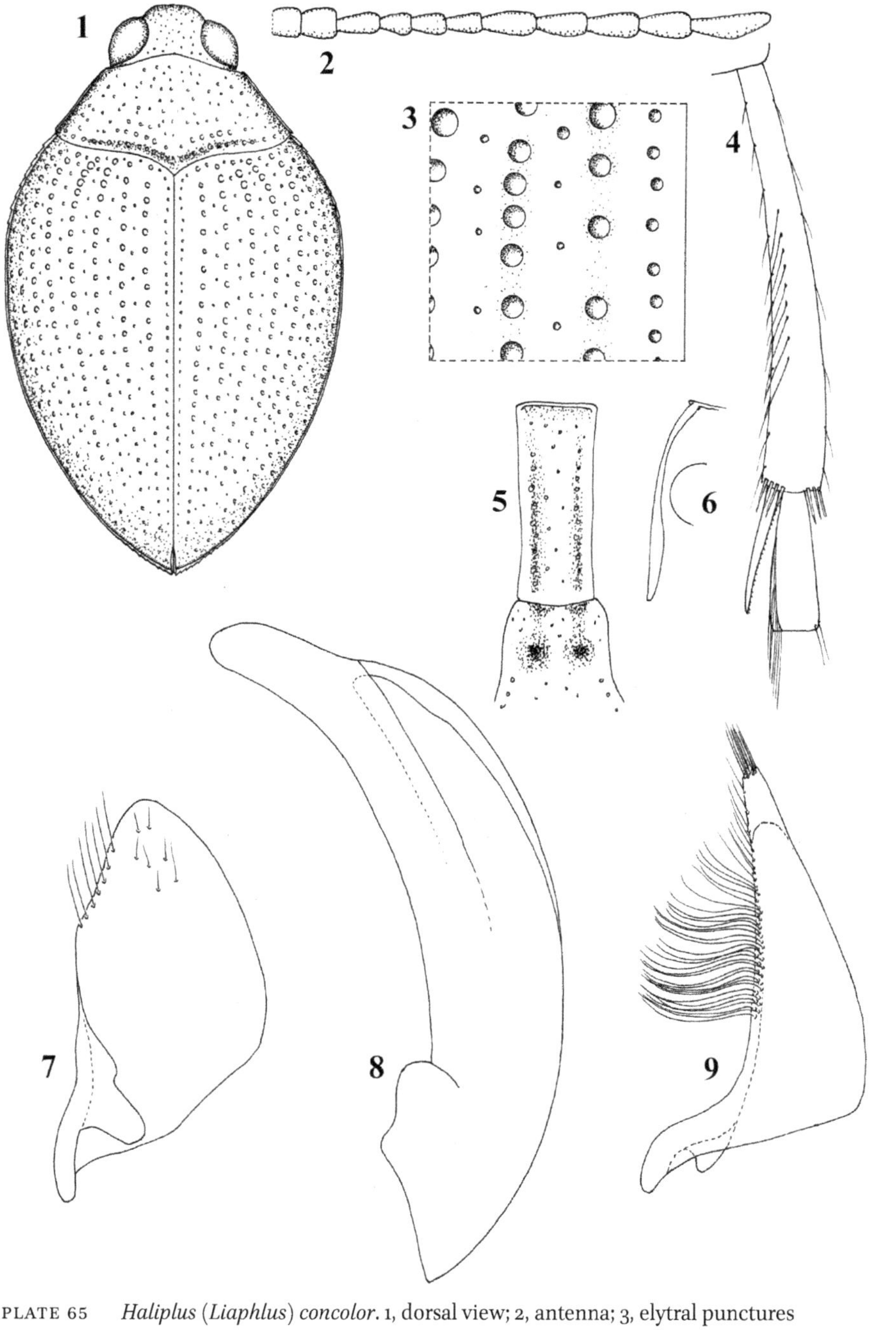

PLATE 65 *Haliplus* (*Liaphlus*) *concolor*. 1, dorsal view; 2, antenna; 3, elytral punctures near base and suture; 4, hind tibia in dorsal view; 5, prosternal and metaventral process; 6, prosternal process in lateral view; 7, left paramere; 8, penis; 9, right paramere

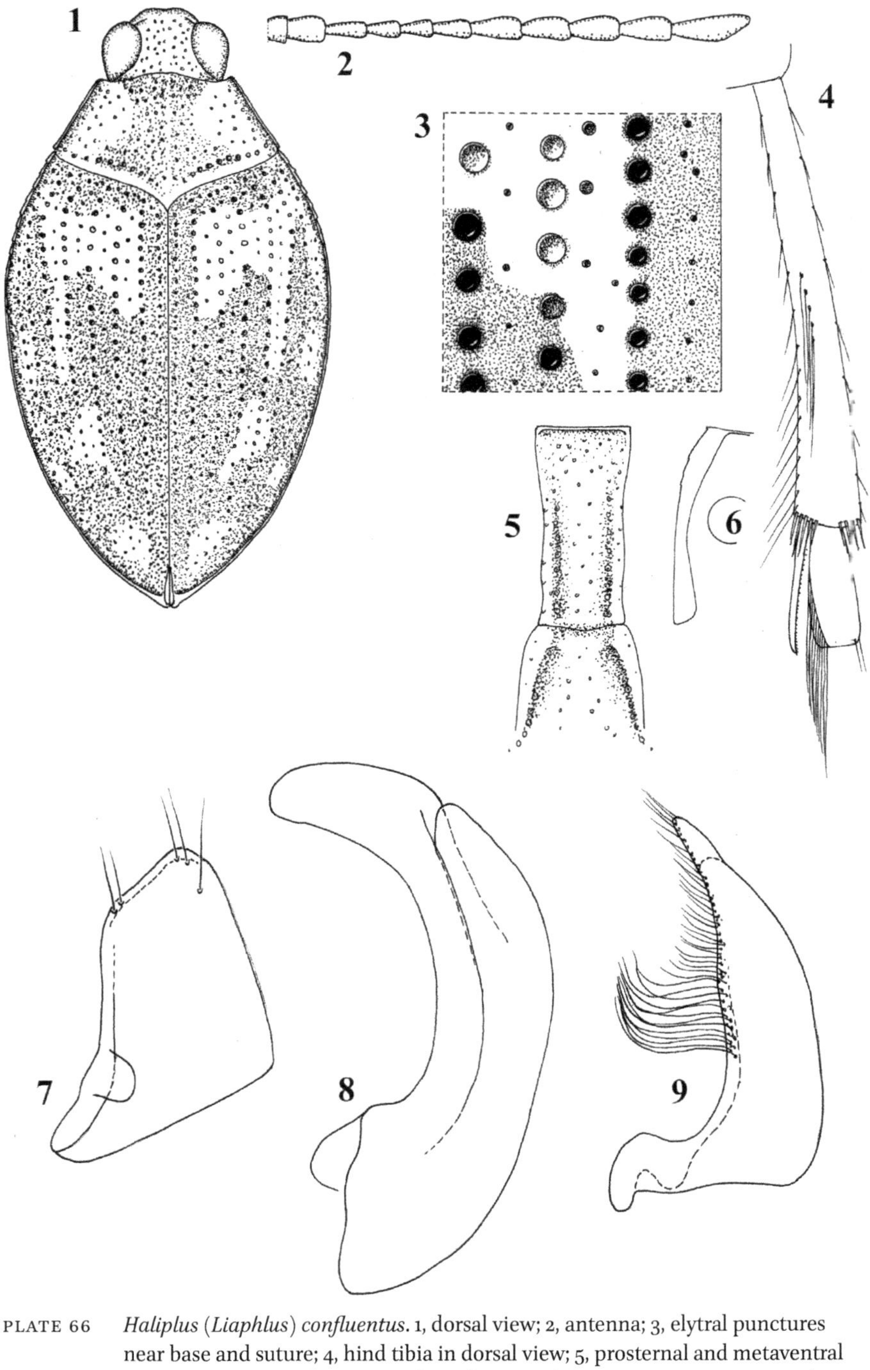

PLATE 66 *Haliplus* (*Liaphlus*) *confluentus*. 1, dorsal view; 2, antenna; 3, elytral punctures near base and suture; 4, hind tibia in dorsal view; 5, prosternal and metaventral process; 6, prosternal process in lateral view; 7, left paramere; 8, penis; 9, right paramere

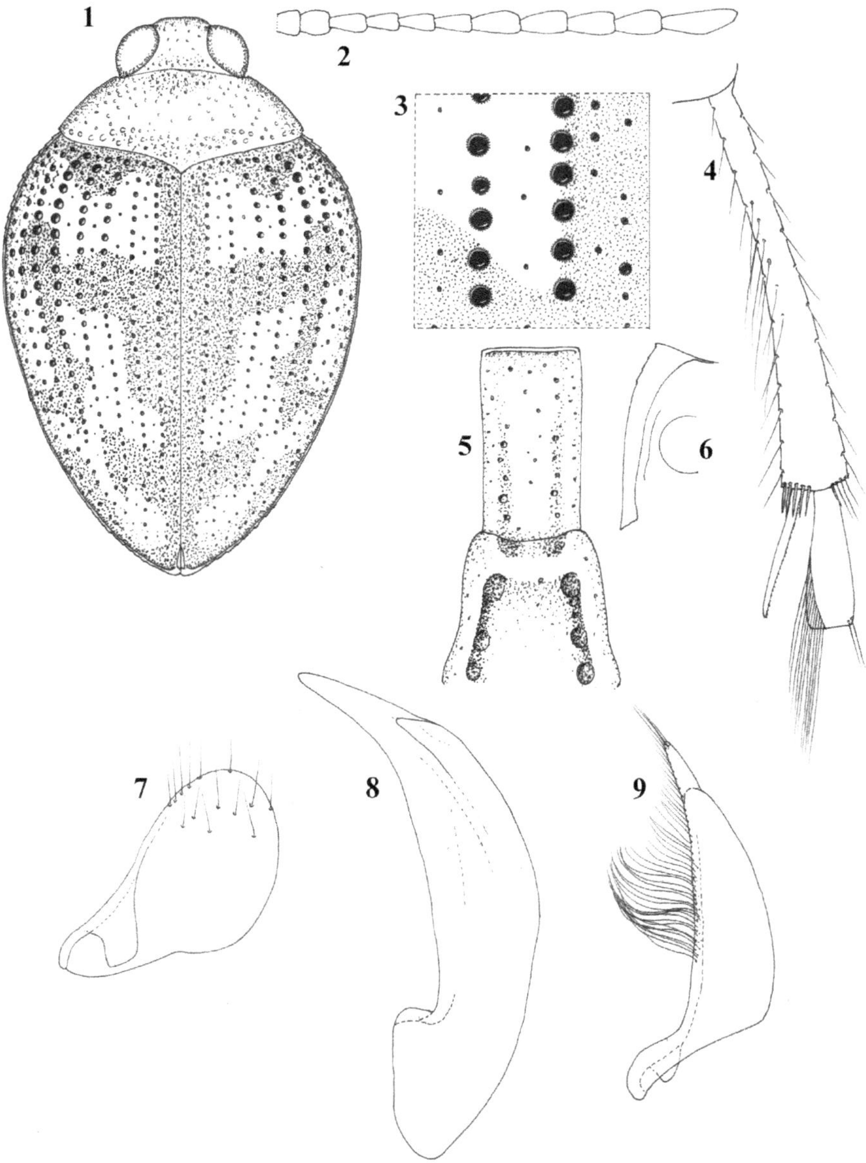

PLATE 67 *Haliplus* (*Liaphlus*) *costaricanus*. 1, dorsal view; 2, antenna; 3, elytral punctures near base and suture; 4, hind tibia in dorsal view; 5, prosternal and metaventral process; 6, prosternal process in lateral view; 7, left paramere; 8, penis; 9, right paramere

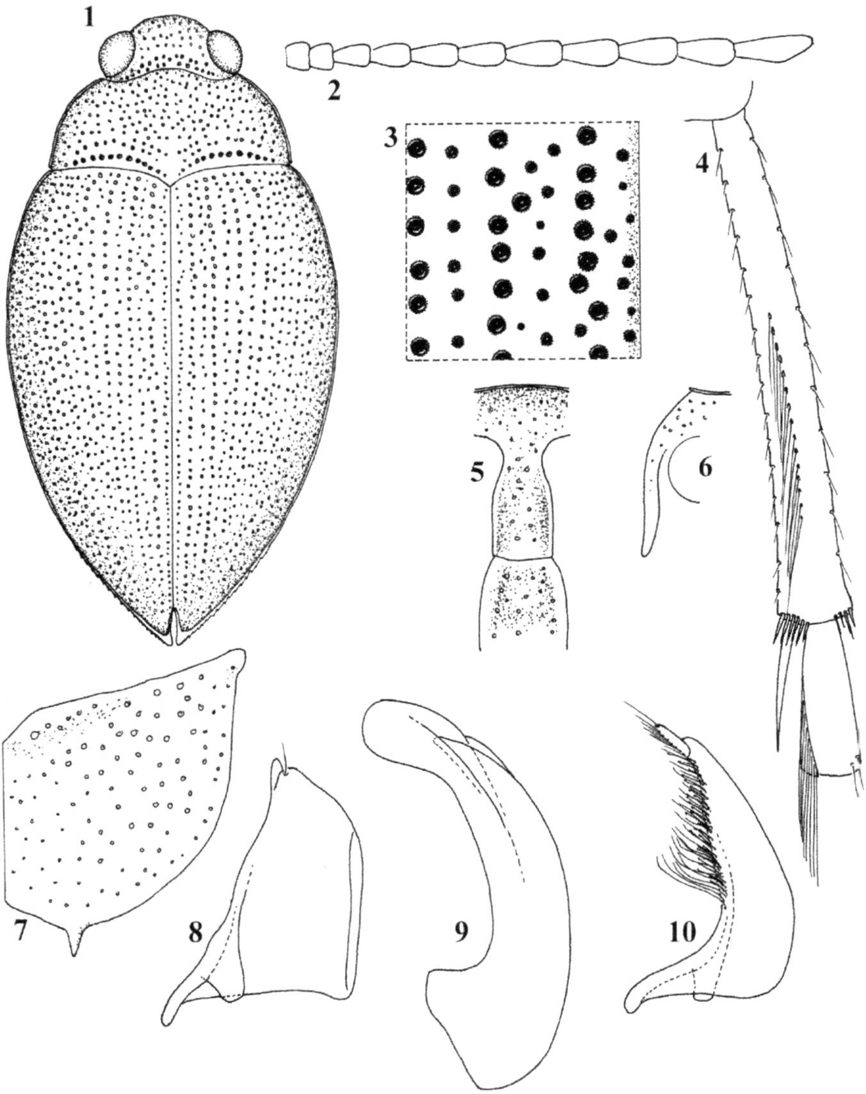

PLATE 68 *Haliplus* (*Liaphlus*) *coxaspinipes*. 1, dorsal view; 2, antenna; 3, elytral punctures near base and suture; 4, hind tibia in dorsal view; 5, prosternal and metaventral process; 6, prosternal process in lateral view; 7, hind coxal plate; 8, left paramere; 9, penis; 10, right paramere

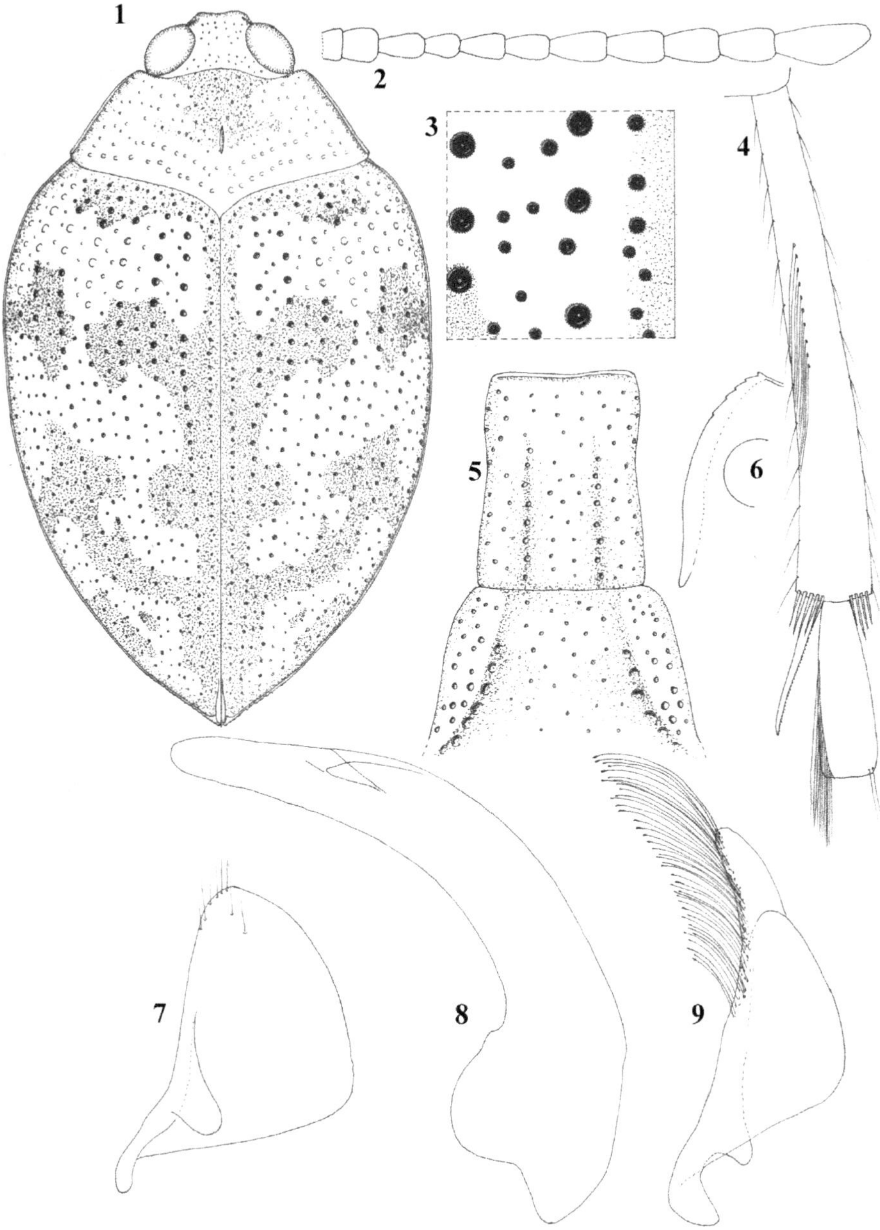

PLATE 69 *Haliplus* (*Liaphlus*) *crassus*. 1, dorsal view; 2, antenna; 3, elytral punctures near base and suture; 4, hind tibia in dorsal view; 5, prosternal and metaventral process; 6, prosternal process in lateral view; 7, left paramere; 8, penis; 9, right paramere

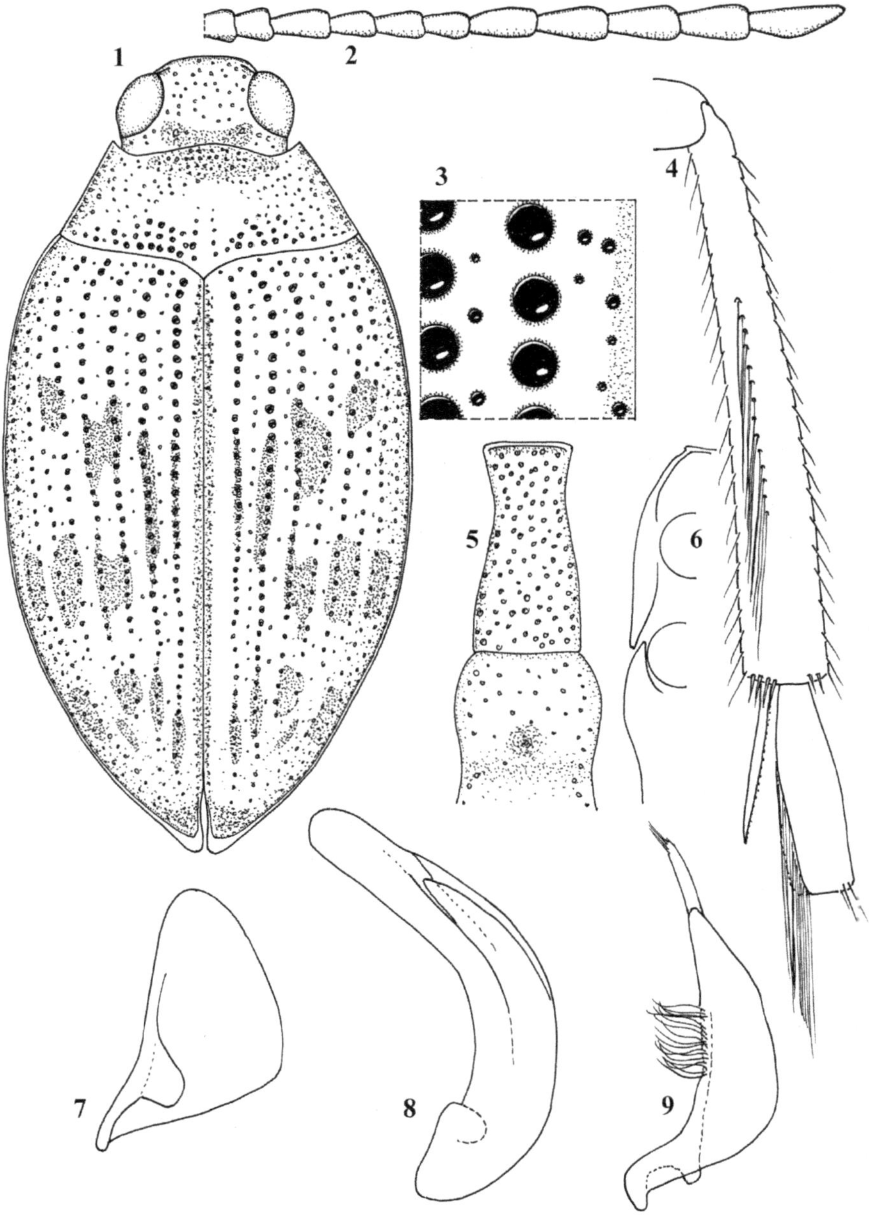

PLATE 70 *Haliplus* (*Liaphlus*) *cribrarius*. 1, dorsal view; 2, antenna; 3, elytral punctures near base and suture; 4, hind tibia in dorsal view; 5, prosternal and metaventral process; 6, prosternal process in lateral view; 7, left paramere; 8, penis; 9, right paramere

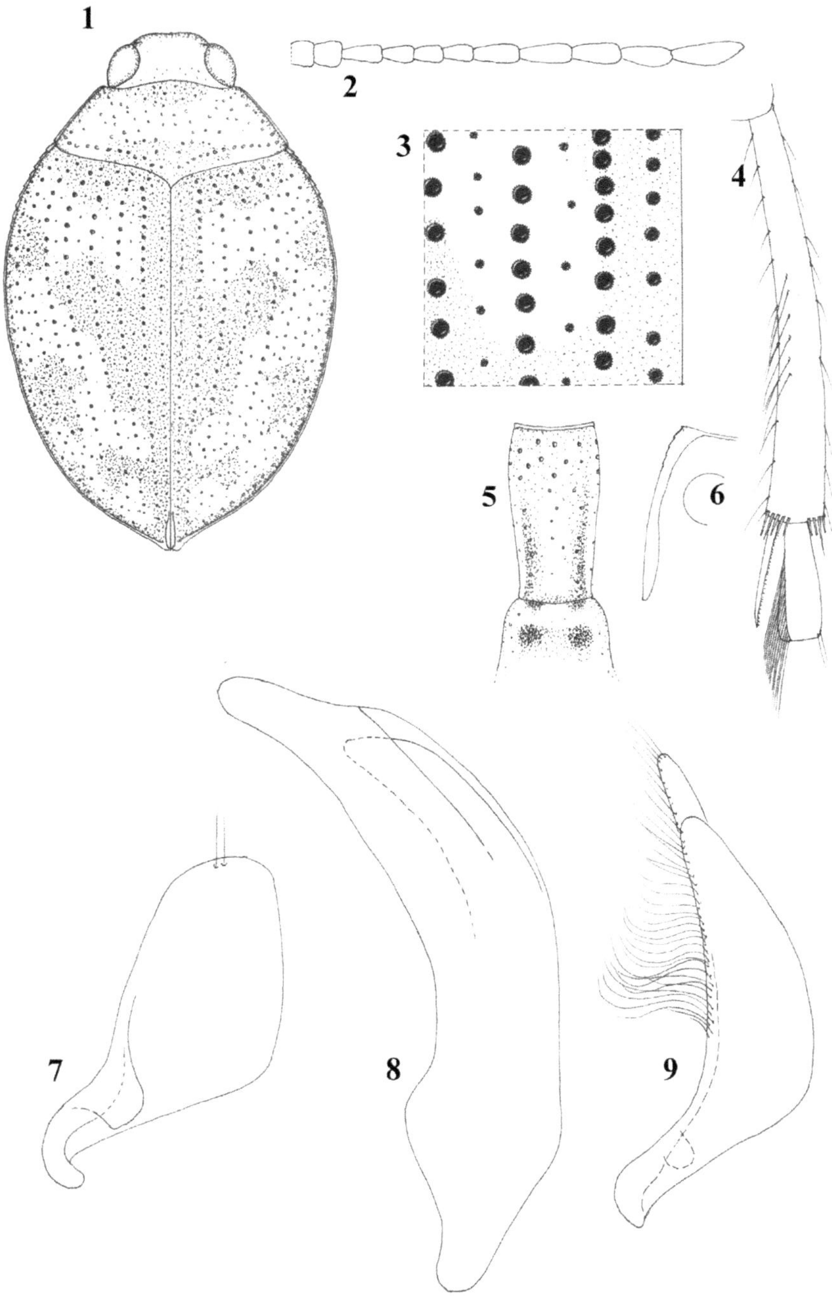

PLATE 71 *Haliplus* (*Liaphlus*) *cubensis*. 1, dorsal view; 2, antenna; 3, elytral punctures near base and suture; 4, hind tibia in dorsal view; 5, prosternal and metaventral process; 6, prosternal process in lateral view; 7, left paramere; 8, penis; 9, right paramere

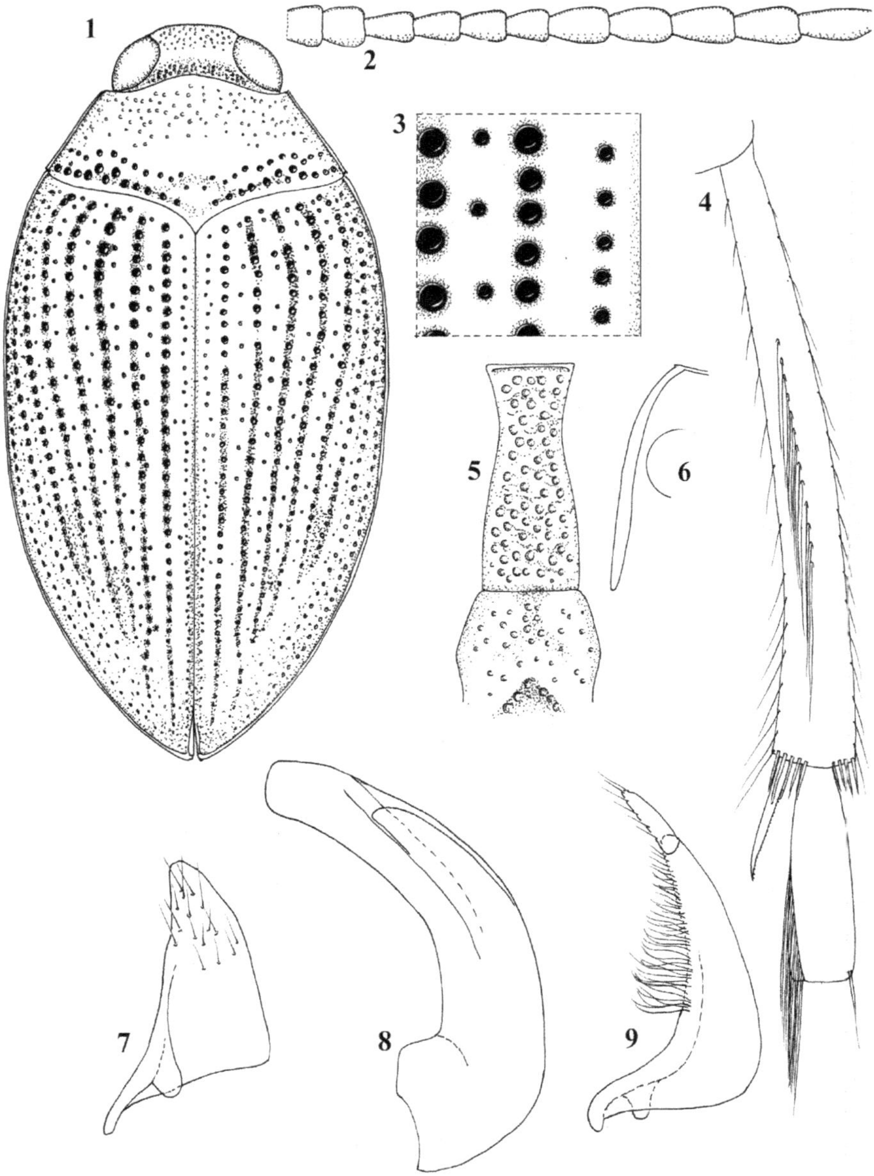

PLATE 72 *Haliplus* (*Liaphlus*) *cylindricus*. 1, dorsal view; 2, antenna; 3, elytral punctures near base and suture; 4, hind tibia in dorsal view; 5, prosternal and metaventral process; 6, prosternal process in lateral view; 7, left paramere; 8, penis; 9, right paramere

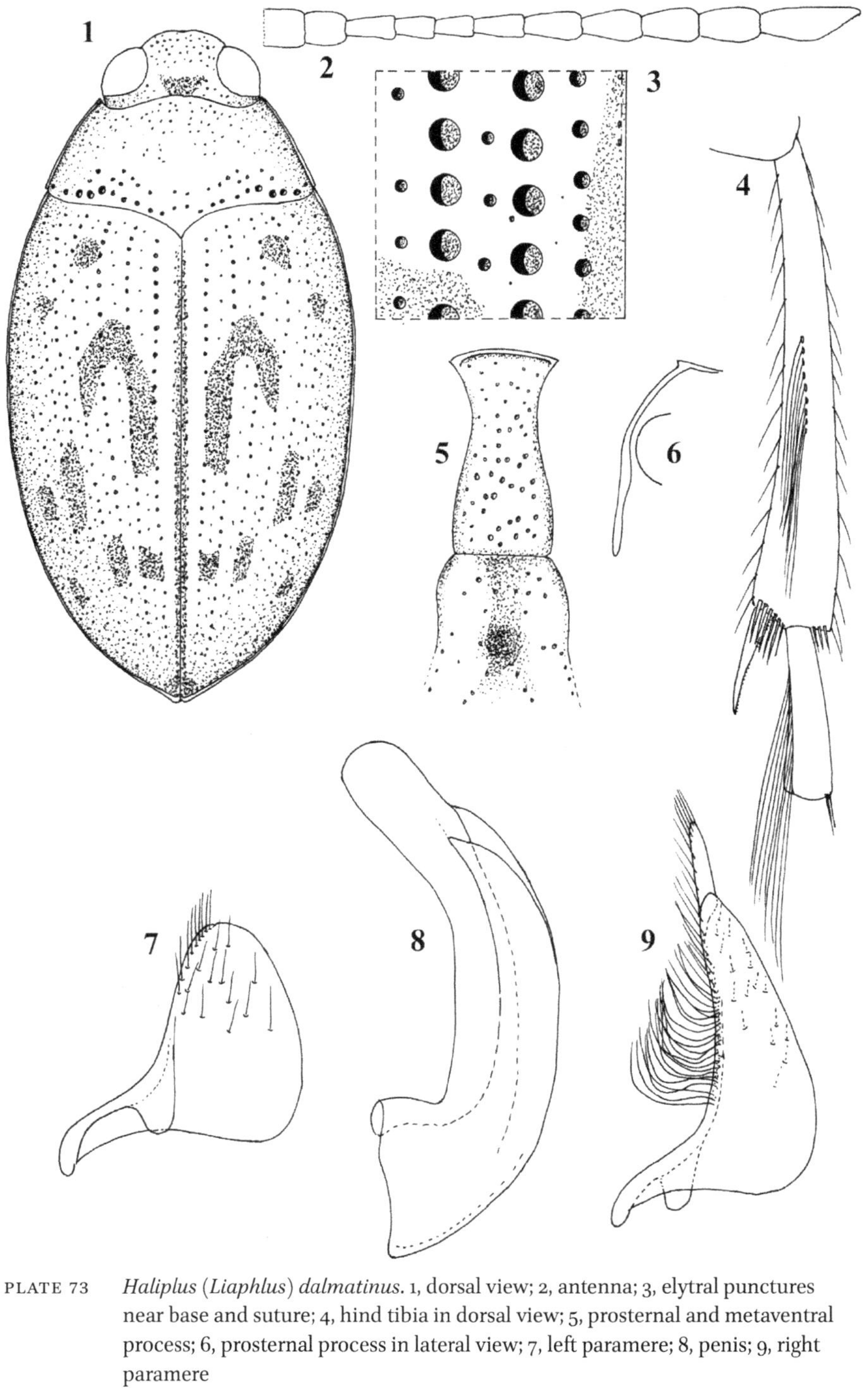

PLATE 73 *Haliplus* (*Liaphlus*) *dalmatinus*. 1, dorsal view; 2, antenna; 3, elytral punctures near base and suture; 4, hind tibia in dorsal view; 5, prosternal and metaventral process; 6, prosternal process in lateral view; 7, left paramere; 8, penis; 9, right paramere

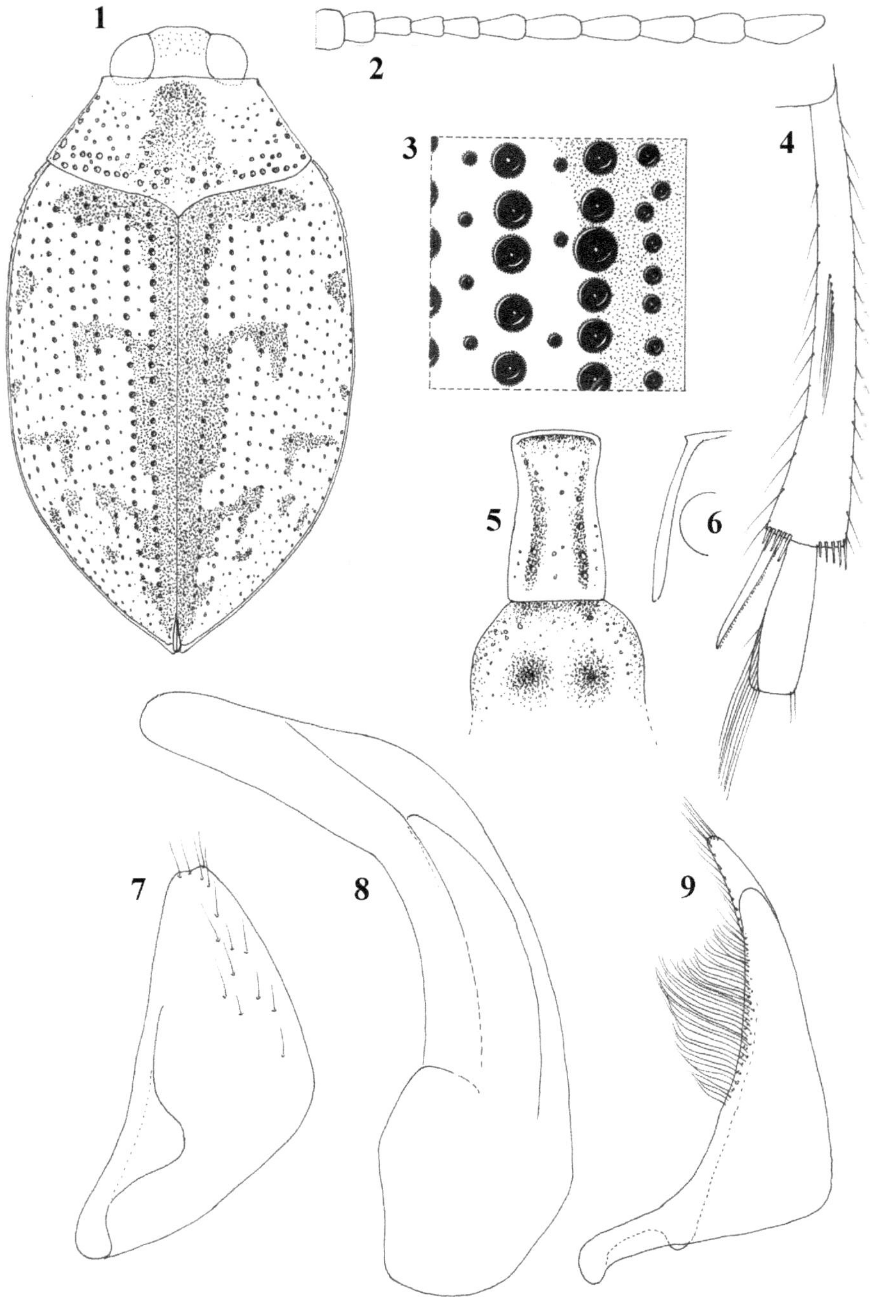

PLATE 74 *Haliplus* (*Liaphlus*) *diopus*. 1, dorsal view; 2, antenna; 3, elytral punctures near base and suture; 4, hind tibia in dorsal view; 5, prosternal and metaventral process; 6, prosternal process in lateral view; 7, left paramere; 8, penis; 9, right paramere

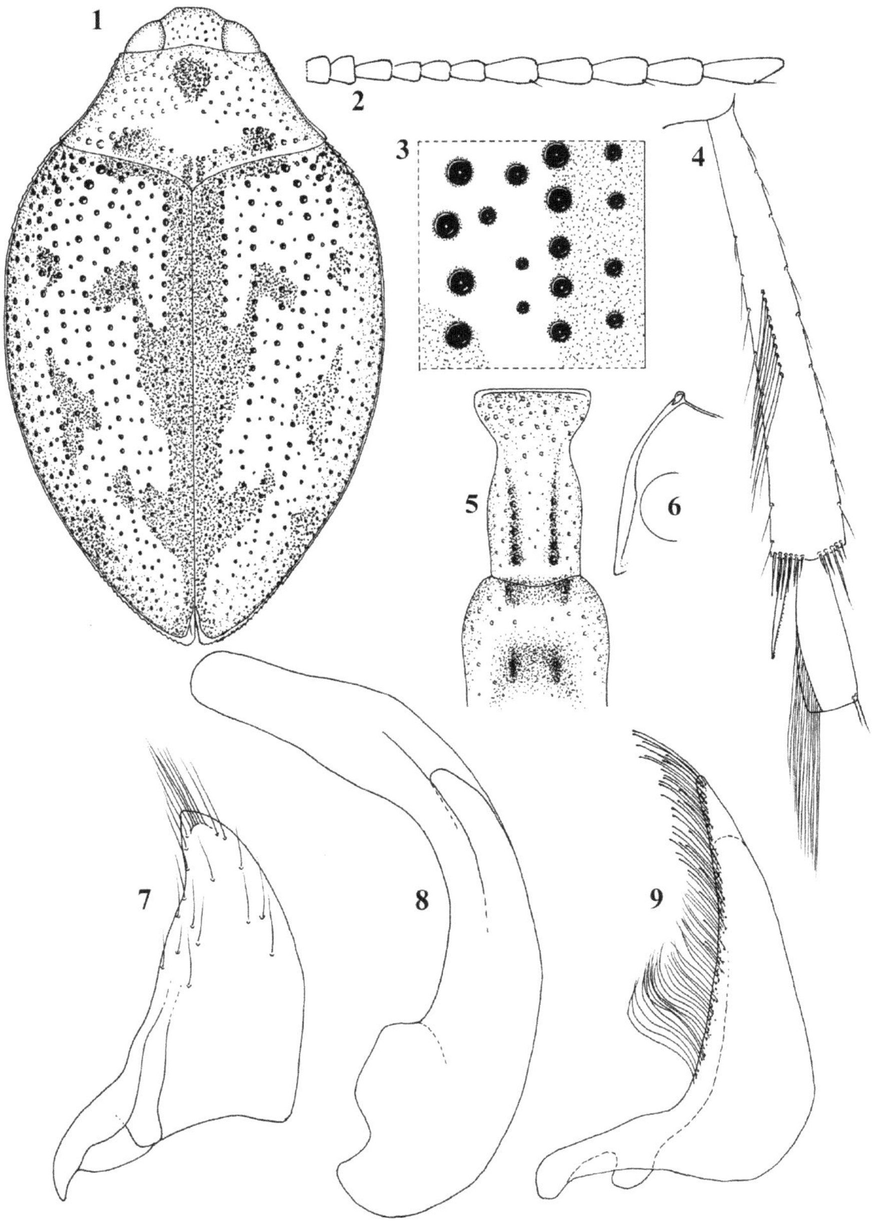

PLATE 75 *Haliplus* (*Liaphlus*) *discessus*. 1, dorsal view; 2, antenna; 3, elytral punctures near base and suture; 4, hind tibia in dorsal view; 5, prosternal and metaventral process; 6, prosternal process in lateral view; 7, left paramere; 8, penis; 9, right paramere

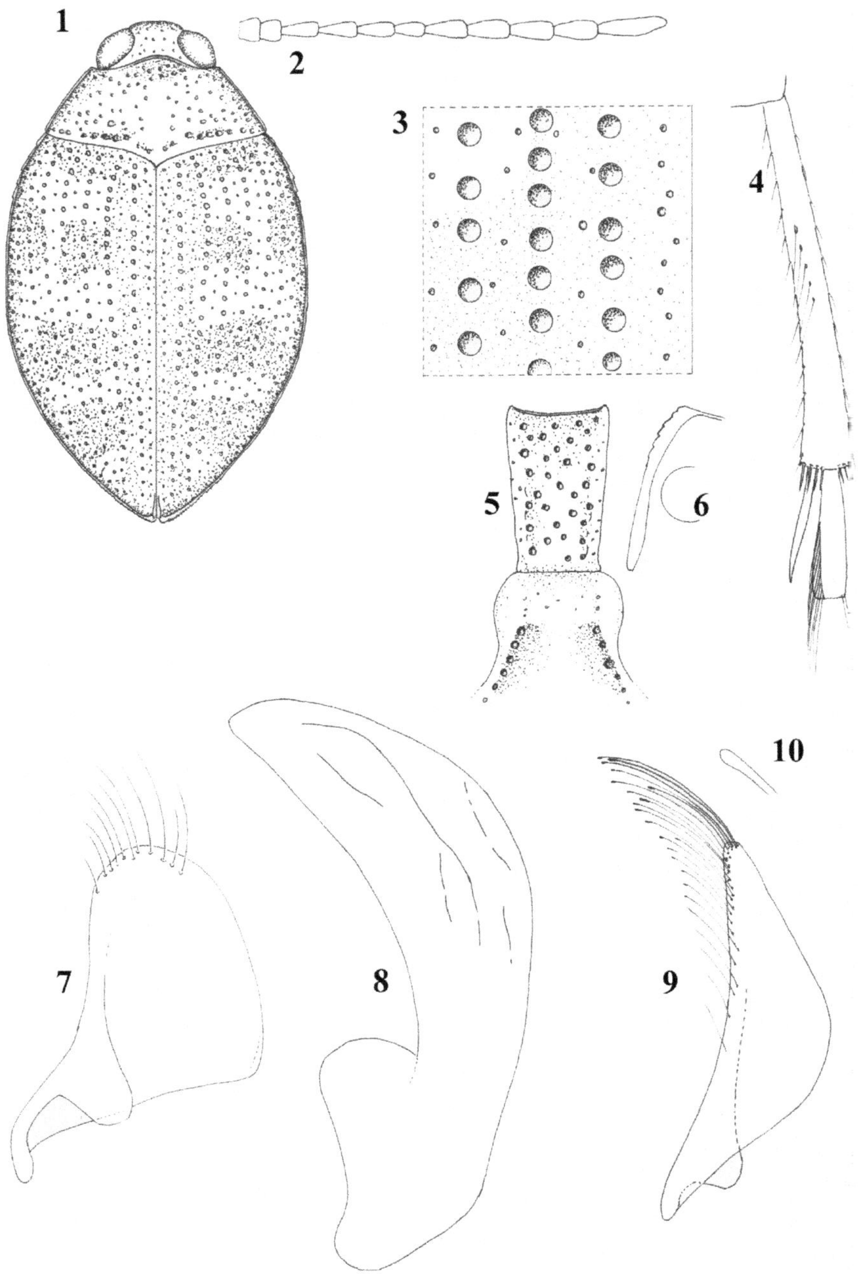

PLATE 76 *Haliplus* (*Liaphlus*) *drechseli*. 1, dorsal view; 2, antenna; 3, elytral punctures near base and suture; 4, hind tibia in dorsal view; 5, prosternal and metaventral process; 6, prosternal process in lateral view; 7, left paramere; 8, penis; 9, right paramere; 10, setae of right paramere

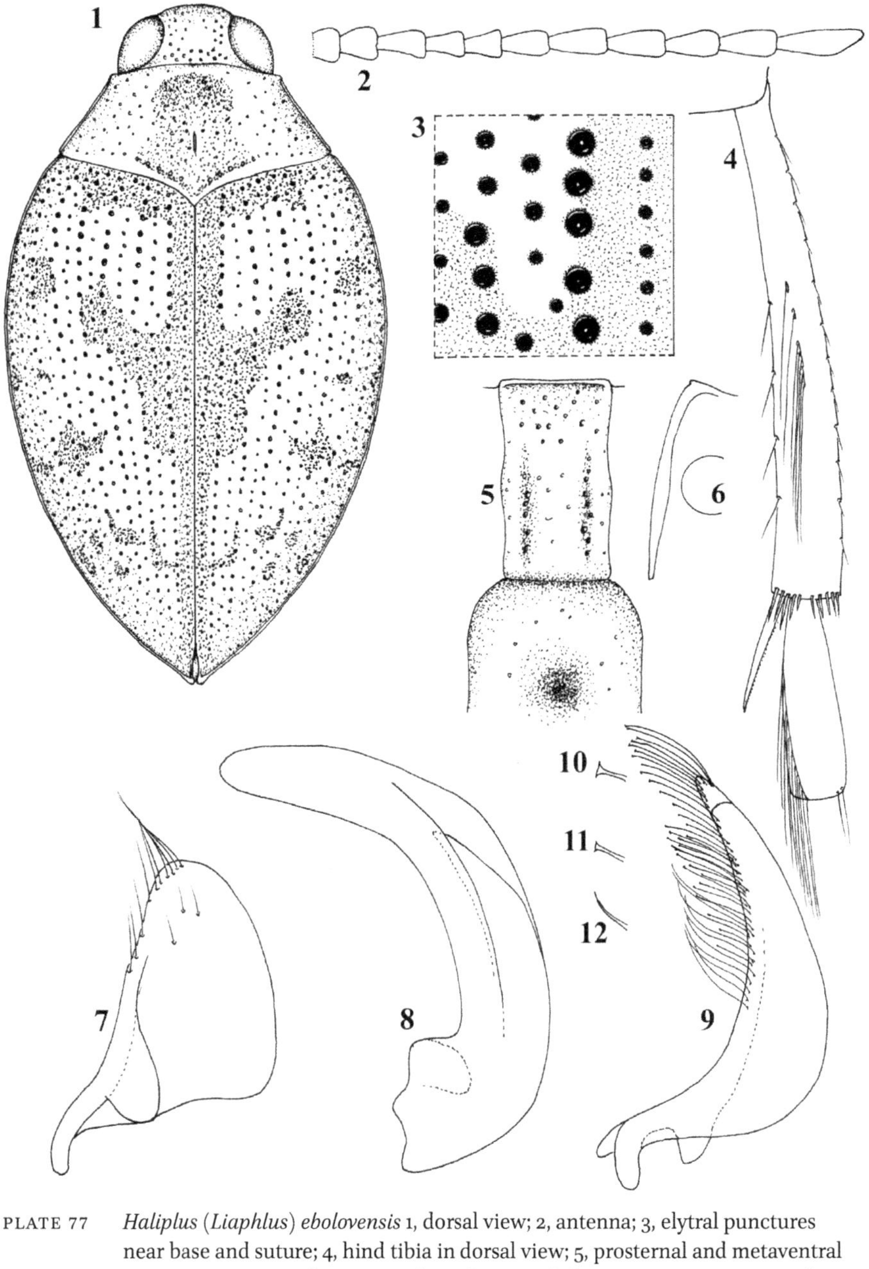

PLATE 77 *Haliplus* (*Liaphlus*) *ebolovensis* 1, dorsal view; 2, antenna; 3, elytral punctures near base and suture; 4, hind tibia in dorsal view; 5, prosternal and metaventral process; 6, prosternal process in lateral view; 7, left paramere; 8, penis; 9, right paramere; 10–12, setae of right paramere

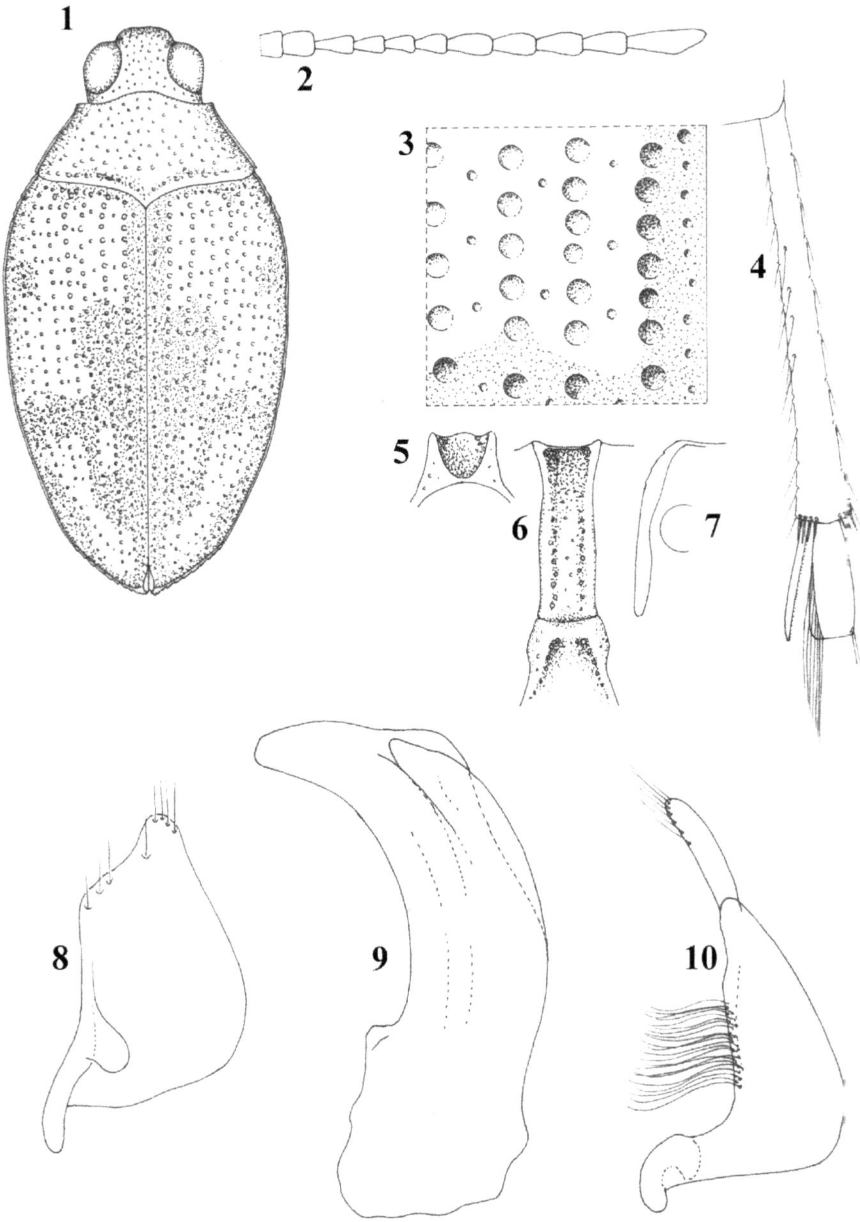

PLATE 78 *Haliplus* (*Liaphlus*) *elsaltous*. 1, dorsal view; 2, antenna; 3, elytral punctures near base and suture; 4, hind tibia in dorsal view; 5, prosternal process in frontal view; 6, prosternal and metaventral process; 7, prosternal process in lateral view; 8, left paramere; 9, penis; 10, right paramere

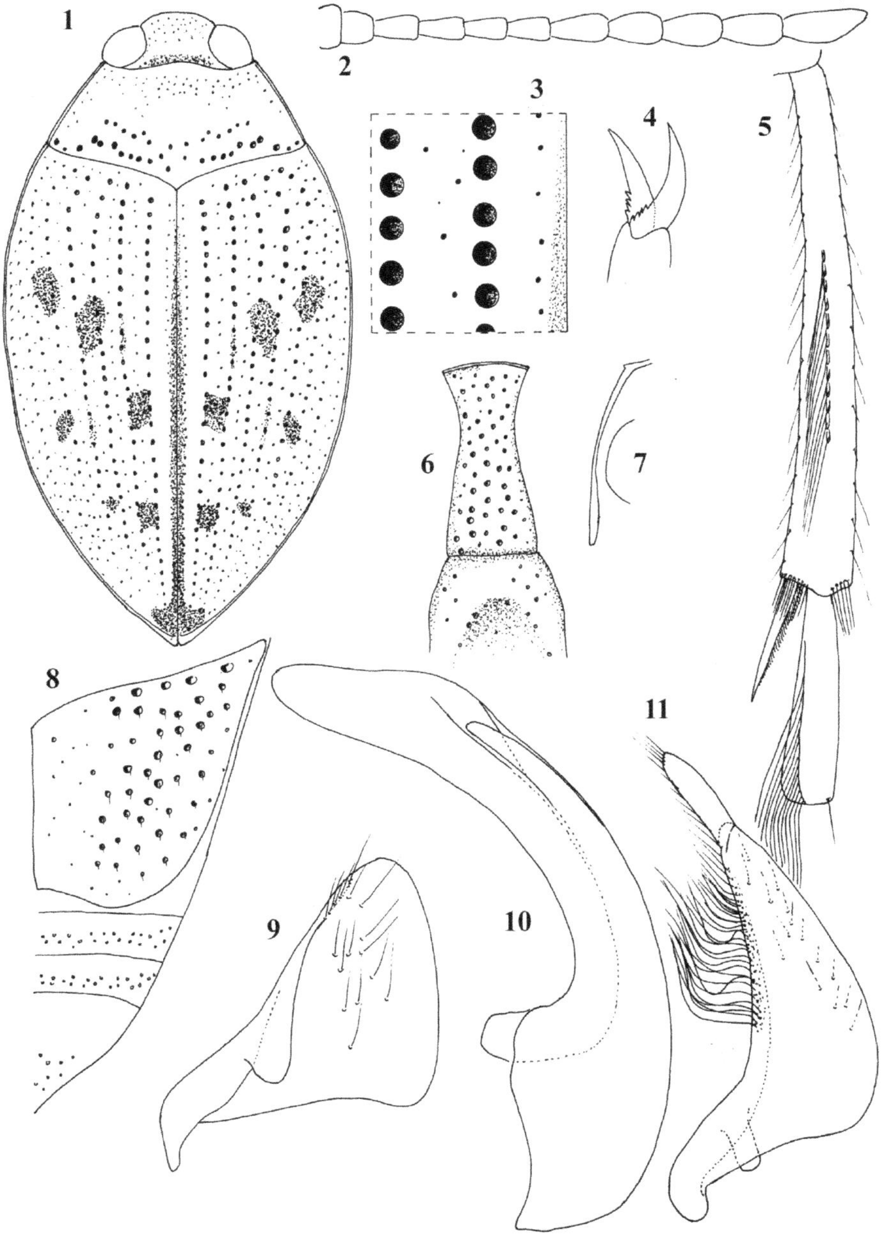

PLATE 79 *Haliplus* (*Liaphlus*) *excoffieri*. 1, dorsal view; 2, antenna; 3, elytral punctures near base and suture; 4, claws of foreleg; 5, hind tibia in dorsal view; 6, prosternal and metaventral process; 7, prosternal process in lateral view; 8, hind coxal plate and sternites; 9, left paramere; 10, penis; 11, right paramere

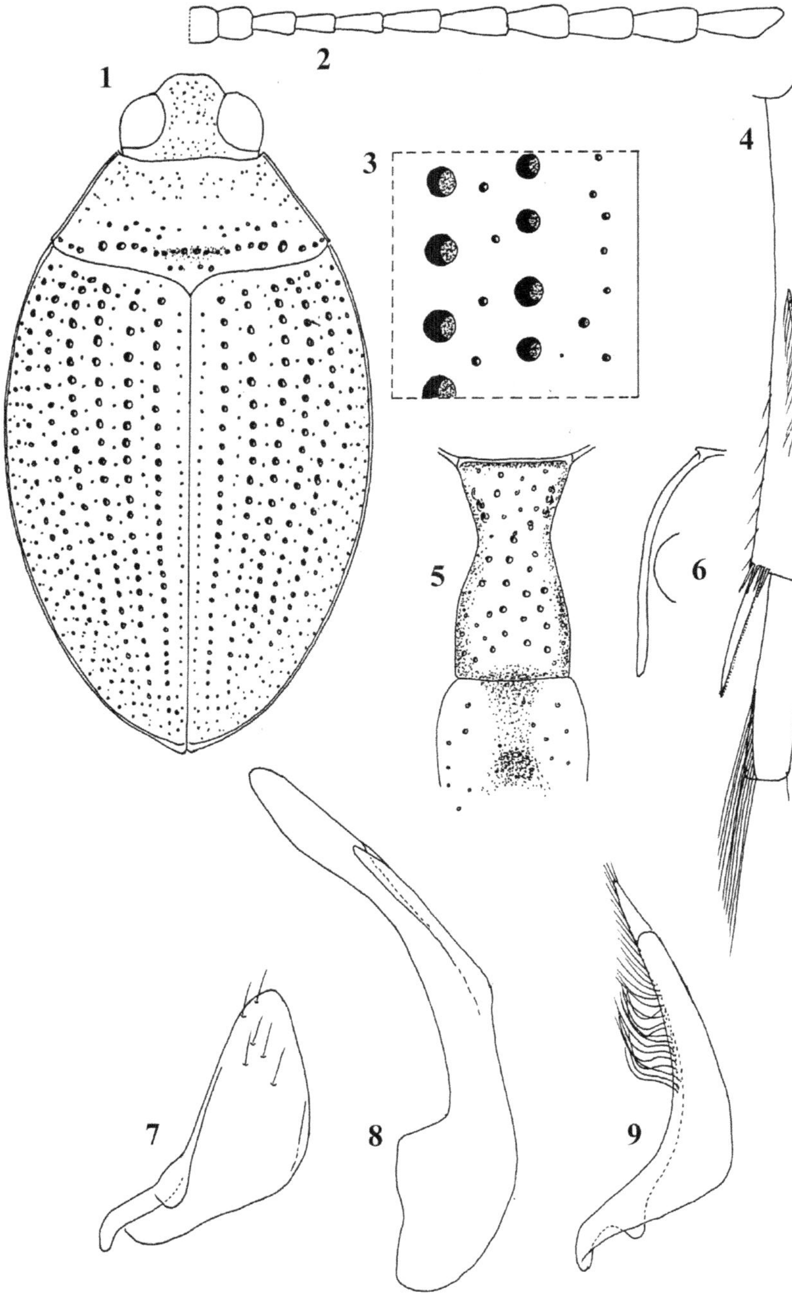

PLATE 80 *Haliplus* (*Liaphlus*) *eximius*. 1, dorsal view; 2, antenna; 3, elytral punctures near base and suture; 4, hind tibia in dorsal view; 5, prosternal and metaventral process; 6, prosternal process in lateral view; 7, left paramere; 8, penis; 9, right paramere

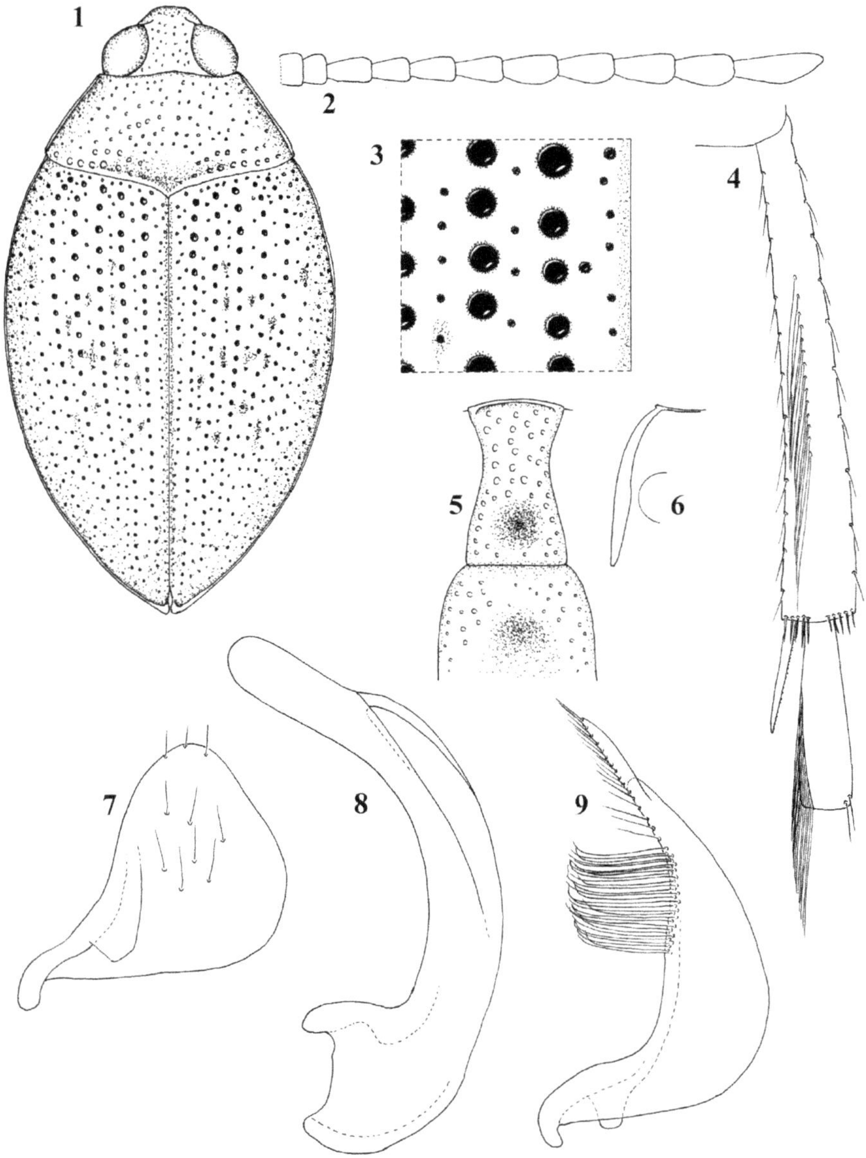

PLATE 81 *Haliplus* (*Liaphlus*) *exsecratus*. 1, dorsal view; 2, antenna; 3, elytral punctures near base and suture; 4, hind tibia in dorsal view; 5, prosternal and metaventral process; 6, prosternal process in lateral view; 7, left paramere; 8, penis; 9, right paramere

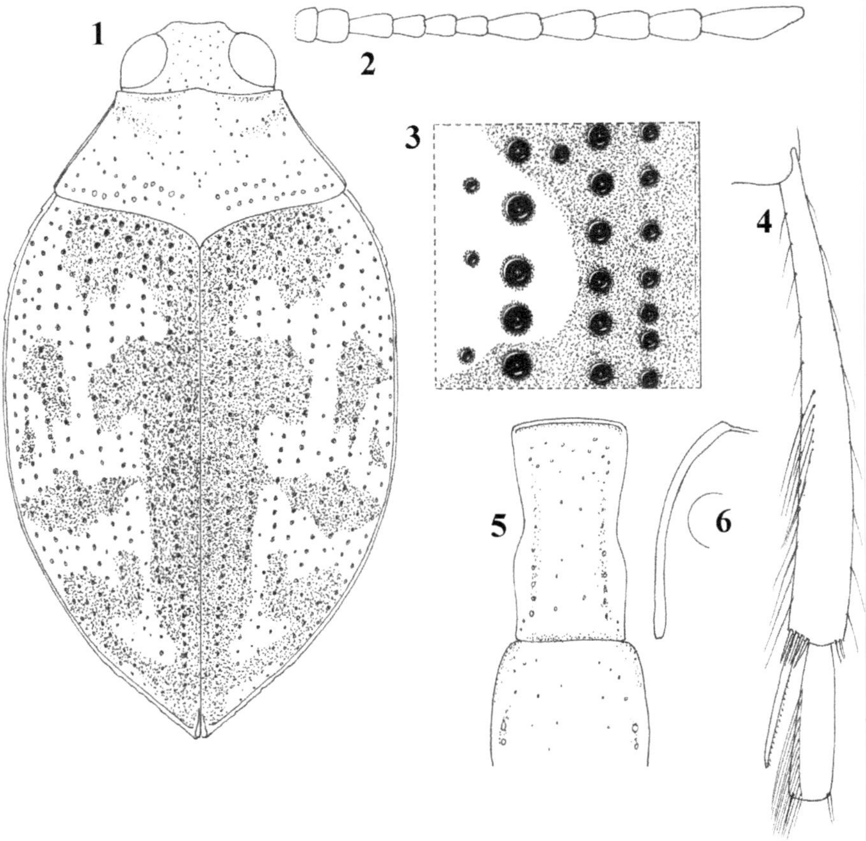

PLATE 82 *Haliplus* (*Liaphlus*) *ferruginipes* female. 1, dorsal view; 2, antenna; 3, elytral punctures near base and suture; 4, hind tibia in dorsal view; 5, prosternal and metaventral process; 6, prosternal process in lateral view

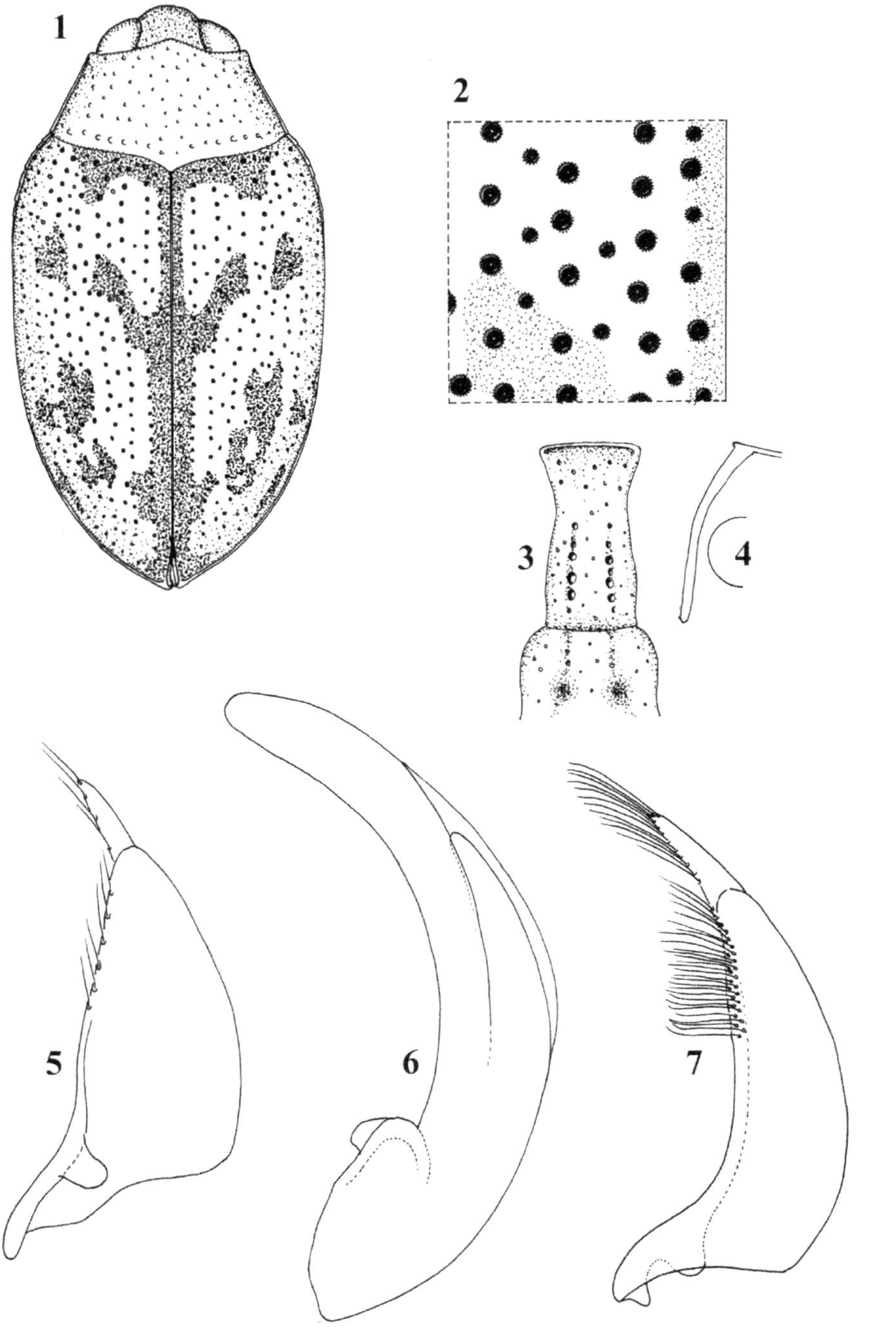

PLATE 83 *Haliplus* (*Liaphlus*) *figuratus*. 1, dorsal view; 2, elytral punctures near base and suture; 3, prosternal and metaventral process; 4, prosternal process in lateral view; 5, left paramere; 6, penis; 7, right paramere

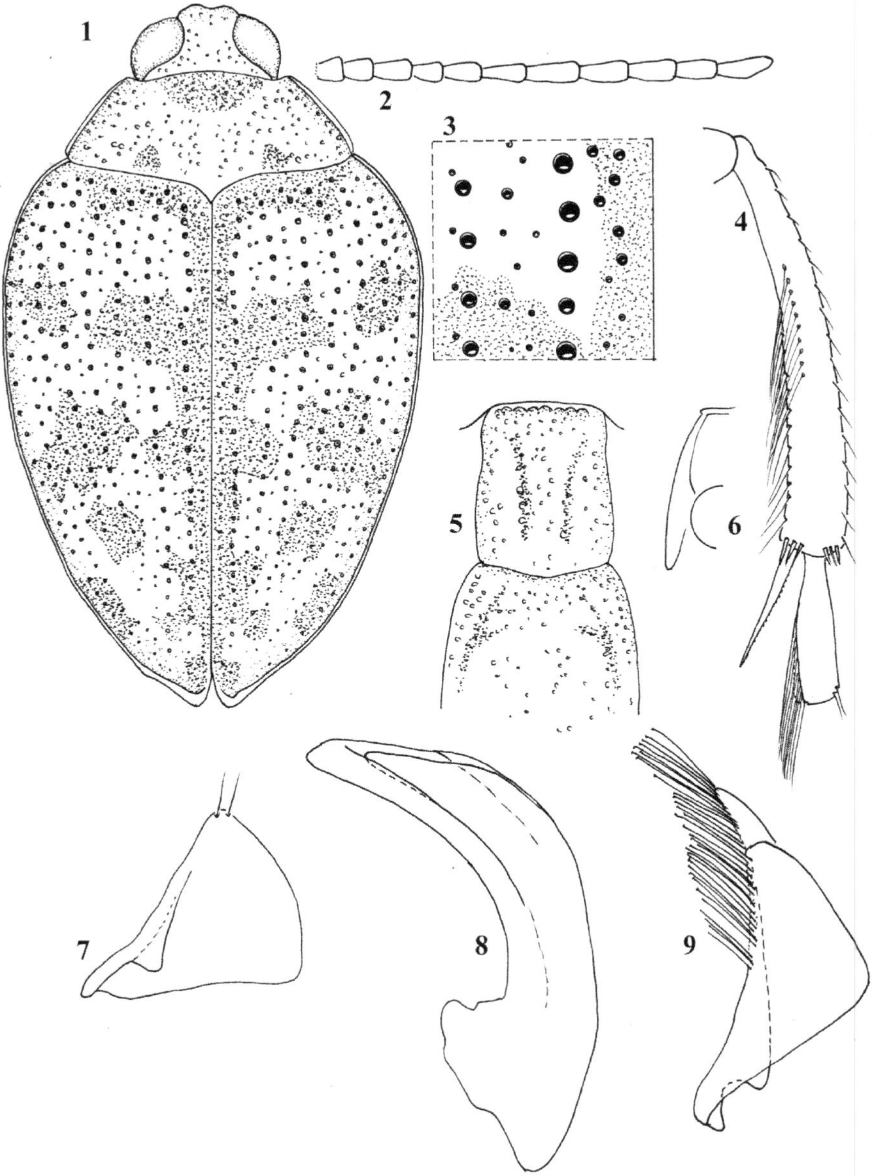

PLATE 84 *Haliplus* (*Liaphlus*) *fiorentini*. 1, dorsal view; 2, antenna; 3, elytral punctures near base and suture; 4, hind tibia in dorsal view; 5, prosternal and metaventral process; 6, prosternal process in lateral view; 7, left paramere; 8, penis; 9, right paramere

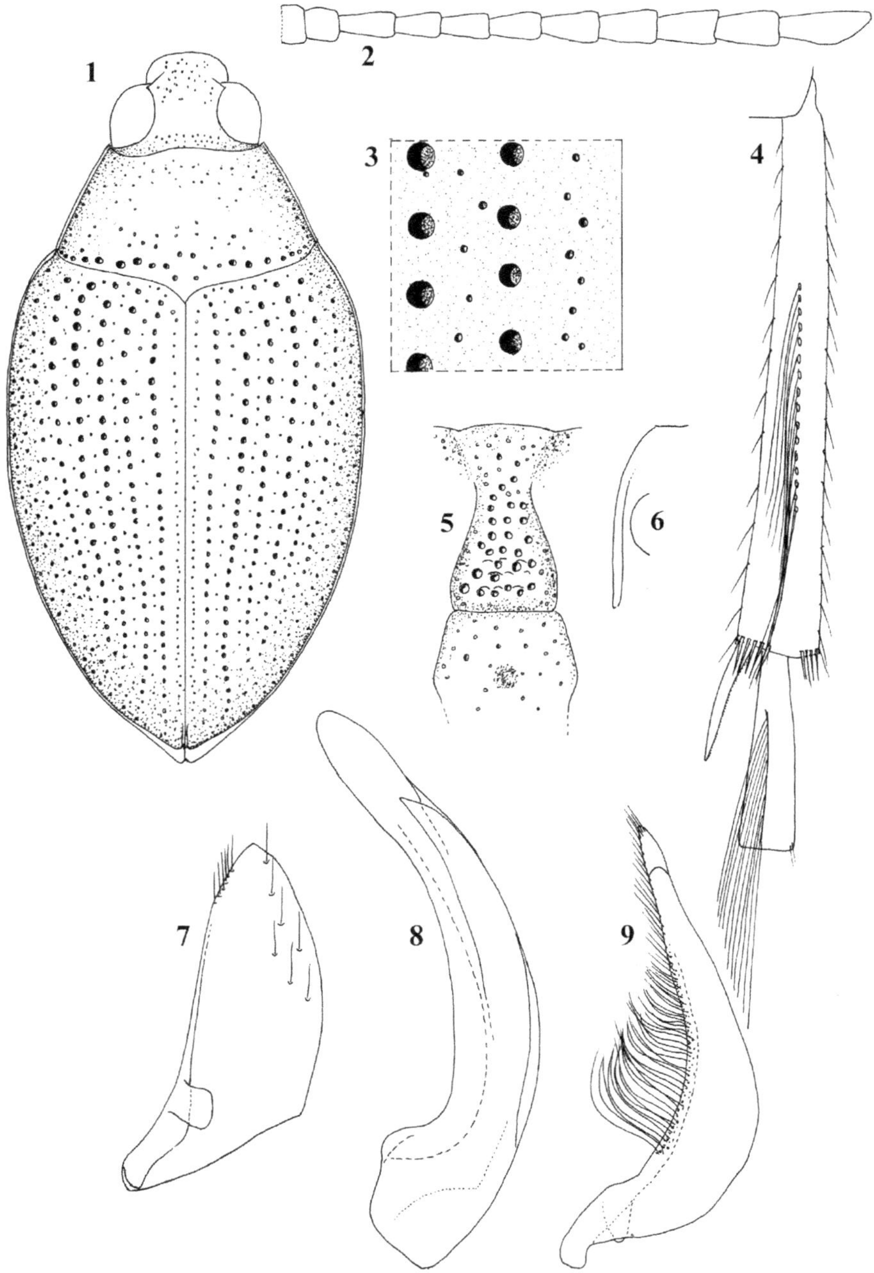

PLATE 85 *Haliplus* (*Liaphlus*) *flavicollis*. 1, dorsal view; 2, antenna; 3, elytral punctures near base and suture; 4, hind tibia in dorsal view; 5, prosternal and metaventral process; 6, prosternal process in lateral view; 7, left paramere; 8, penis; 9, right paramere

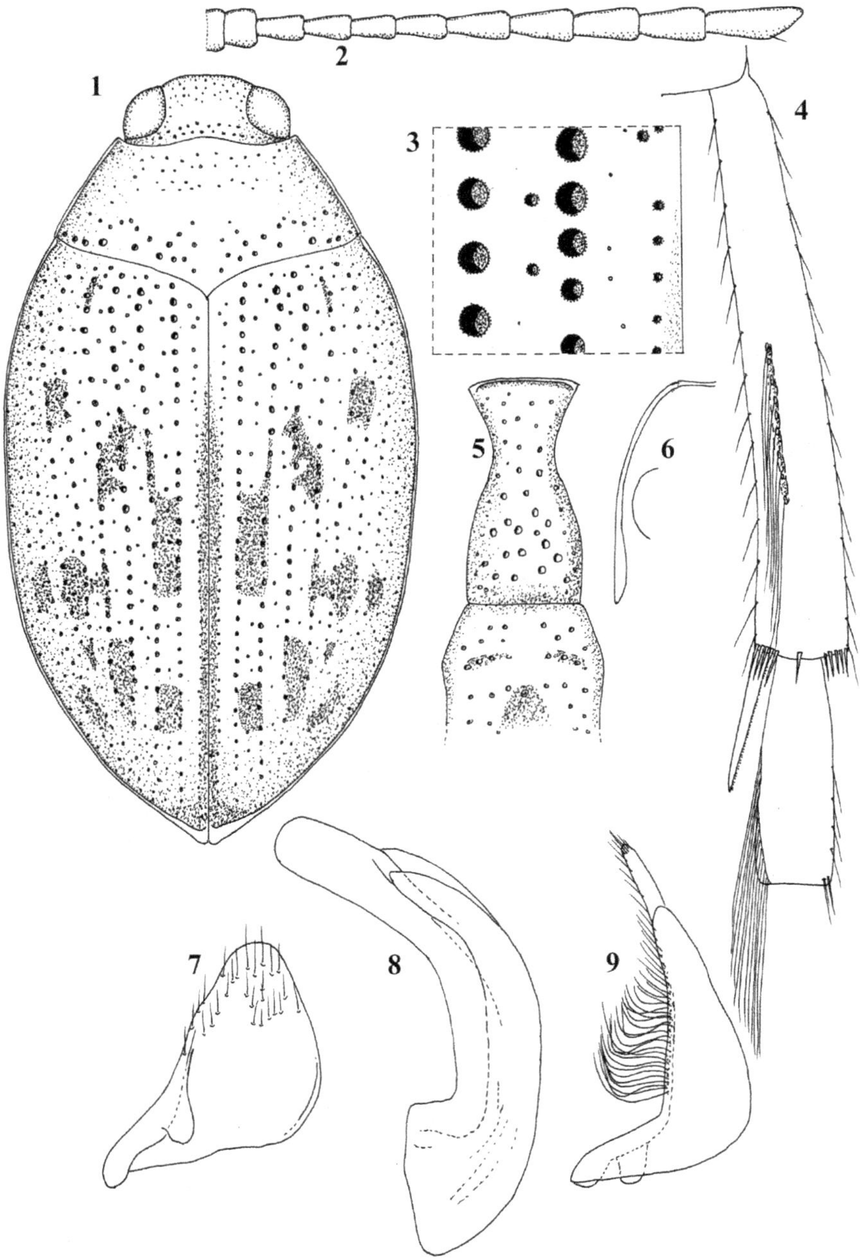

PLATE 86 *Haliplus* (*Liaphlus*) *fulvus*. 1, dorsal view; 2, antenna; 3, elytral punctures near base and suture; 4, hind tibia in dorsal view; 5, prosternal and metaventral process; 6, prosternal process in lateral view; 7, left paramere; 8, penis; 9, right paramere

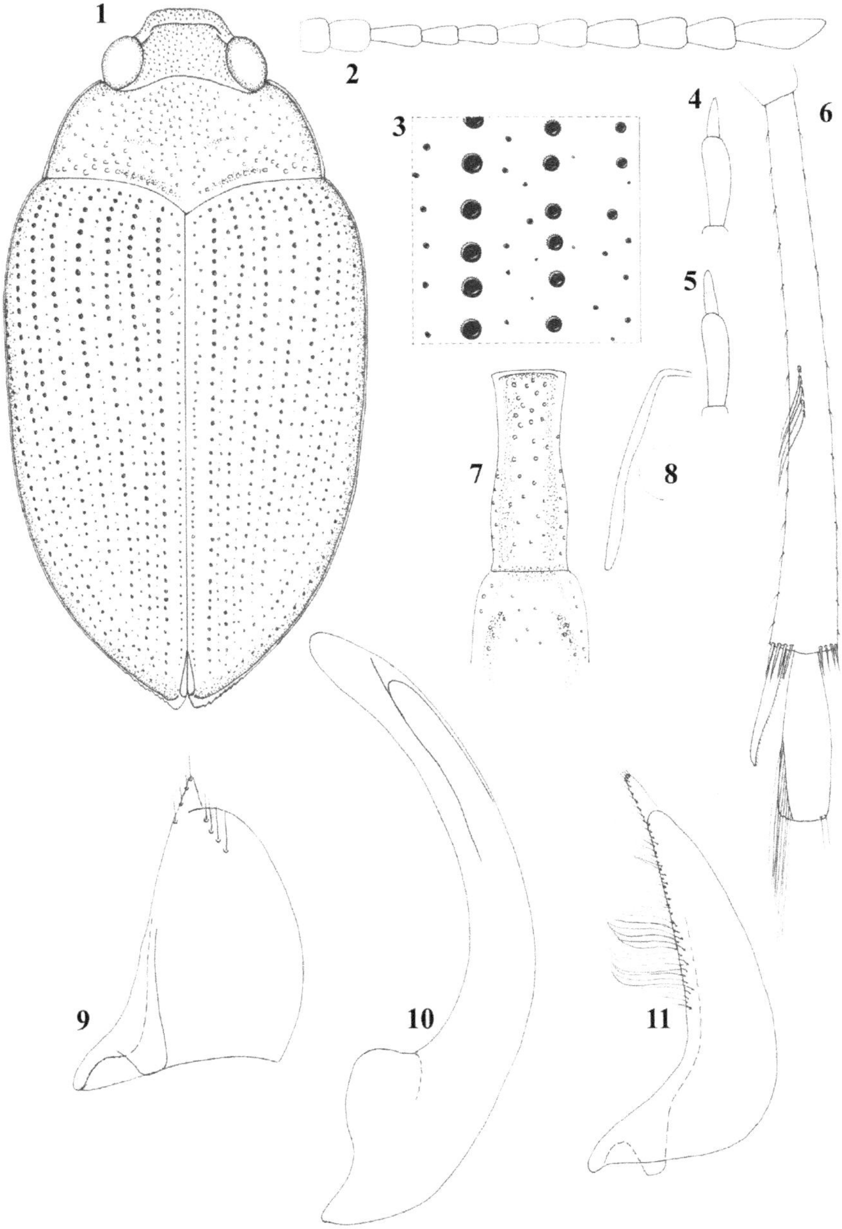

PLATE 87 *Haliplus* (*Liaphlus*) *fuscipennis*. 1, dorsal view; 2, antenna; 3, elytral punctures near base and suture; 4, labial palpus; 5, maxillary palpus; 6, hind tibia in dorsal view; 7, prosternal and metaventral process; 8, prosternal process in lateral view; 9, left paramere; 10, penis; 11, right paramere

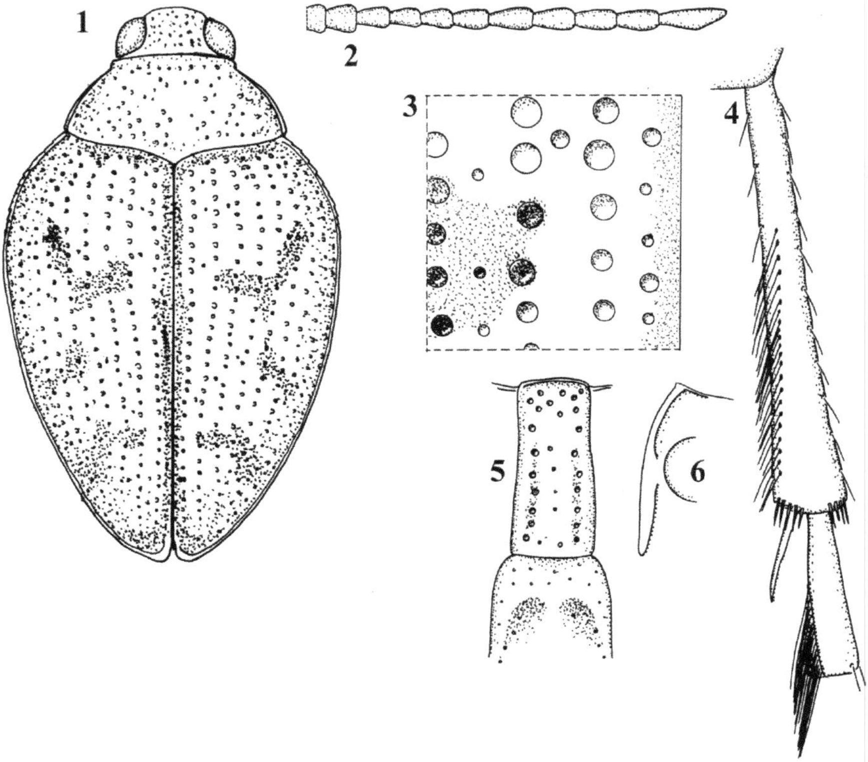

PLATE 88 *Haliplus* (*Liaphlus*) *gamma* female. 1, dorsal view; 2, antenna; 3, elytral punctures near base and suture; 4, hind tibia in dorsal view; 5, prosternal and metaventral process; 6, prosternal process in lateral view

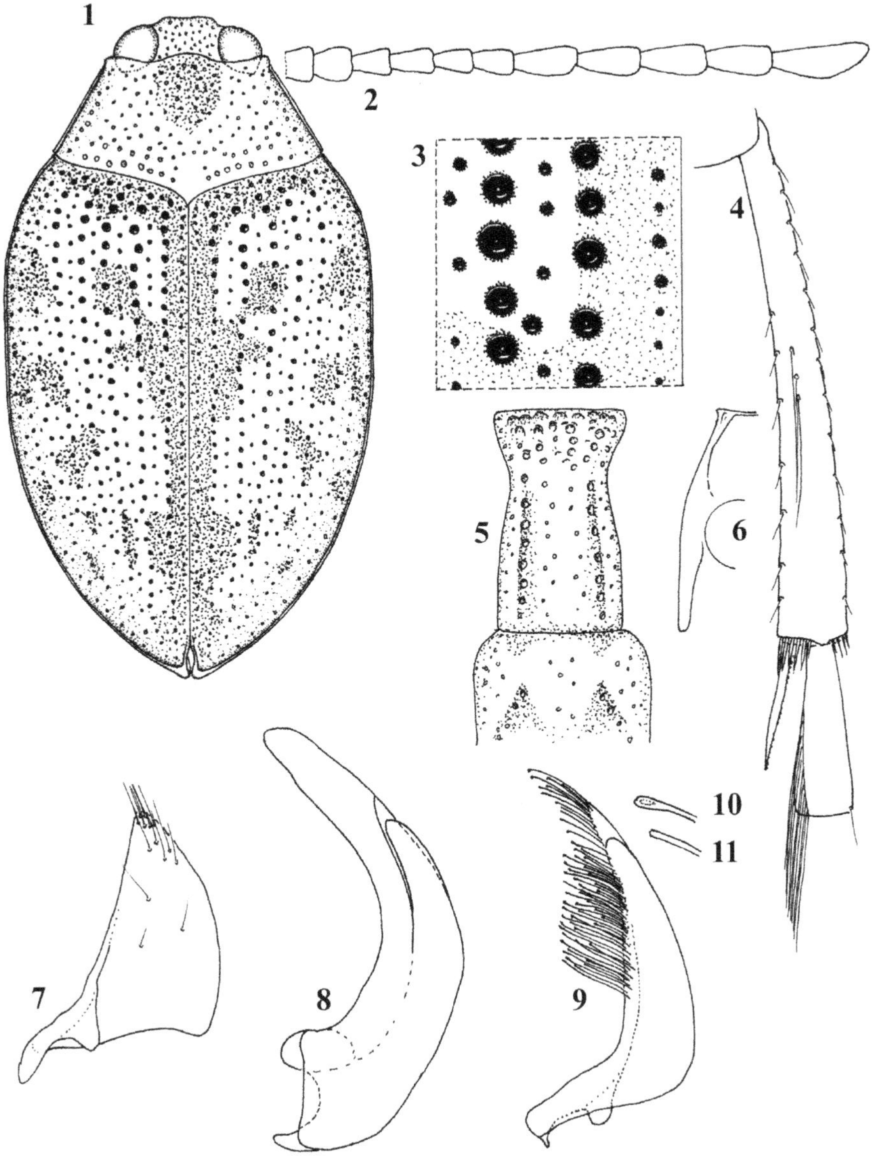

PLATE 89 *Haliplus* (*Liaphlus*) *garambanus*. 1, dorsal view; 2, antenna; 3, elytral punctures near base and suture; 4, hind tibia in dorsal view; 5, prosternal and metaventral process; 6, prosternal process in lateral view; 7, left paramere; 8, penis; 9, right paramere; 10, 11, setae of right paramere

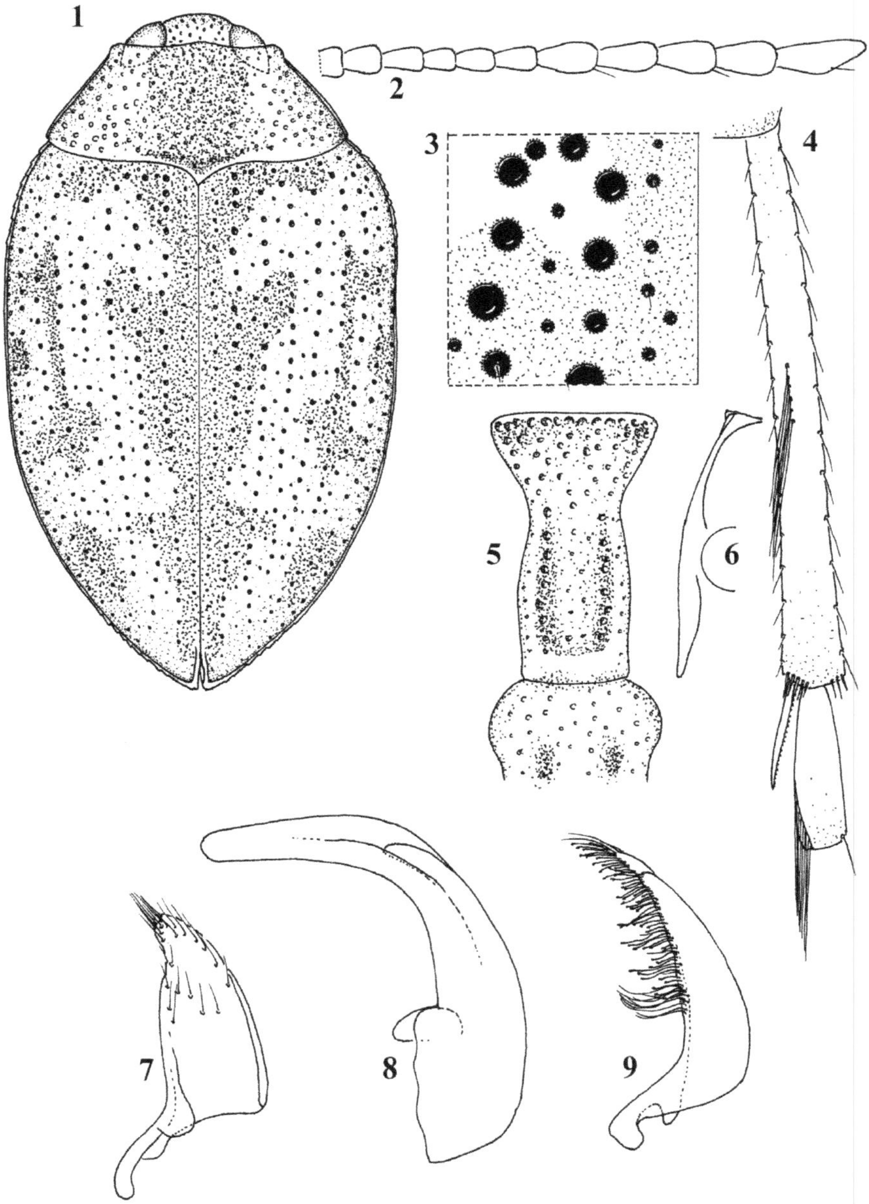

PLATE 90 *Haliplus* (*Liaphlus*) *genieri*. 1, dorsal view; 2, antenna; 3, elytral punctures near base and suture; 4, hind tibia in dorsal view; 5, prosternal and metaventral process; 6, prosternal process in lateral view; 7, left paramere; 8, penis; 9, right paramere

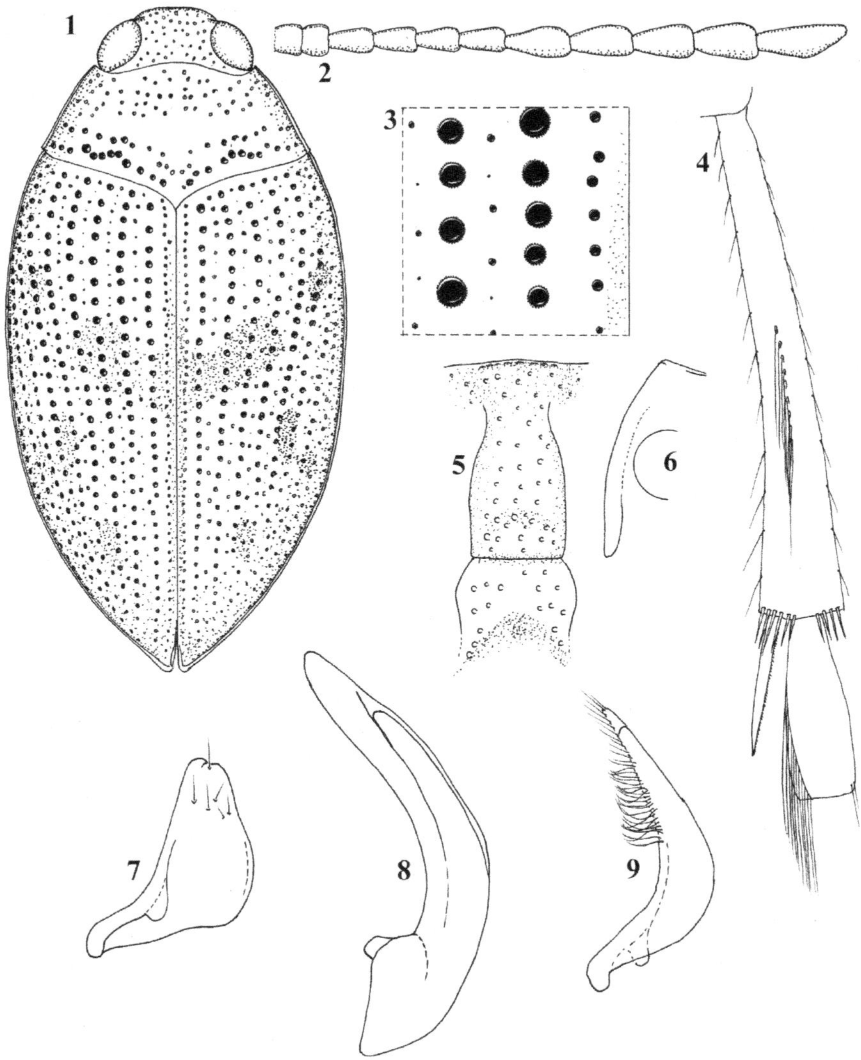

PLATE 91 *Haliplus* (*Liaphlus*) *gracilis*. 1, dorsal view; 2, antenna; 3, elytral punctures near base and suture; 4, hind tibia in dorsal view; 5, prosternal and metaventral process; 6, prosternal process in lateral view; 7, left paramere; 8, penis; 9, right paramere

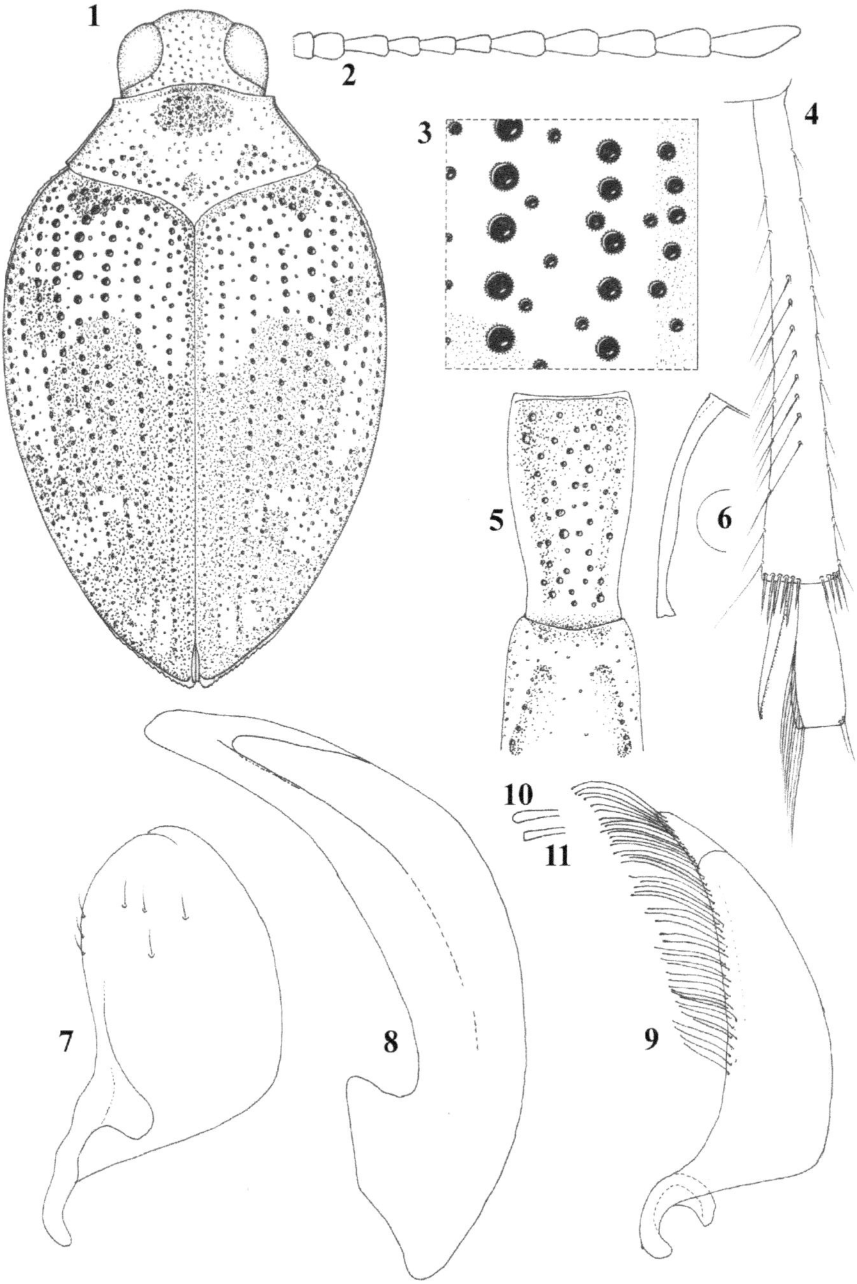

PLATE 92 *Haliplus* (*Liaphlus*) *gravidoides*. 1, dorsal view; 2, antenna; 3, elytral punctures near base and suture; 4, hind tibia in dorsal view; 5, prosternal and metaventral process; 6, prosternal process in lateral view; 7, left paramere; 8, penis; 9, right paramere; 10, 11, setae of right paramere

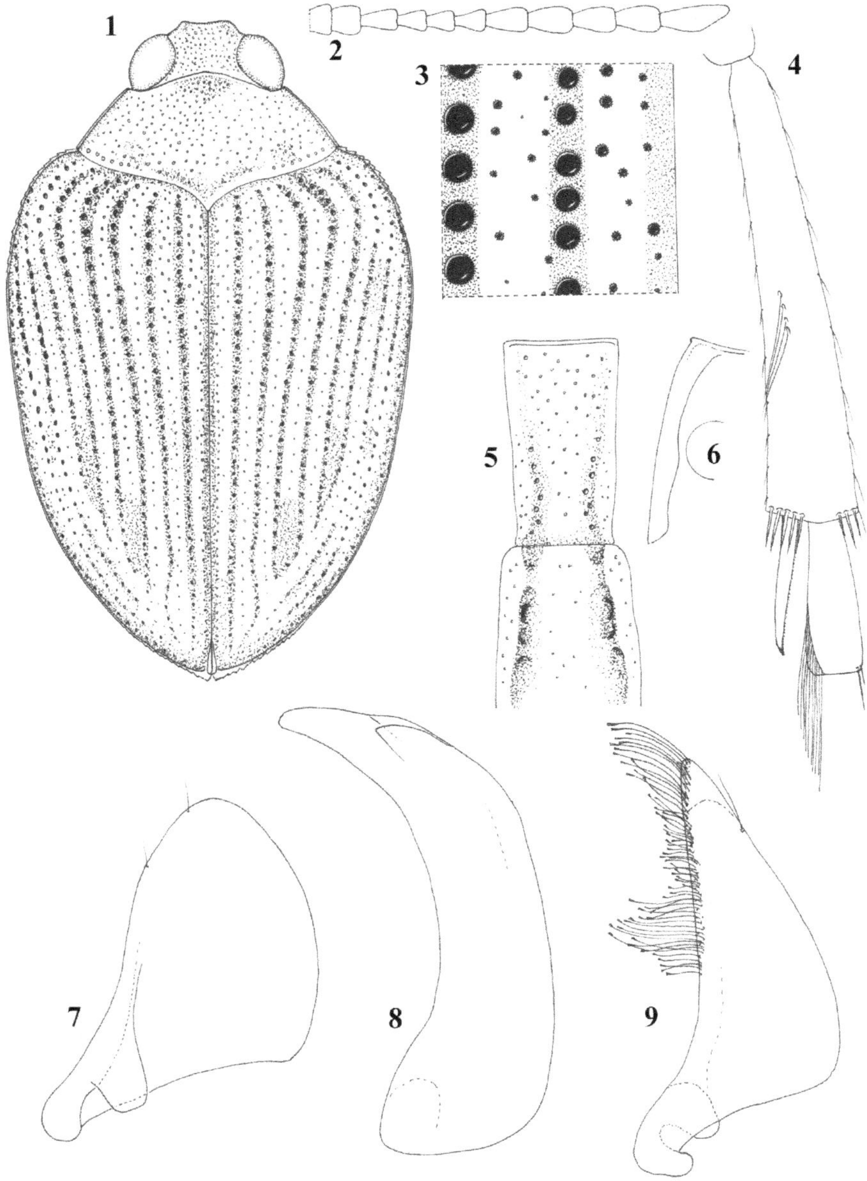

PLATE 93 *Haliplus* (*Liaphlus*) *gravidus*. 1, dorsal view; 2, antenna; 3, elytral punctures near base and suture; 4, hind tibia in dorsal view; 5, prosternal and metaventral process; 6, prosternal process in lateral view; 7, left paramere; 8, penis; 9, right paramere

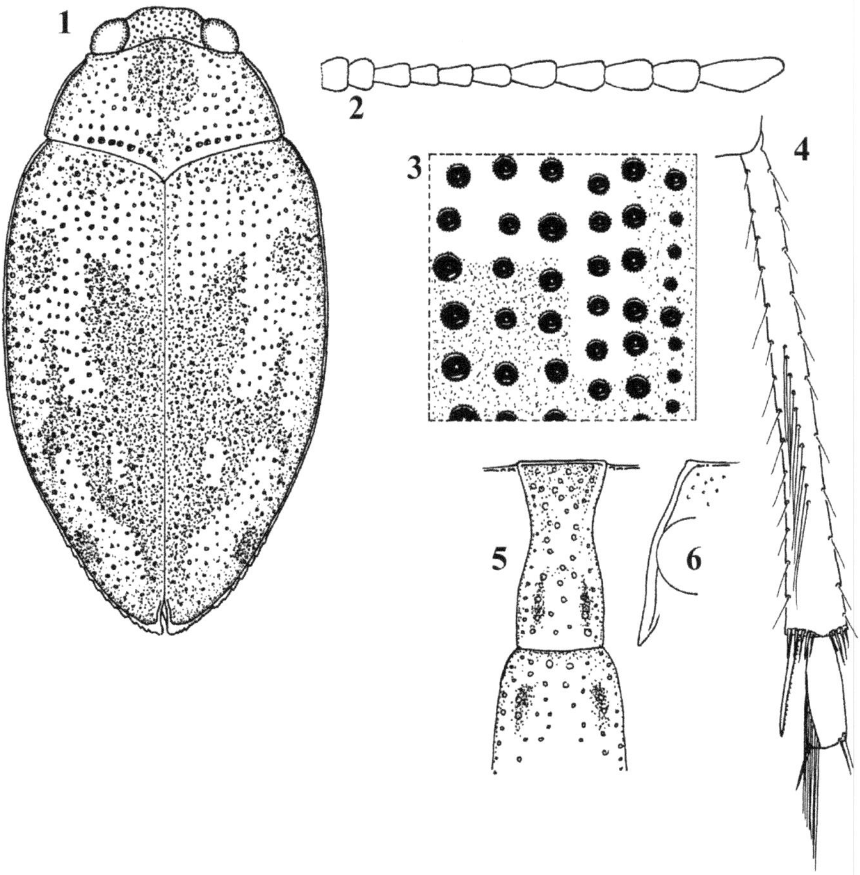

PLATE 94 *Haliplus* (*Liaphlus*) *guignoti* female. 1, dorsal view; 2, antenna; 3, elytral punctures near base and suture; 4, hind tibia in dorsal view; 5, prosternal and metaventral process; 6, prosternal process in lateral view

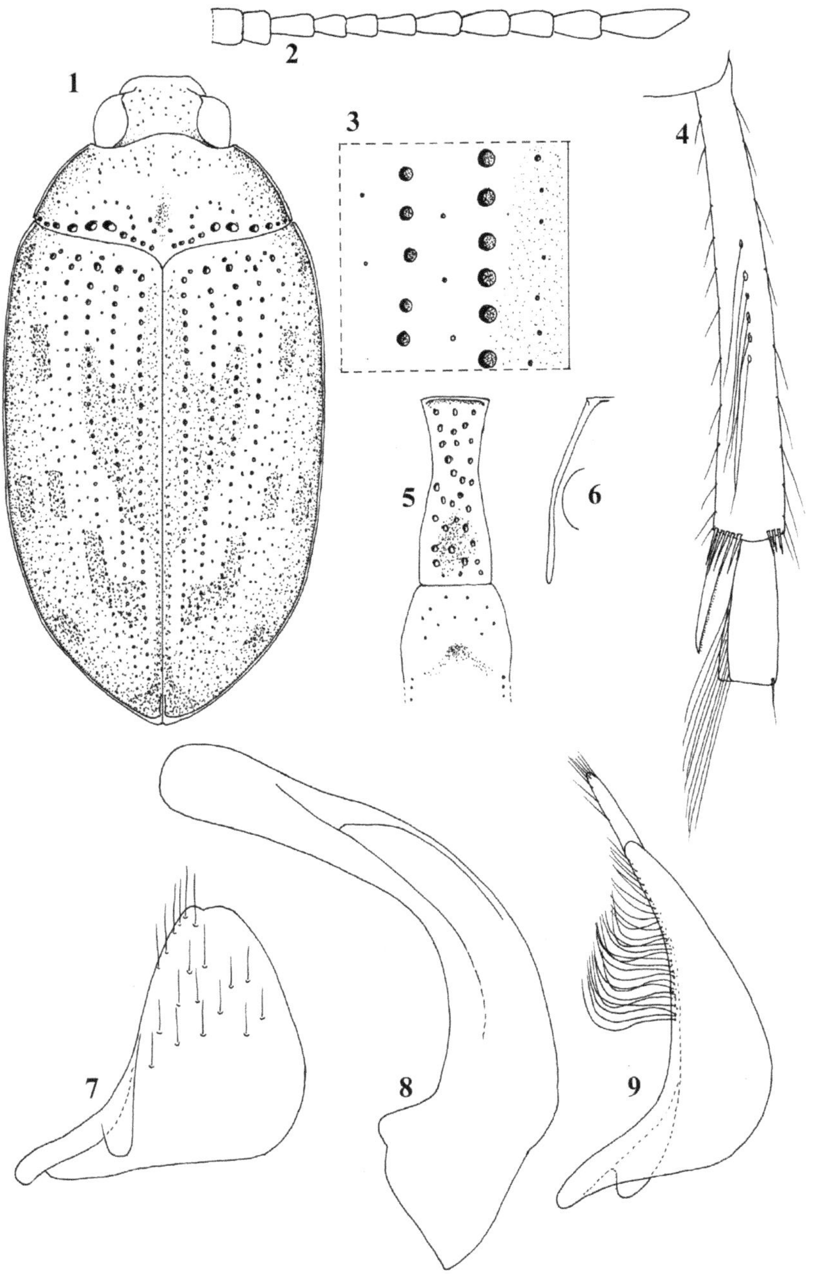

PLATE 95 *Haliplus* (*Liaphlus*) *guttatus*. 1, dorsal view; 2, antenna; 3, elytral punctures near base and suture; 4, hind tibia in dorsal view; 5, prosternal and metaventral process; 6, prosternal process in lateral view; 7, left paramere; 8, penis; 9, right paramere

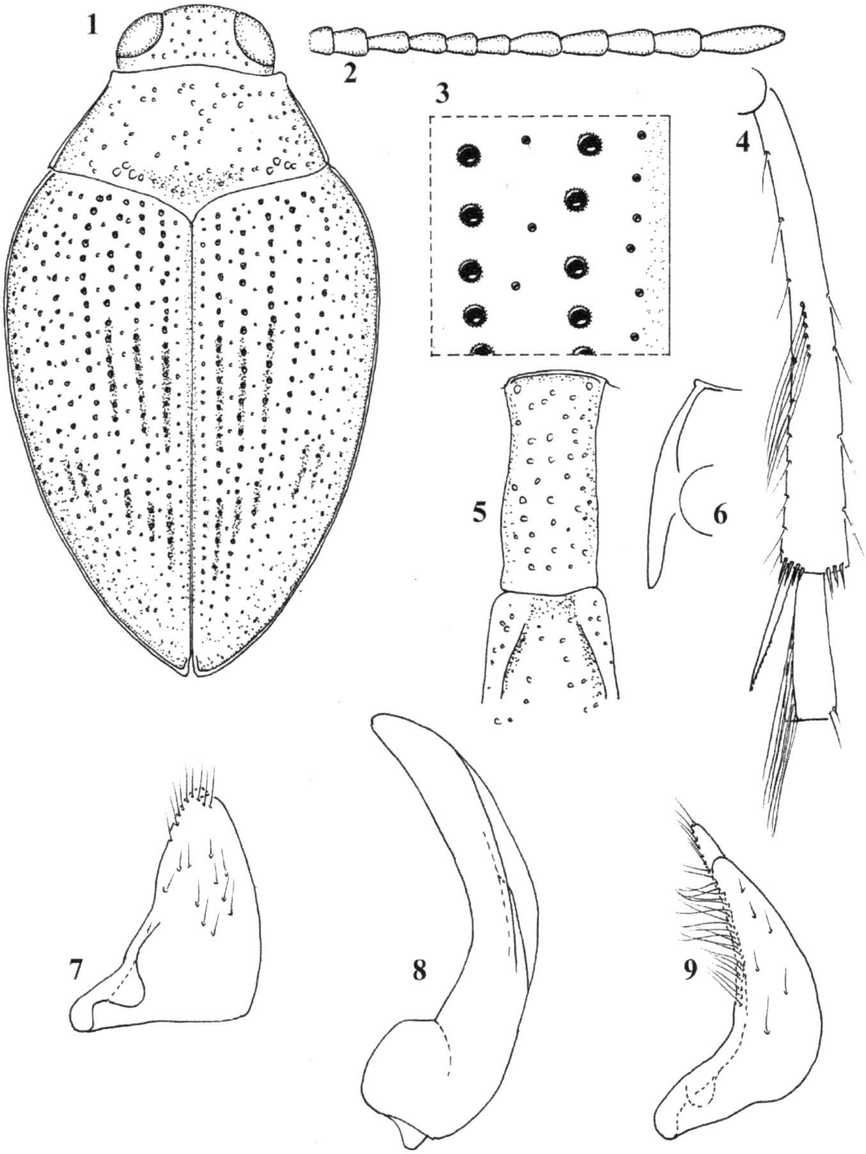

PLATE 96 *Haliplus* (*Liaphlus*) *halsei*. 1, dorsal view; 2, antenna; 3, elytral punctures near base and suture; 4, hind tibia in dorsal view; 5, prosternal and metaventral process; 6, prosternal process in lateral view; 7, left paramere; 8, penis; 9, right paramere

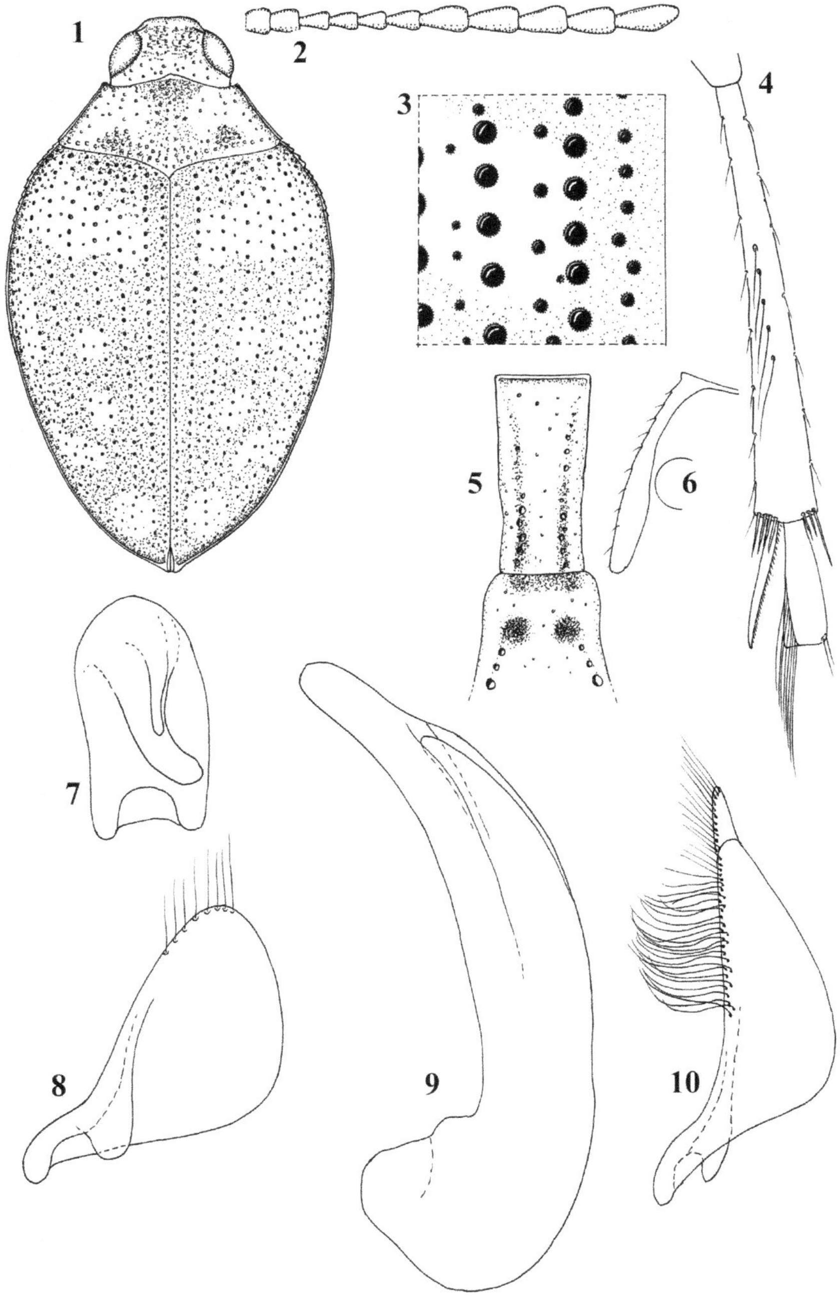

PLATE 97 *Haliplus* (*Liaphlus*) *havaniensis*. 1, dorsal view; 2, antenna; 3, elytral punctures near base and suture; 4, hind tibia in dorsal view; 5, prosternal and metaventral process; 6, prosternal process in lateral view; 7, penis in frontal view; 8, left paramere; 9, penis; 10, right paramere

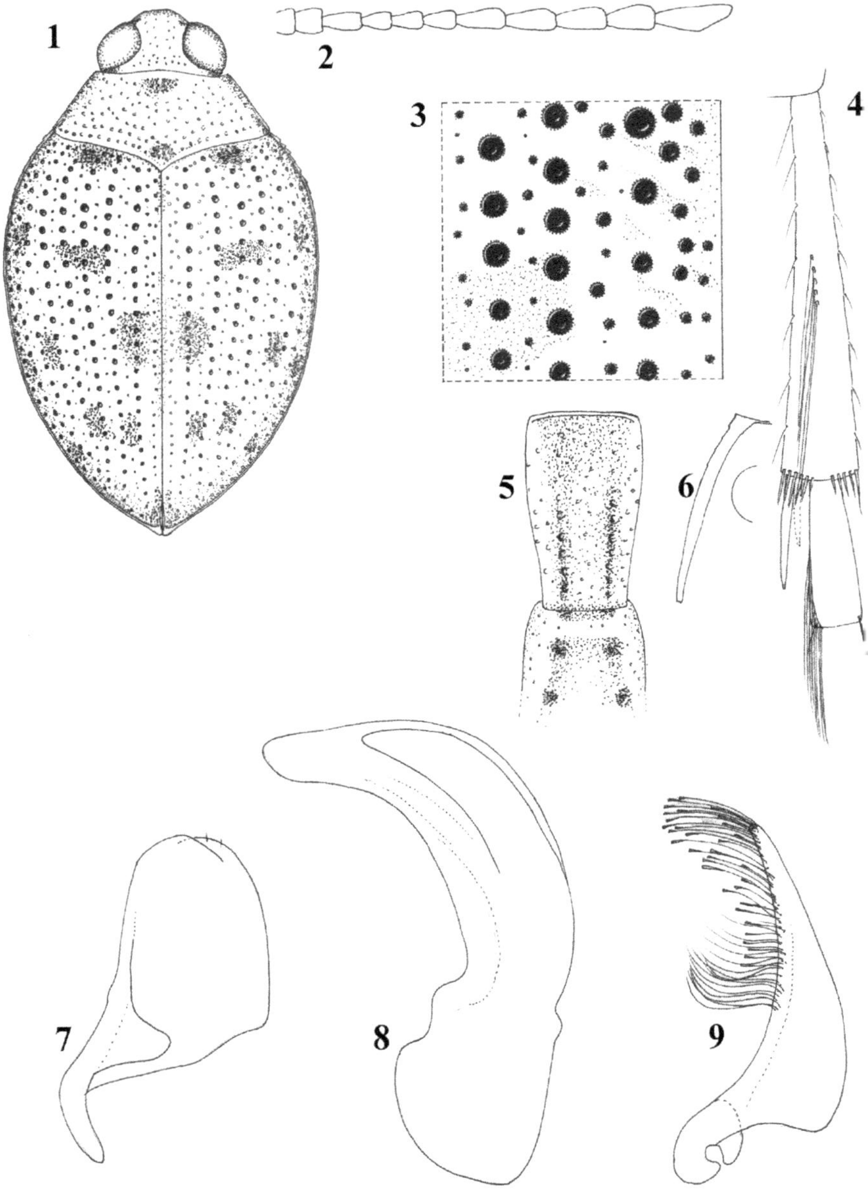

PLATE 98 *Haliplus* (*Liaphlus*) *heppneri*. 1, dorsal view; 2, antenna; 3, elytral punctures near base and suture; 4, hind tibia in dorsal view; 5, prosternal and metaventral process; 6, prosternal process in lateral view; 7, left paramere; 8, penis; 9, right paramere

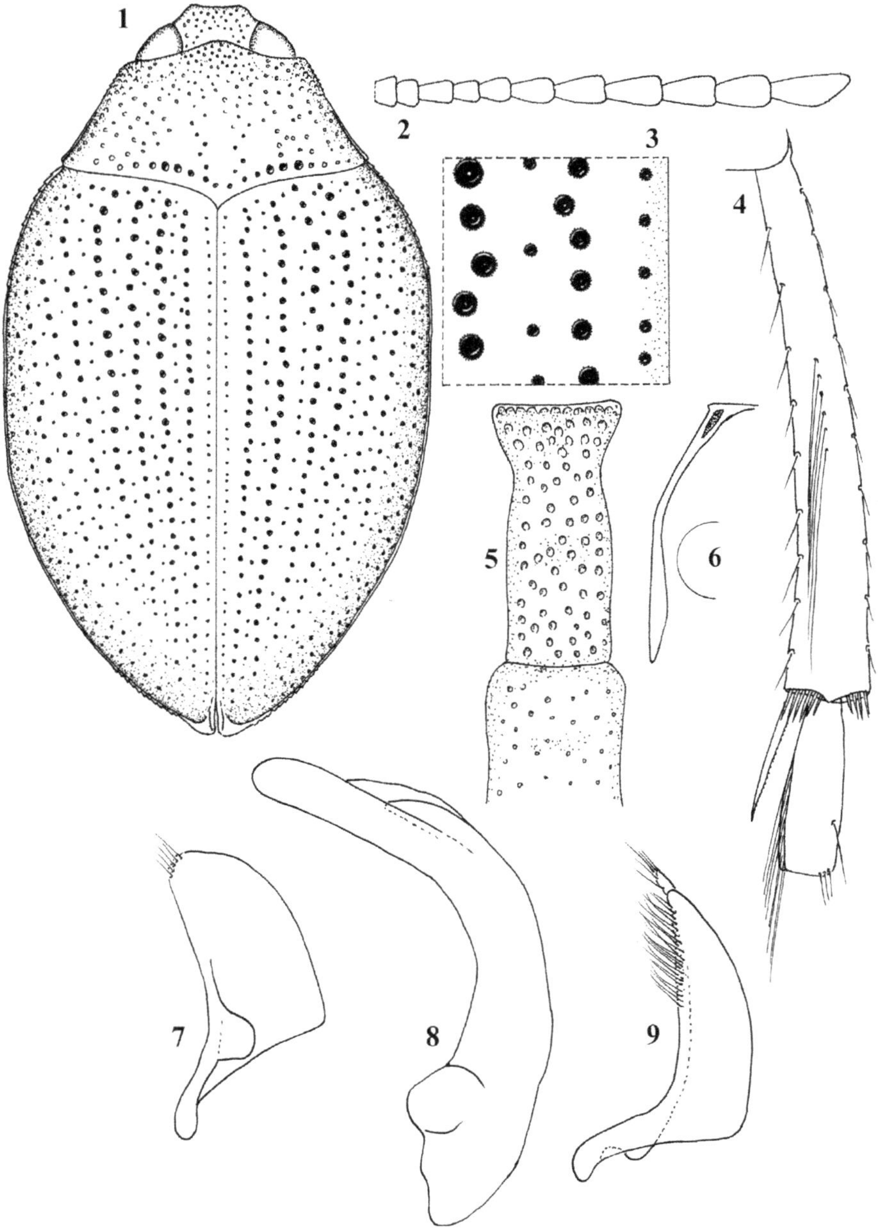

PLATE 99 *Haliplus* (*Liaphlus*) *incrassatus*. 1, dorsal view; 2, antenna; 3, elytral punctures near base and suture; 4, hind tibia in dorsal view; 5, prosternal and metaventral process; 6, prosternalprocess in lateral view; 7, left paramere; 8, penis; 9, right paramere

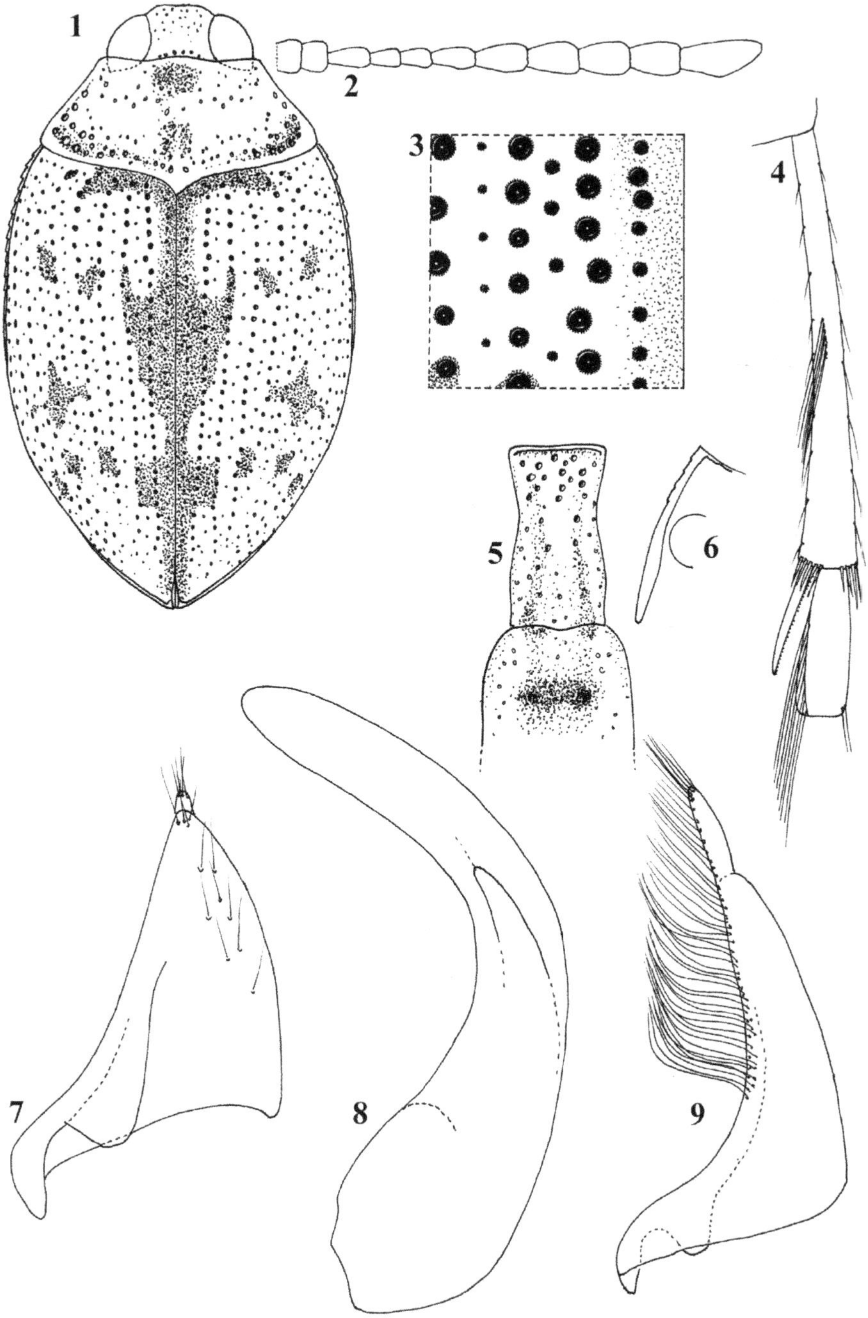

PLATE 100 *Haliplus* (*Liaphlus*) *indicus*. 1, dorsal view; 2, antenna; 3, elytral punctures near base and suture; 4, hind tibia in dorsal view; 5, prosternal and metaventral process; 6, prosternal process in lateral view; 7, left paramere; 8, penis; 9, right paramere

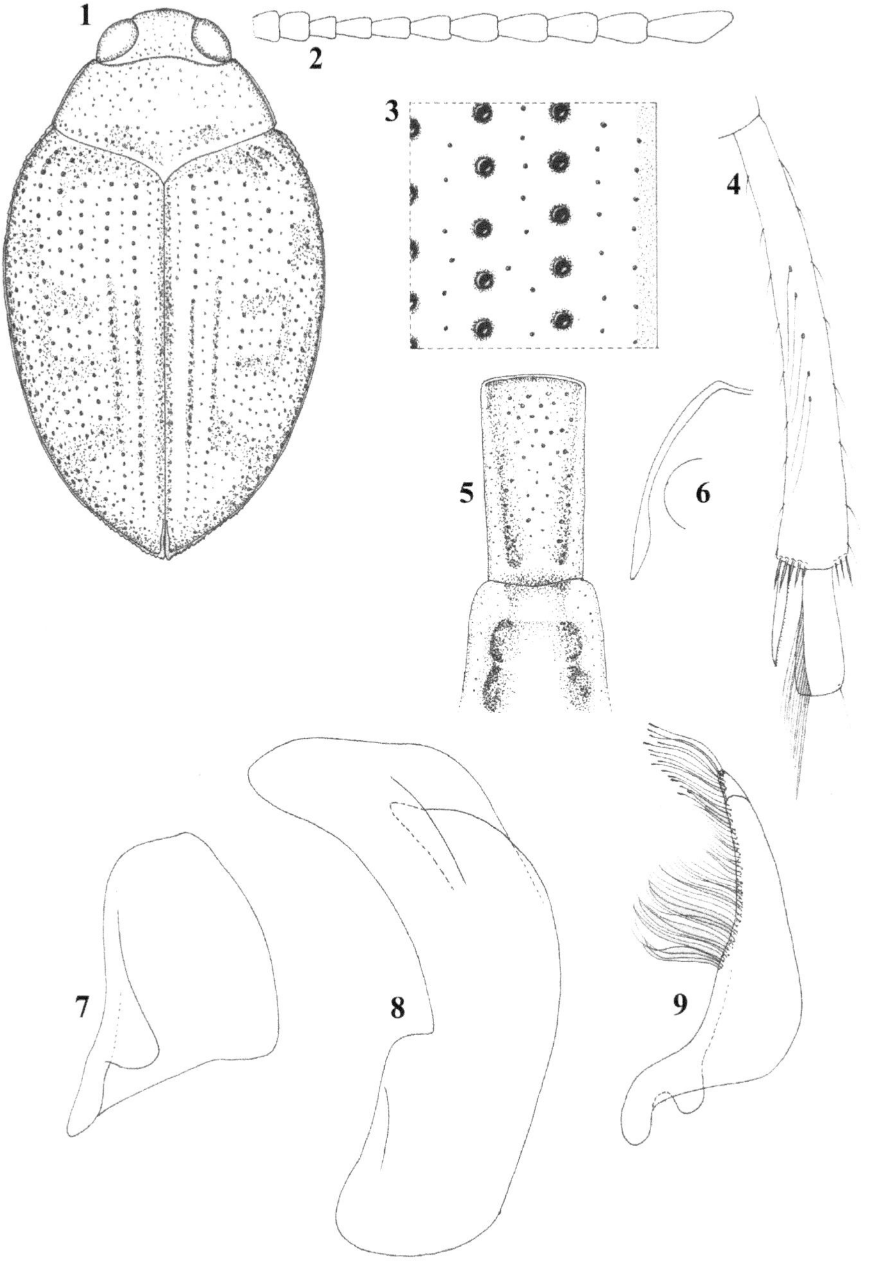

PLATE 101 *Haliplus* (*Liaphlus*) *indistinctus*. 1, dorsal view; 2, antenna; 3, elytral punctures near base and suture; 4, hind tibia in dorsal view; 5, prosternal and metaventral process; 6, prosternal process in lateral view; 7, left paramere; 8, penis; 9, right paramere

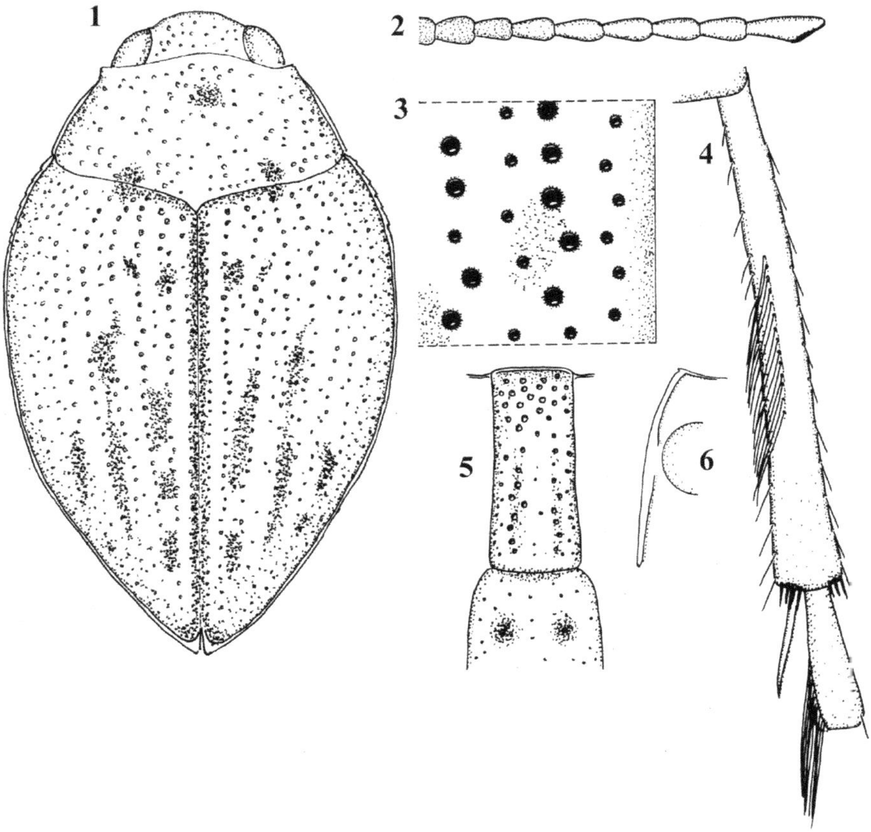

PLATE 102 *Haliplus* (*Liaphlus*) *insularis* female. 1, dorsal view; 2, antenna; 3, elytral punctures near base and suture; 4, hind tibia in dorsal view; 5, prosternal and metaventral process; 6, prosternal process in lateral view

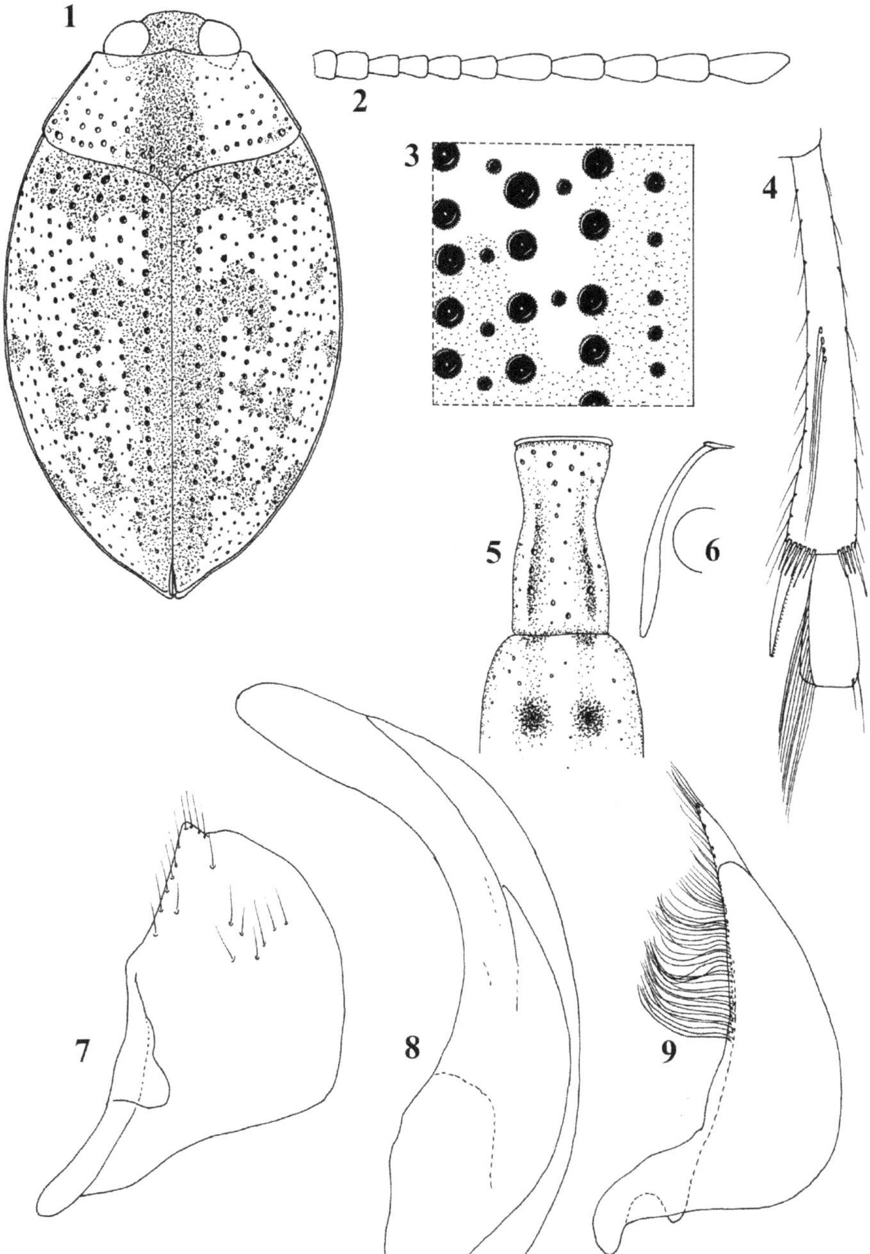

PLATE 103 *Haliplus* (*Liaphlus*) *javanicus*. 1, dorsal view; 2, antenna; 3, elytral punctures near base and suture; 4, hind tibia in dorsal view; 5, prosternal and metaventral process; 6, prosternal process in lateral view; 7, left paramere; 8, penis; 9, right paramere

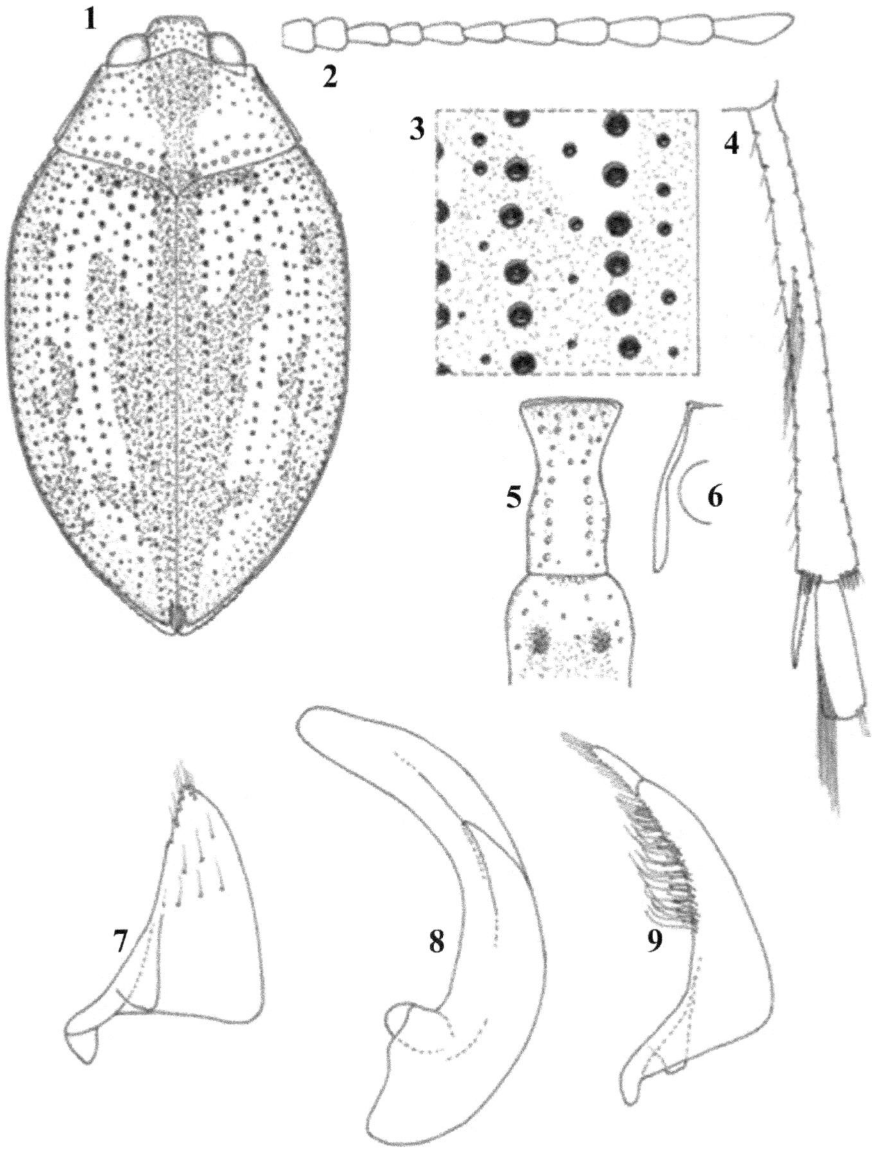

PLATE 104 *Haliplus* (*Liaphlus*) *kantneri*. 1, dorsal view; 2, antenna; 3, elytral punctures near base and suture; 4, hind tibia in dorsal view; 5, prosternal and metaventral process; 6, prosternal process in lateral view; 7, left paramere; 8, penis; 9, right paramere

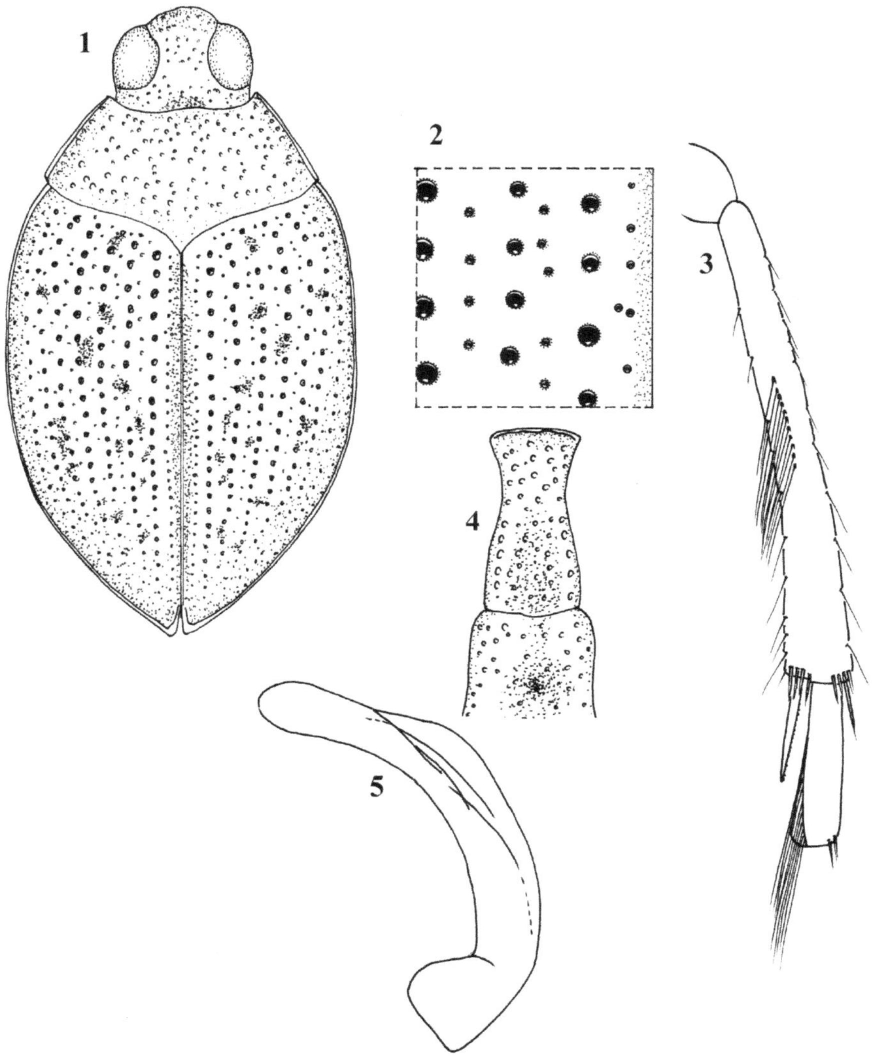

PLATE 105 *Haliplus* (*Liaphlus*) *kapuri*. 1, dorsal view; 2, elytral punctures near base and suture; 3, hind tibia in dorsal view; 4, prosternal and metaventral process; 5, penis

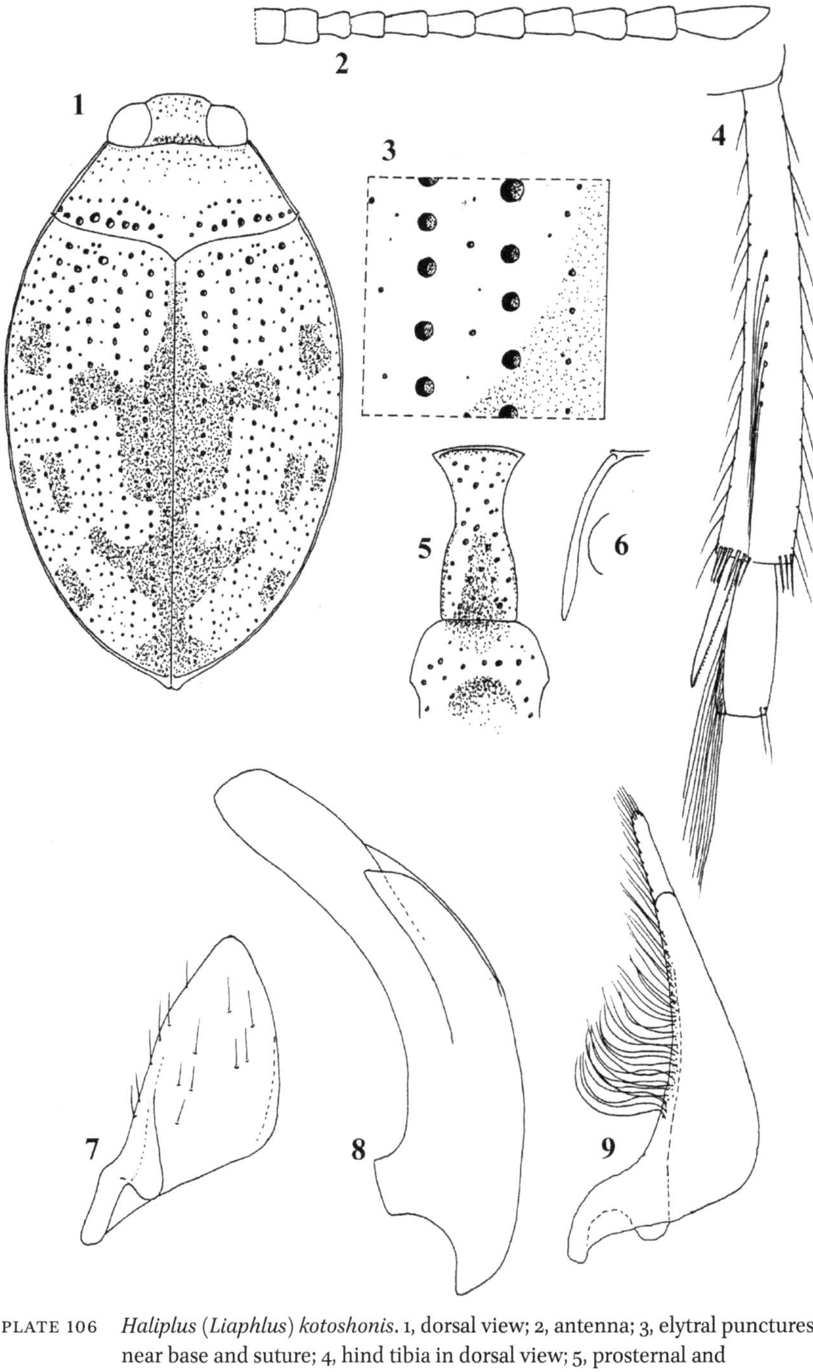

PLATE 106 *Haliplus* (*Liaphlus*) *kotoshonis*. 1, dorsal view; 2, antenna; 3, elytral punctures near base and suture; 4, hind tibia in dorsal view; 5, prosternal and metaventral process; 6, prosternal process in lateral view; 7, left paramere; 8, penis; 9, right paramere

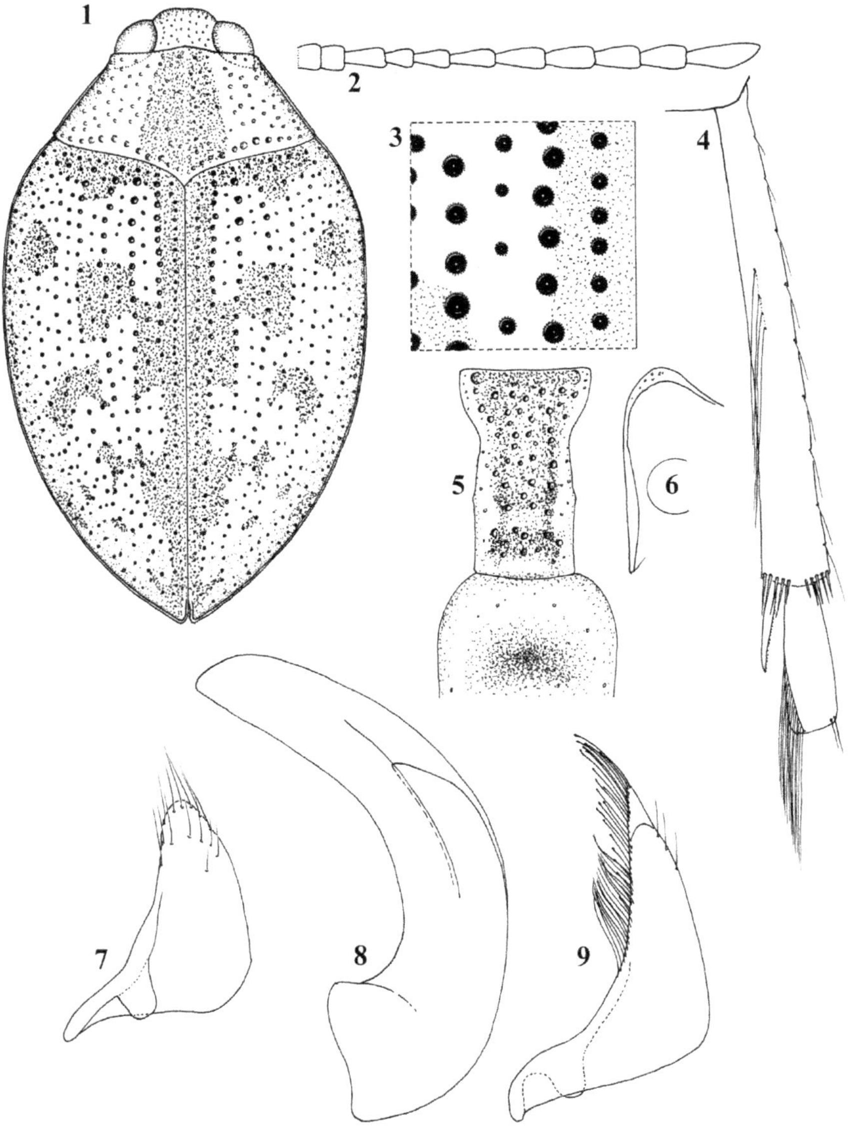

PLATE 107 *Haliplus* (*Liaphlus*) *lamottei*. 1, dorsal view; 2, antenna; 3, elytral punctures near base and suture; 4, hind tibia in dorsal view; 5, prosternal and metaventral process; 6, prosternal process in lateral view; 7, left paramere; 8, penis; 9, right paramere

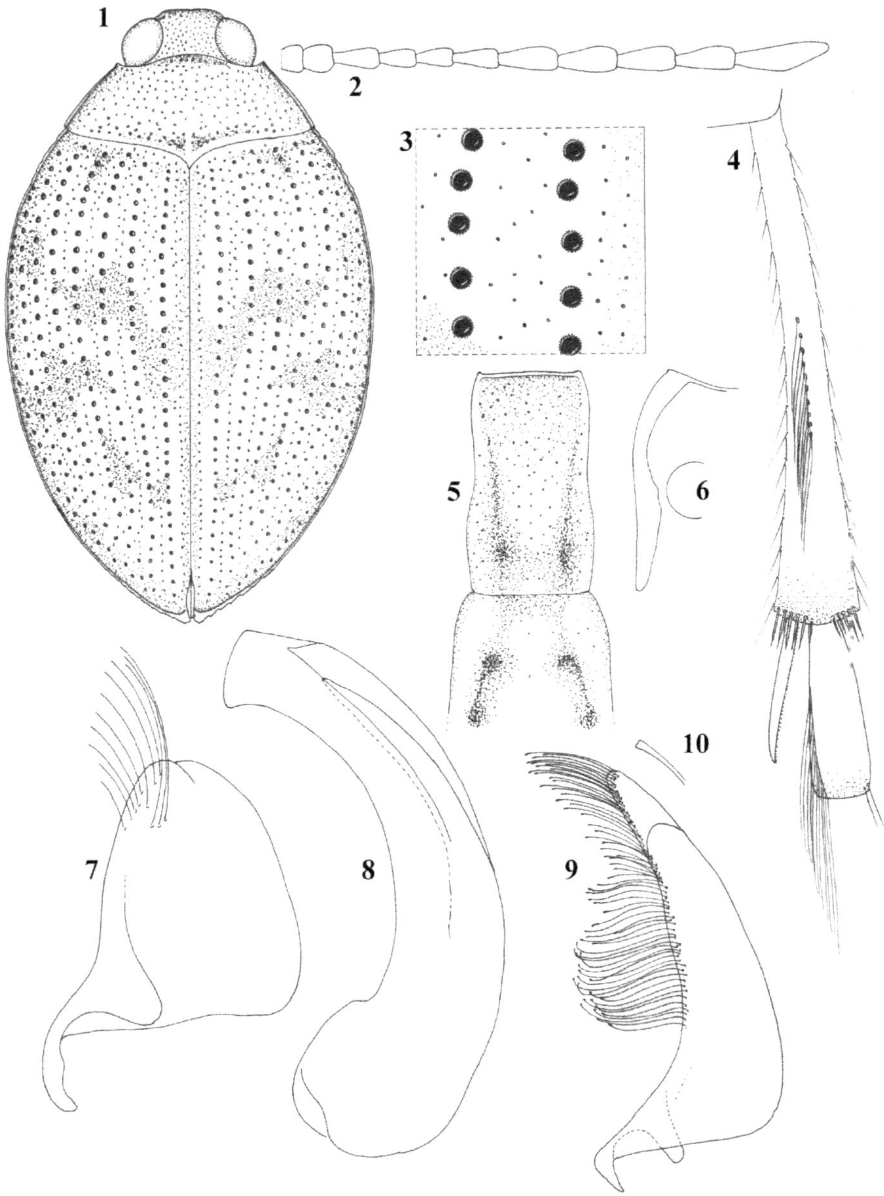

PLATE 108 *Haliplus* (*Liaphlus*) *langleyi*. 1, dorsal view; 2, antenna; 3, elytral punctures near base and suture; 4, hind tibia in dorsal view; 5, prosternal and metaventral process; 6, prosternal process in lateral view; 7, left paramere; 8, penis; 9, right paramere; 10, setae of right paramere

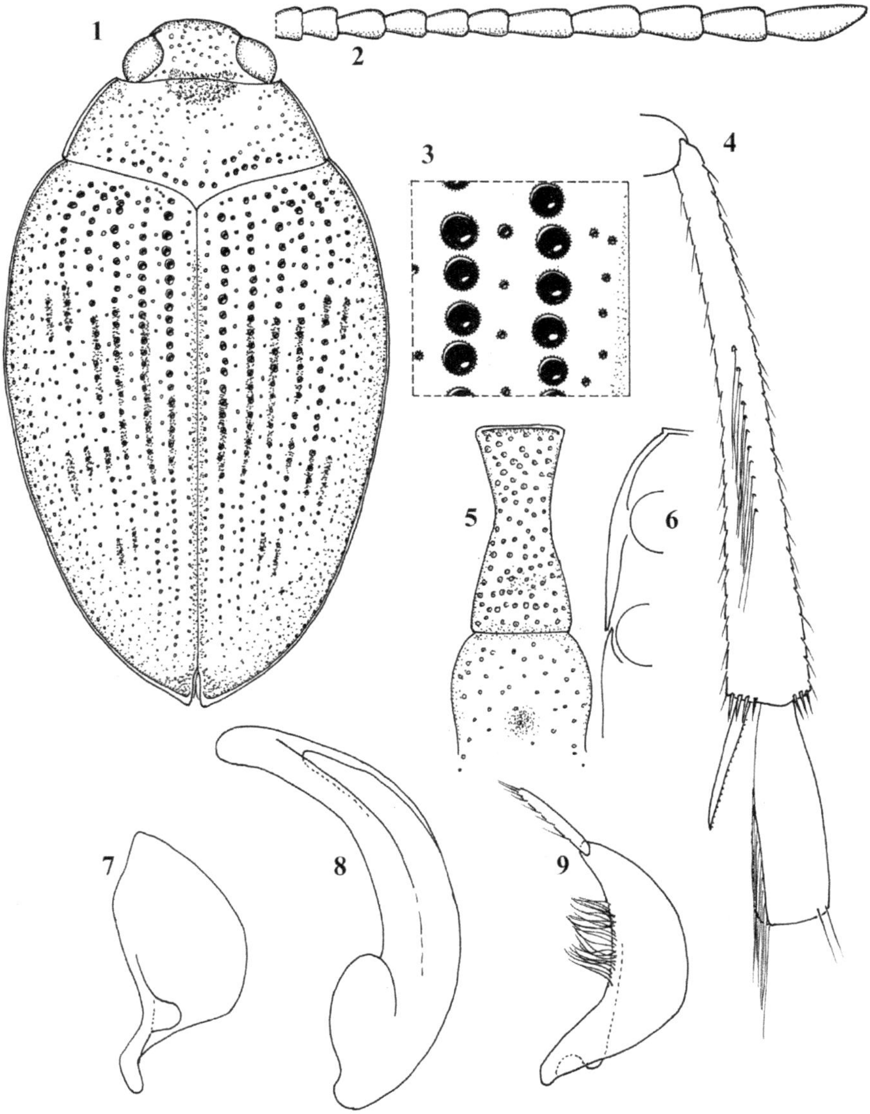

PLATE 109 *Haliplus* (*Liaphlus*) *leechi*. 1, dorsal view; 2, antenna; 3, elytral punctures near base and suture; 4, hind tibia in dorsal view; 5, prosternal and metaventral process; 6, prosternal process in lateral view; 7, left paramere; 8, penis; 9, right paramere

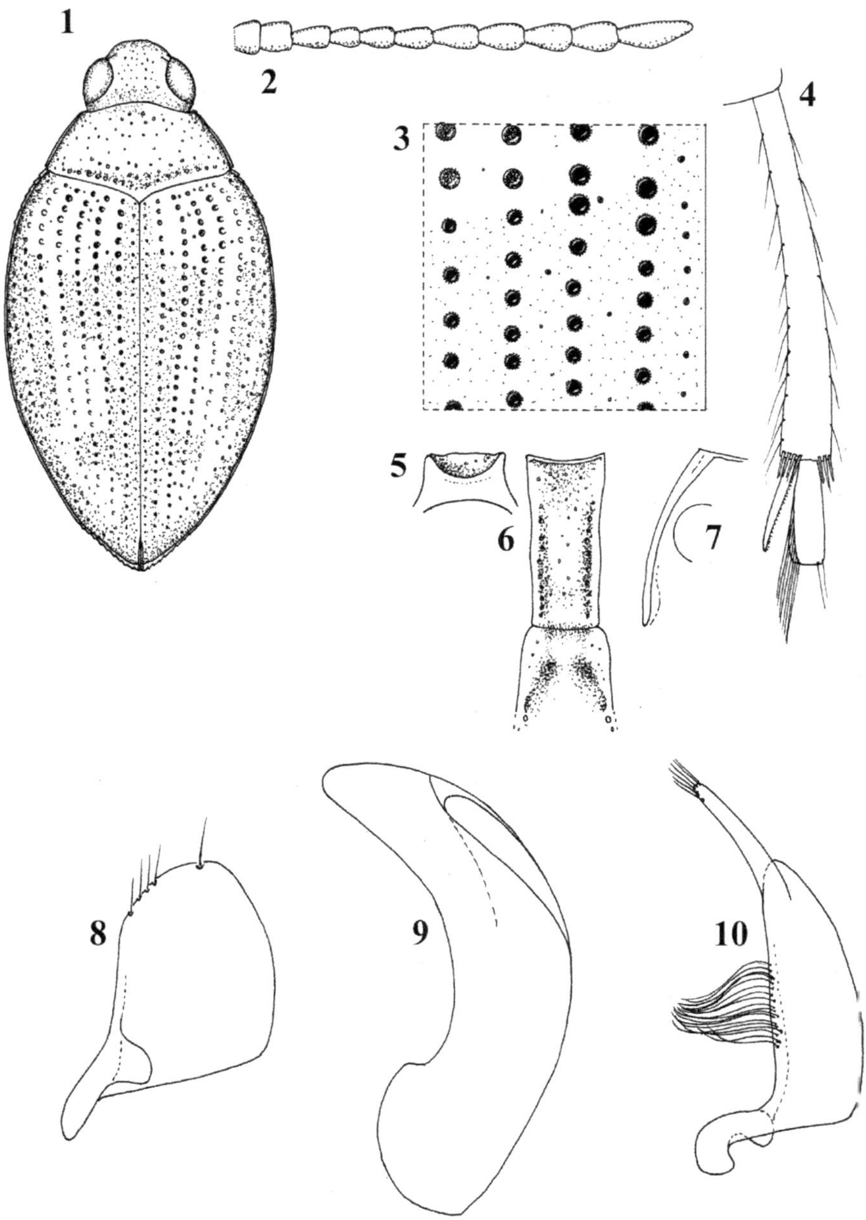

PLATE 110 *Haliplus* (*Liaphlus*) *lewisii*. 1, dorsal view; 2, antenna; 3, elytral punctures near base and suture; 4, hind tibia in dorsal view; 5, prosternal process in frontal view; 6, prosternal and metaventral process in ventral view; 7, prosternal process in lateral view; 8, left paramere; 9, penis; 10, right paramere

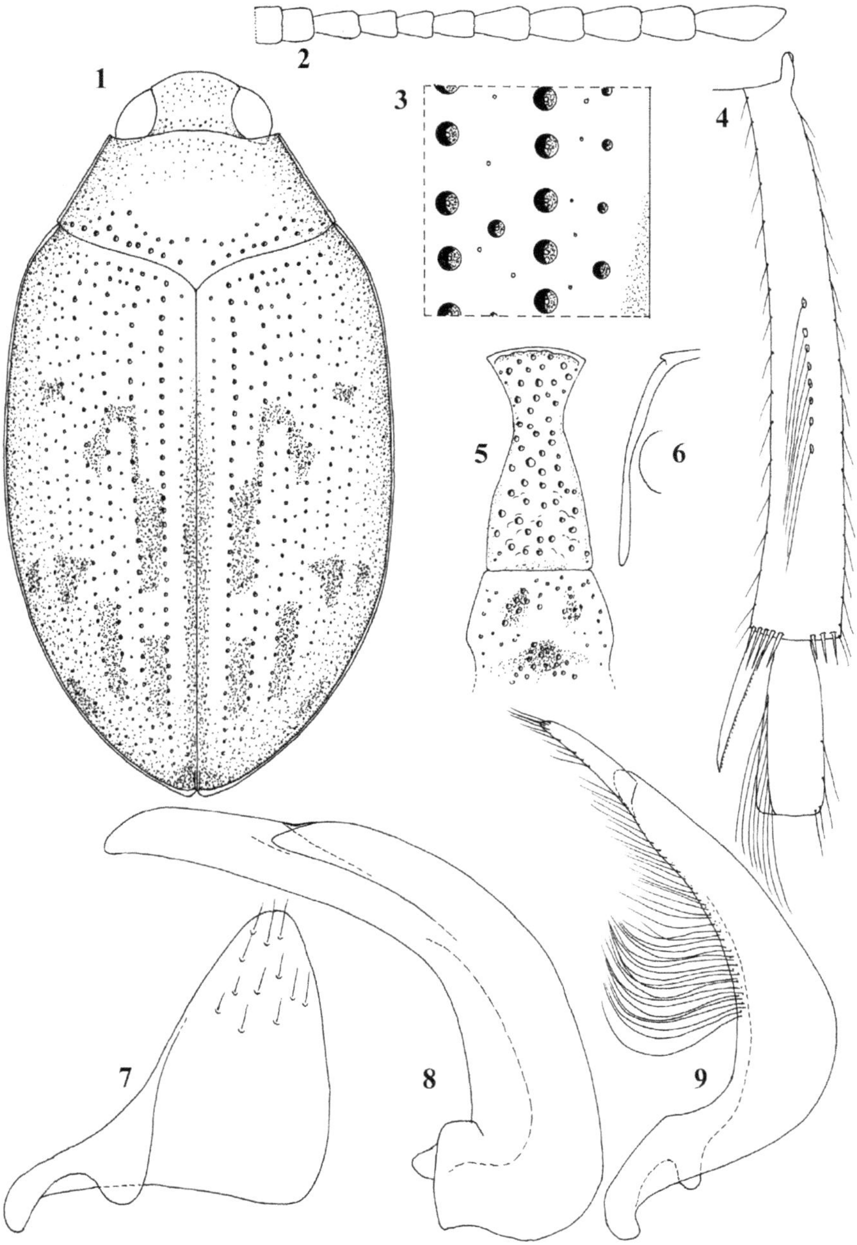

PLATE 111 *Haliplus* (*Liaphlus*) *maculatus*. 1, dorsal view; 2, antenna; 3, elytral punctures near base and suture; 4, hind tibia in dorsal view; 5, prosternal and metaventral process; 6, prosternal process in lateral view; 7, left paramere; 8, penis; 9, right paramere

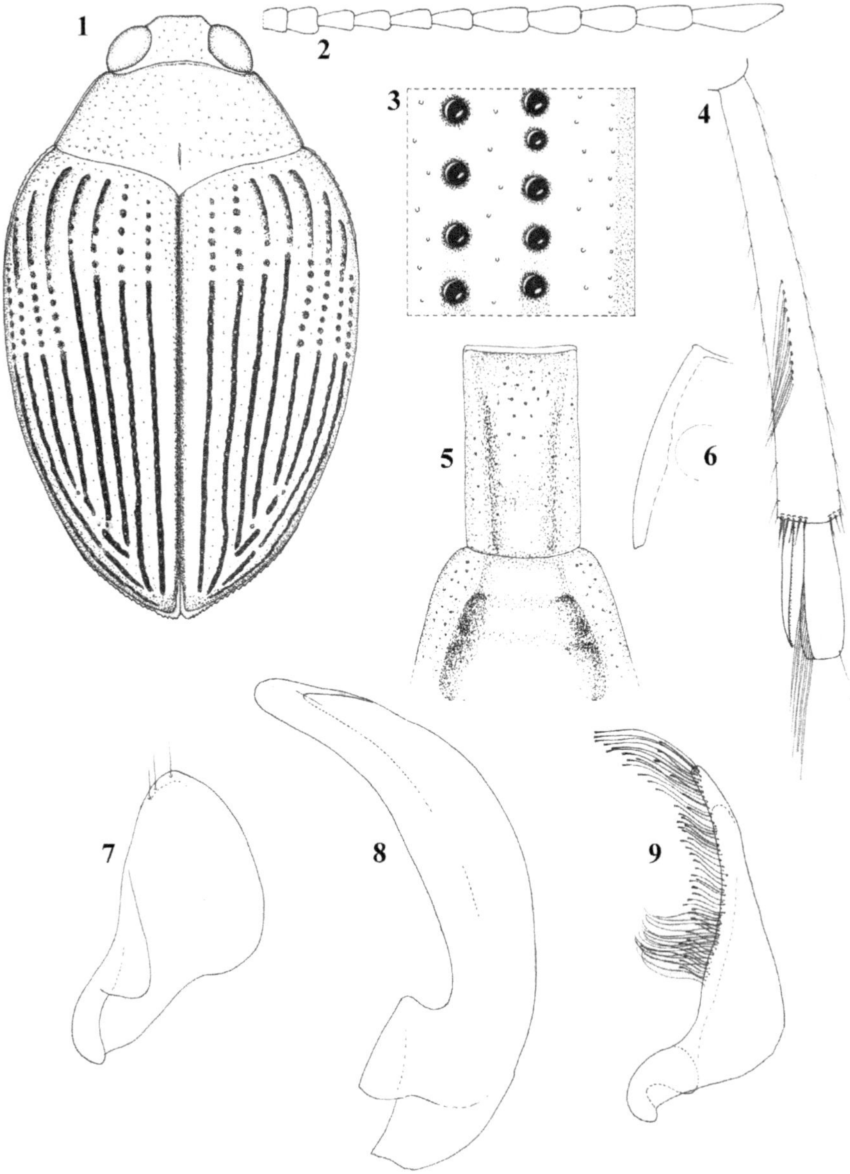

PLATE 112 *Haliplus* (*Liaphlus*) *maculicollis*. 1, dorsal view; 2, antenna; 3, elytral punctures near base and suture; 4, hind tibia in dorsal view; 5, prosternal and metaventral process; 6, prosternal process in lateral view; 7, left paramere; 8, penis; 9, right paramere

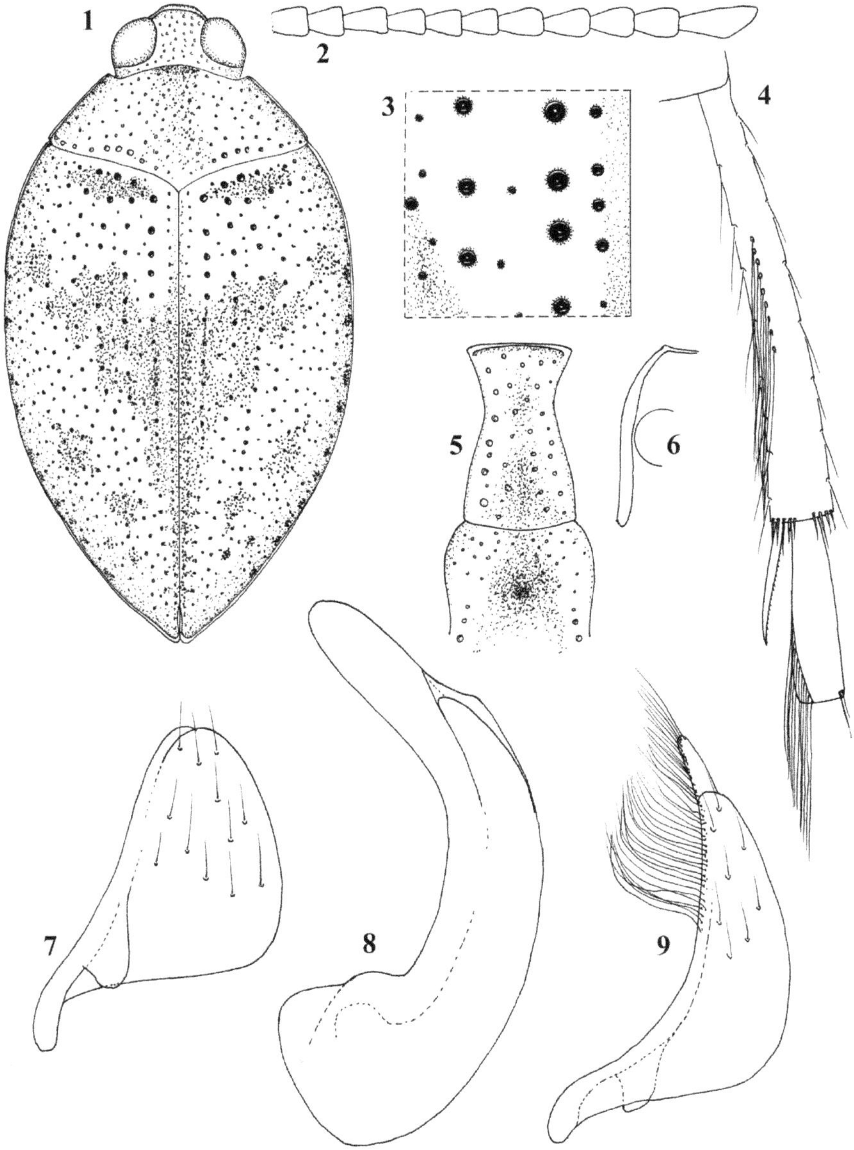

PLATE 113 *Haliplus* (*Liaphlus*) *maculipennis*. 1, dorsal view; 2, antenna; 3, elytral punctures near base and suture; 4, hind tibia in dorsal view; 5, prosternal and metaventral process; 6, prosternal process in lateral view; 7, left paramere; 8, penis; 9, right paramere

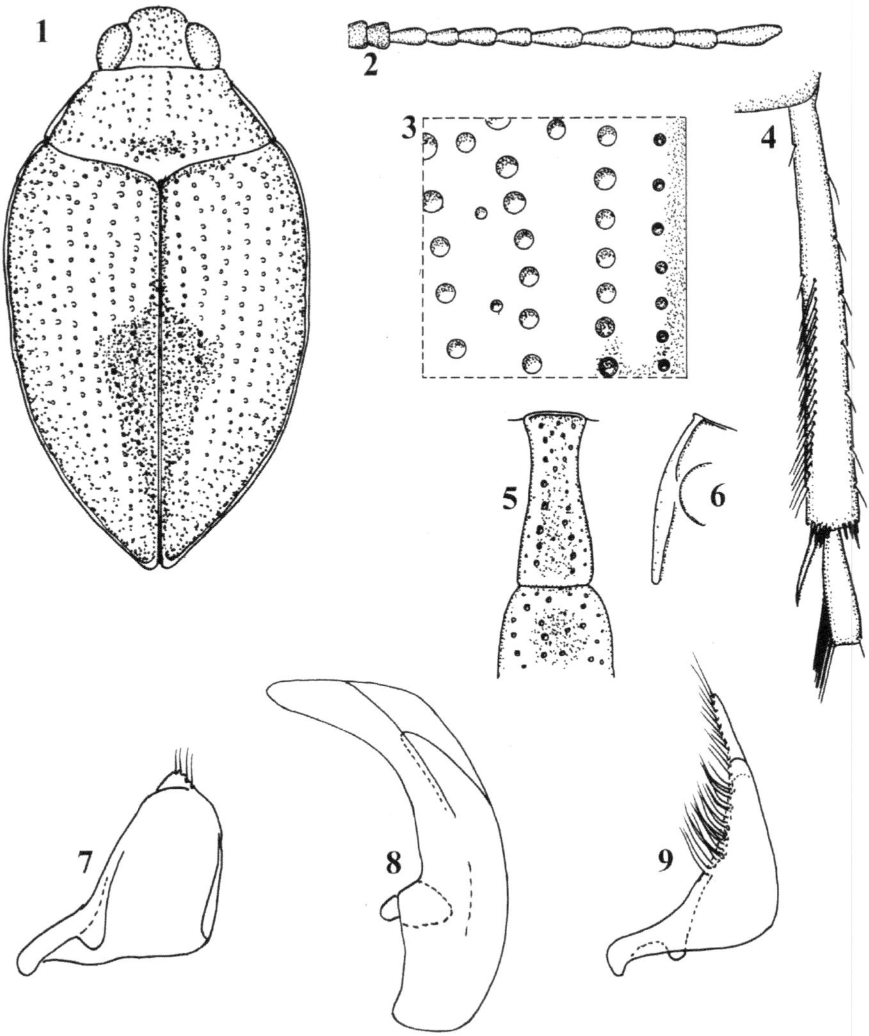

PLATE 114 *Haliplus* (*Liaphlus*) *madagascariensis*. 1, dorsal view; 2, antenna; 3, elytral punctures near base and suture; 4, hind tibia in dorsal view; 5, prosternal and metaventral process; 6, prosternal process in lateral view; 7, left paramere; 8, penis; 9, right paramere

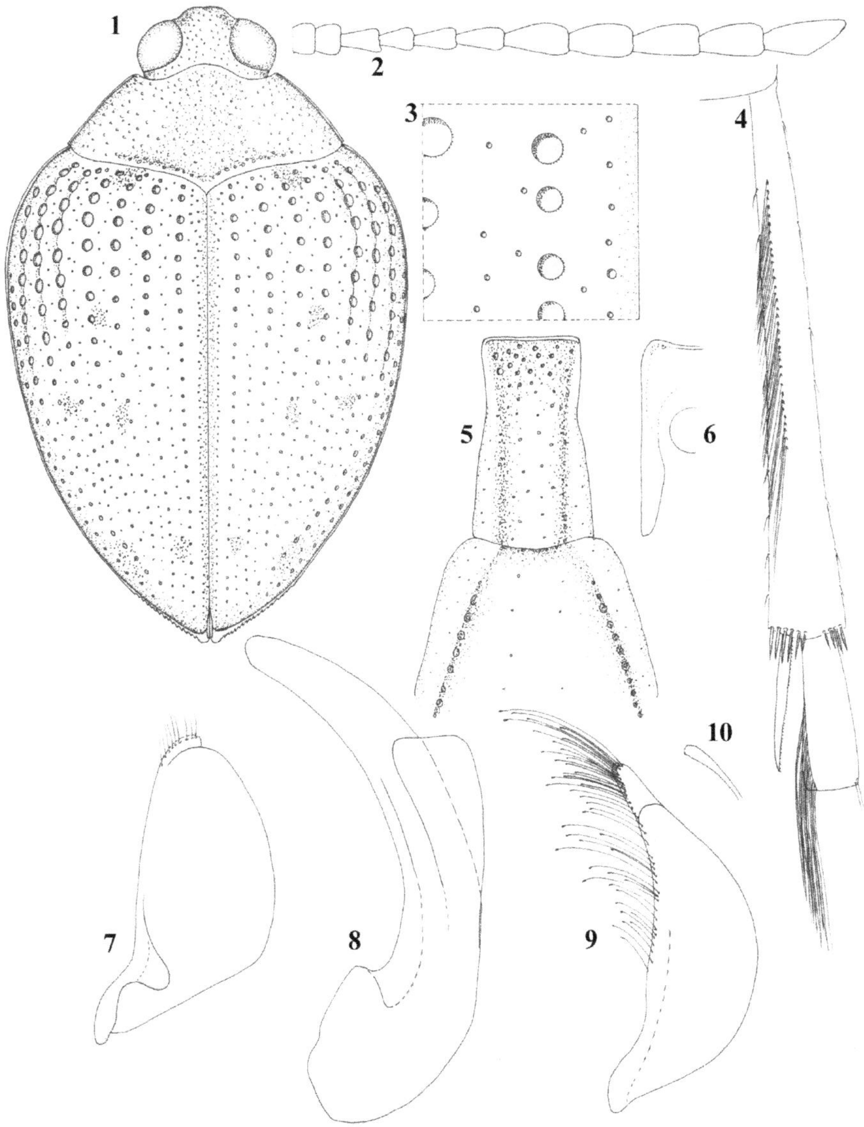

PLATE 115 *Haliplus* (*Liaphlus*) *megapunctatus*. 1, dorsal view; 2, antenna; 3, elytral punctures near base and suture; 4, hind tibia in dorsal view; 5, prosternal and metaventral process; 6, prosternal process in lateral view; 7, left paramere; 8, penis; 9, right paramere; 10 setae of right paramere

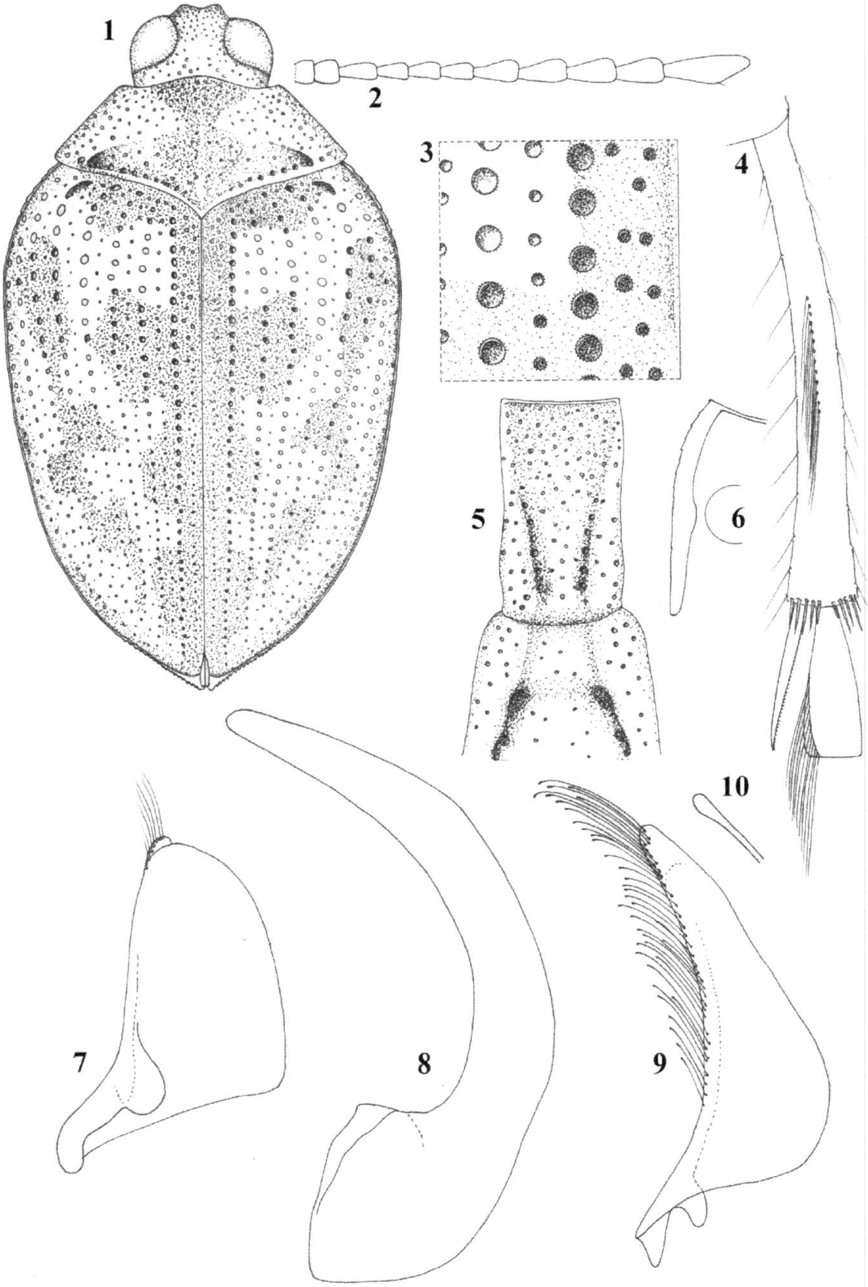

PLATE 116 *Haliplus* (*Liaphlus*) *mesoamericanus*. 1, dorsal view; 2, antenna; 3, elytral punctures near base and suture; 4, hind tibia in dorsal view; 5, prosternal and metaventral process; 6, prosternal process in lateral view; 7, left paramere; 8, penis; 9, right paramere; 10, setae of right paramere

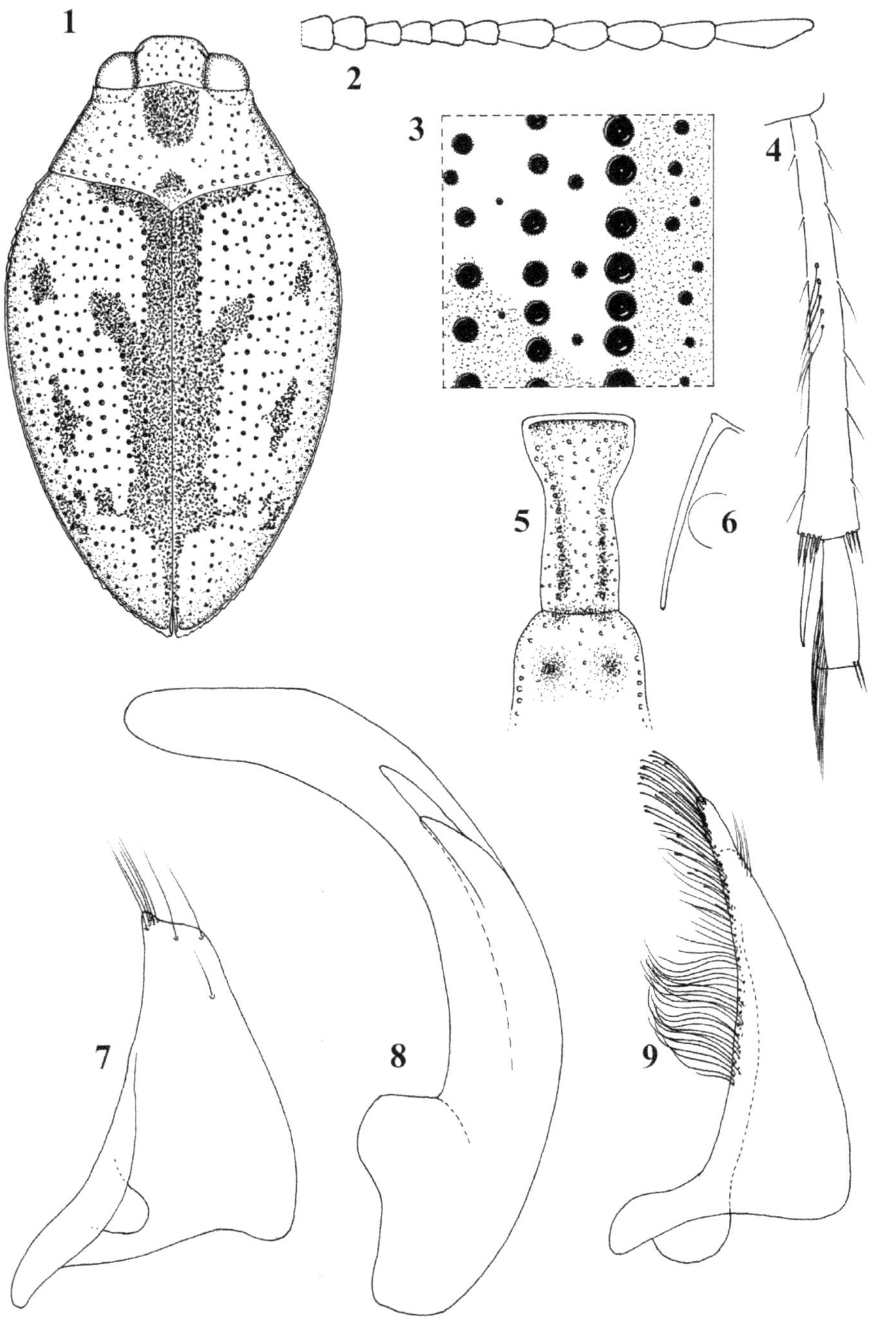

PLATE 117 *Haliplus* (*Liaphlus*) *methneri*. 1, dorsal view; 2, antenna; 3, elytral punctures near base and suture; 4, hind tibia in dorsal view; 5, prosternal and metaventral process; 6, prosternal process in lateral view; 7, left paramere; 8, penis; 9, right paramere

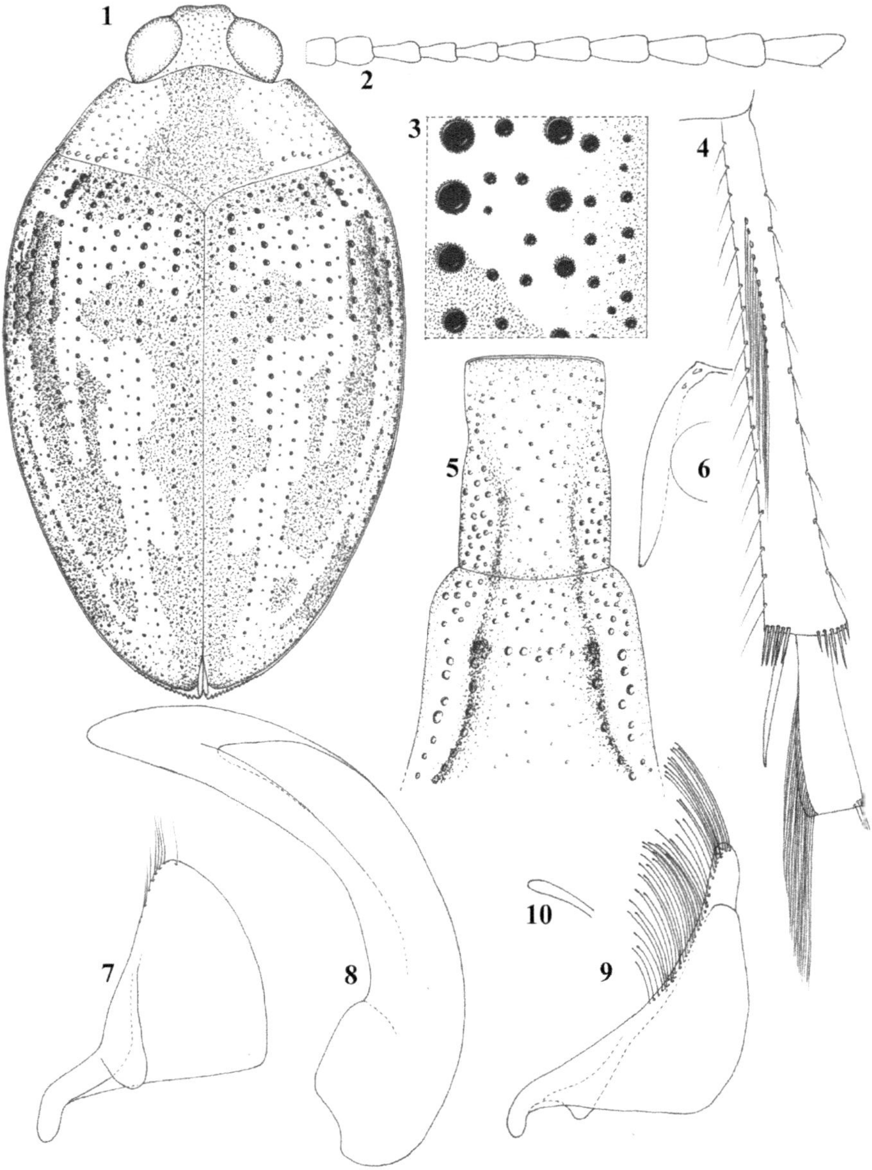

PLATE 118 *Haliplus* (*Liaphlus*) *mexicanus*. 1, dorsal view; 2, antenna; 3, elytral punctures near base and suture; 4, hind tibia in dorsal view; 5, prosternal and metaventral process; 6, prosternal process in lateral view; 7, left paramere; 8, penis; 9, right paramere; 10, setae of right paramere

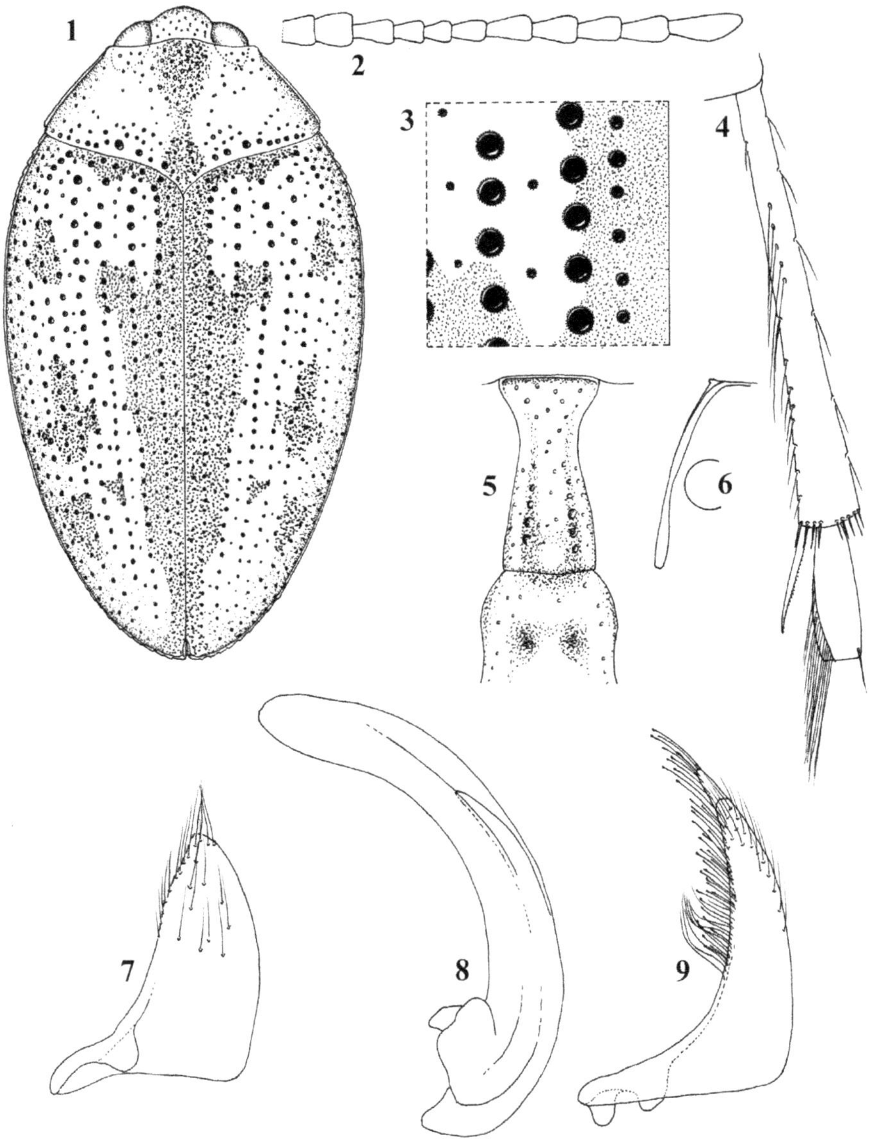

PLATE 119 *Haliplus* (*Liaphlus*) *mimulus*. 1, dorsal view; 2, antenna; 3, elytral punctures near base and suture; 4, hind tibia in dorsal view; 5, prosternal and metaventral process; 6, prosternal process in lateral view; 7, left paramere; 8, penis; 9, right paramere

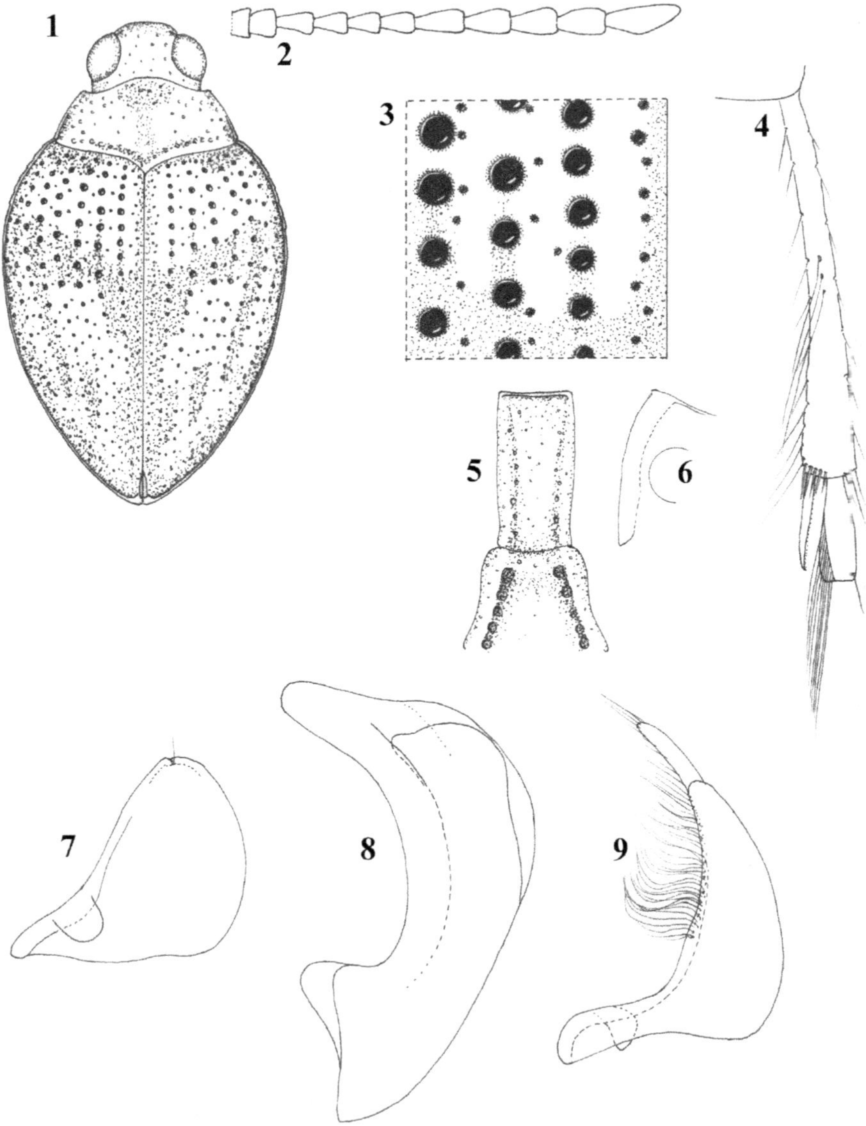

PLATE 120 *Haliplus* (*Liaphlus*) *minimus*. 1, dorsal view; 2, antenna; 3, elytral punctures near base and suture; 4, hind tibia in dorsal view; 5, prosternal and metaventral process; 6, prosternal process in lateral view; 7, left paramere; 8, penis; 9, right paramere

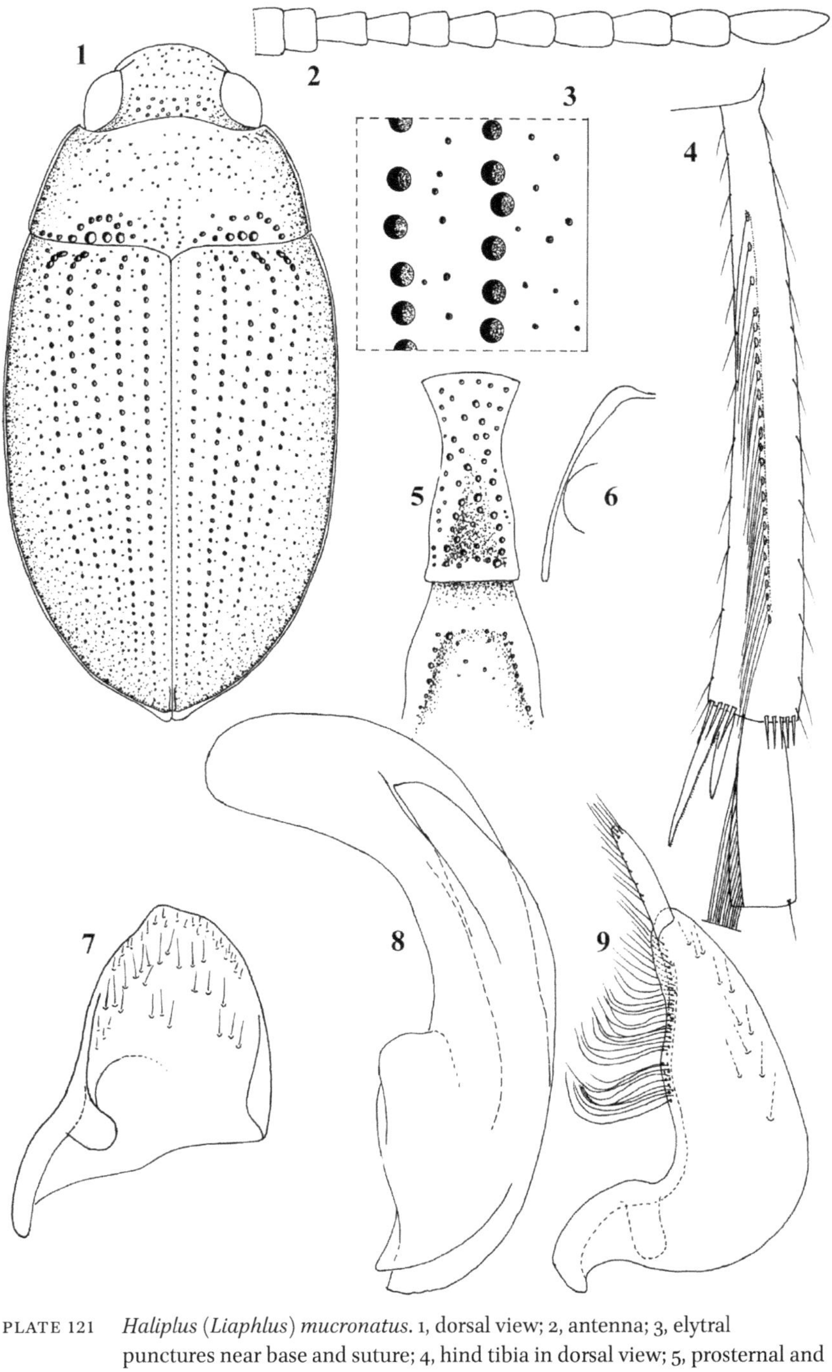

PLATE 121 *Haliplus* (*Liaphlus*) *mucronatus*. 1, dorsal view; 2, antenna; 3, elytral punctures near base and suture; 4, hind tibia in dorsal view; 5, prosternal and metaventral process; 6, prosternal process in lateral view; 7, left paramere; 8, penis; 9, right paramere

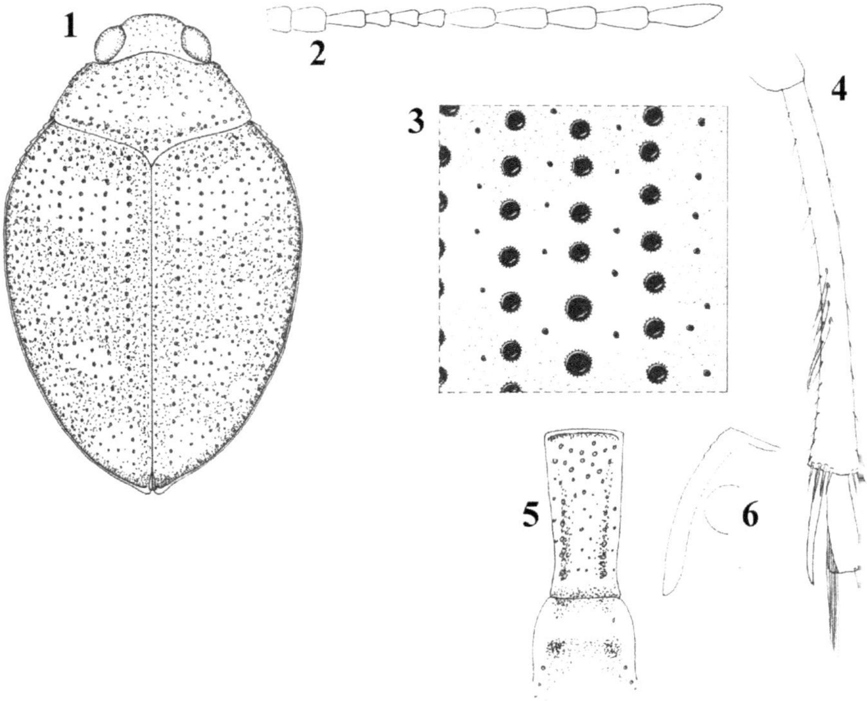

PLATE 122 *Haliplus* (*Liaphlus*) *nanus* female. 1, dorsal view; 2, antenna; 3, elytral punctures near base and suture; 4, hind tibia in dorsal view; 5, prosternal and metaventral process; 6, prosternal process in lateral view

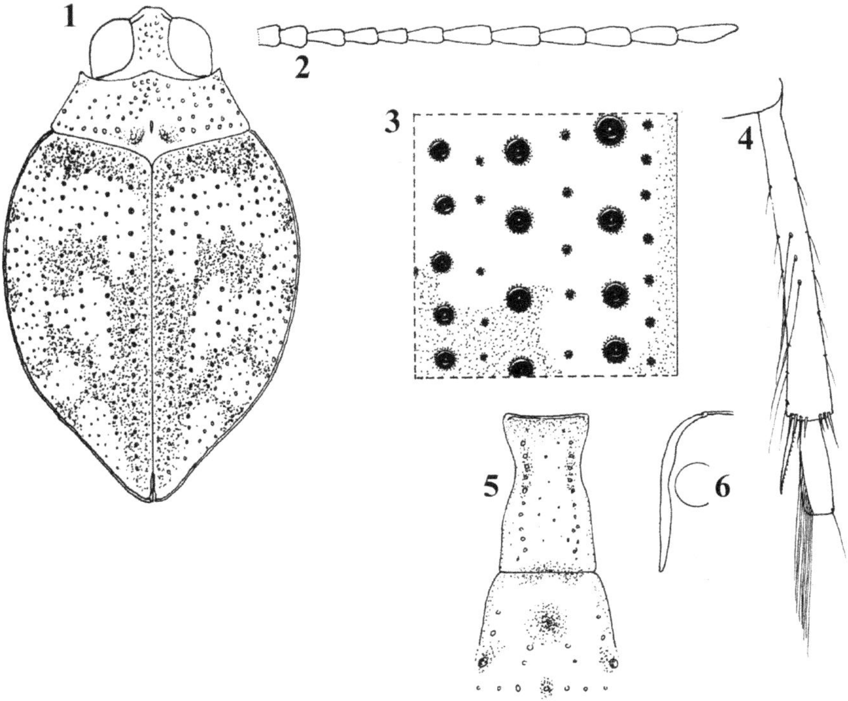

PLATE 123 *Haliplus* (*Liaphlus*) *napolovi* female. 1, dorsal view; 2, antenna; 3, elytral punctures near base and suture; 4, hind tibia in dorsal view; 5, prosternal and metaventral process; 6, prosternal process in lateral view

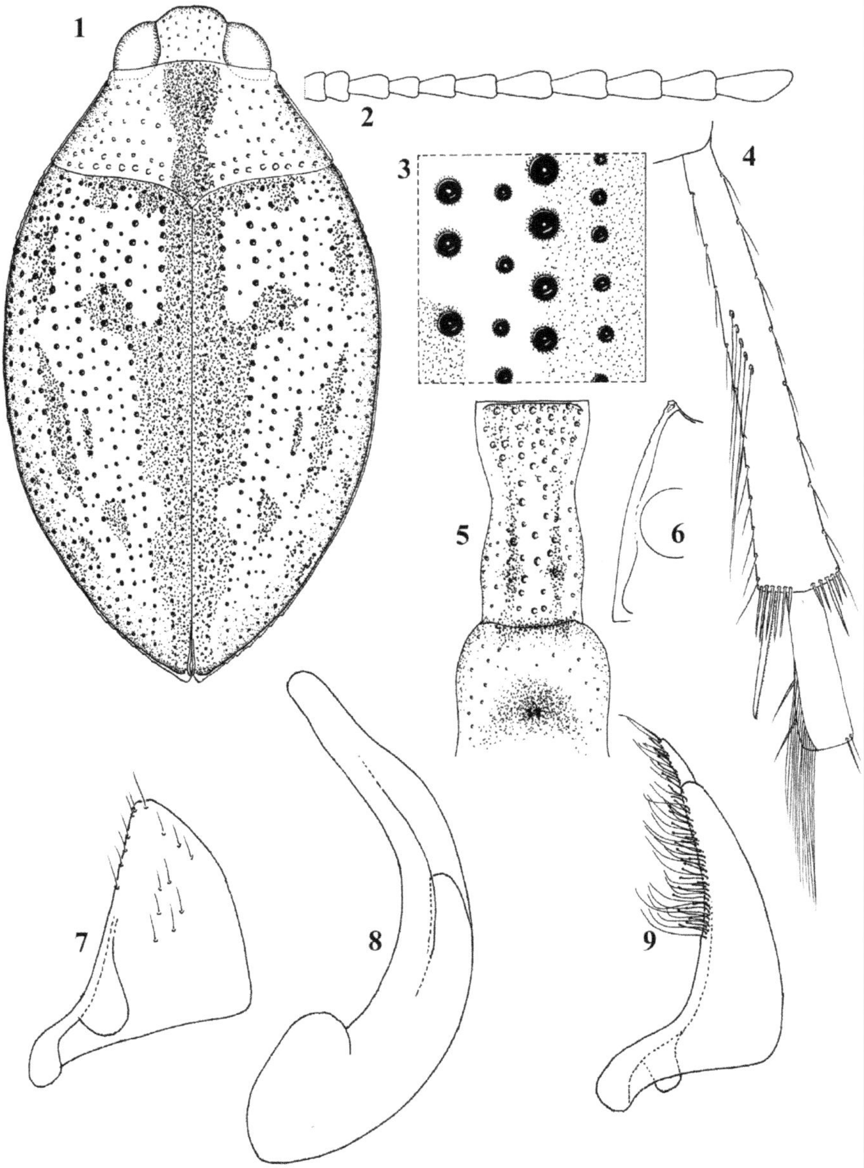

PLATE 124 *Haliplus* (*Liaphlus*) *natalensis*. 1, dorsal view; 2, antenna; 3, elytral punctures near base and suture; 4, hind tibia in dorsal view; 5, prosternal and metaventral process; 6, prosternal process in lateral view; 7, left paramere; 8, penis; 9, right paramere

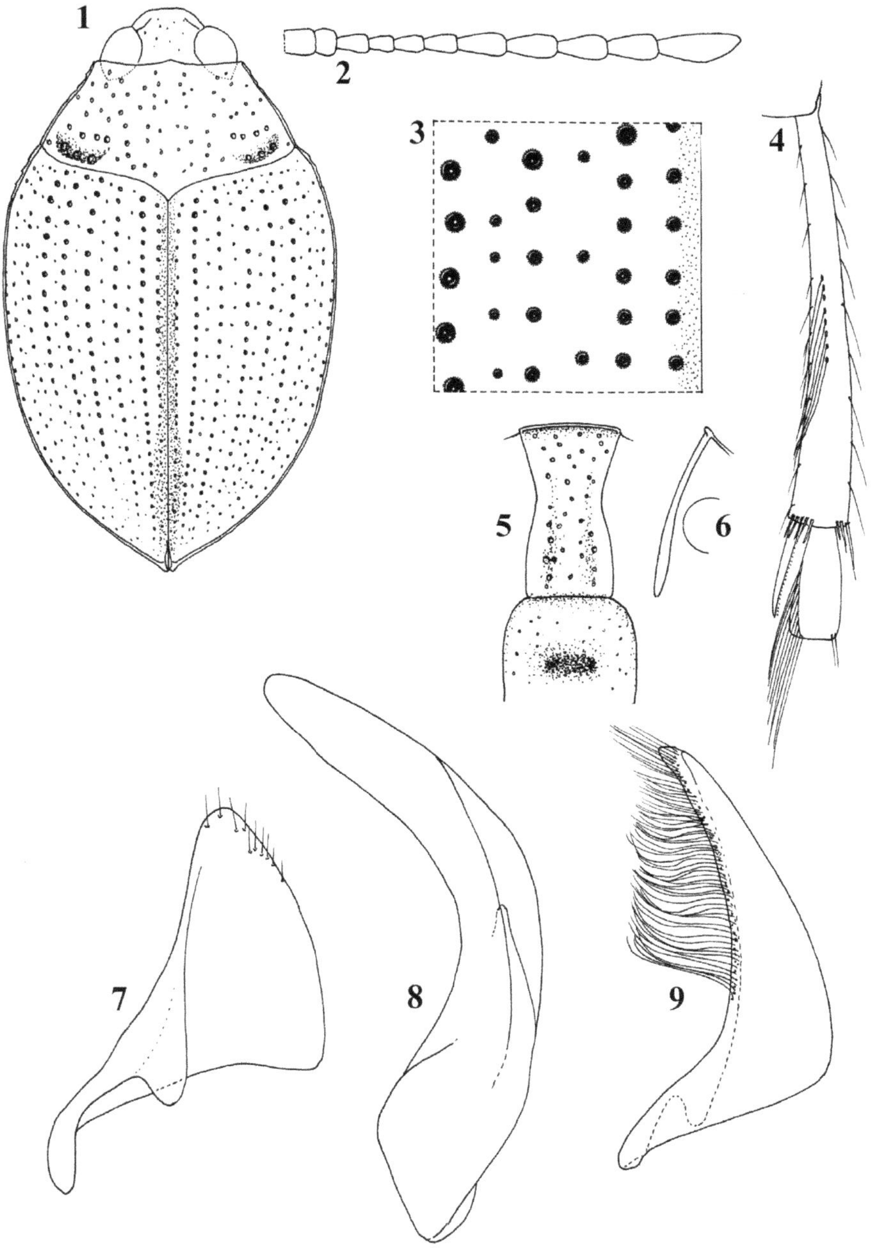

PLATE 125 *Haliplus* (*Liaphlus*) *nedungaduensis*. 1, dorsal view; 2, antenna; 3, elytral punctures near base and suture; 4, hind tibia in dorsal view; 5, prosternal and metaventral process; 6, prosternal process in lateral view; 7, left paramere; 8, penis; 9, right paramere

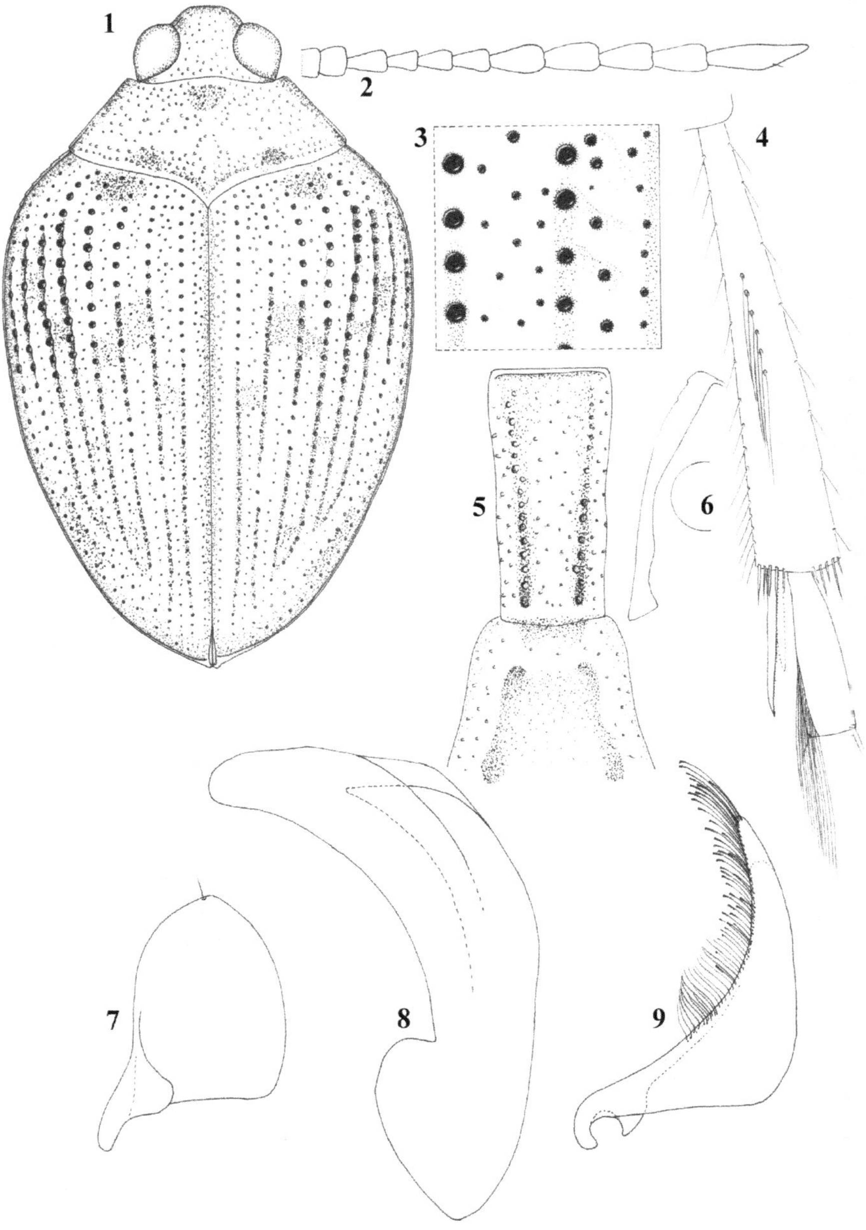

PLATE 126 *Haliplus* (*Liaphlus*) *nieseri*. 1, dorsal view; 2, antenna; 3, elytral punctures near base and suture; 4, hind tibia in dorsal view; 5, prosternal and metaventral process; 6, prosternal process in lateral view; 7, left paramere; 8, penis; 9, right paramere

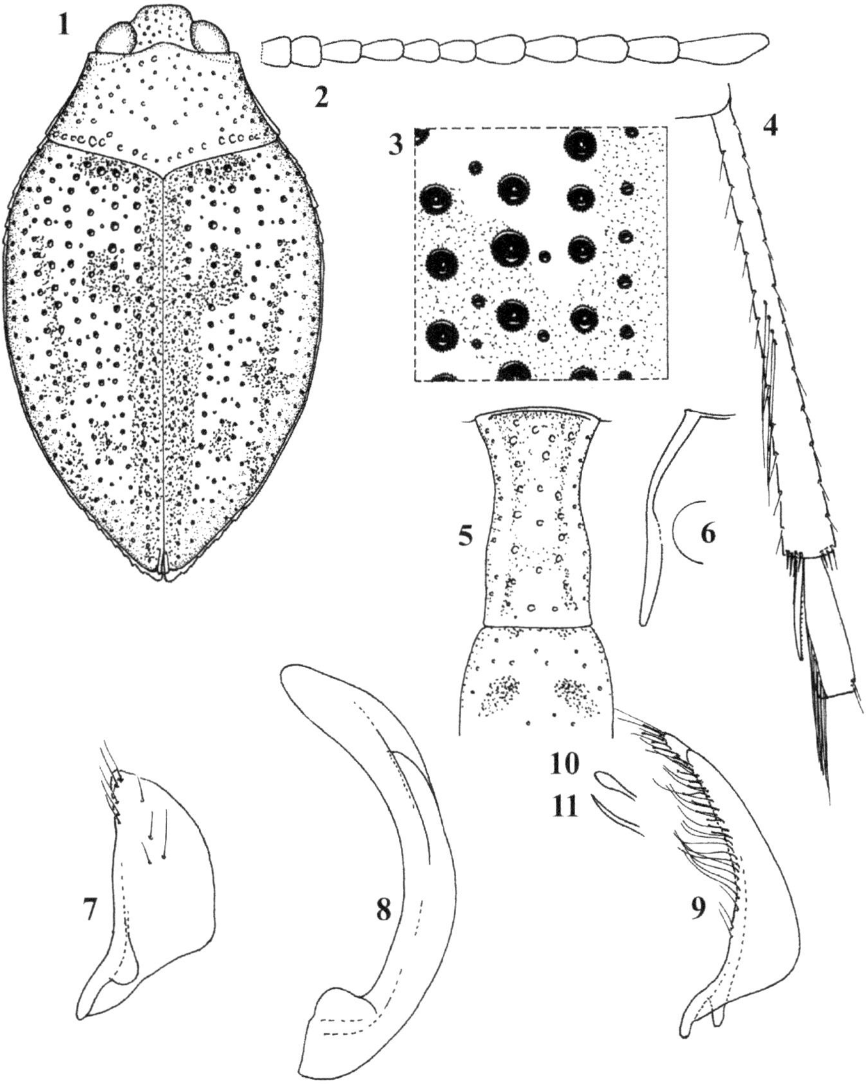

PLATE 127 *Haliplus* (*Liaphlus*) *nigerianus*. 1, dorsal view; 2, antenna; 3, elytral punctures near base and suture; 4, hind tibia in dorsal view; 5, prosternal and metaventral process; 6, prosternal process in lateral view; 7, left paramere; 8, penis; 9, right paramere; 10, 11, setae of right paramere

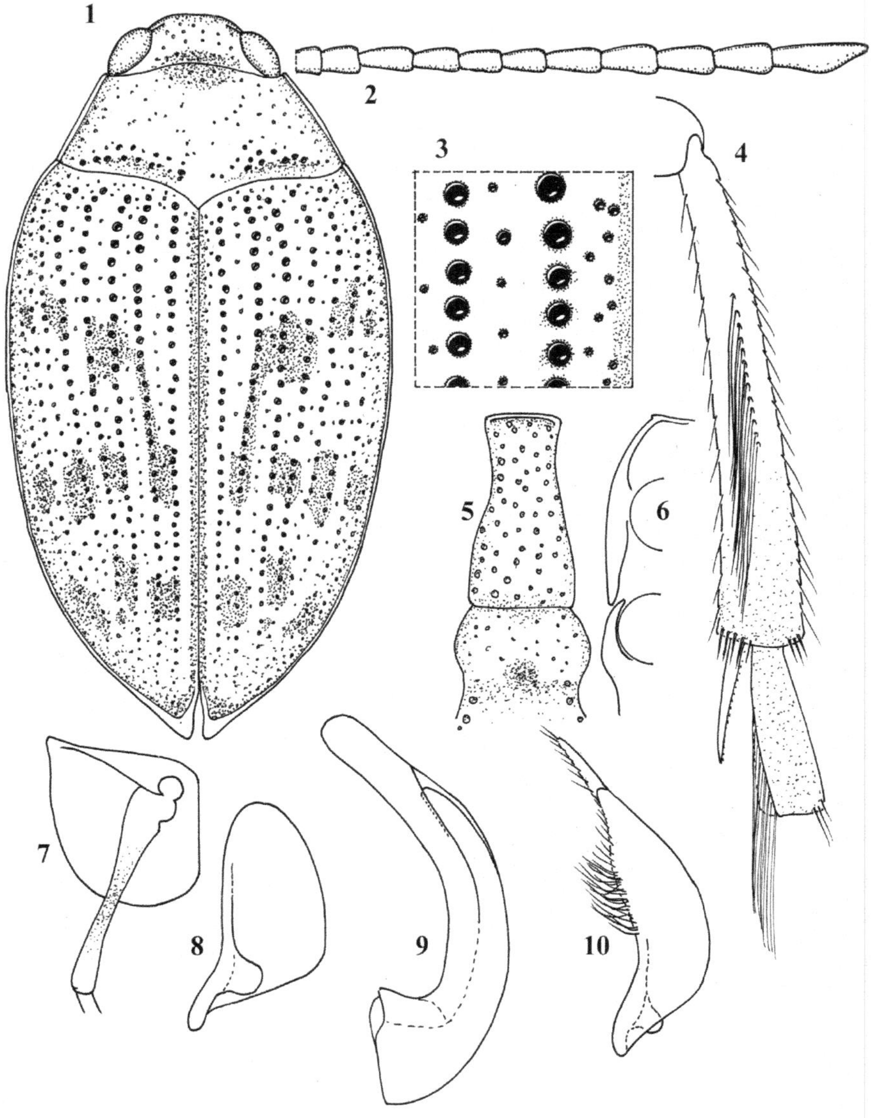

PLATE 128 *Haliplus* (*Liaphlus*) *nitens*. 1, dorsal view; 2, antenna; 3, elytral punctures near base and suture; 4, hind tibia in dorsal view; 5, prosternal and metaventral process; 6, prosternal process in lateral view; 7, hind coxal plate and femur; 8, left paramere; 9, penis; 10, right paramere

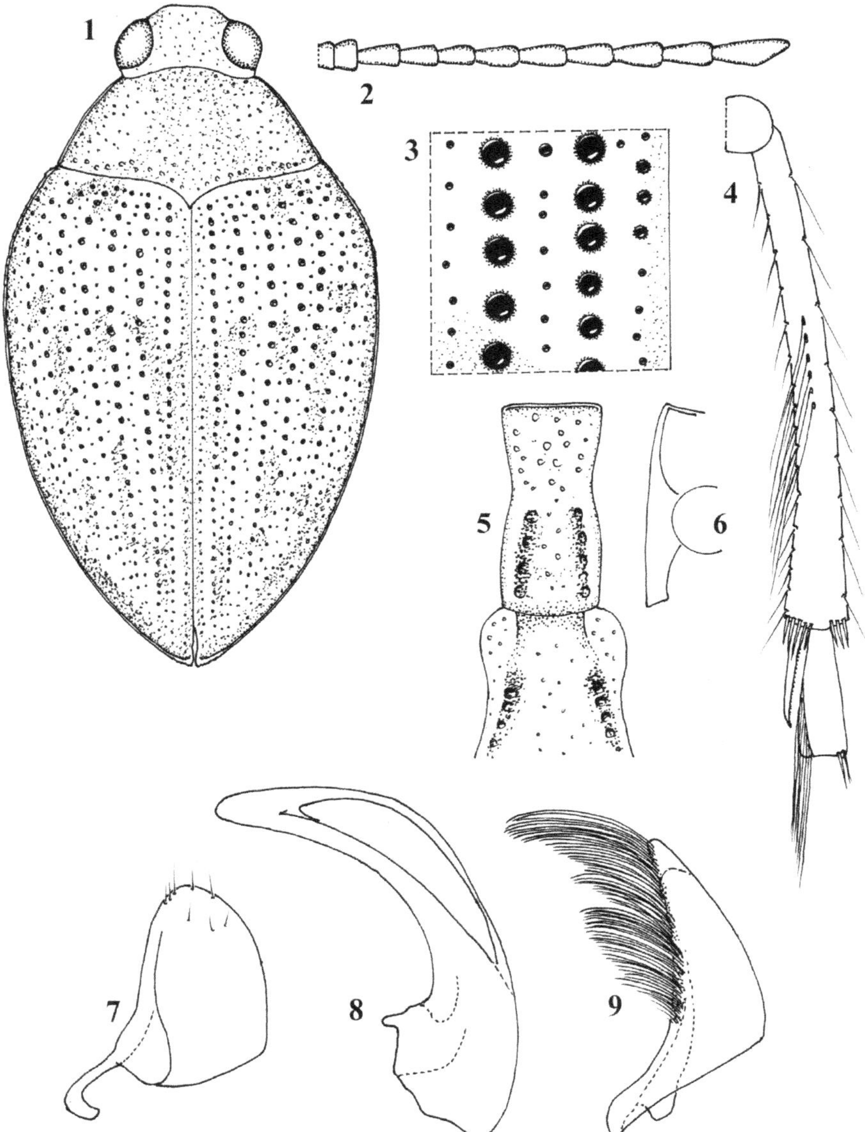

PLATE 129 *Haliplus* (*Liaphlus*) *novateutonianus*. 1, dorsal view; 2, antenna; 3, elytral punctures near base and suture; 4, hind tibia in dorsal view; 5, prosternal and metaventral process; 6, prosternal process in lateral view; 7, left paramere; 8, penis; 9, right paramere

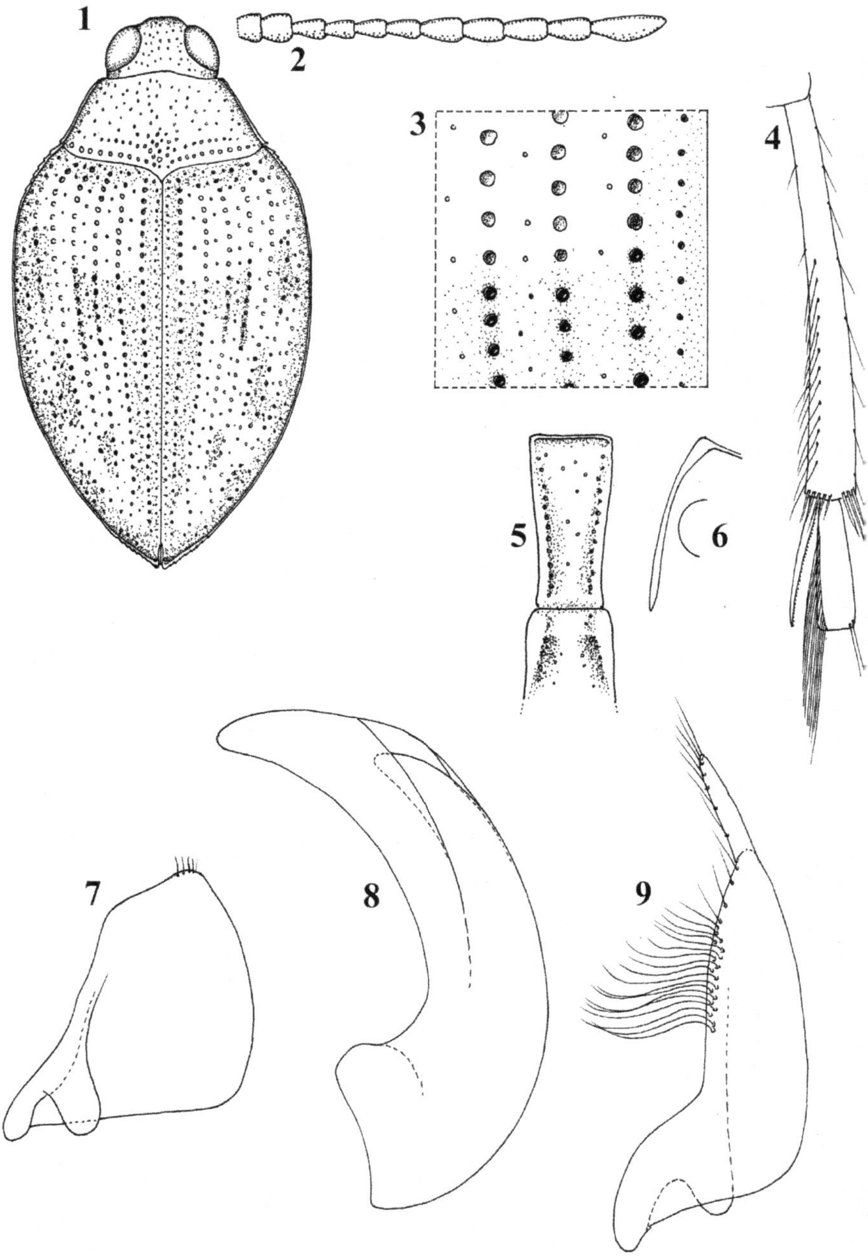

PLATE 130 *Haliplus* (*Liaphlus*) *oklahomensis*. 1, dorsal view; 2, antenna; 3, elytral punctures near base and suture; 4, hind tibia in dorsal view; 5, prosternal and metaventral process; 6, prosternal process in lateral view; 7, left paramere; 8, penis; 9, right paramere

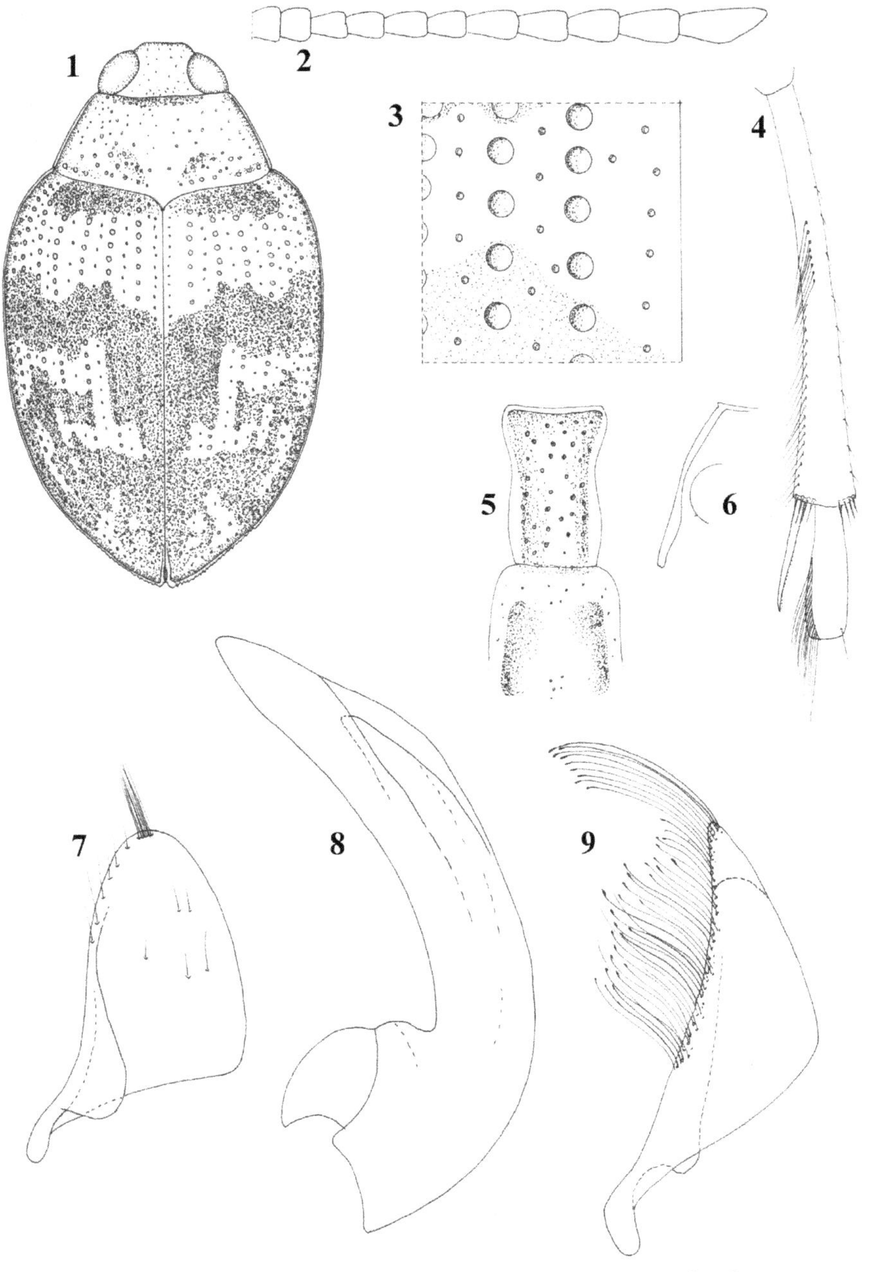

PLATE 131 *Haliplus* (*Liaphlus*) *ornatipennis*. 1, dorsal view; 2, antenna; 3, elytral punctures near base and suture; 4, hind tibia in dorsal view; 5, prosternal and metaventral process; 6, prosternal process in lateral view; 7, left paramere; 8, penis; 9, right paramere

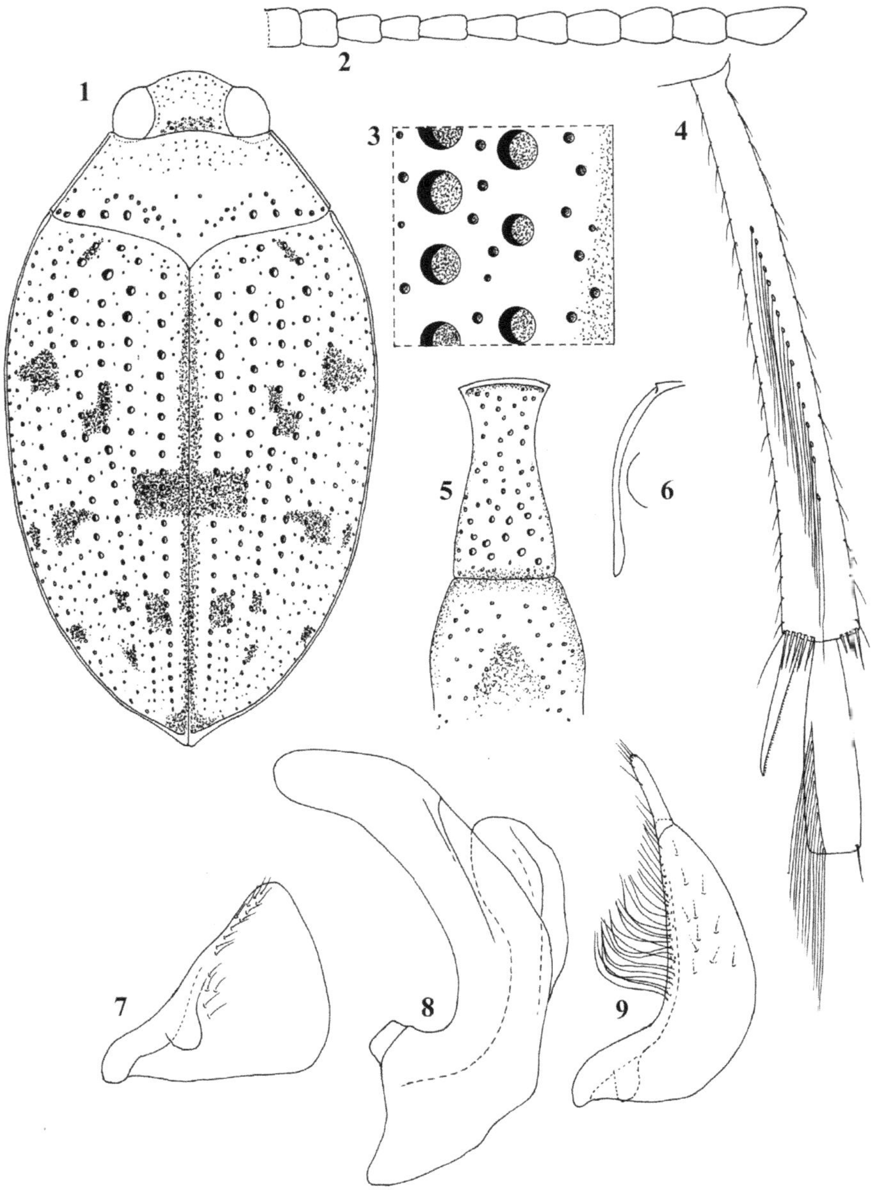

PLATE 132 *Haliplus* (*Liaphlus*) *ovalis*. 1, dorsal view; 2, antenna; 3, elytral punctures near base and suture; 4, hind tibia in dorsal view; 5, prosternal and metaventral process; 6, prosternal process in lateral view; 7, left paramere; 8, penis; 9, right paramere

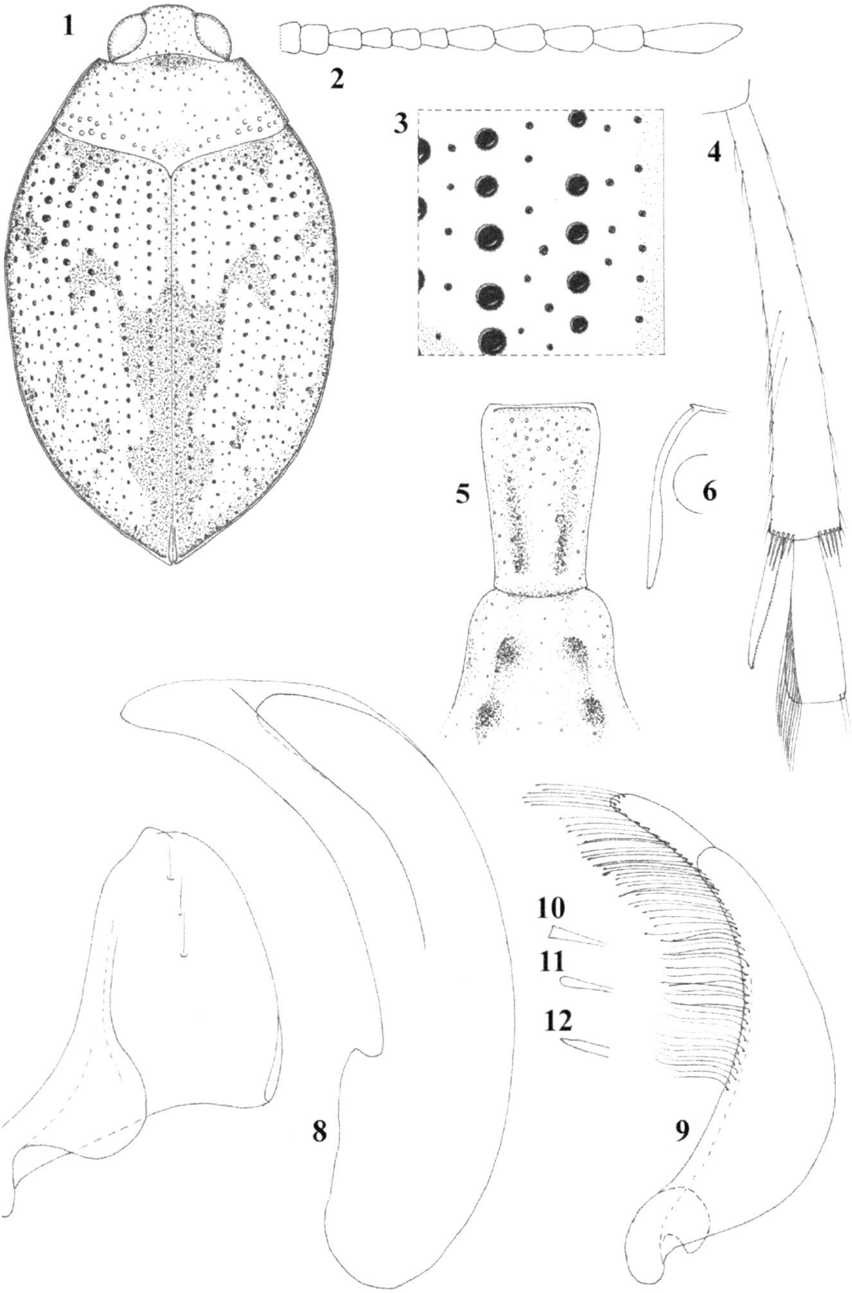

PLATE 133 *Haliplus* (*Liaphlus*) *panamanus*. 1, dorsal view; 2, antenna; 3, elytral punctures near base and suture; 4, hind tibia in dorsal view; 5, prosternal and metaventral process; 6, prosternal process in lateral view; 7, left paramere; 8, penis; 9, right paramere; 10–12 setae of right paramere

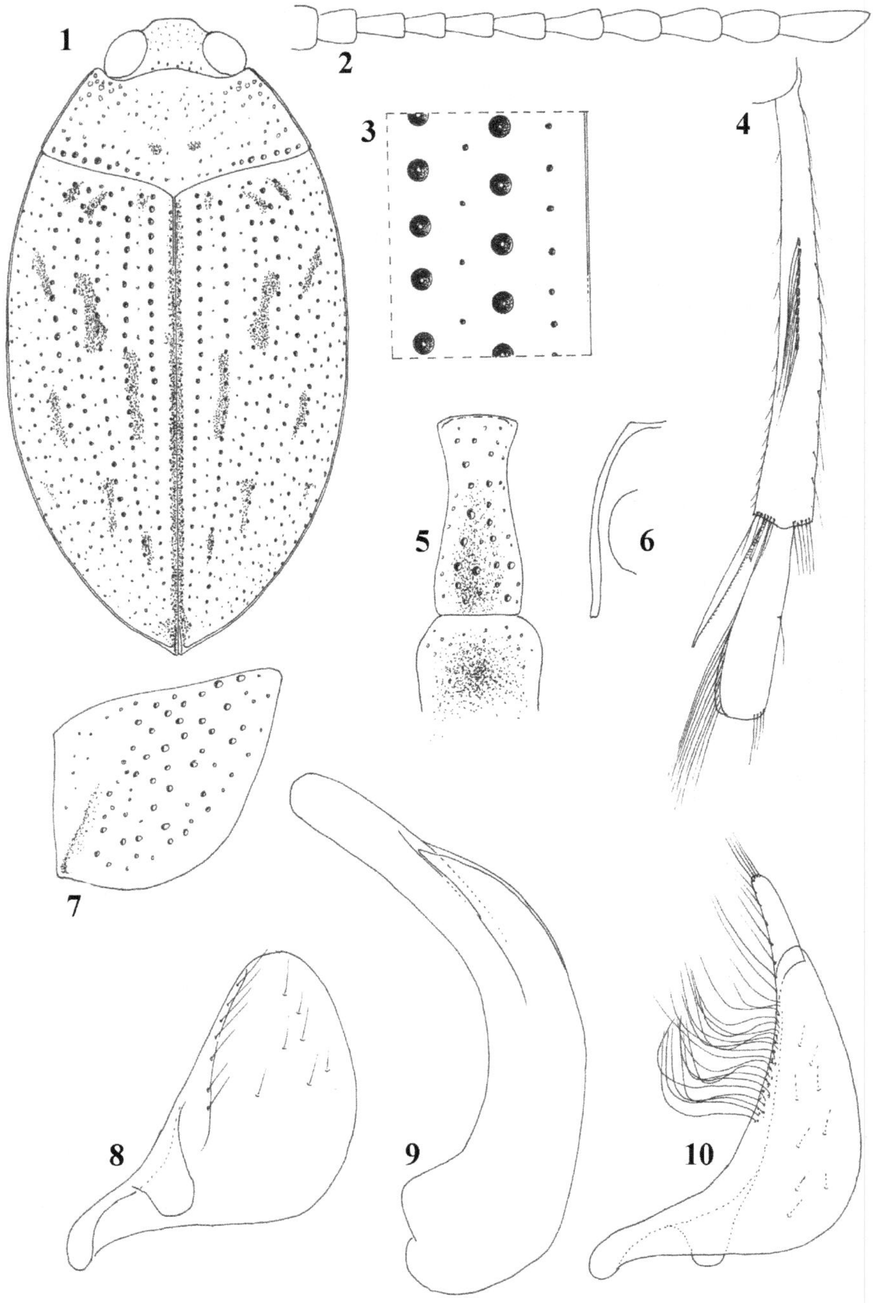

PLATE 134 *Haliplus* (*Liaphlus*) *perroti*. 1, dorsal view; 2, antenna; 3, elytral punctures near base and suture; 4, hind tibia in dorsal view; 5, prosternal and metaventral process; 6, prosternal process in lateral view; 7, hind coxal plate; 8, left paramere; 9, penis; 10, right paramere

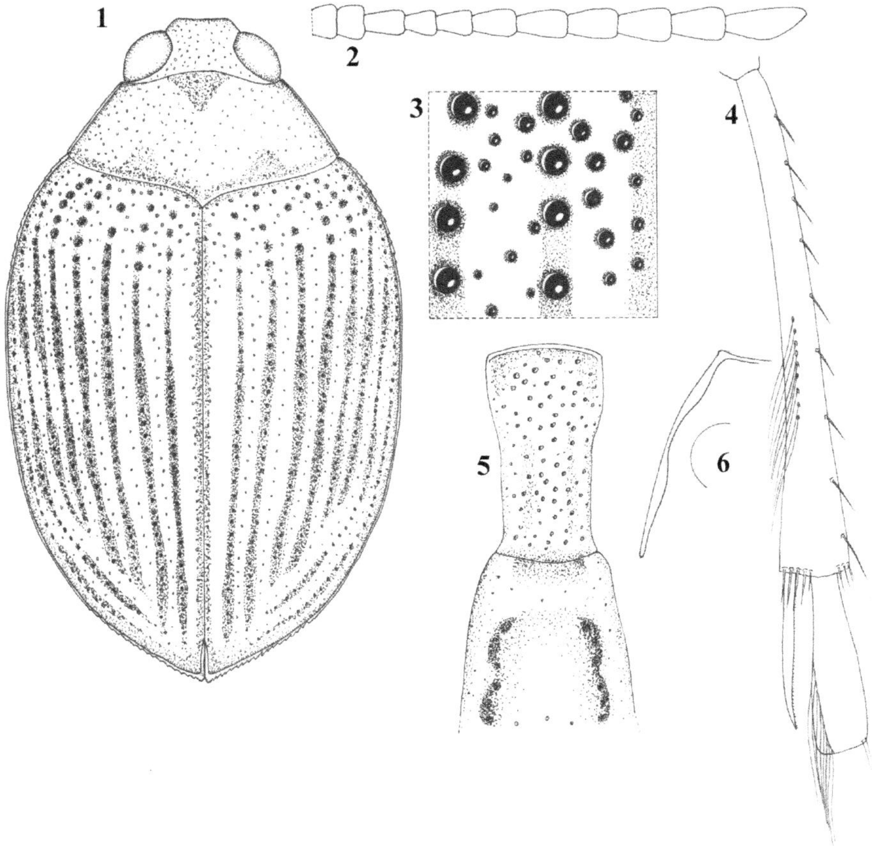

PLATE 135 *Haliplus* (*Liaphlus*) *peruanus* female. 1, dorsal view; 2, antenna; 3, elytral punctures near base and suture; 4, hind tibia in dorsal view; 5, prosternal and metaventral process; 6, prosternal process in lateral view

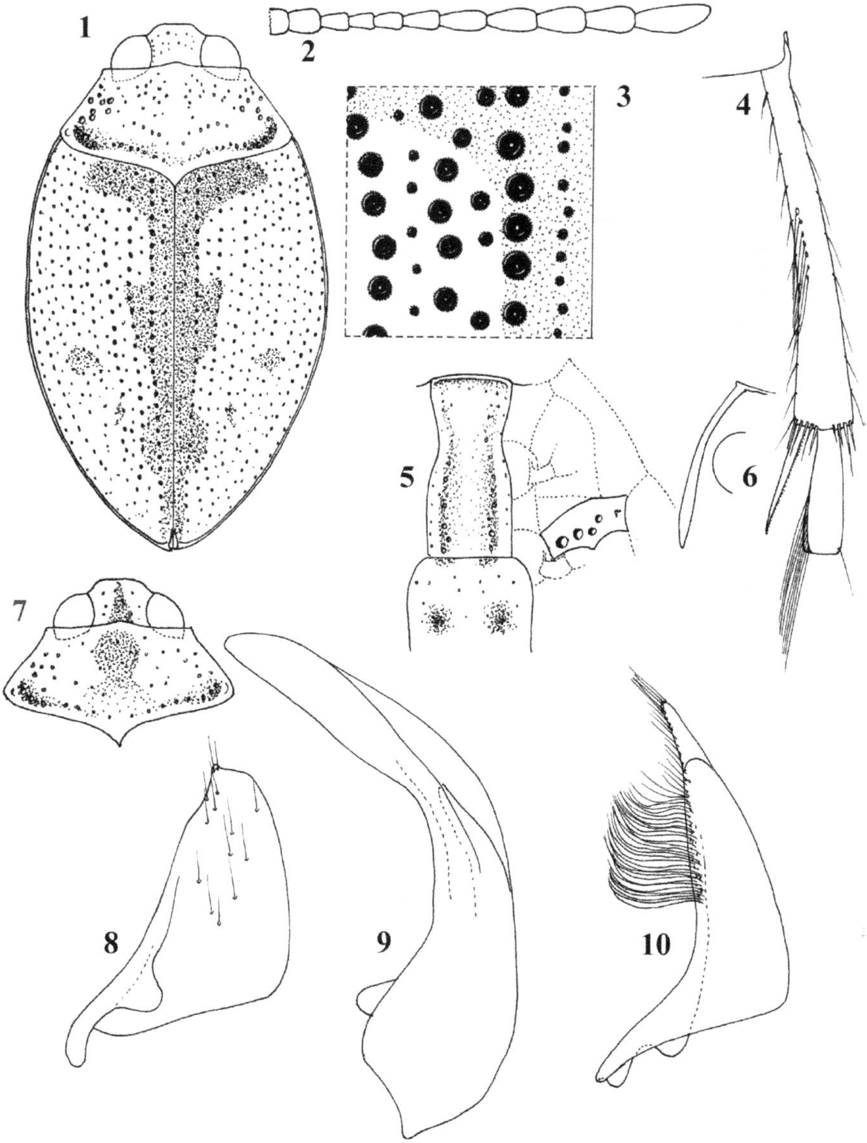

PLATE 136 *Haliplus* (*Liaphlus*) *philippinus*. 1, dorsal view; 2, antenna; 3, elytral punctures near base and suture; 4, hind tibia in dorsal view; 5, prosternal and metaventral process and mesepimeron; 6, prosternal process in lateral view; 7, head and pronotum with possible maculation; 8, left paramere; 9, penis;10, right paramere

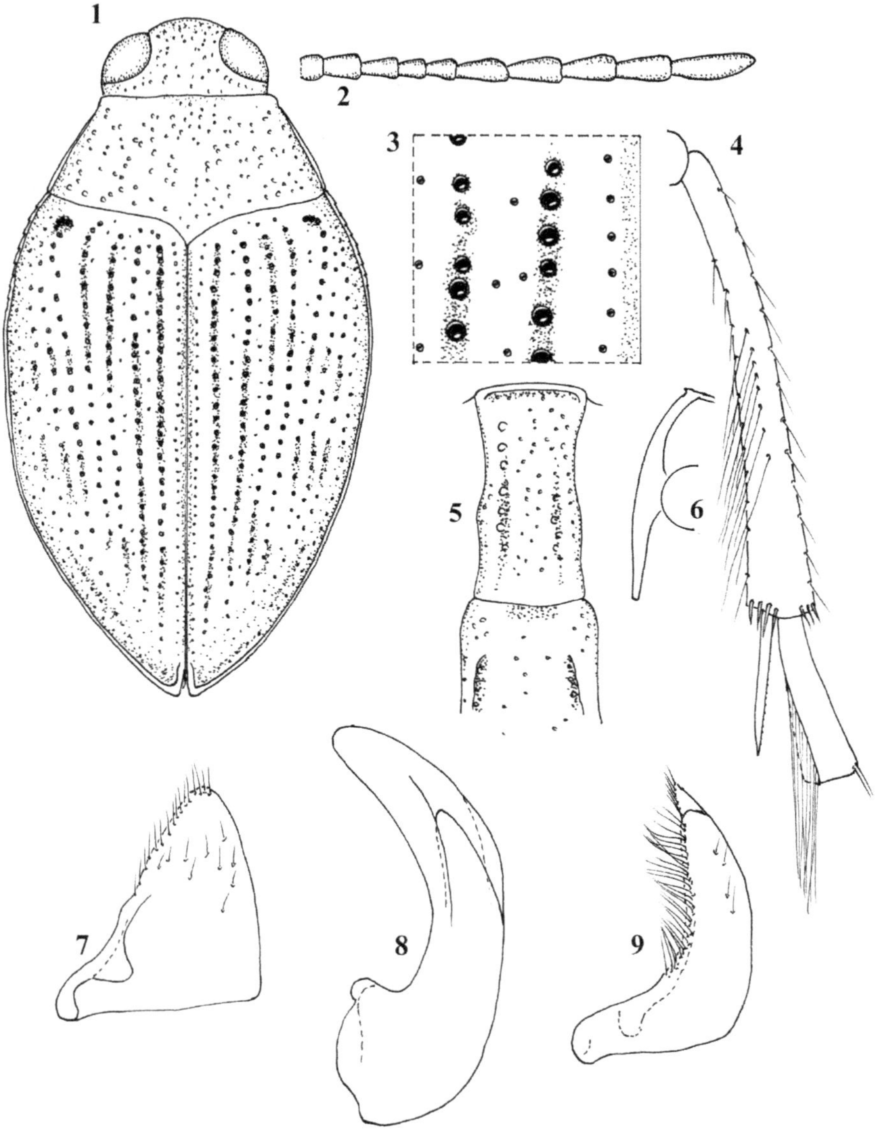

PLATE 137 *Haliplus* (*Liaphlus*) *pilbaraensis*. 1, dorsal view; 2, antenna; 3, elytral punctures near base and suture; 4, hind tibia in dorsal view; 5, prosternal and metaventral process; 6, prosternal process in lateral view; 7, left paramere; 8, penis; 9, right paramere

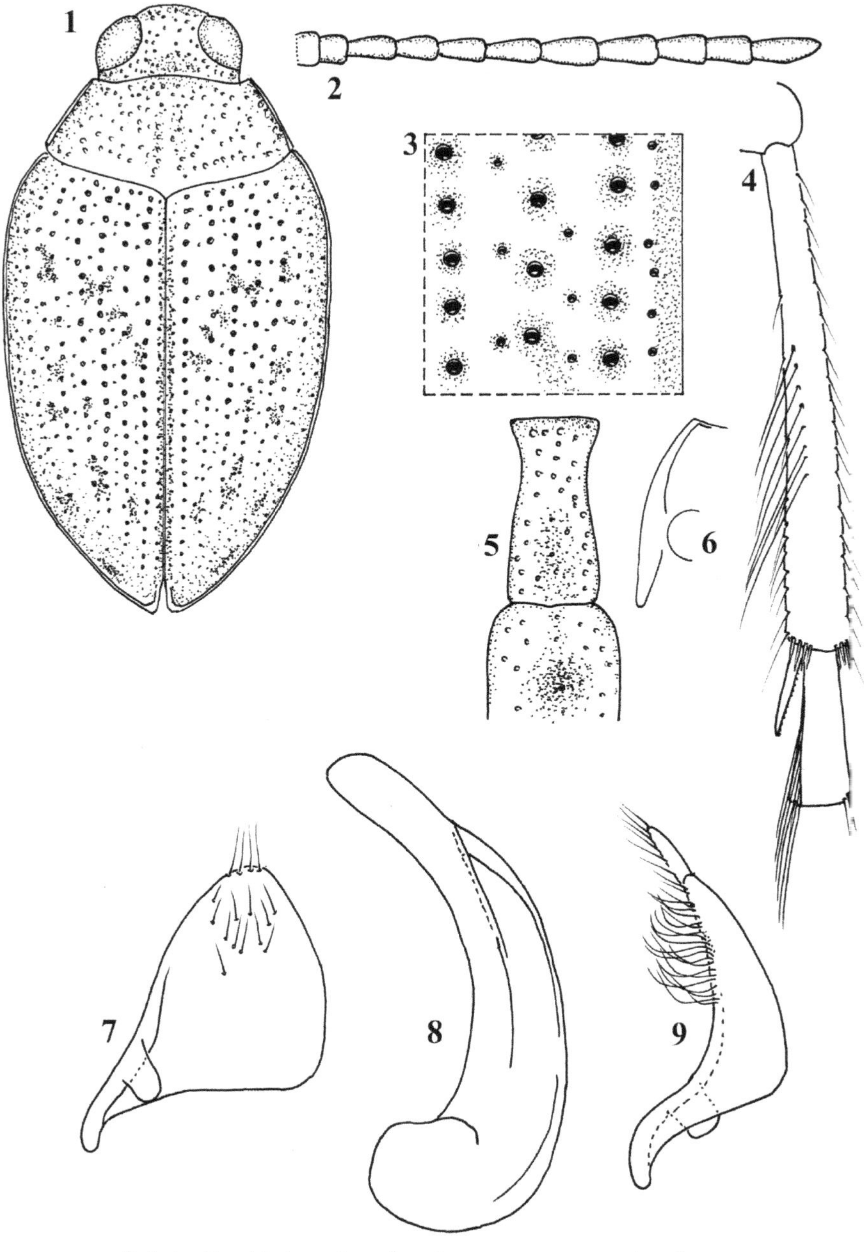

PLATE 138 *Haliplus* (*Liaphlus*) *pruthii*. 1, dorsal view; 2, antenna; 3, elytral punctures near base and suture; 4, hind tibia in dorsal view; 5, prosternal and metaventral process; 6, prosternal process in lateral view; 7, left paramere; 8, penis; 9, right paramere

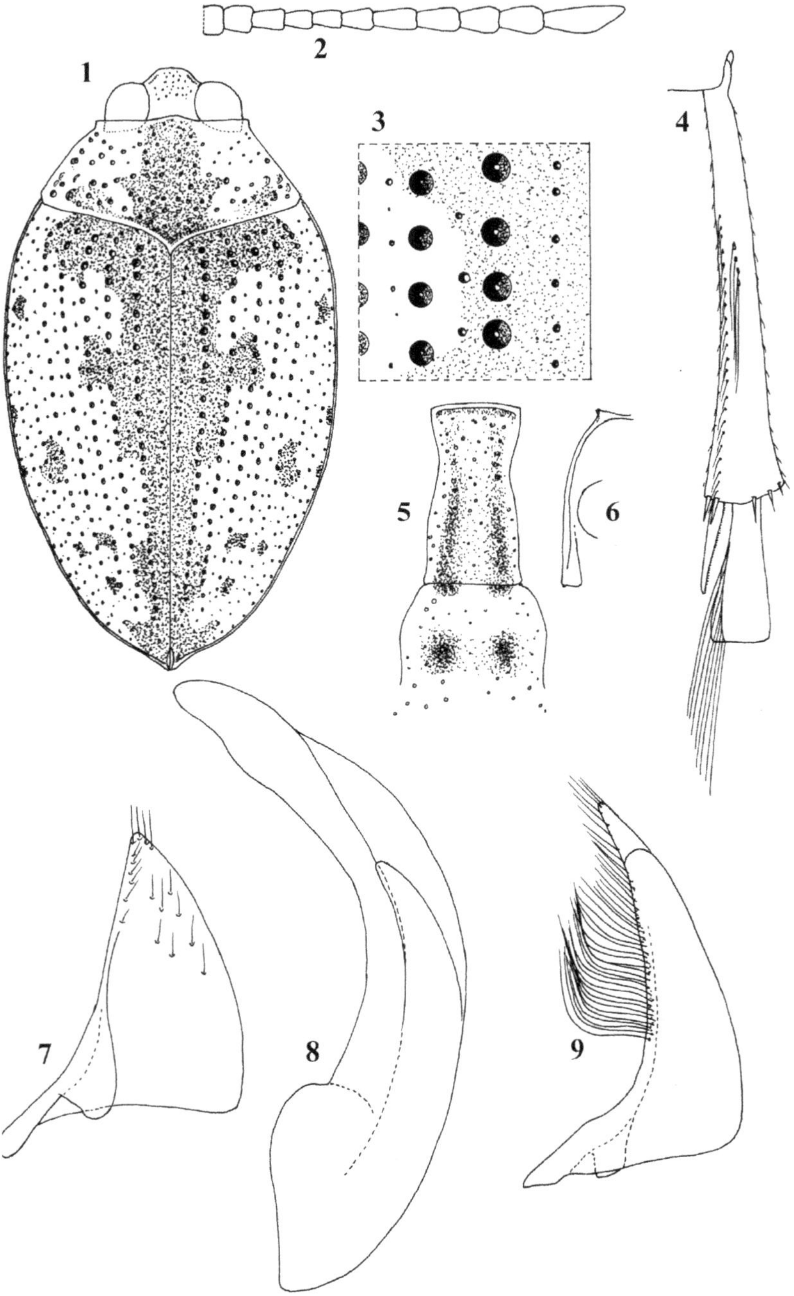

PLATE 139 *Haliplus* (*Liaphlus*) *pulchellus*. 1, dorsal view; 2, antenna; 3, elytral punctures near base and suture; 4, hind tibia in dorsal view; 5, prosternal and metaventral process; 6, prosternal process in lateral view; 7, left paramere; 8, penis; 9, right paramere

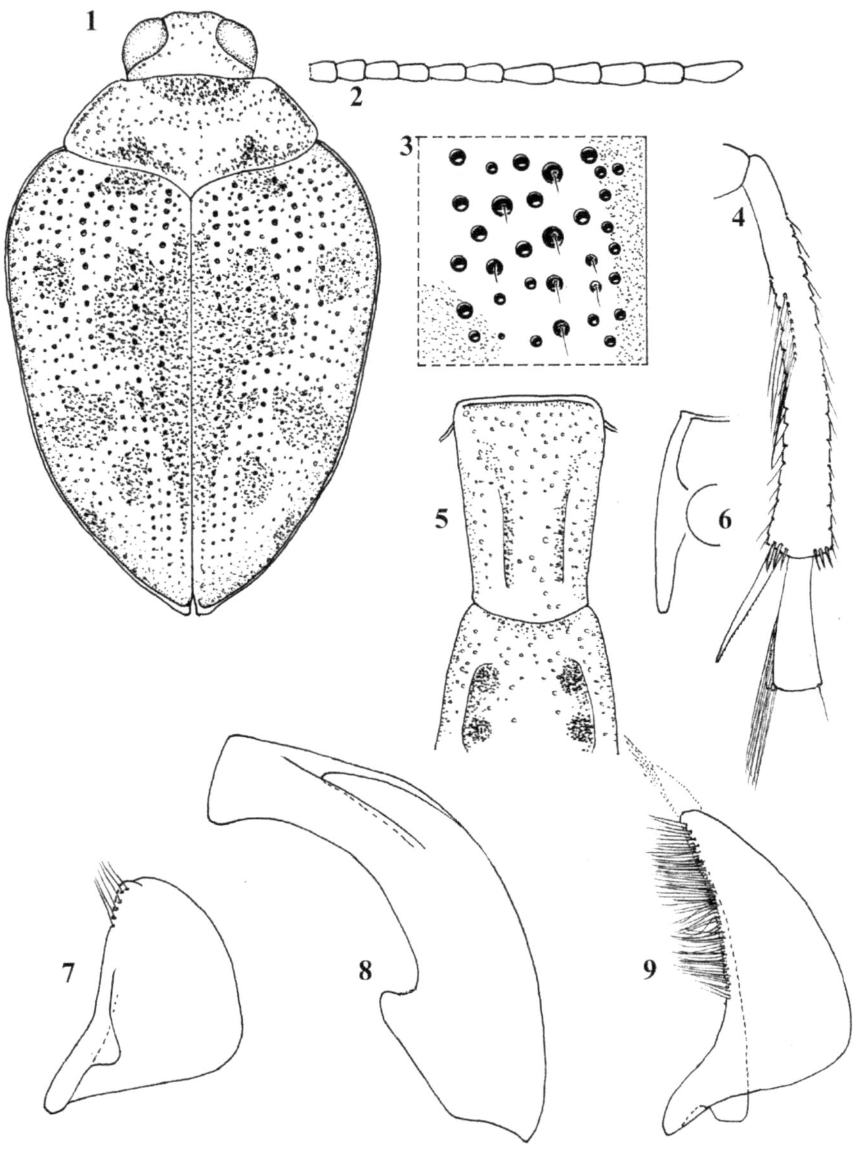

PLATE 140 *Haliplus* (*Liaphlus*) *regili.* 1, dorsal view; 2, antenna; 3, elytral punctures near base and suture; 4, hind tibia in dorsal view; 5, prosternal and metaventral process; 6, prosternal process in lateral view; 7, left paramere; 8, penis; 9, right paramere

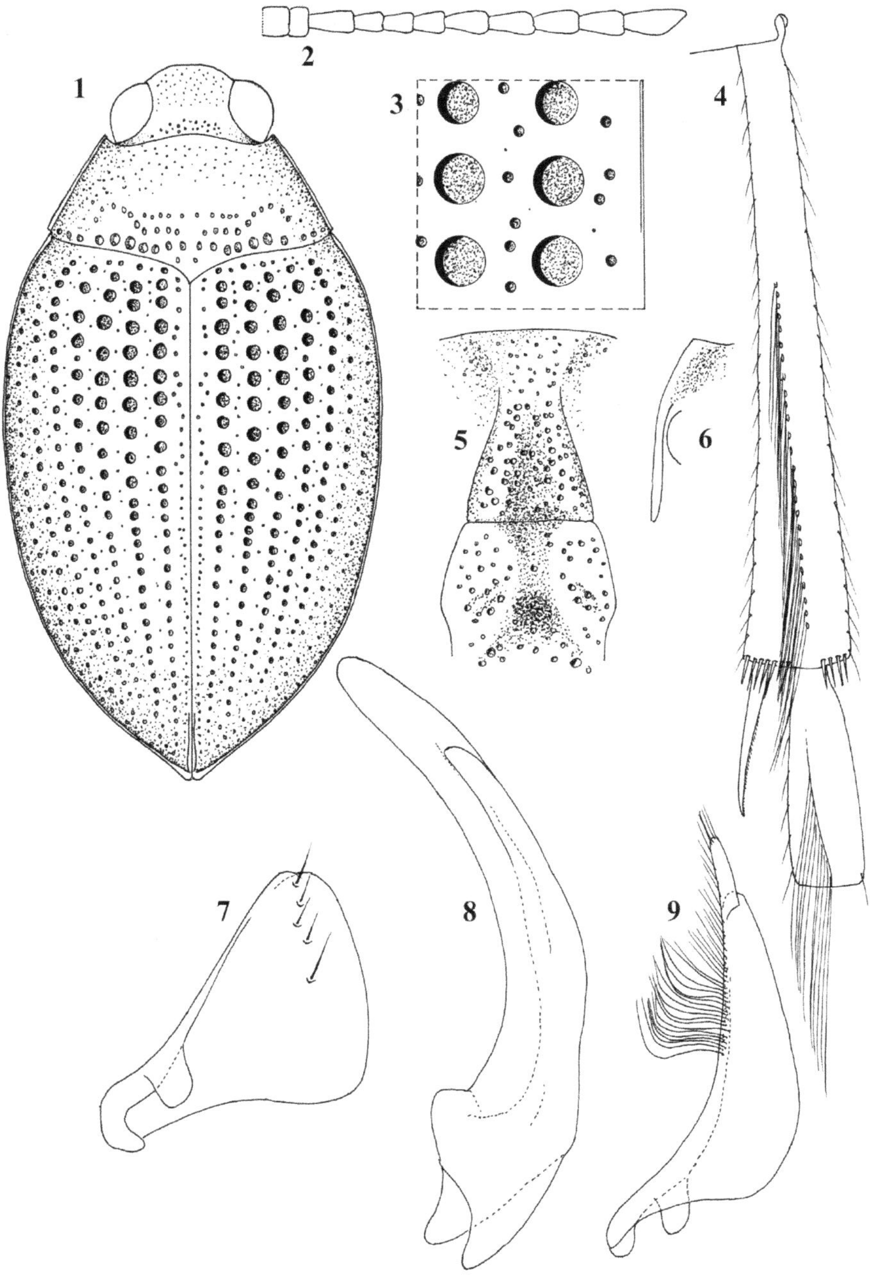

PLATE 141 *Haliplus* (*Liaphlus*) *rubidus*. 1, dorsal view; 2, antenna; 3, elytral punctures near base and suture; 4, hind tibia in dorsal view; 5, prosternal and metaventral process; 6, prosternal process in lateral view; 7, left paramere; 8, penis; 9, right paramere

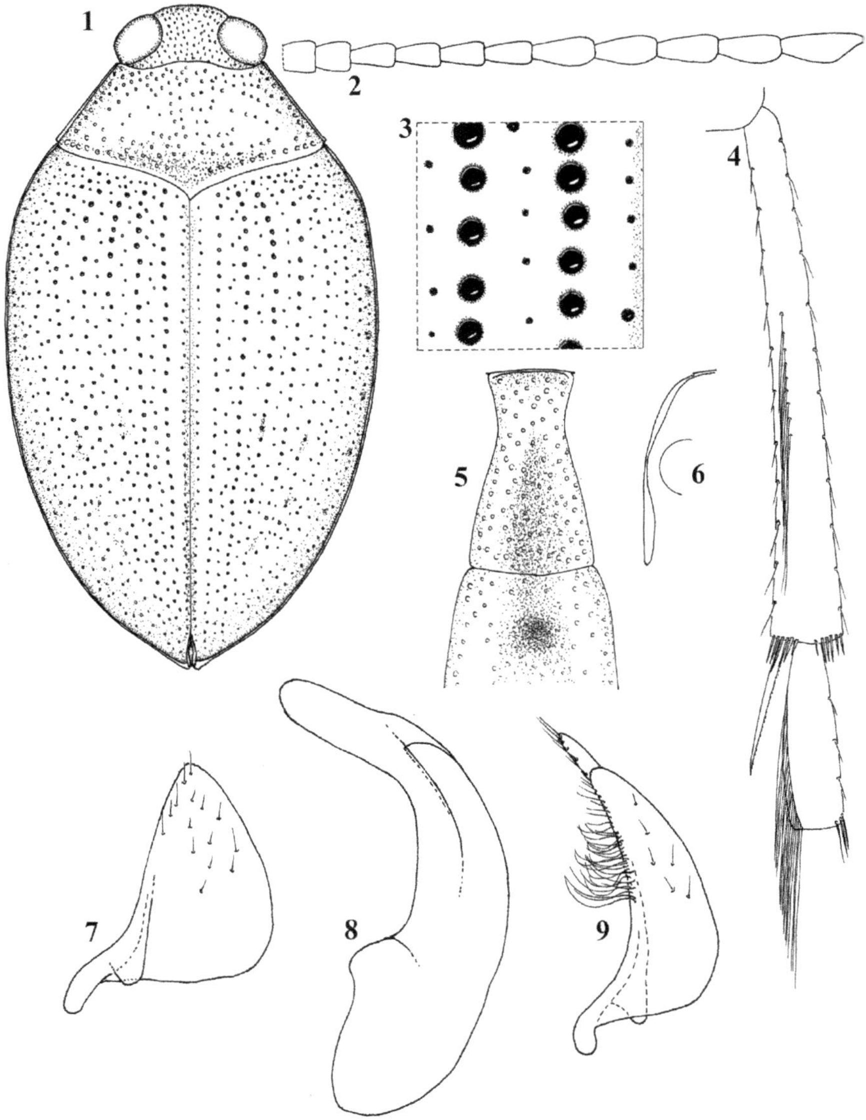

PLATE 142 *Haliplus* (*Liaphlus*) *rufescens*. 1, dorsal view; 2, antenna; 3, elytral punctures near base and suture; 4, hind tibia in dorsal view; 5, prosternal and metaventral process; 6, prosternal process in lateral view; 7, left paramere; 8, penis; 9, right paramere

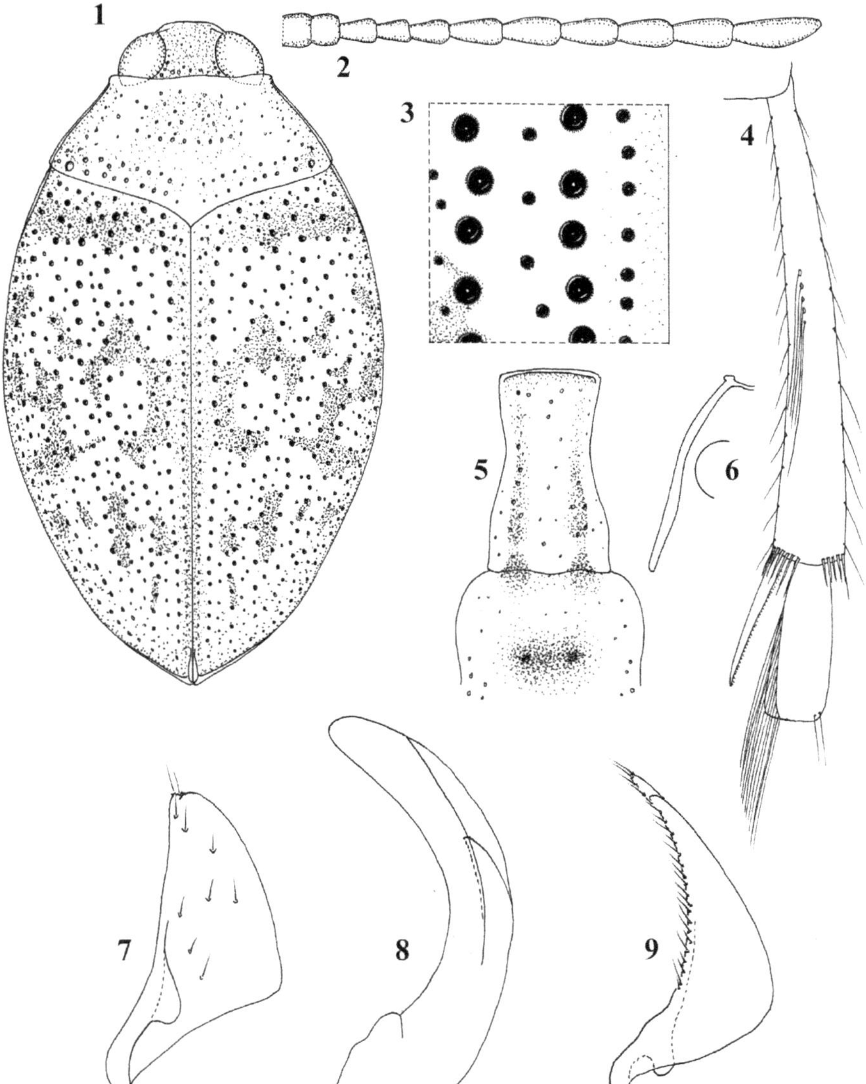

PLATE 143 *Haliplus* (*Liaphlus*) *samosirensis*. 1, dorsal view; 2, antenna; 3, elytral punctures near base and suture; 4, hind tibia in dorsal view; 5, prosternal and metaventral process; 6, prosternal process in lateral view; 7, left paramere; 8, penis; 9, right paramere

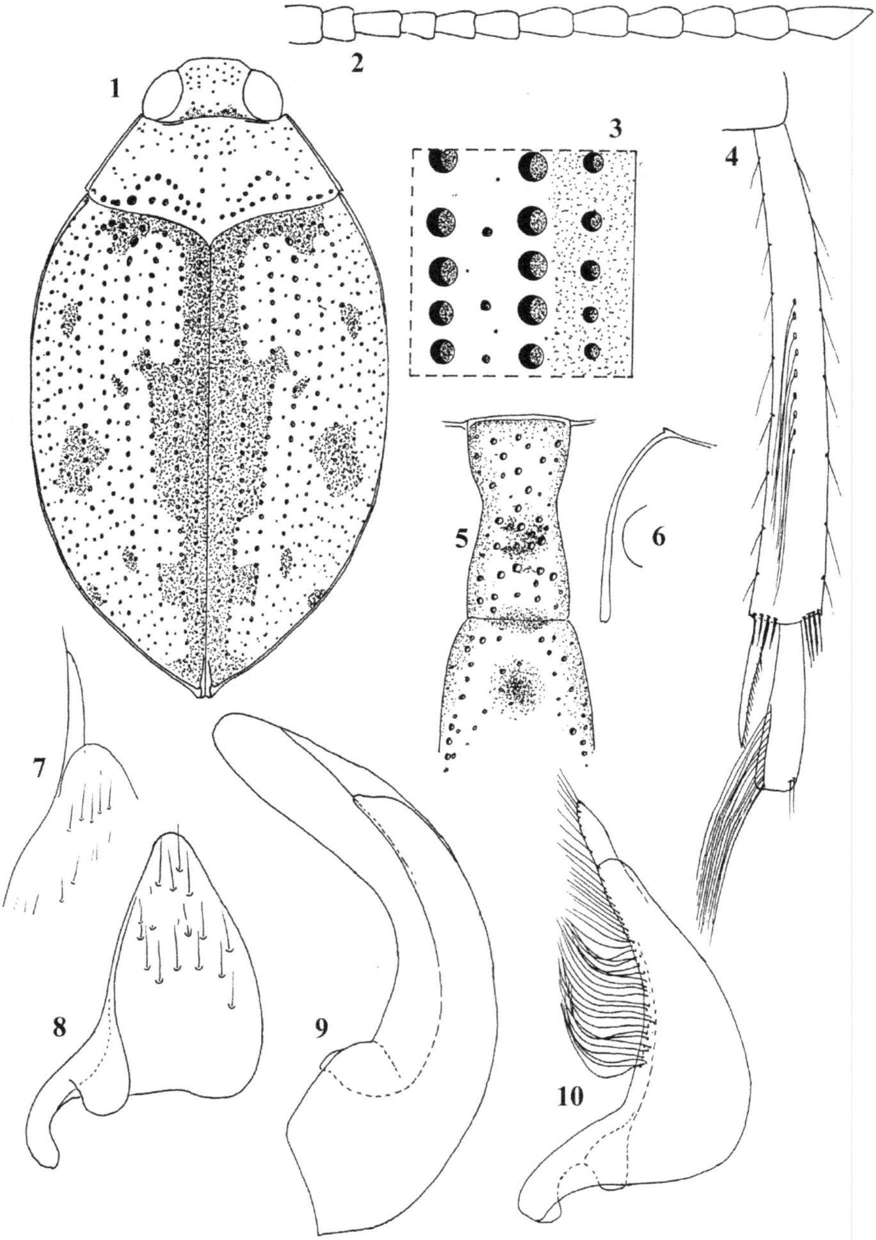

PLATE 144 *Haliplus* (*Liaphlus*) *sharpi*. 1, dorsal view; 2, antenna; 3, elytral punctures near base and suture; 4, hind tibia in dorsal view; 5, prosternal and metaventral process; 6, prosternal process in lateral view; 7, 8, left paramere; 9, penis; 10, right paramere

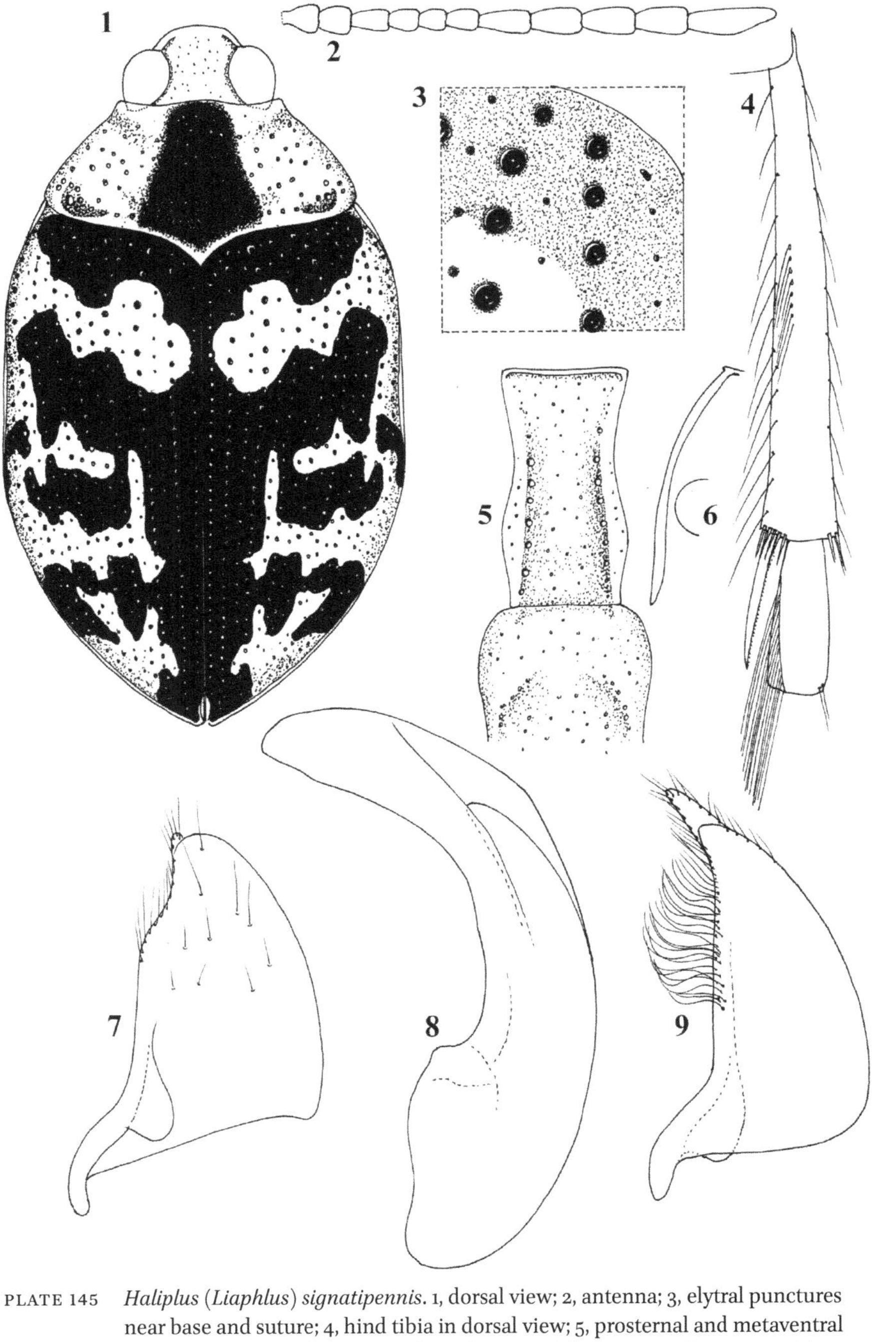

PLATE 145 *Haliplus* (*Liaphlus*) *signatipennis*. 1, dorsal view; 2, antenna; 3, elytral punctures near base and suture; 4, hind tibia in dorsal view; 5, prosternal and metaventral process; 6, prosternal process in lateral view; 7, left paramere; 8, penis; 9, right paramere

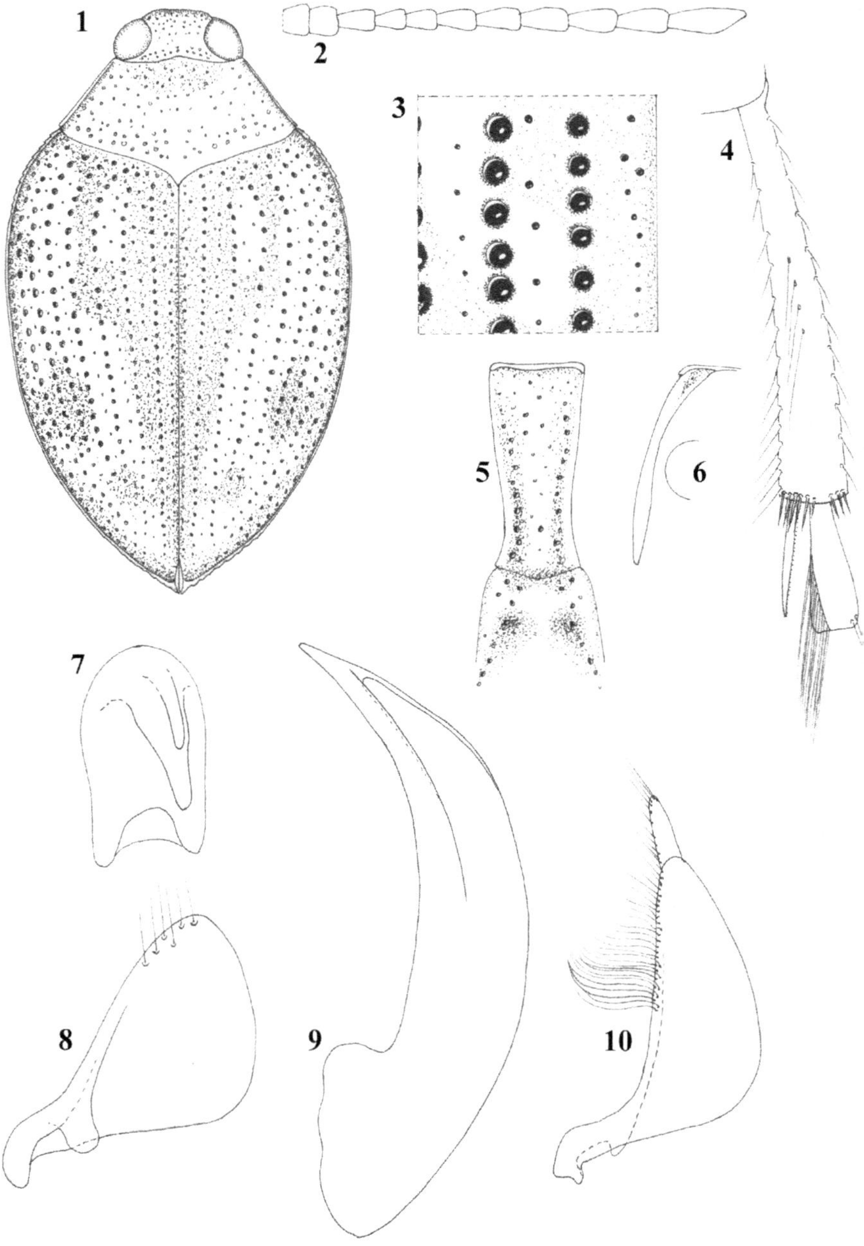

PLATE 146 *Haliplus* (*Liaphlus*) *signatus*. 1, dorsal view; 2, antenna; 3, elytral punctures near base and suture; 4, hind tibia in dorsal view; 5, prosternal and metaventral process; 6, prosternal process in lateral view; 7, left paramer in frontal view; 8, left paramere; 9, penis; 10, right paramere

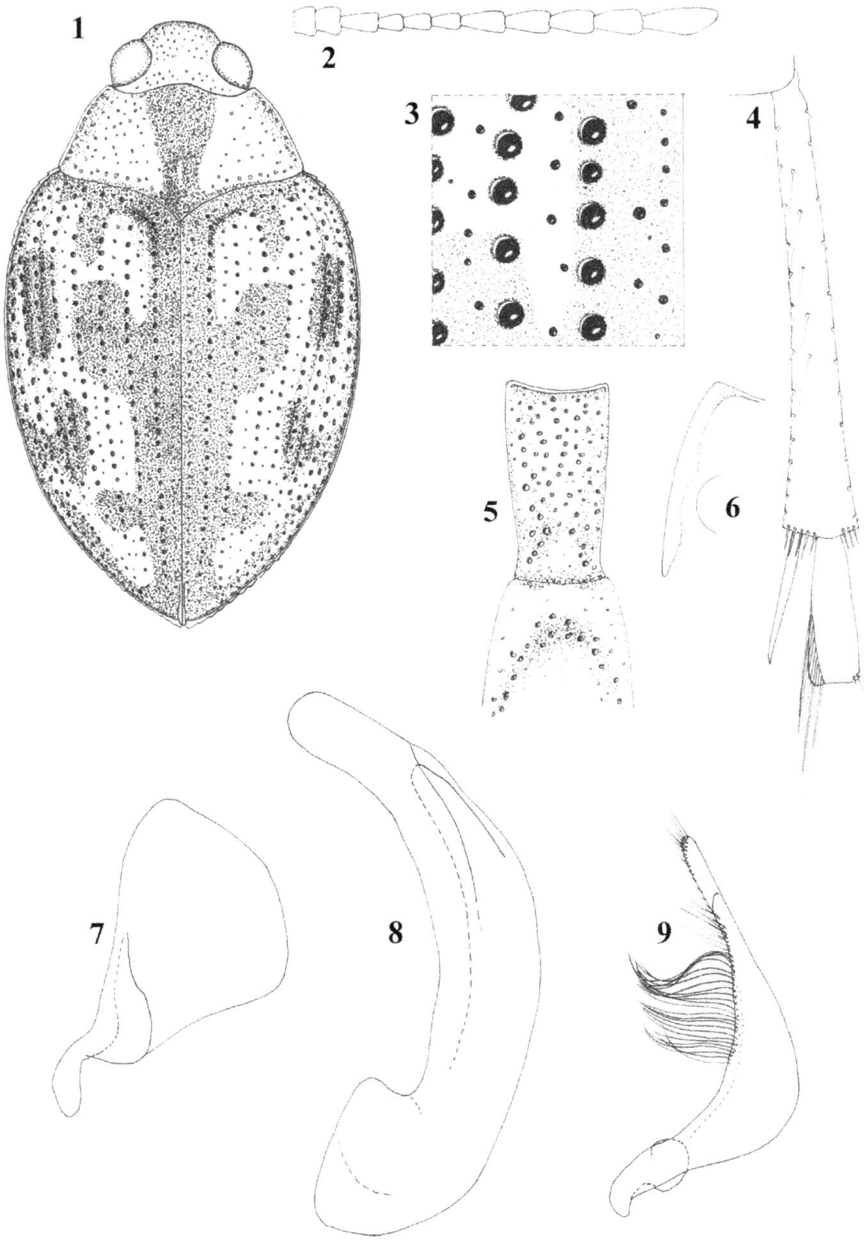

PLATE 147 *Haliplus* (*Liaphlus*) *solitarius*. 1, dorsal view; 2, antenna; 3, elytral punctures near base and suture; 4, hind tibia in dorsal view; 5, prosternal and metaventral process; 6, prosternal process in lateral view; 7, left paramere; 8, penis; 9, right paramere

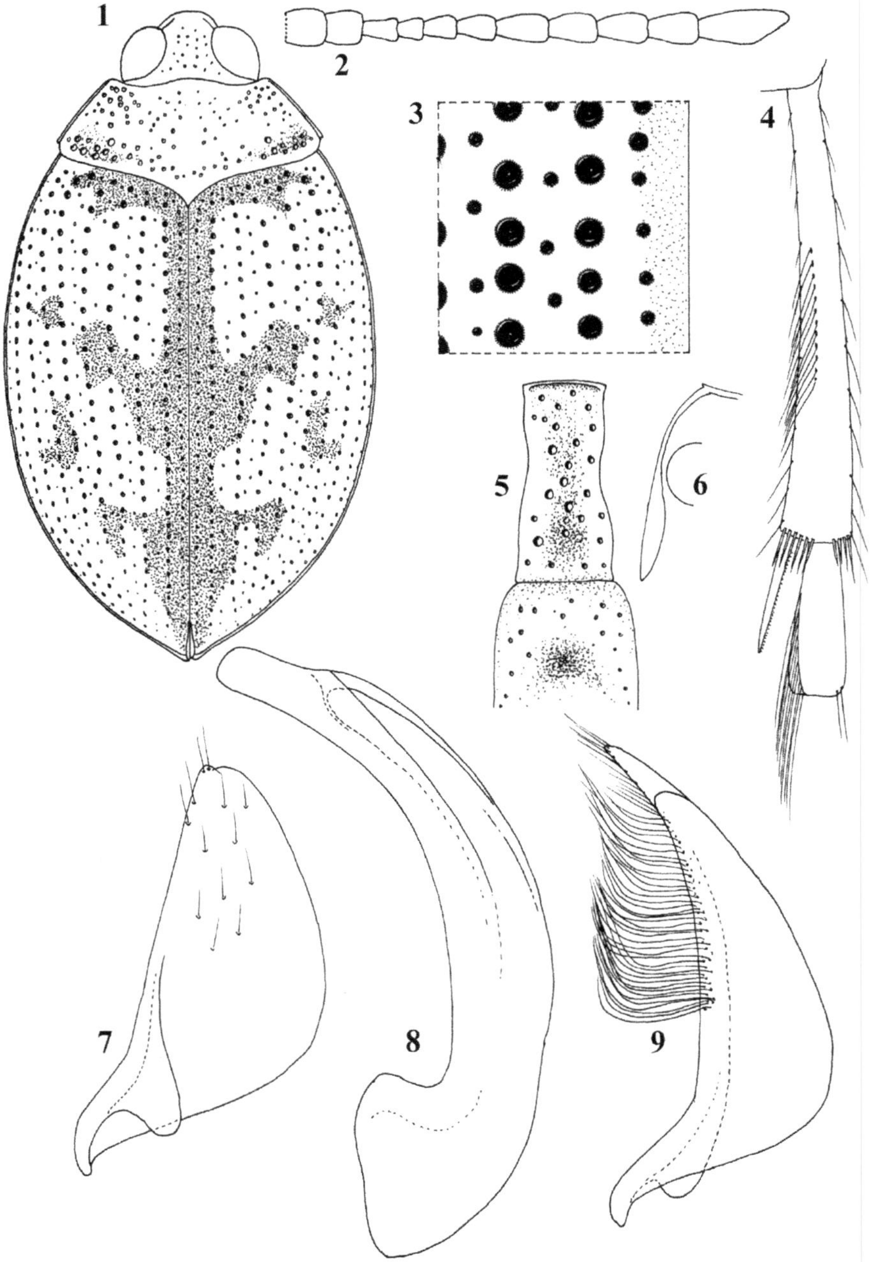

PLATE 148 *Haliplus* (*Liaphlus*) *srilankanus*. 1, dorsal view; 2, antenna; 3, elytral punctures near base and suture; 4, hind tibia in dorsal view; 5, prosternal and metaventral process; 6, prosternal process in lateral view; 7, left paramere; 8, penis; 9, right paramere

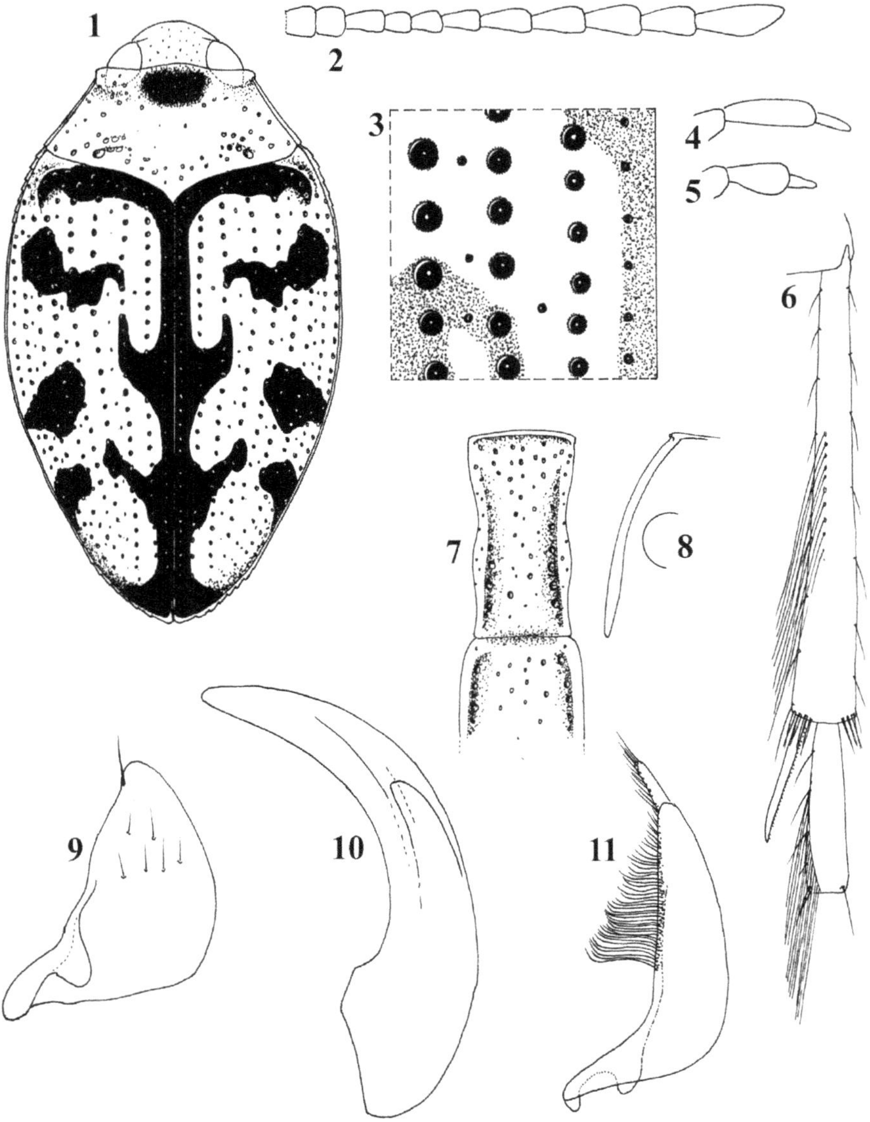

PLATE 149 *Haliplus* (*Liaphlus*) *stepheni*. 1, dorsal view; 2, antenna; 3, elytral punctures near base and suture; 4, maxillary palpus; 5, labial palpus; 6, hind tibia in dorsal view; 7, prosternal and metaventral process; 8, prosternal process in lateral view; 9, left paramere; 10, penis; 11, right paramere

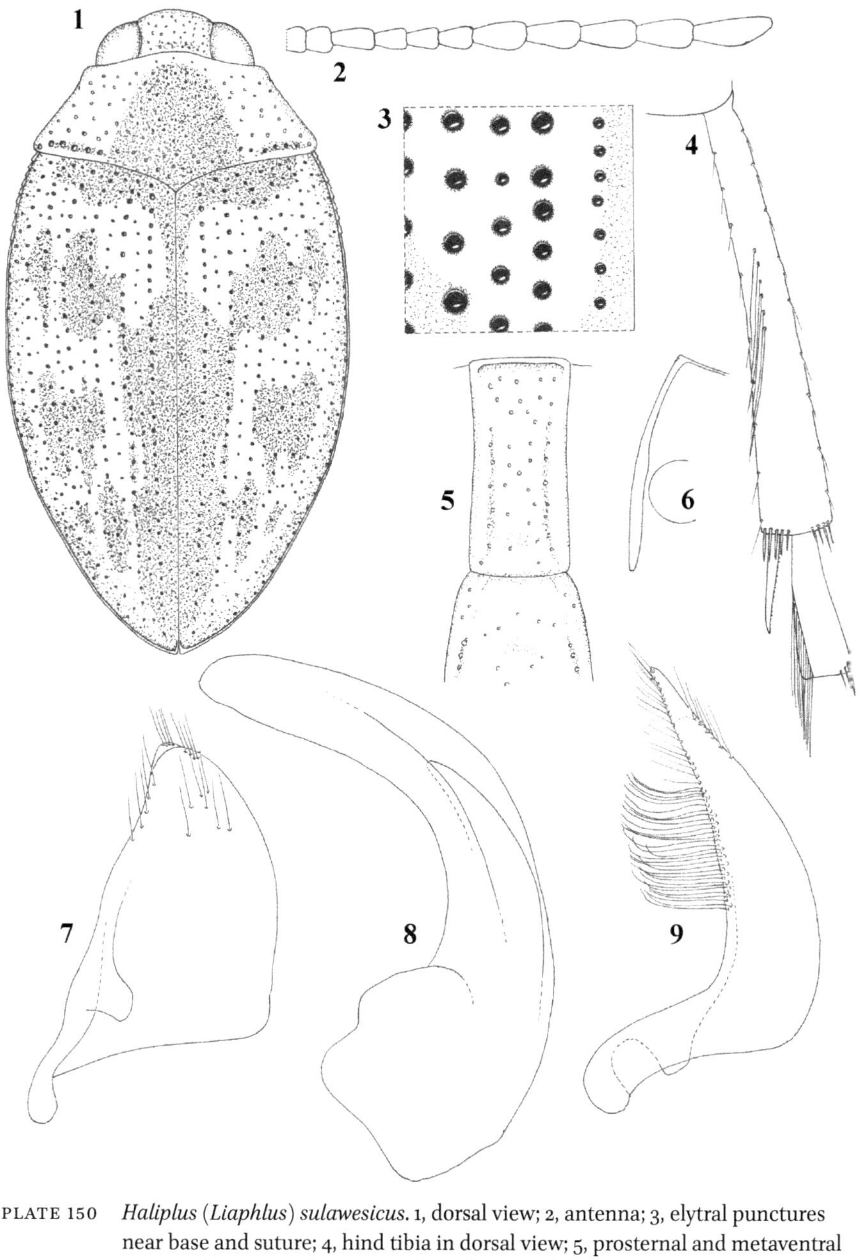

PLATE 150 *Haliplus* (*Liaphlus*) *sulawesicus*. 1, dorsal view; 2, antenna; 3, elytral punctures near base and suture; 4, hind tibia in dorsal view; 5, prosternal and metaventral process; 6, prosternal process in lateral view; 7, left paramere; 8, penis; 9, right paramere

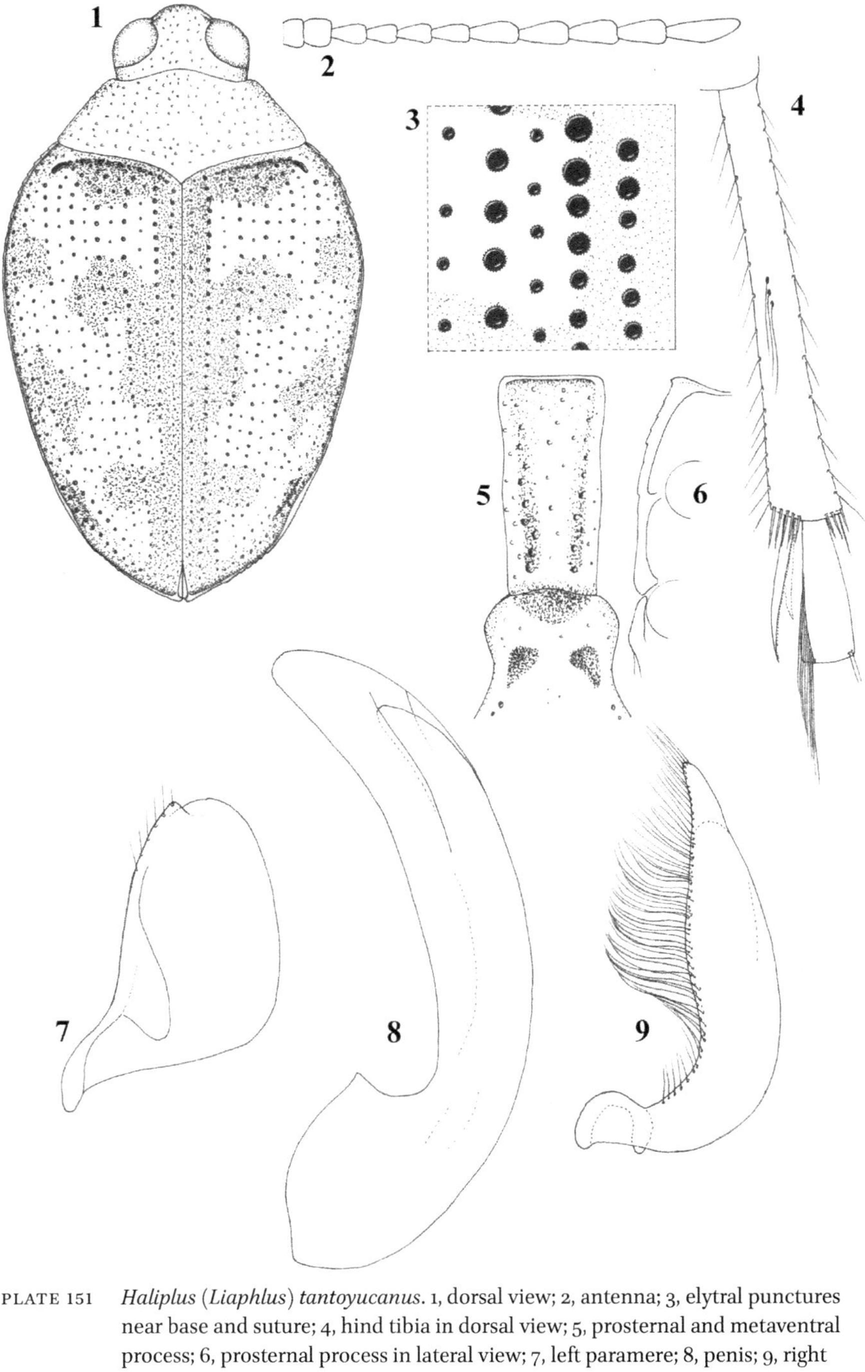

PLATE 151 *Haliplus* (*Liaphlus*) *tantoyucanus*. 1, dorsal view; 2, antenna; 3, elytral punctures near base and suture; 4, hind tibia in dorsal view; 5, prosternal and metaventral process; 6, prosternal process in lateral view; 7, left paramere; 8, penis; 9, right paramere

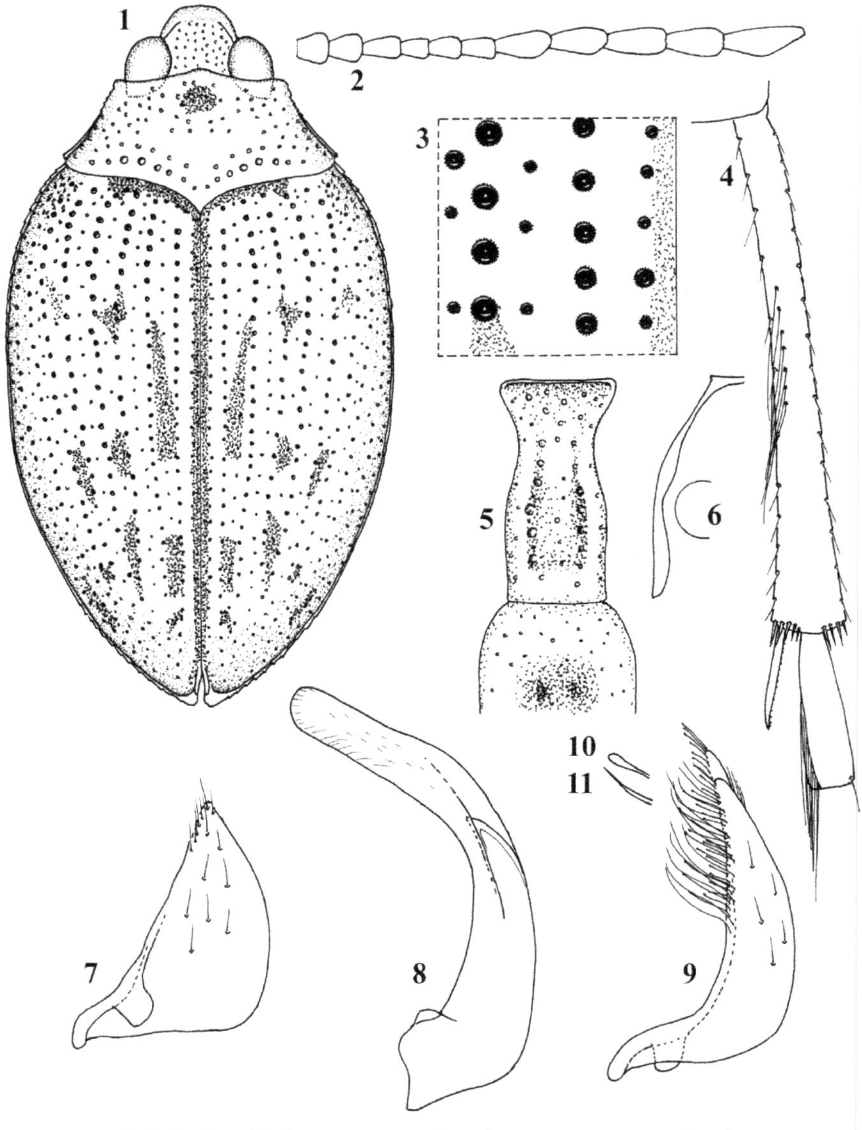

PLATE 152 *Haliplus* (*Liaphlus*) *tanzanianus*. 1, dorsal view; 2, antenna; 3, elytral punctures near base and suture; 4, hind tibia in dorsal view; 5, prosternal and metaventral process; 6, prosternal process in lateral view; 7, left paramere; 8, penis; 9, right paramere; 10, 11, setae of right paramere

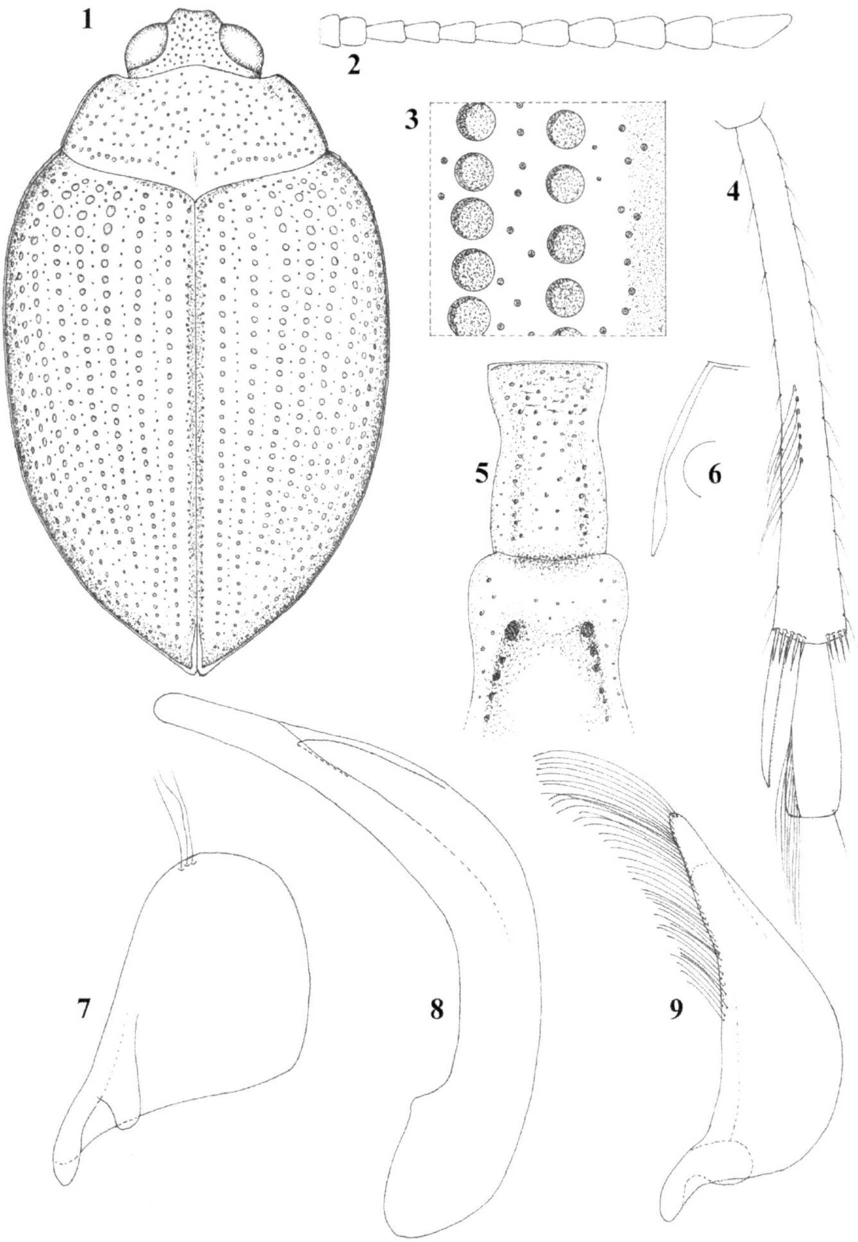

PLATE 153 *Haliplus* (*Liaphlus*) *testaceus*. 1, dorsal view; 2, antenna; 3, elytral punctures near base and suture; 4, hind tibia in dorsal view; 5, prosternal and metaventral process; 6, prosternal process in lateral view; 7, left paramere; 8, penis; 9, right paramere

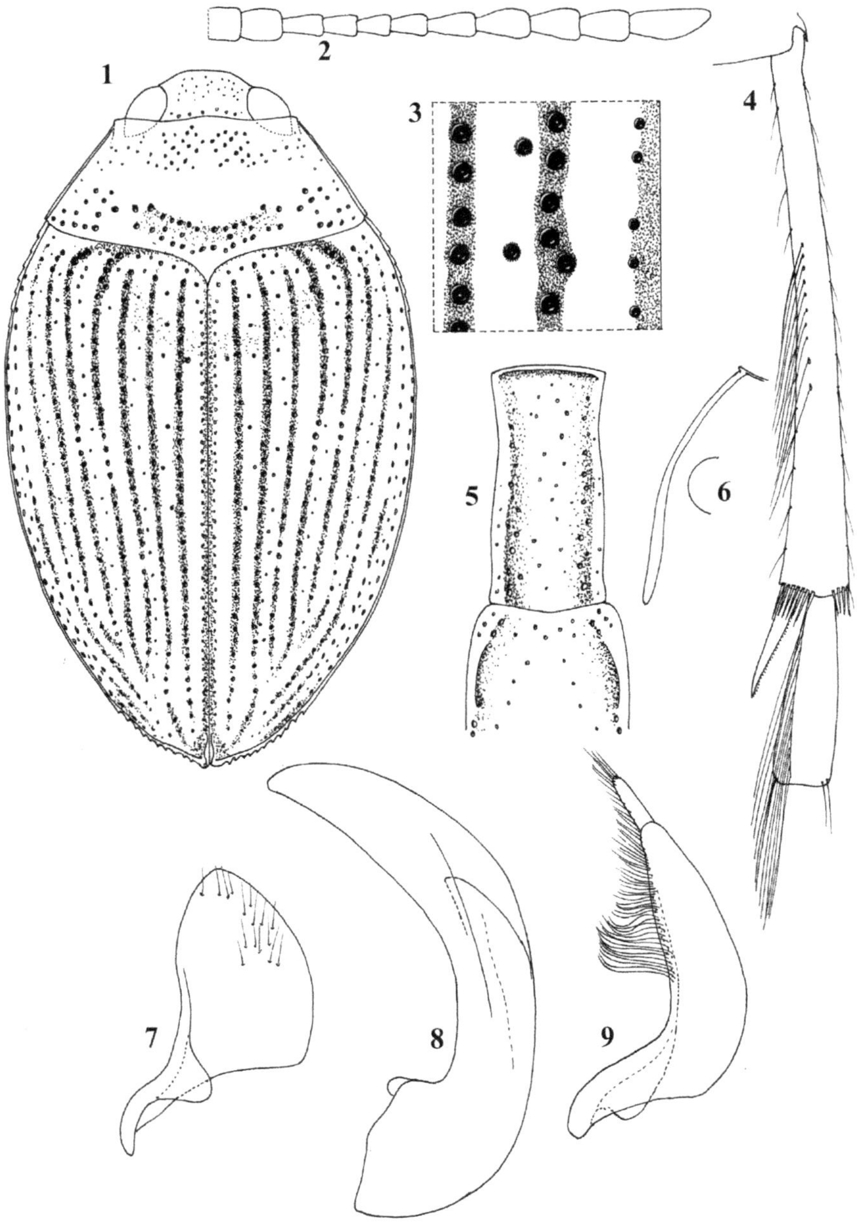

PLATE 154 *Haliplus* (*Liaphlus*) *testudo*. 1, dorsal view; 2, antenna; 3, elytral punctures near base and suture; 4, hind tibia in dorsal view; 5, prosternal and metaventral process; 6, prosternal process in lateral view; 7, left paramere; 8, penis; 9, right paramere

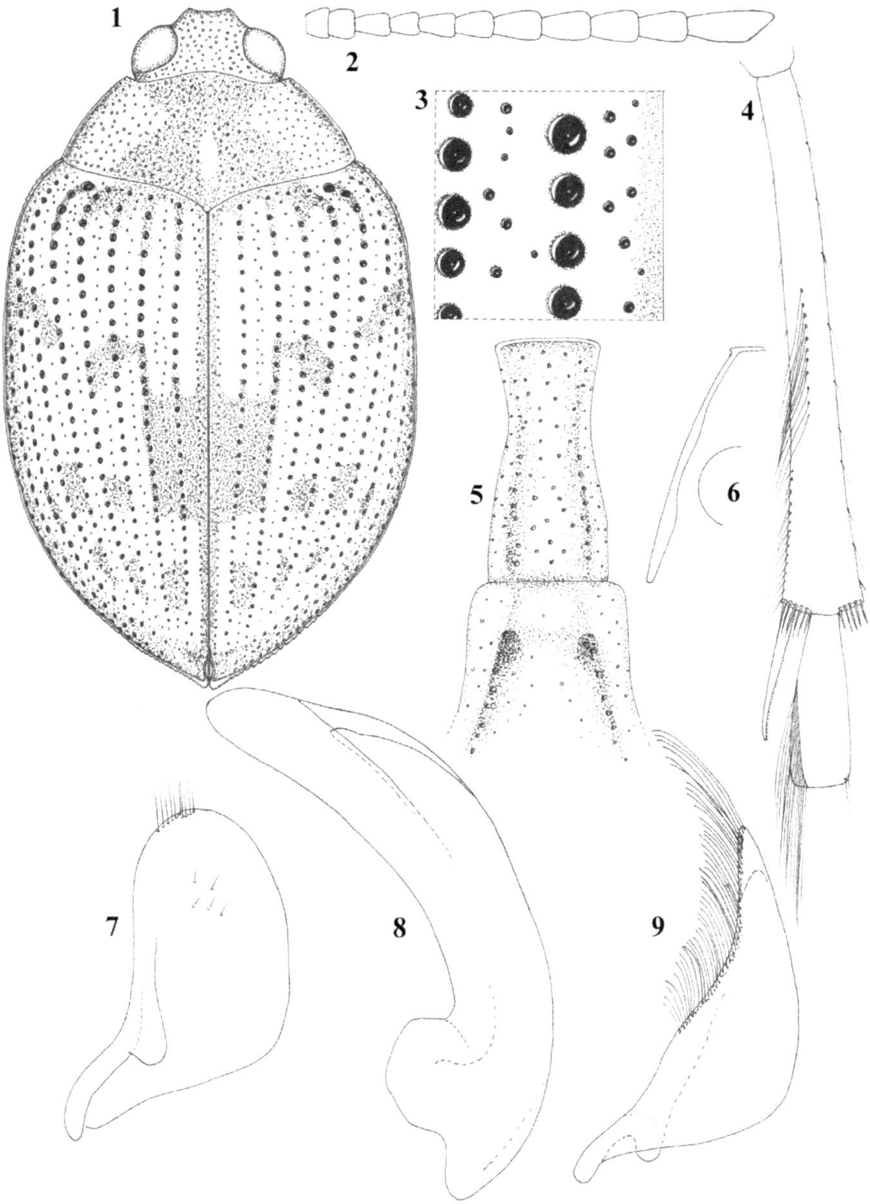

PLATE 155 *Haliplus* (*Liaphlus*) *thoracicus*. 1, dorsal view; 2, antenna; 3, elytral punctures near base and suture; 4, hind tibia in dorsal view; 5, prosternal and metaventral process; 6, prosternal process in lateral view; 7, left paramere; 8, penis; 9, right paramere

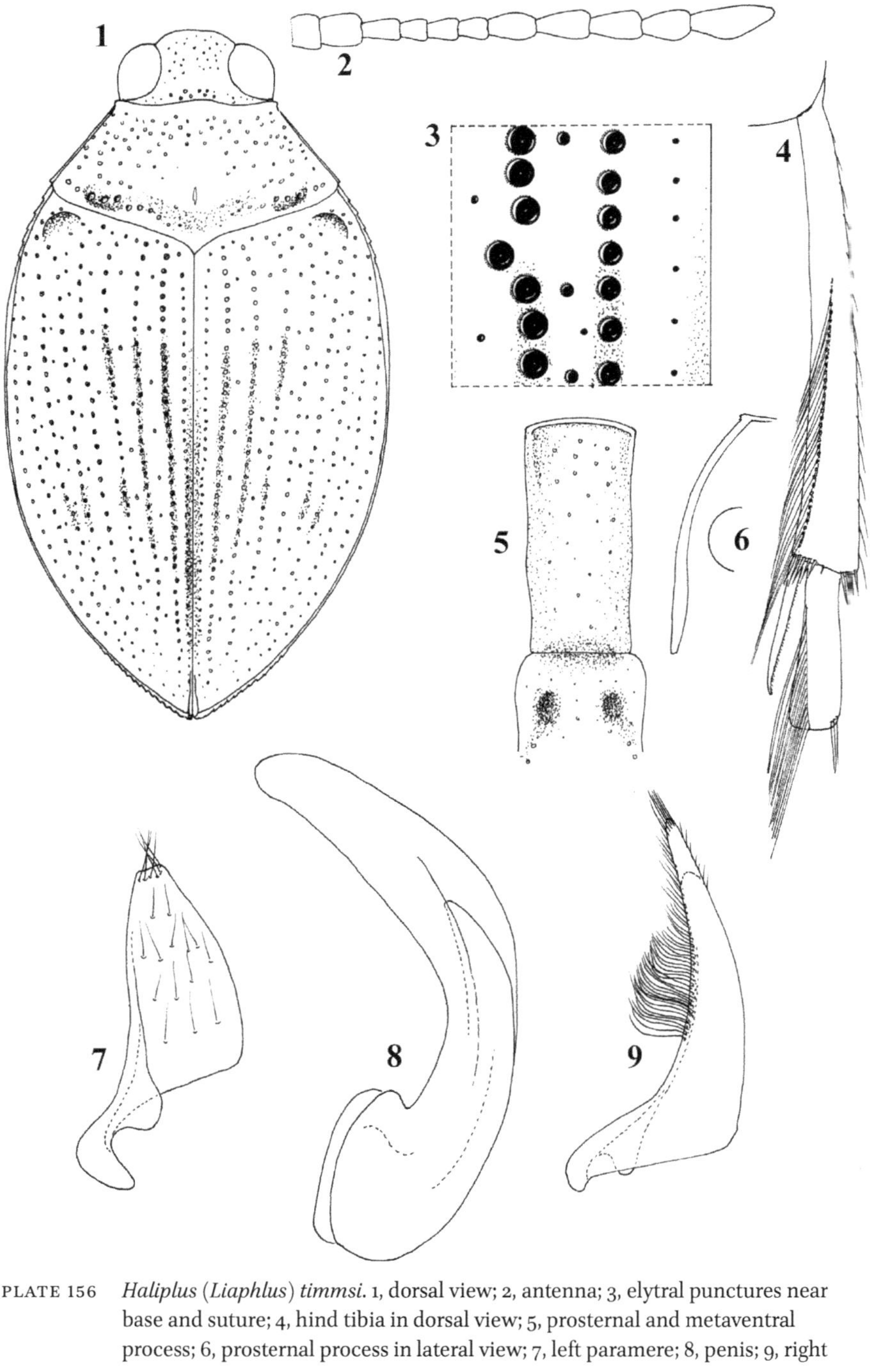

PLATE 156 *Haliplus* (*Liaphlus*) *timmsi*. 1, dorsal view; 2, antenna; 3, elytral punctures near base and suture; 4, hind tibia in dorsal view; 5, prosternal and metaventral process; 6, prosternal process in lateral view; 7, left paramere; 8, penis; 9, right paramere

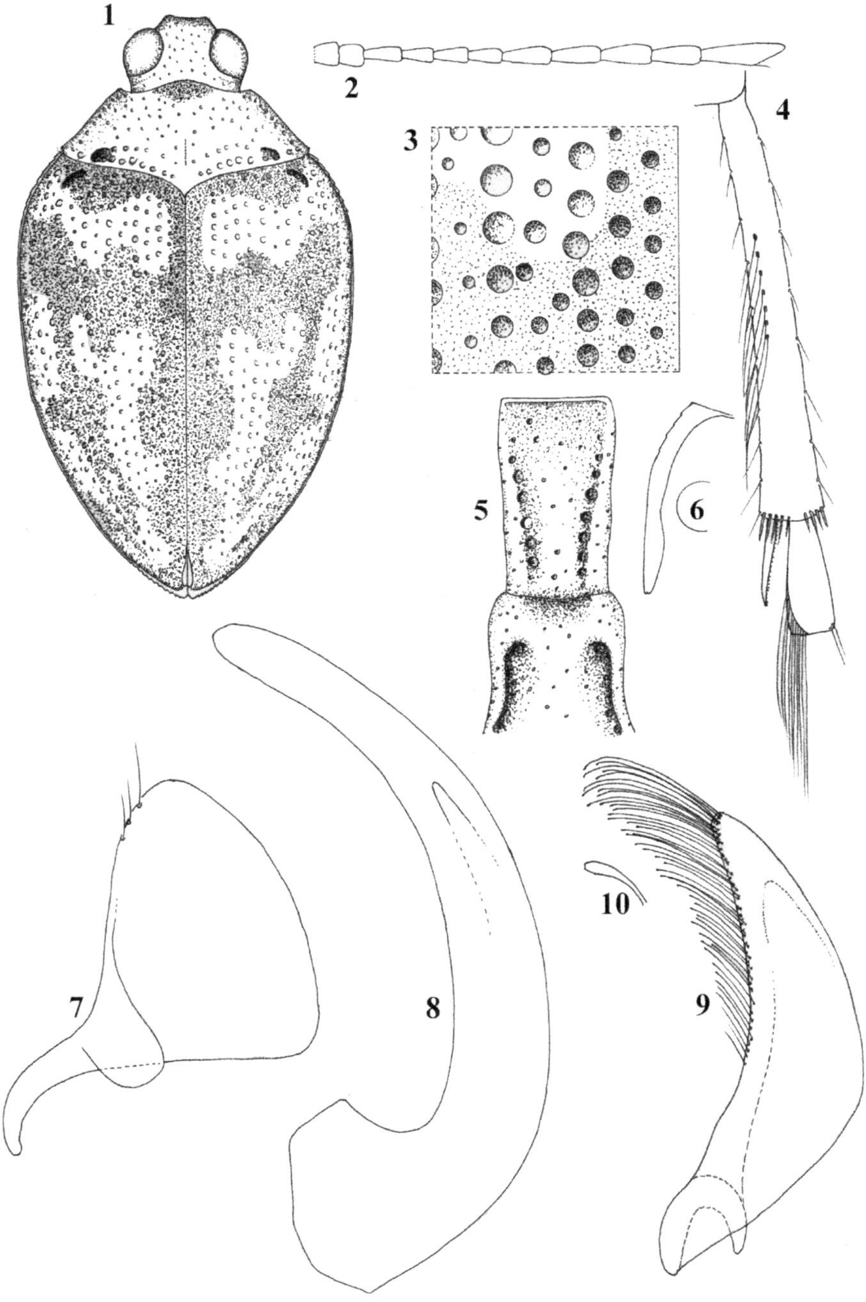

PLATE 157 *Haliplus* (*Liaphlus*) *tocumenus*. 1, dorsal view; 2, antenna; 3, elytral punctures near base and suture; 4, hind tibia in dorsal view; 5, prosternal and metaventral process; 6, prosternal process in lateral view; 7, left paramere; 8, penis; 9, right paramere; 10, setae of right paramere

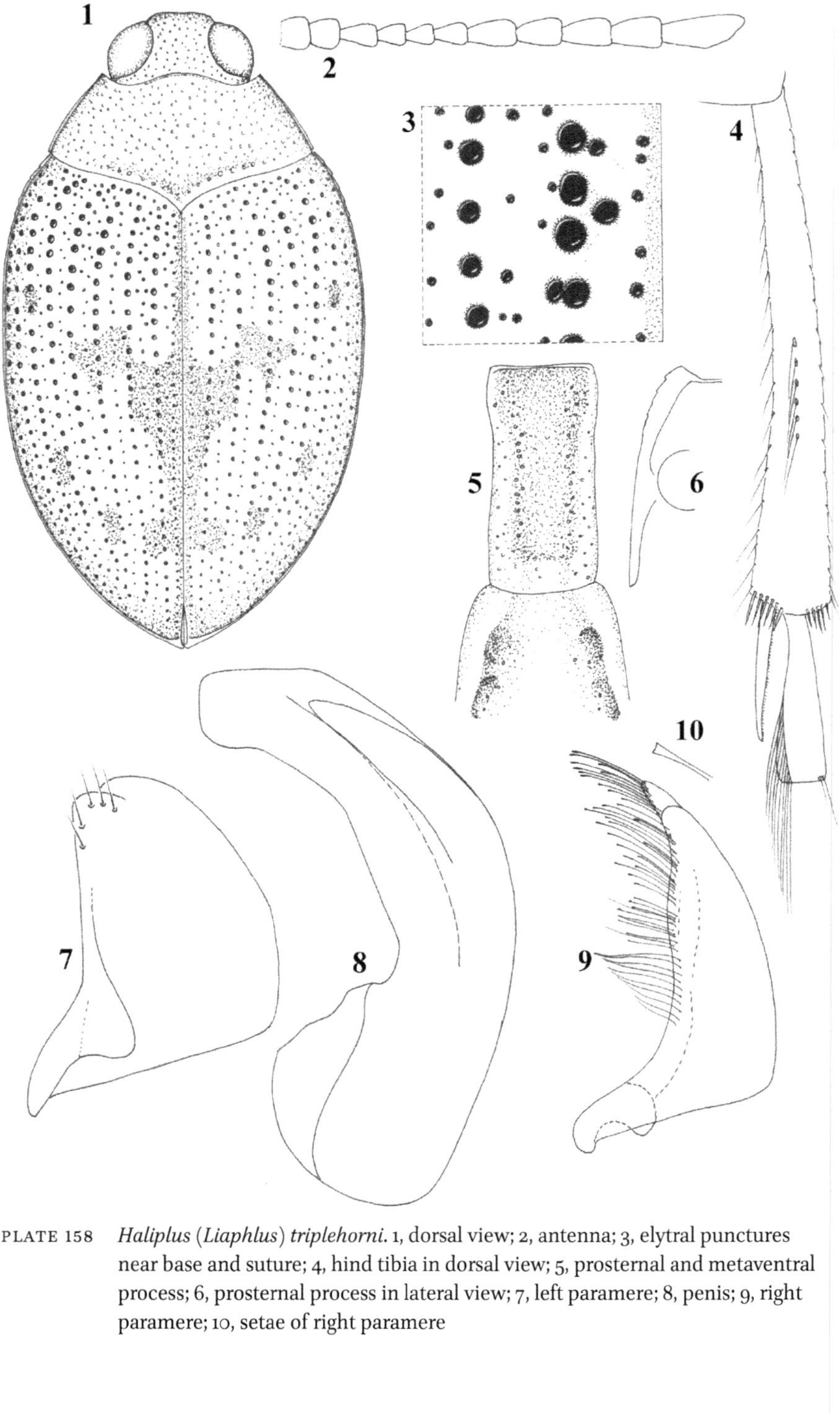

PLATE 158 *Haliplus* (*Liaphlus*) *triplehorni*. 1, dorsal view; 2, antenna; 3, elytral punctures near base and suture; 4, hind tibia in dorsal view; 5, prosternal and metaventral process; 6, prosternal process in lateral view; 7, left paramere; 8, penis; 9, right paramere; 10, setae of right paramere

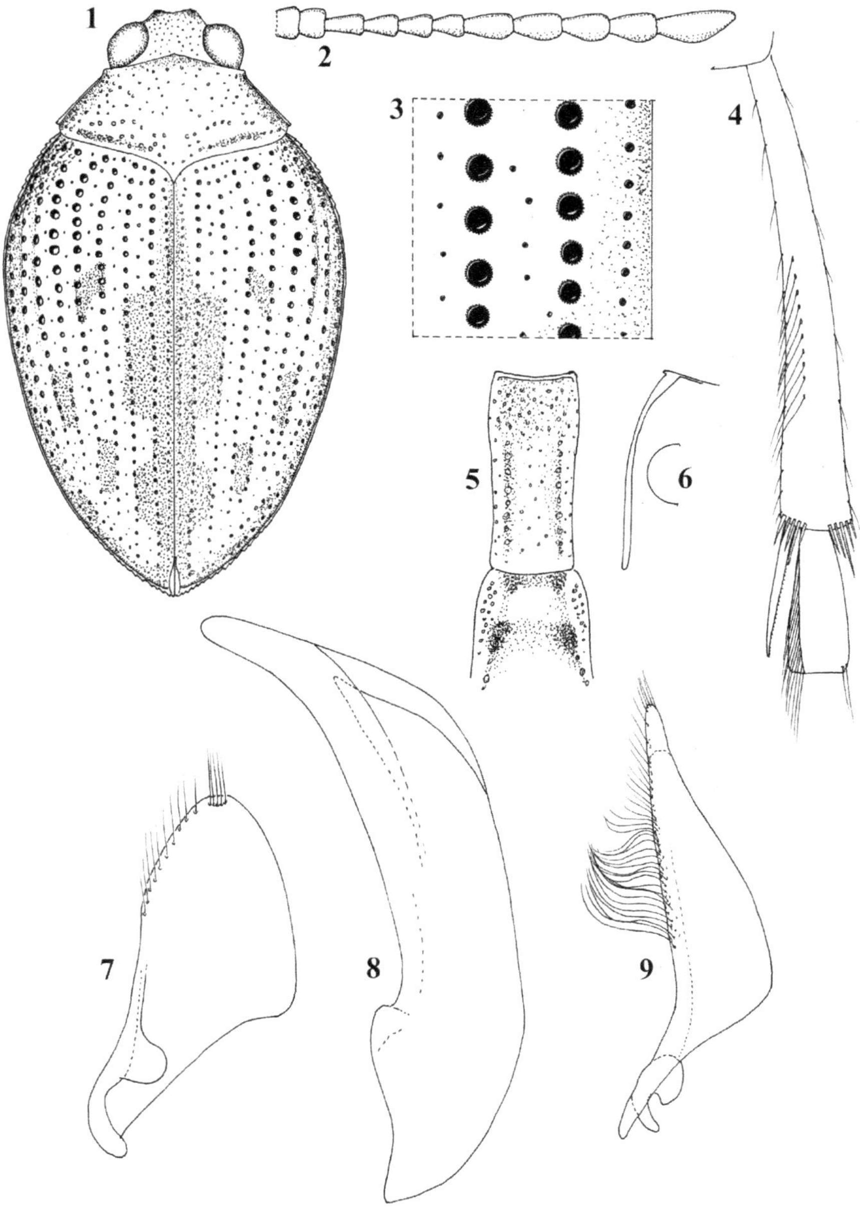

PLATE 159 *Haliplus* (*Liaphlus*) *tumidus*. 1, dorsal view; 2, antenna; 3, elytral punctures near base and suture; 4, hind tibia in dorsal view; 5, prosternal and metaventral process; 6, prosternal process in lateral view; 7, left paramere; 8, penis; 9, right paramere

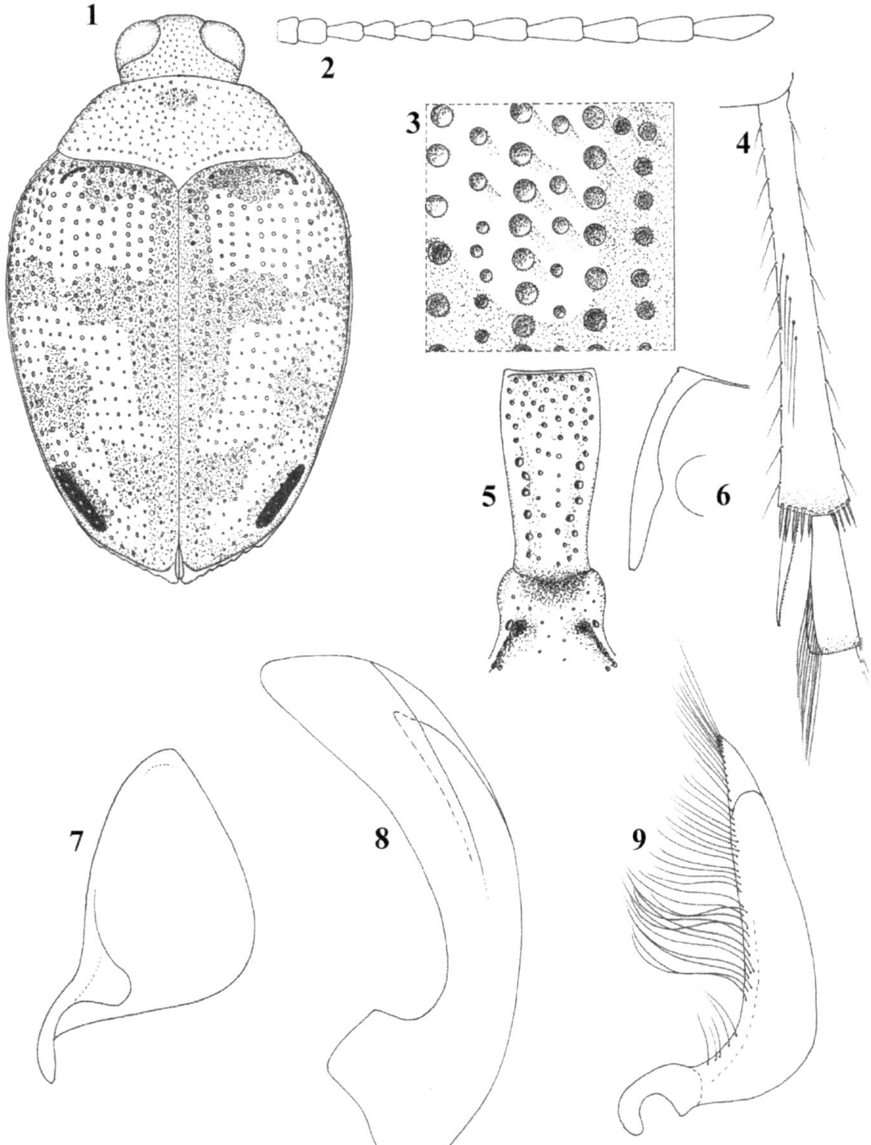

PLATE 160 *Haliplus* (*Liaphlus*) *unicarinatus*. 1, dorsal view; 2, antenna; 3, elytral punctures near base and suture; 4, hind tibia in dorsal view; 5, prosternal and metaventral process; 6, prosternal process in lateral view; 7, left paramere; 8, penis; 9, right paramere

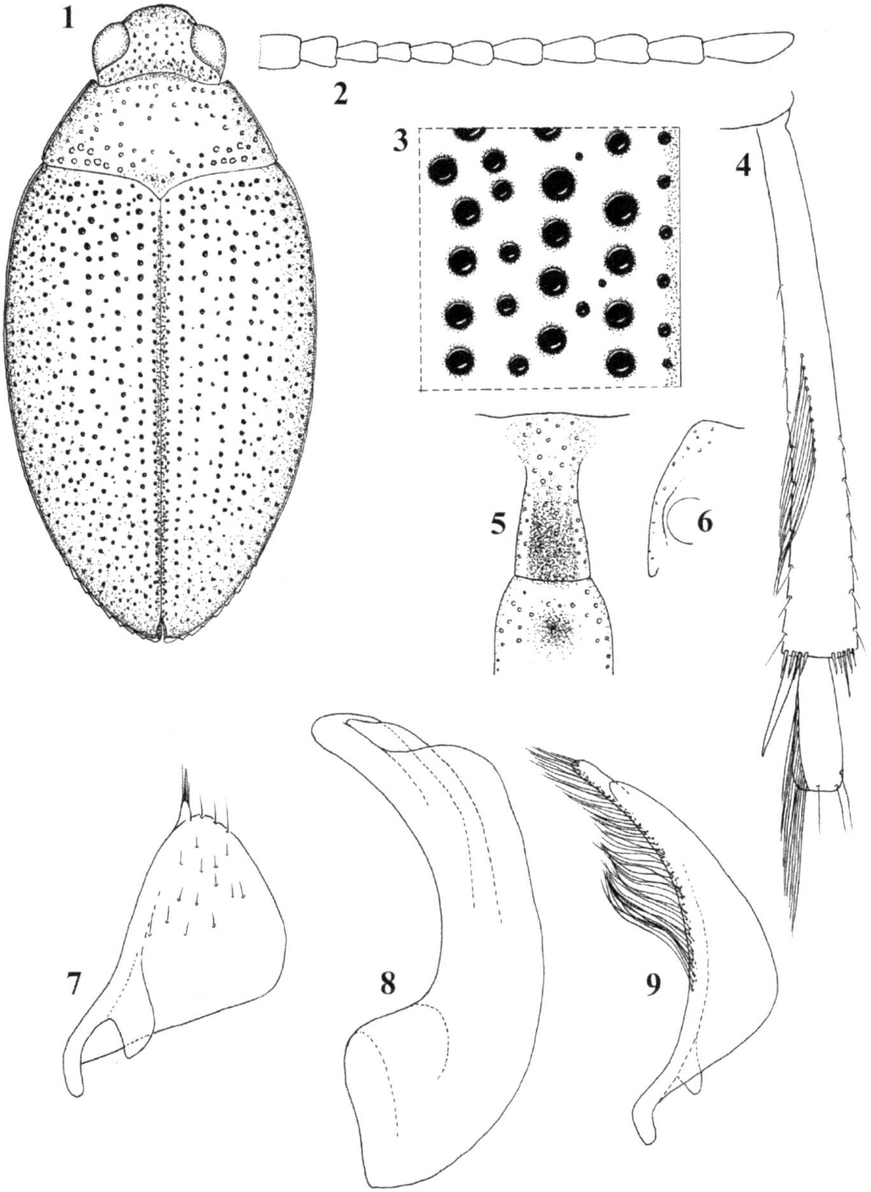

PLATE 161 *Haliplus* (*Liaphlus*) *uniformis*. 1, dorsal view; 2, antenna; 3, elytral punctures near base and suture; 4, hind tibia in dorsal view; 5, prosternal and metaventral process; 6, prosternal process in lateral view; 7, left paramere; 8, penis; 9, right paramere

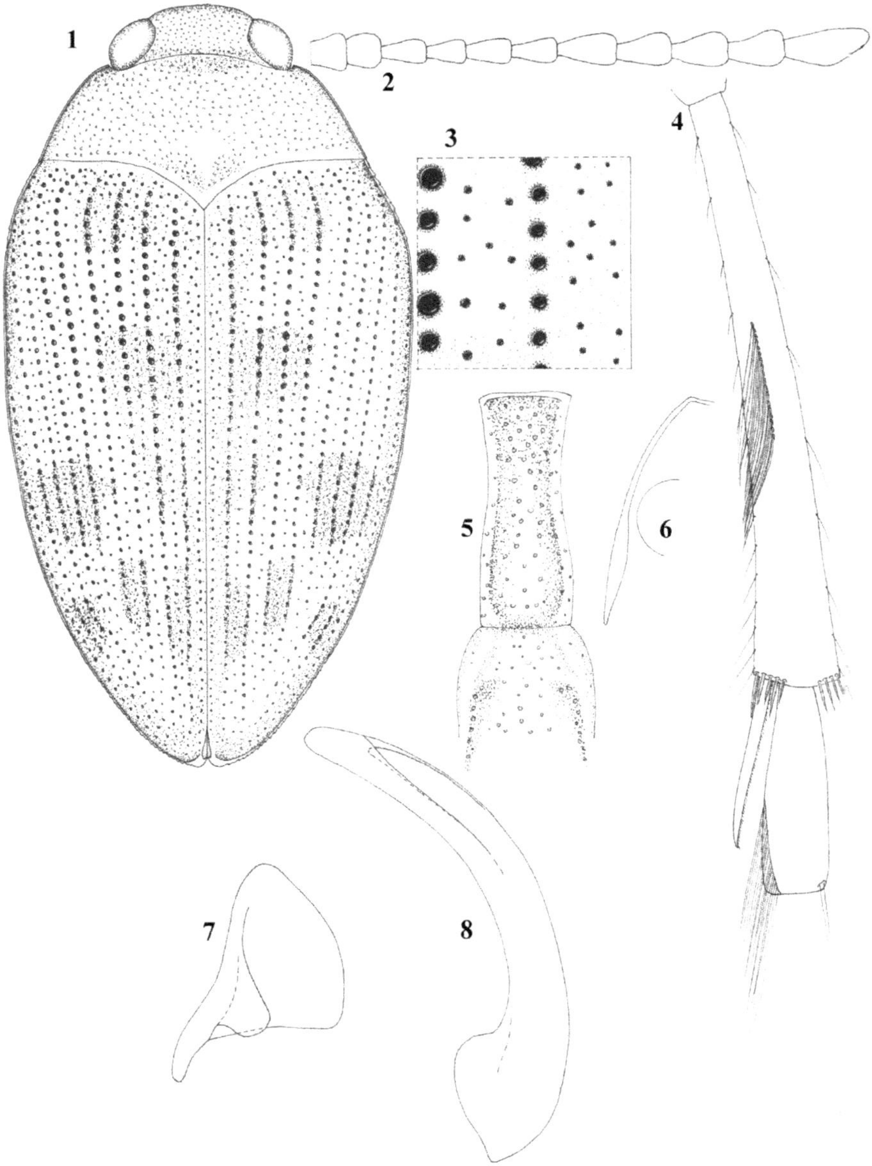

PLATE 162 *Haliplus* (*Liaphlus*) *valdiviensis*. 1, dorsal view; 2, antenna; 3, elytral punctures near base and suture; 4, hind tibia in dorsal view; 5, prosternal and metaventral process; 6, prosternal process in lateral view; 7, left paramere; 8, penis

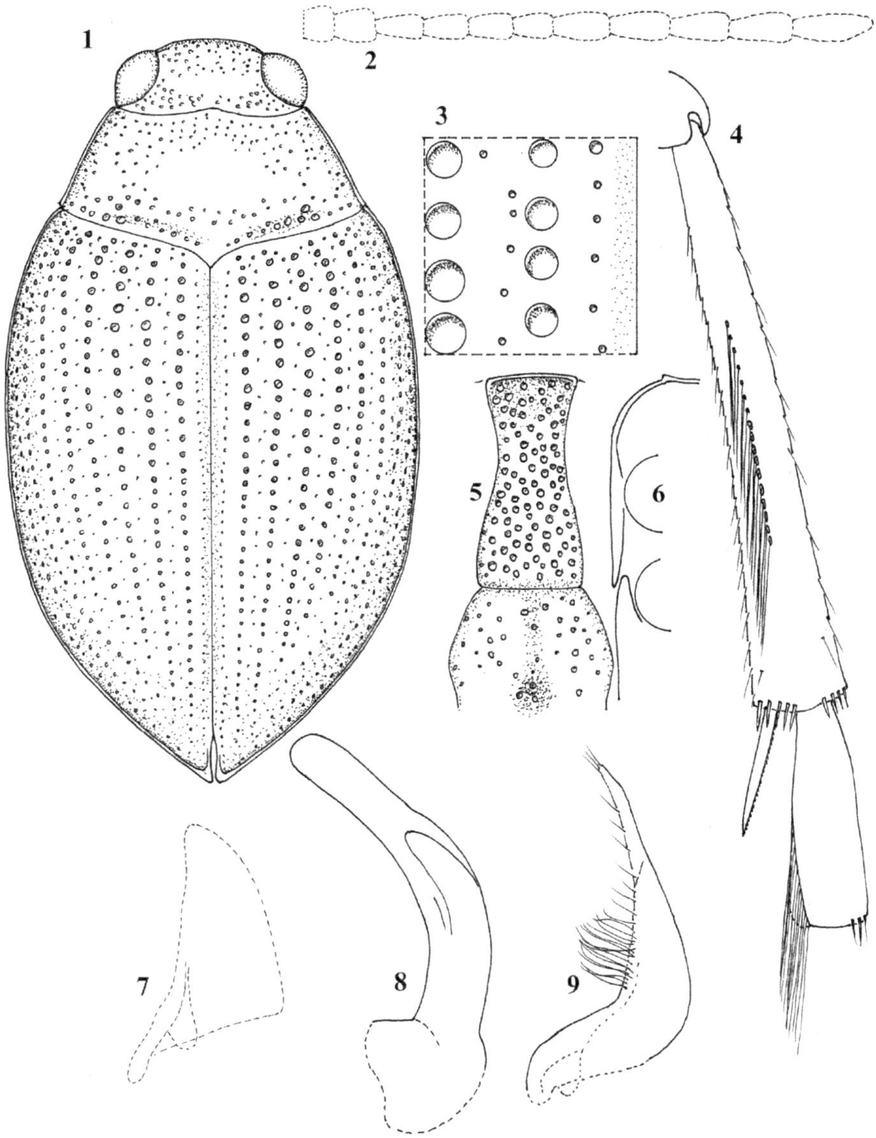

PLATE 163 *Haliplus* (*Liaphlus*) *vancouverensis*. 1, dorsal view; 2, antenna; 3, elytral punctures near base and suture; 4, hind tibia in dorsal view; 5, prosternal and metaventral process; 6, prosternal process in lateral view; 7, left paramere; 8, penis; 9, right paramere

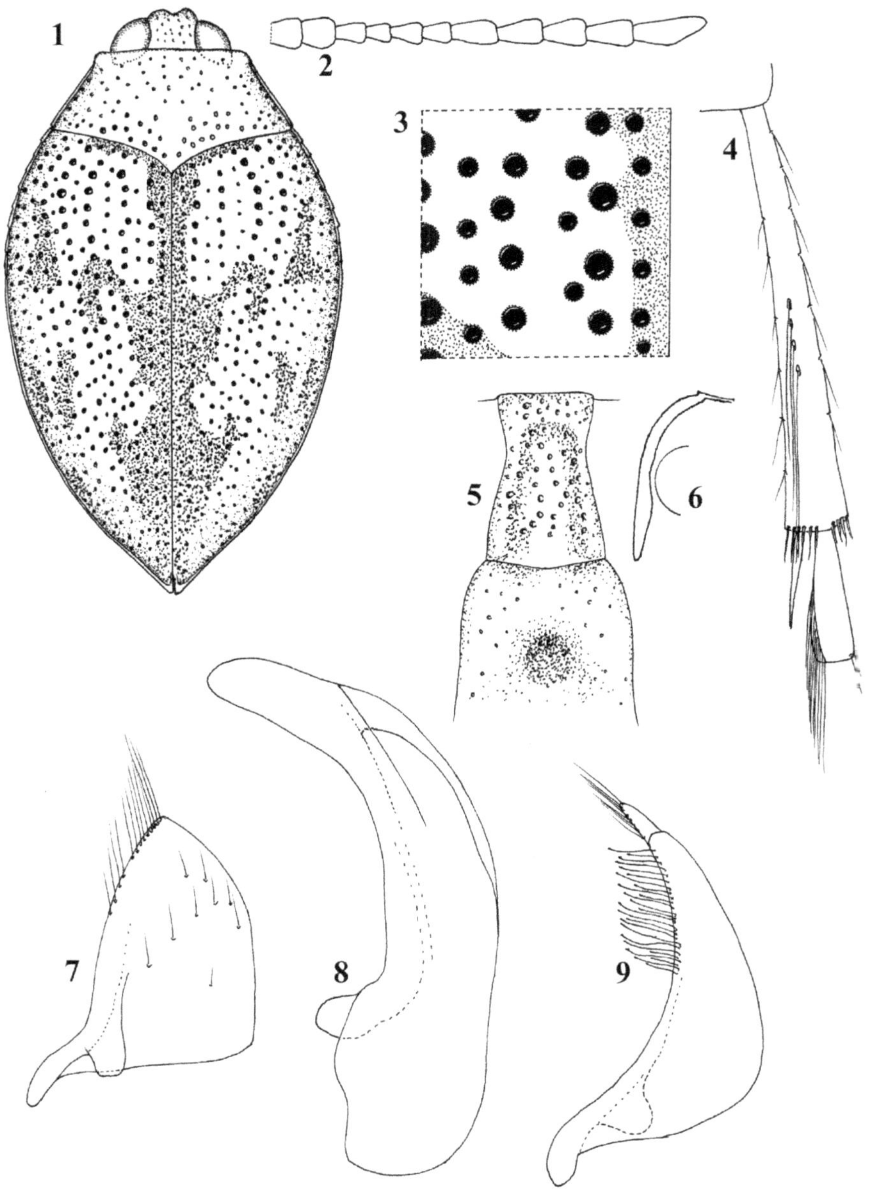

PLATE 164 *Haliplus* (*Liaphlus*) *varicator*. 1, dorsal view; 2, antenna; 3, elytral punctures near base and suture; 4, hind tibia in dorsal view; 5, prosternal and metaventral process; 6, prosternal process in lateral view; 7, left paramere; 8, penis; 9, right paramere

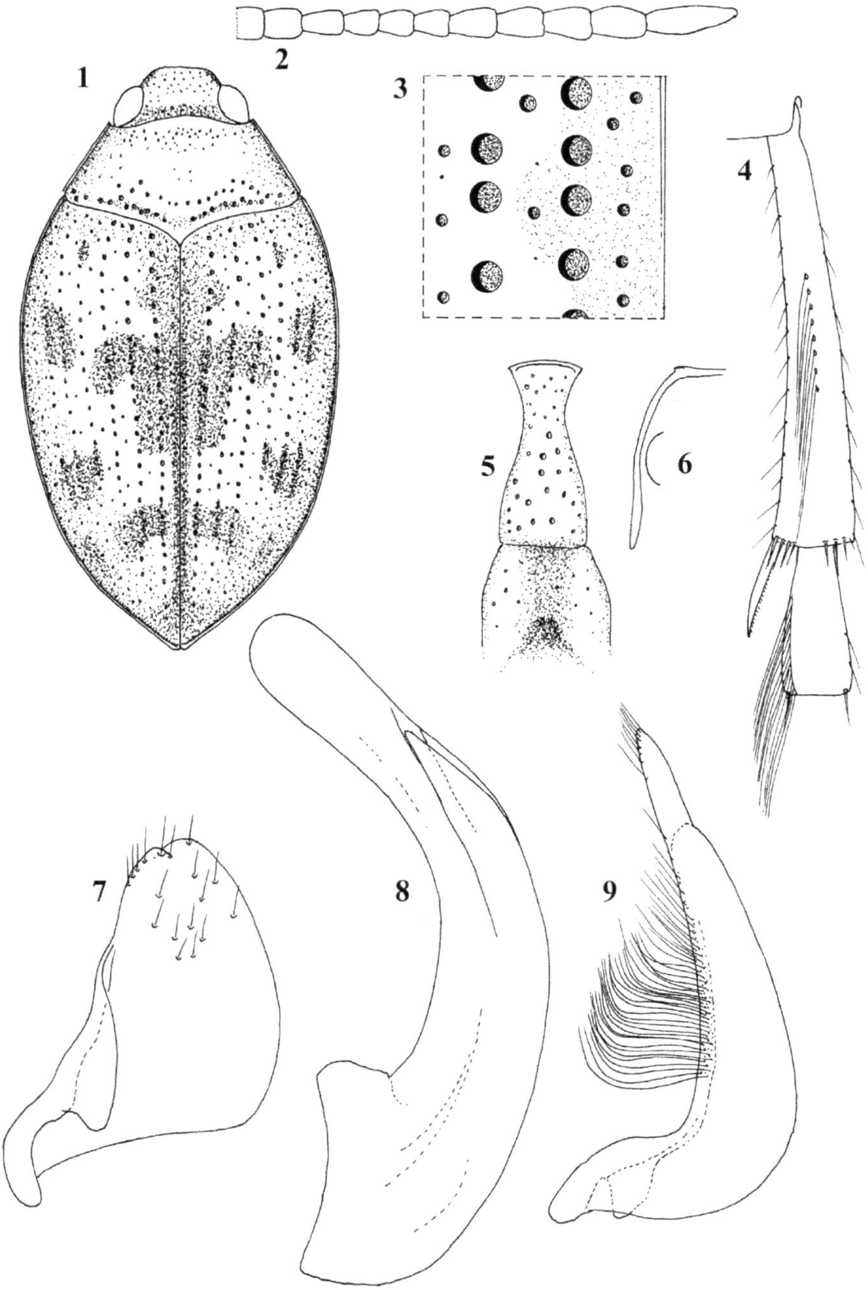

PLATE 165 *Haliplus* (*Liaphlus*) *variegatus*. 1, dorsal view; 2, antenna; 3, elytral punctures near base and suture; 4, hind tibia in dorsal view; 5, prosternal and metaventral process; 6, prosternal process in lateral view; 7, left paramere; 8, penis; 9, right paramere

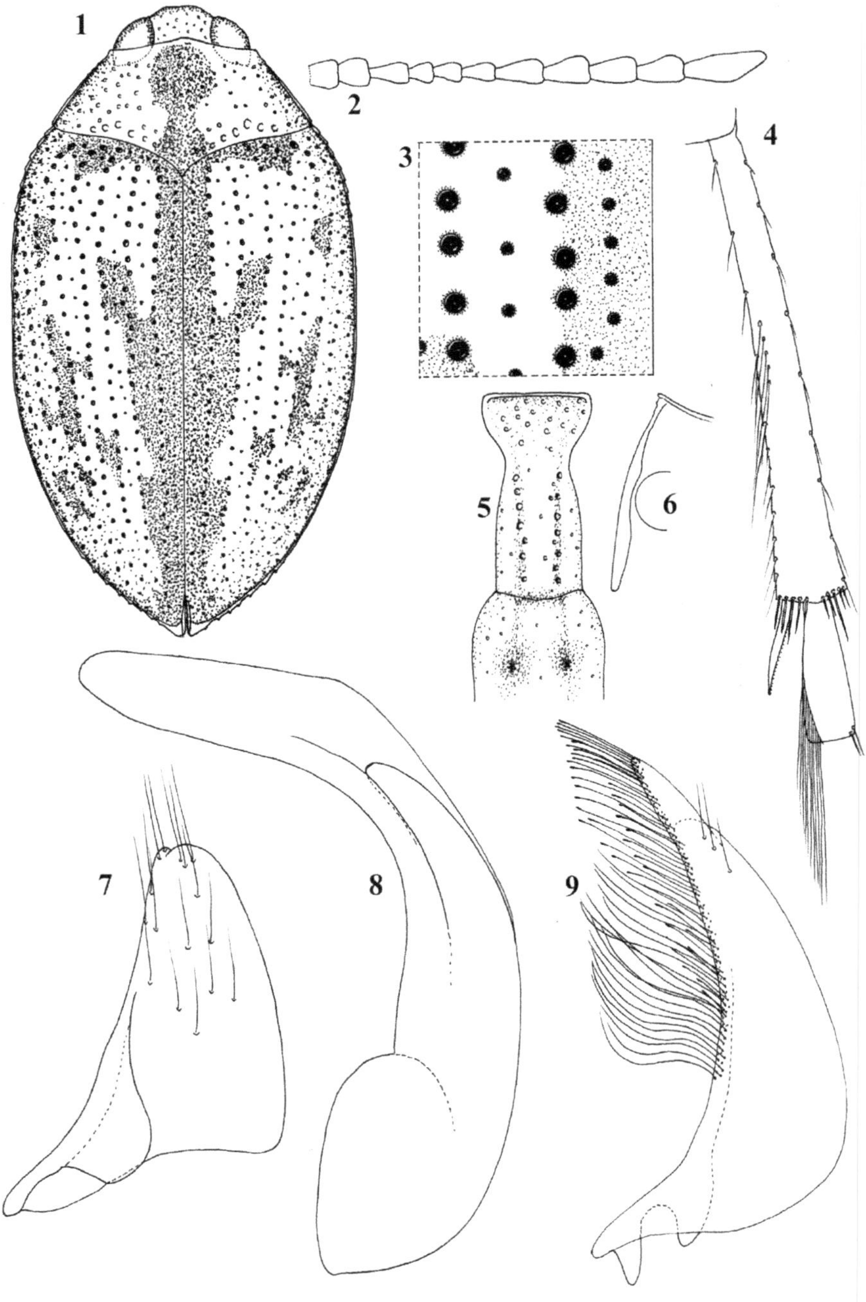

PLATE 166 *Haliplus* (*Liaphlus*) *venustus*. 1, dorsal view; 2, antenna; 3, elytral punctures near base and suture; 4, hind tibia in dorsal view; 5, prosternal and metaventral process; 6, prosternal process in lateral view; 7, left paramere; 8, penis; 9, right paramere

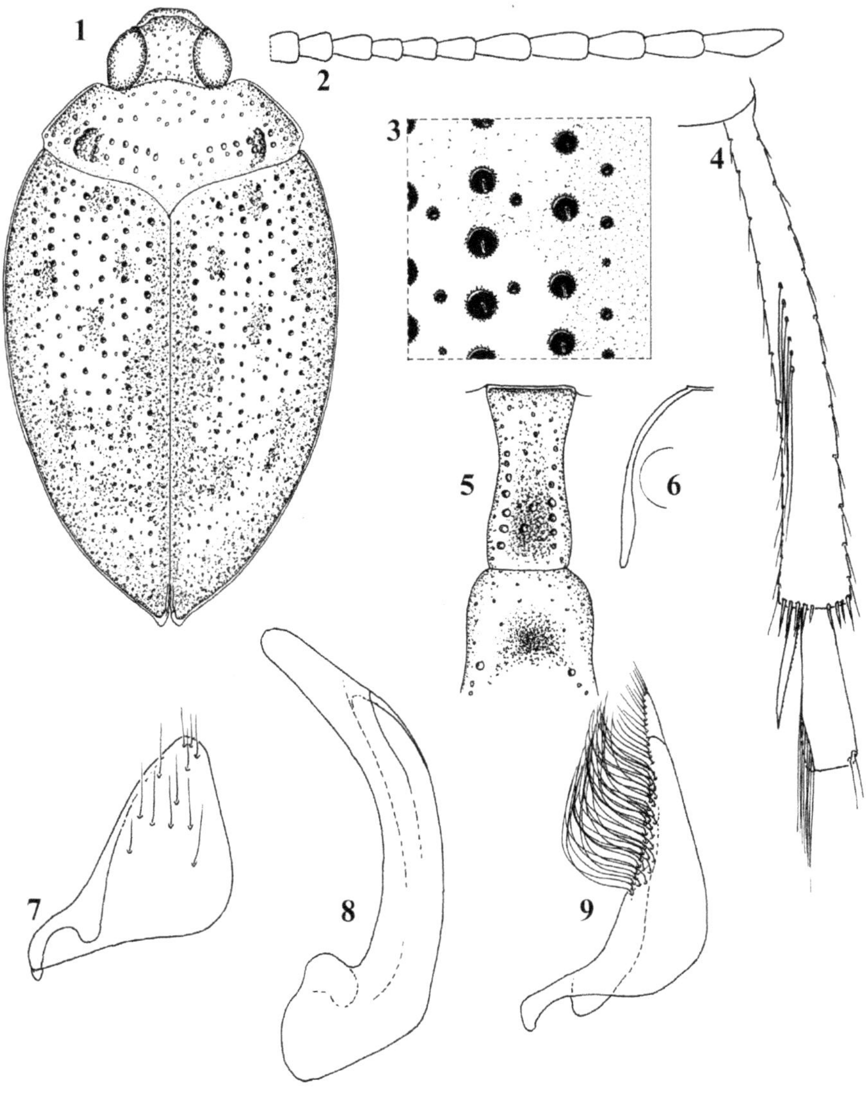

PLATE 167 *Haliplus* (*Liaphlus*) *volpei*. 1, dorsal view; 2, antenna; 3, elytral punctures near base and suture; 4, hind tibia in dorsal view; 5, prosternal and metaventral process; 6, prosternal process in lateral view; 7, left paramere; 8, penis; 9, right paramere

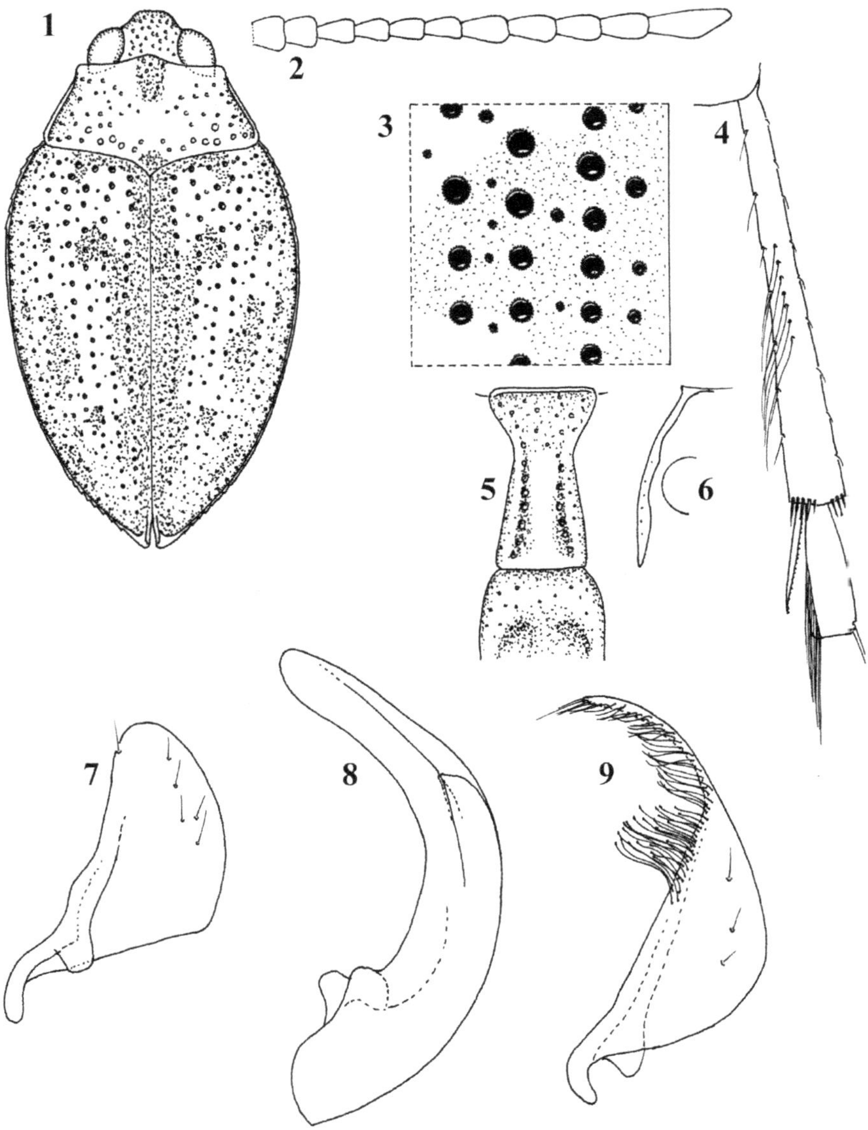

PLATE 168 *Haliplus* (*Liaphlus*) *watsoni*. 1, dorsal view; 2, antenna; 3, elytral punctures near base and suture; 4, hind tibia in dorsal view; 5, prosternal and metaventral process; 6, prosternal process in lateral view; 7, left paramere; 8, penis; 9, right paramere

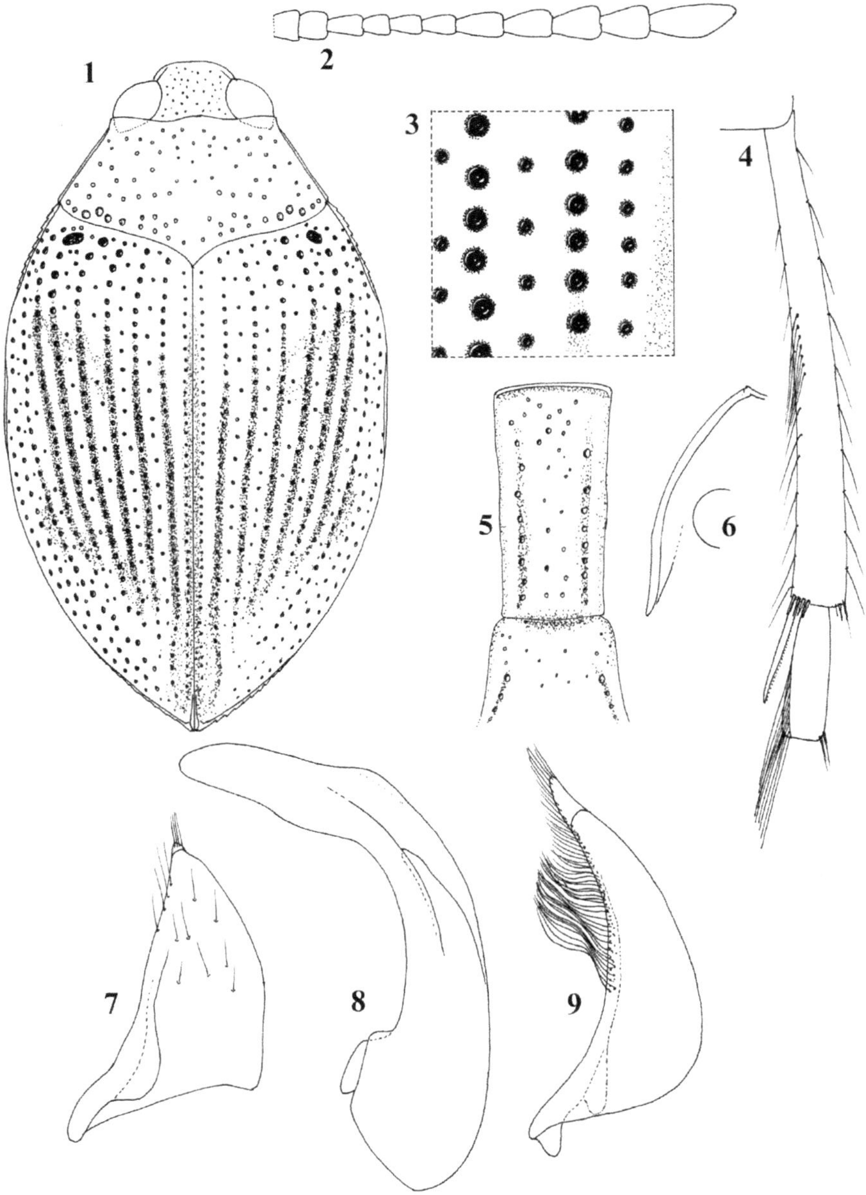

PLATE 169 *Haliplus* (*Liaphlus*) *wattsi*. 1, dorsal view; 2, antenna; 3, elytral punctures near base and suture; 4, hind tibia in dorsal view; 5, prosternal and metaventral process; 6, prosternal process in lateral view; 7, left paramere; 8, penis; 9, right paramere

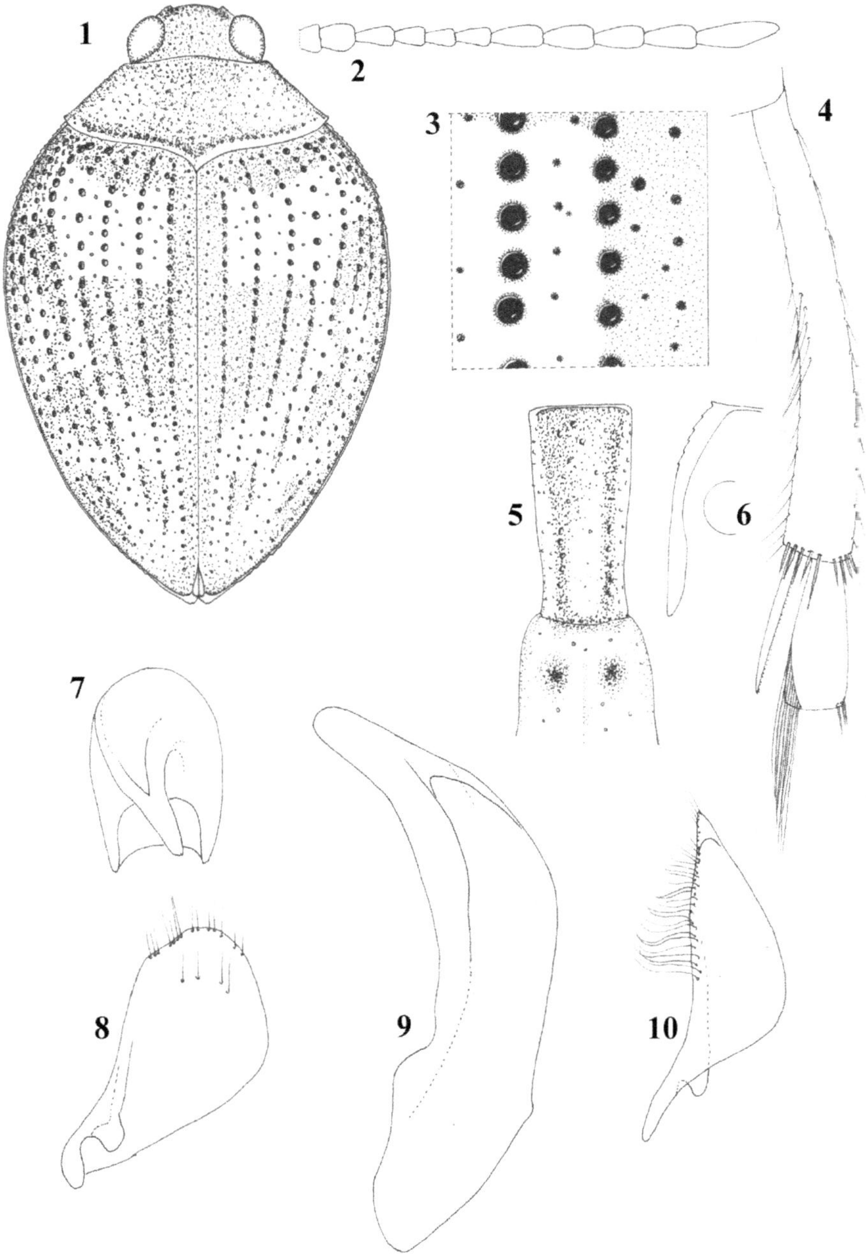

PLATE 170 *Haliplus* (*Liaphlus*) *youngi*. 1, dorsal view; 2, antenna; 3, elytral punctures near base and suture; 4, hind tibia in dorsal view; 5, prosternal and metaventral process; 6, prosternal process in lateral view; 7, left paramer in frontal view; 8, left paramere; 9, penis; 10, right paramere

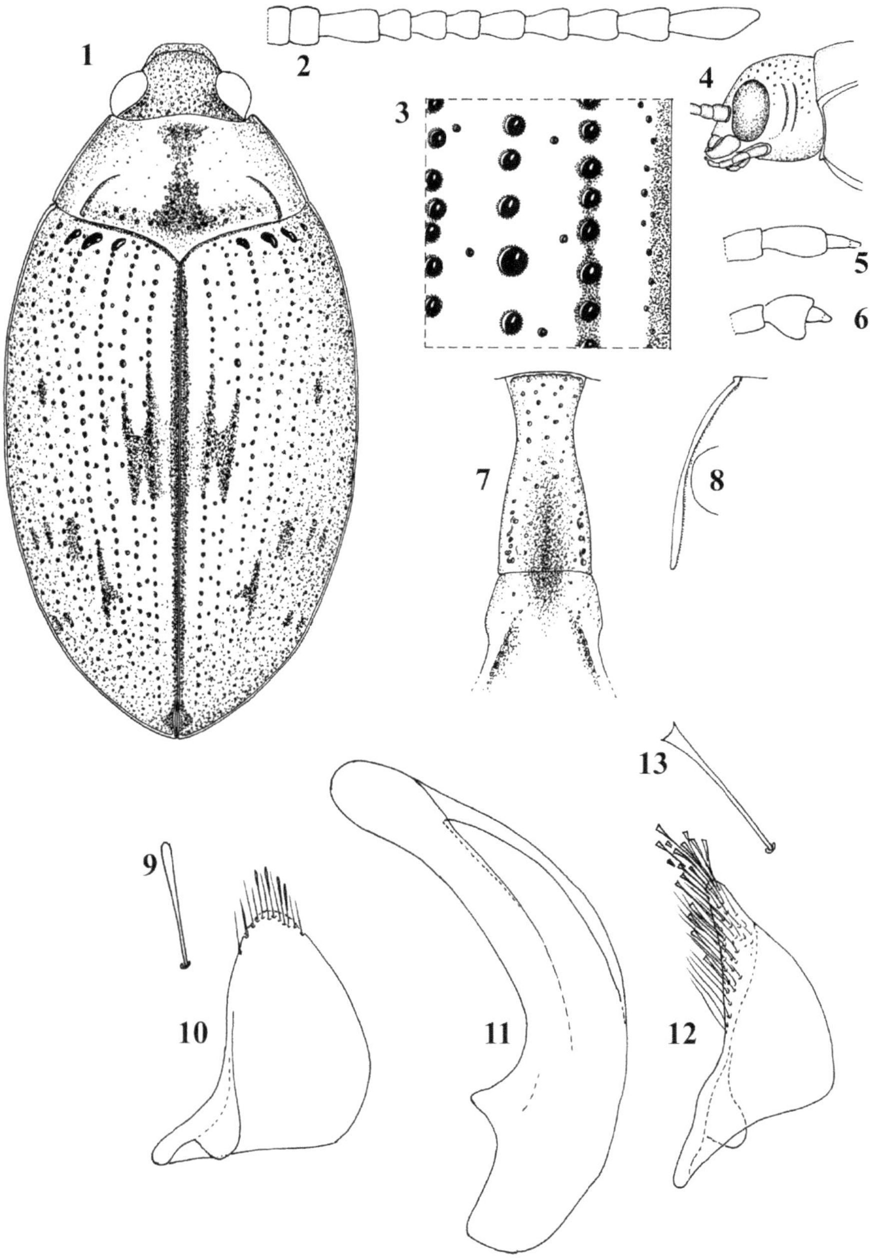

PLATE 171 *Haliplus* (*Neohaliplus*) *lineatocollis*. 1, dorsal view; 2, antenna; 3, elytral punctures near base and suture; head in lateral view; 5, maxillary palpus; 6, labial palpus; 7, prosternal and metaventral process; 8, prosternal process in lateral view; 9, setae of left paramere; 10, left paramere; 11, penis; 12, right paramere; 13, setae of right paramere

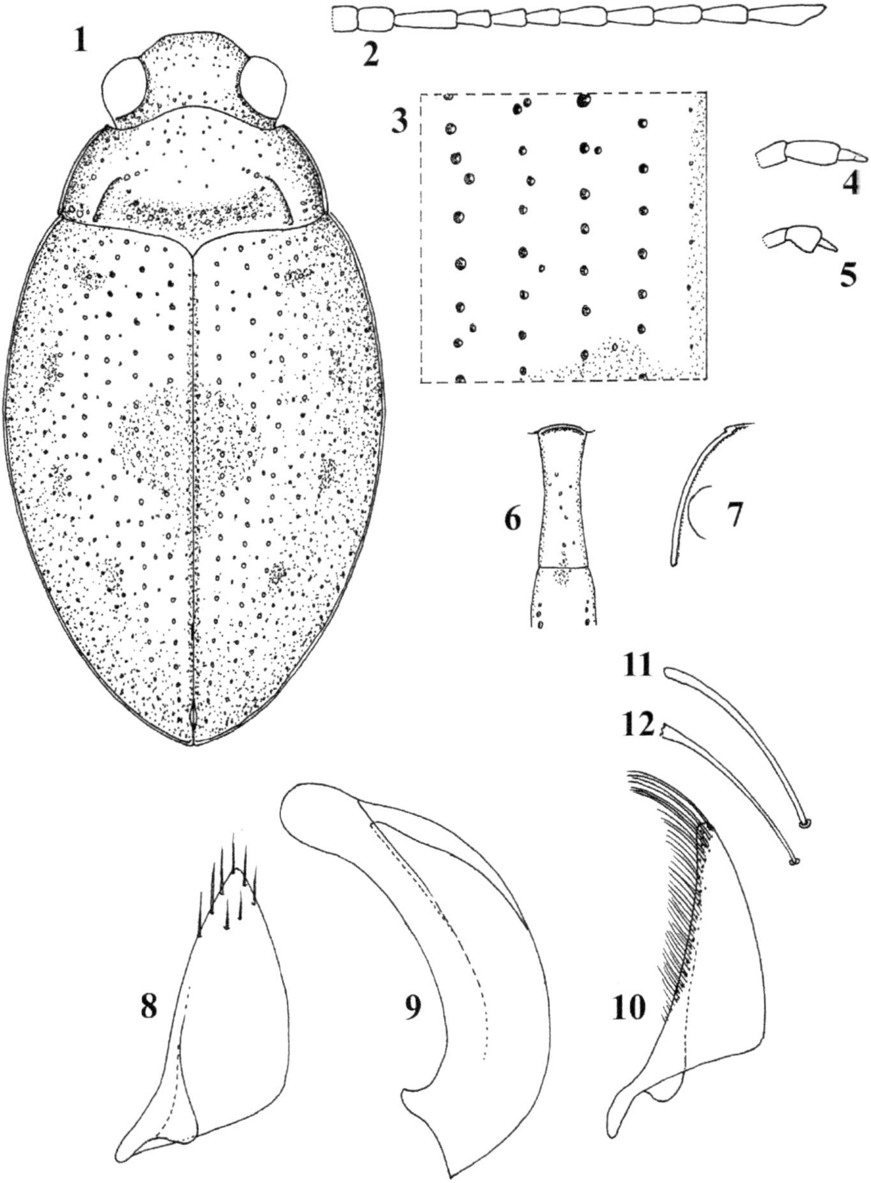

PLATE 172 *Haliplus* (*Neohaliplus*) *ruficeps*. 1, dorsal view; 2, antenna; 3, elytral punctures near base and suture; 4, maxillary palpus; 5, labial palpus; 6, prosternal and metaventral process; 7, prosternal process in lateral view; 8, left paramere; 9, penis; 10, right paramere; 11, 12, setae of right paramere

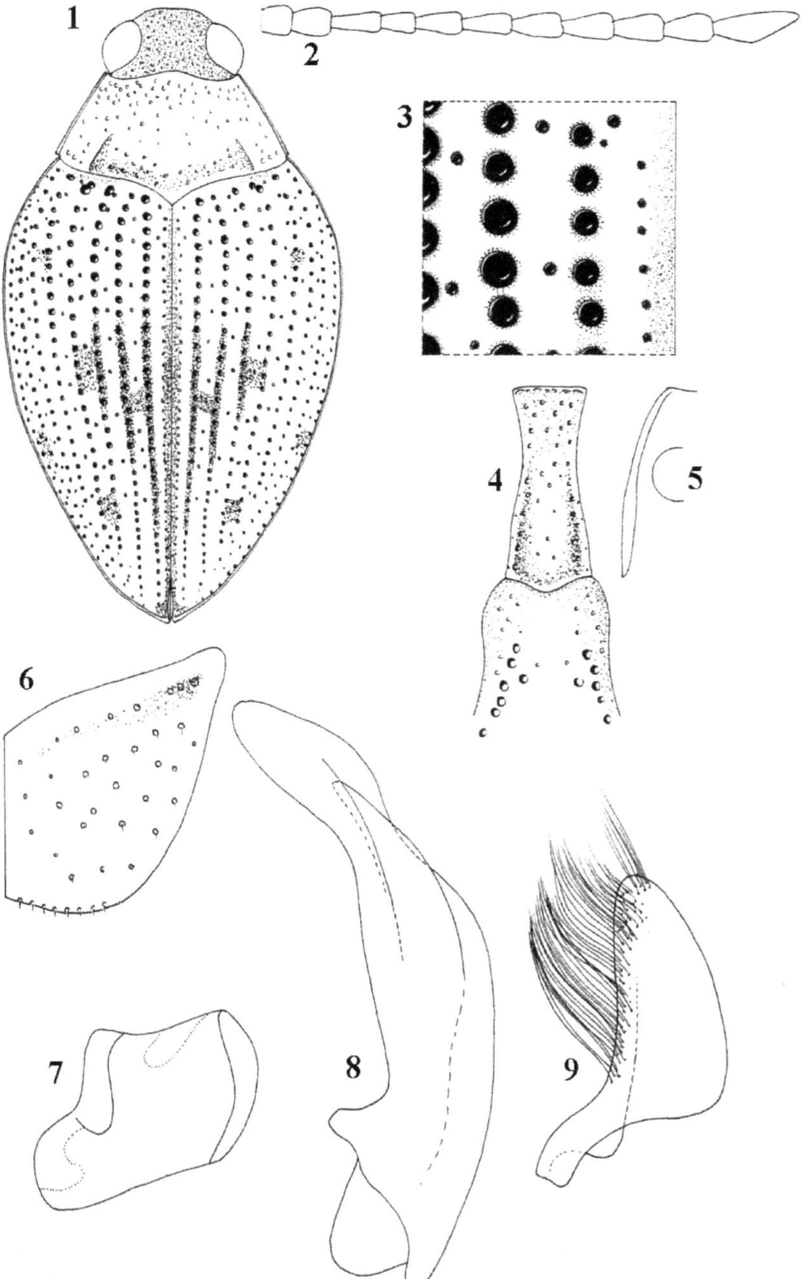

PLATE 173 *Haliplus* (*Nipponiplus*) *japonicus*. 1, dorsal view; 2, antenna; 3, elytral punctures near base and suture; 4, prosternal and metaventral process; 5, prosternal process in lateral view; 6; hind coxal plate; 7, left paramere; 8, penis; 9, right paramere

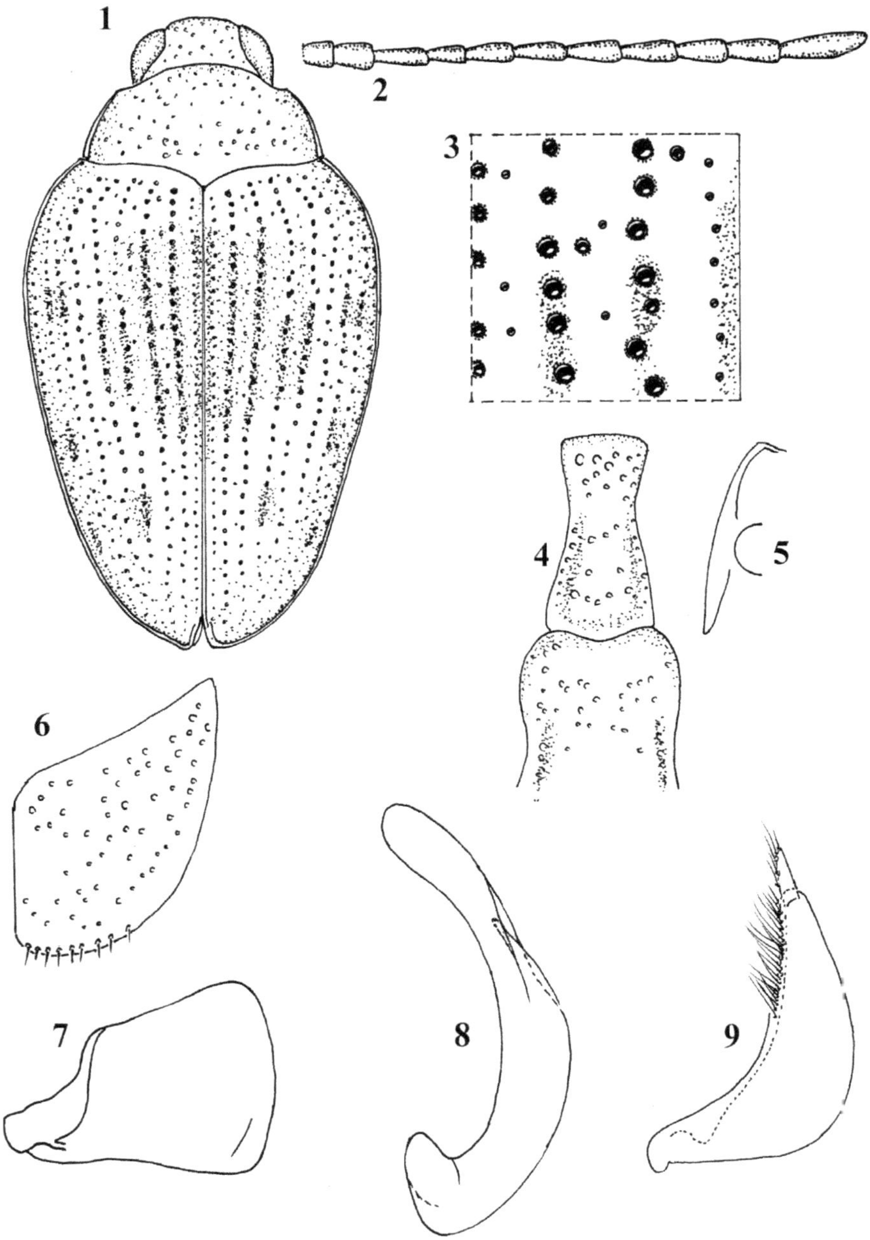

PLATE 174 *Haliplus* (*Nipponiplus*) *latreillei*. 1, dorsal view; 2, antenna; 3, elytral punctures near base and suture; 4, prosternal and metaventral process; 5, prosternal process in lateral view; 6; hind coxal plate; 7, left paramere; 8, penis; 9, right paramere

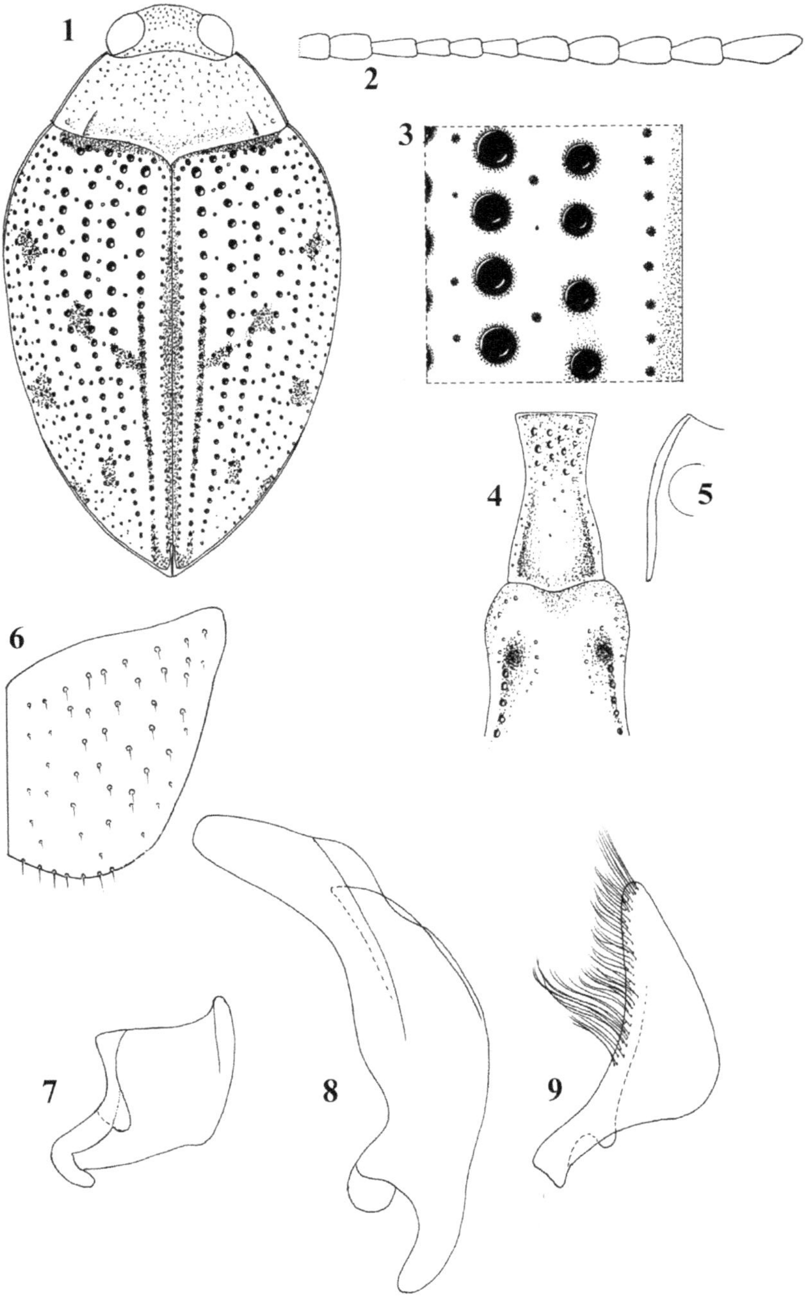

PLATE 175 *Haliplus* (*Nipponiplus*) *regimbarti*. 1, dorsal view; 2, antenna; 3, elytral punctures near base and suture; 4, prosternal and metaventral process; 5, prosternal process in lateral view; 6; hind coxal plate; 7, left paramere; 8, penis; 9, right paramere

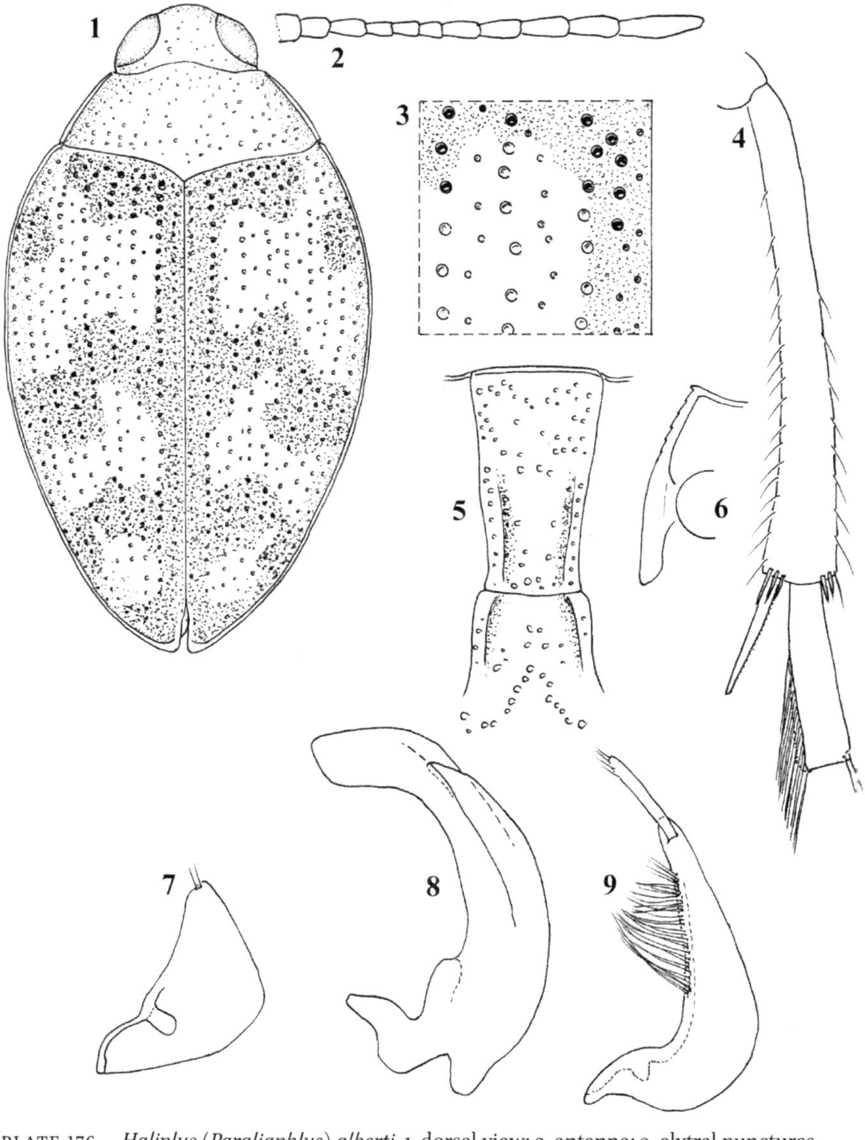

PLATE 176 *Haliplus* (*Paraliaphlus*) *alberti*. 1, dorsal view; 2, antenna; 3, elytral punctures near base and suture; 4, hind tibia in dorsal view; 5, prosternal and metaventral process; 6, prosternal process in lateral view; 7, left paramere; 8, penis; 9, right paramere

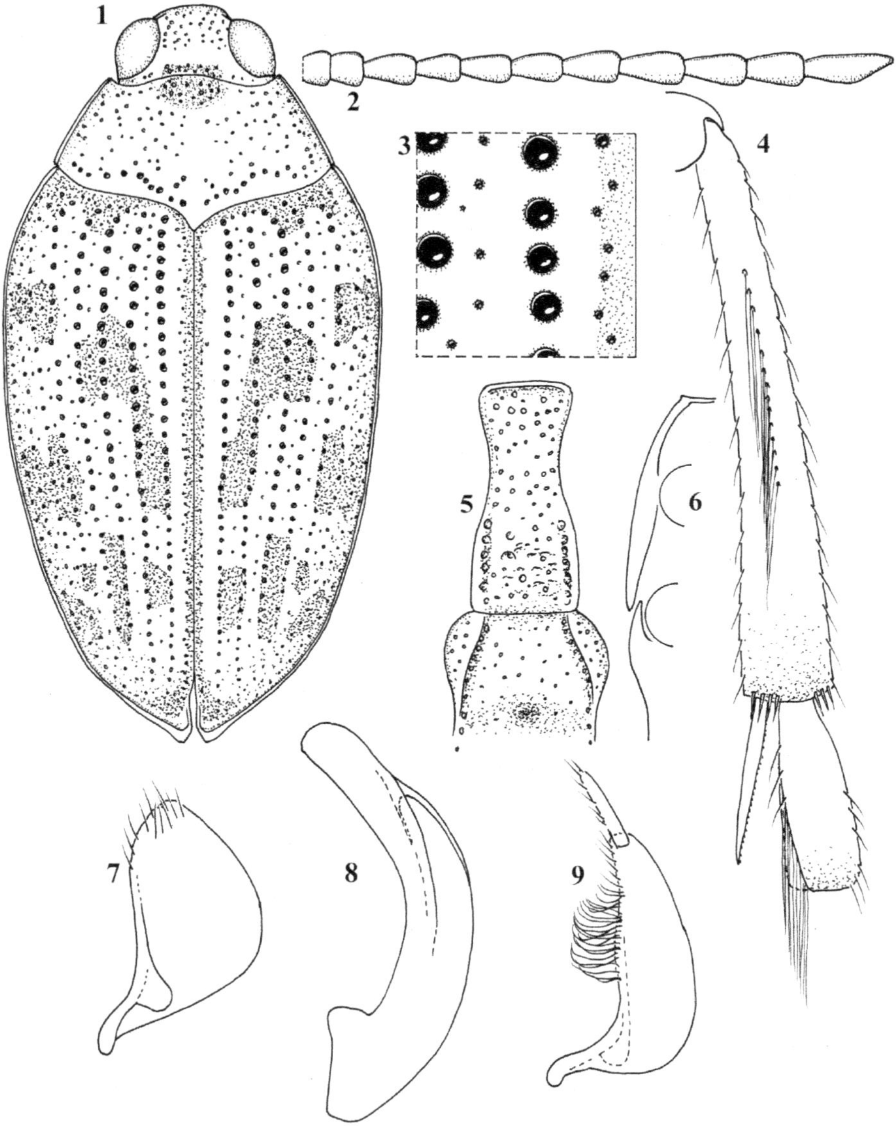

PLATE 177 *Haliplus* (*Paraliaphlus*) *apostolicus*. 1, dorsal view; 2, antenna; 3, elytral punctures near base and suture; 4, hind tibia in dorsal view; 5, prosternal and metaventral process; 6, prosternal process in lateral view; 7, left paramere; 8, penis; 9, right paramere

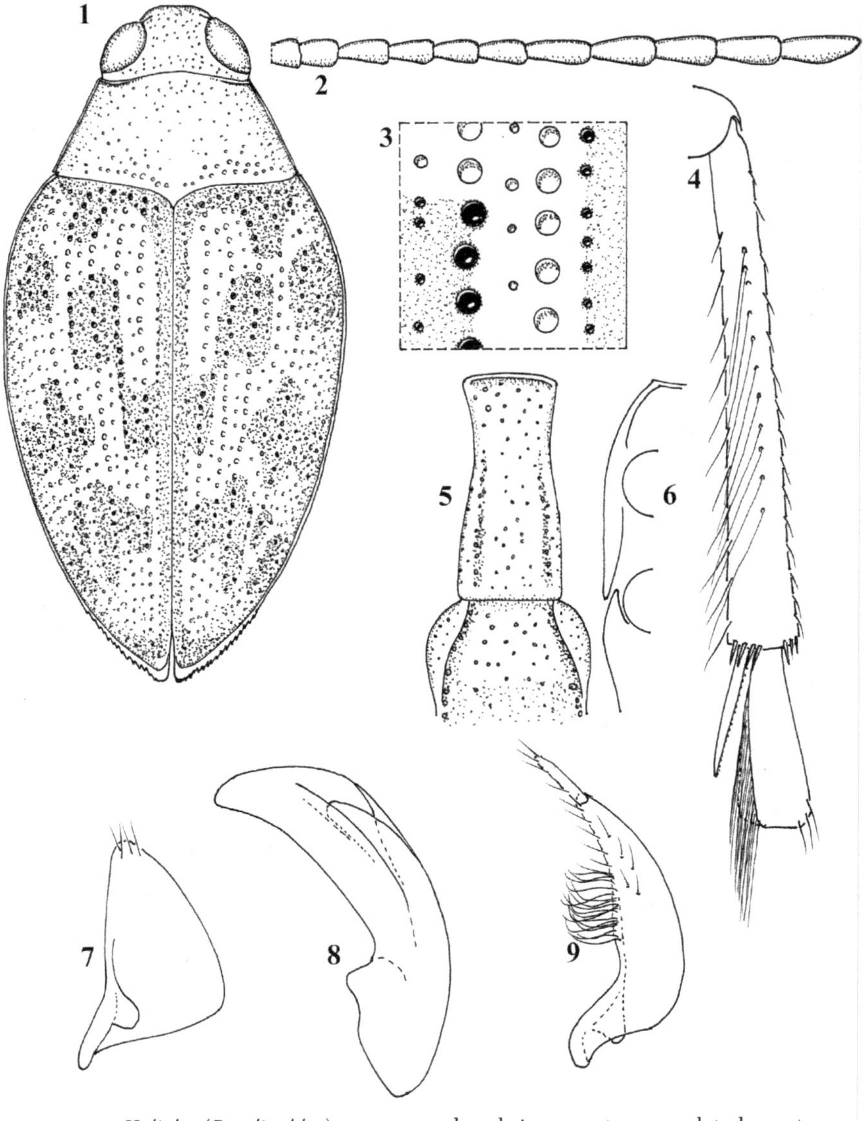

PLATE 178 *Haliplus* (*Paraliaphlus*) *connexus*. 1, dorsal view; 2, antenna; 3, elytral punctures near base and suture; 4, hind tibia in dorsal view; 5, prosternal and metaventral process; 6, prosternal process in lateral view; 7, left paramere; 8, penis; 9, right paramere

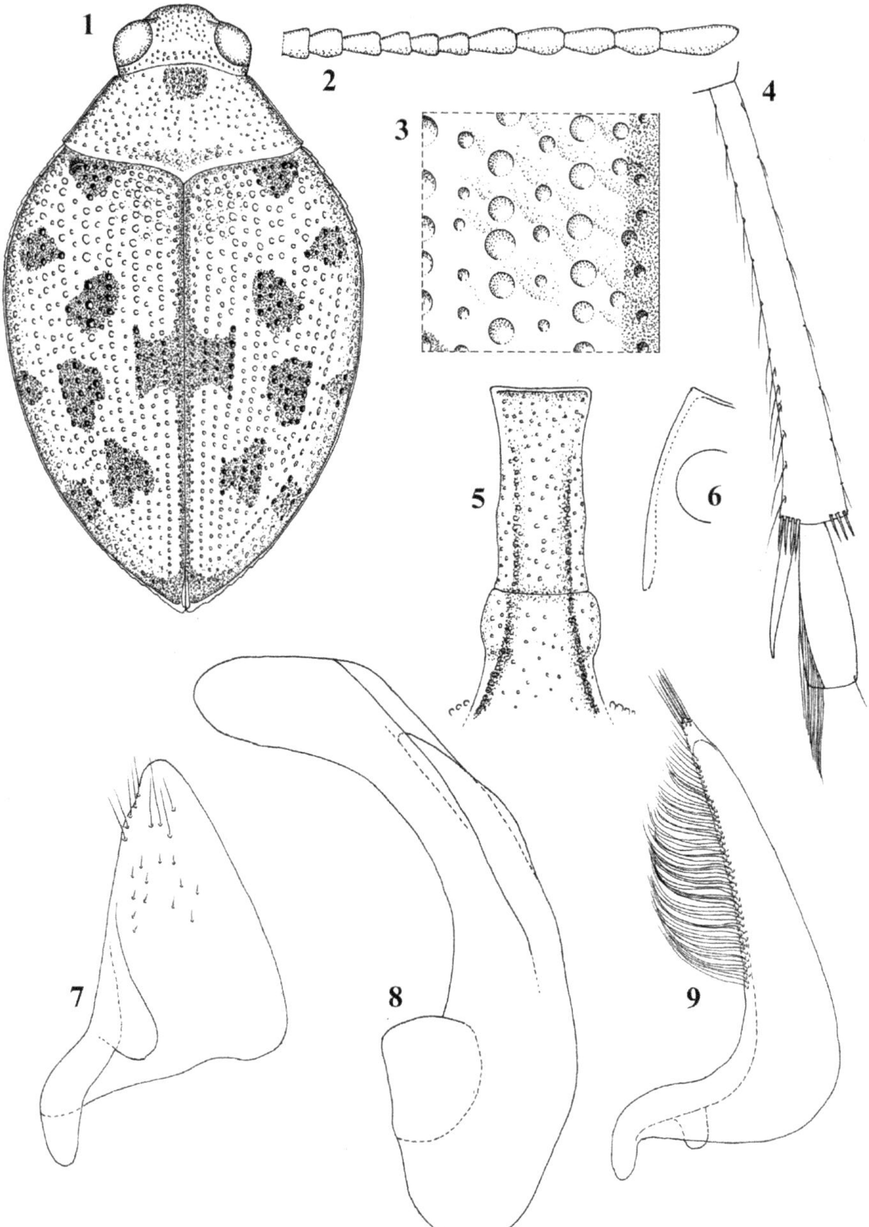

PLATE 179 *Haliplus* (*Paraliaphlus*) *deceptus*. 1, dorsal view; 2, antenna; 3, elytral punctures near base and suture; 4, hind tibia in dorsal view; 5, prosternal and metaventral process; 6, prosternal process in lateral view; 7, left paramere; 8, penis; 9, right paramere

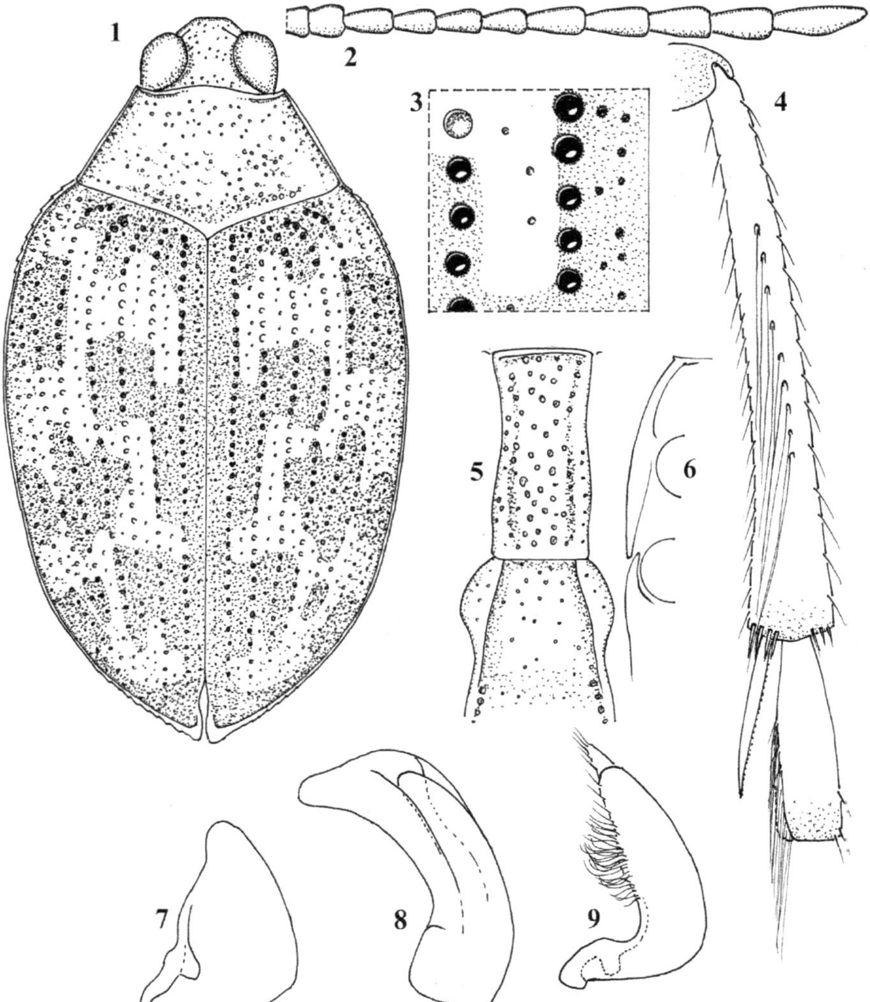

PLATE 180 *Haliplus* (*Paraliaphlus*) *fasciatus*. 1, dorsal view; 2, antenna; 3, elytral punctures near base and suture; 4, hind tibia in dorsal view; 5, prosternal and metaventral process; 6, prosternal process in lateral view; 7, left paramere; 8, penis; 9, right paramere

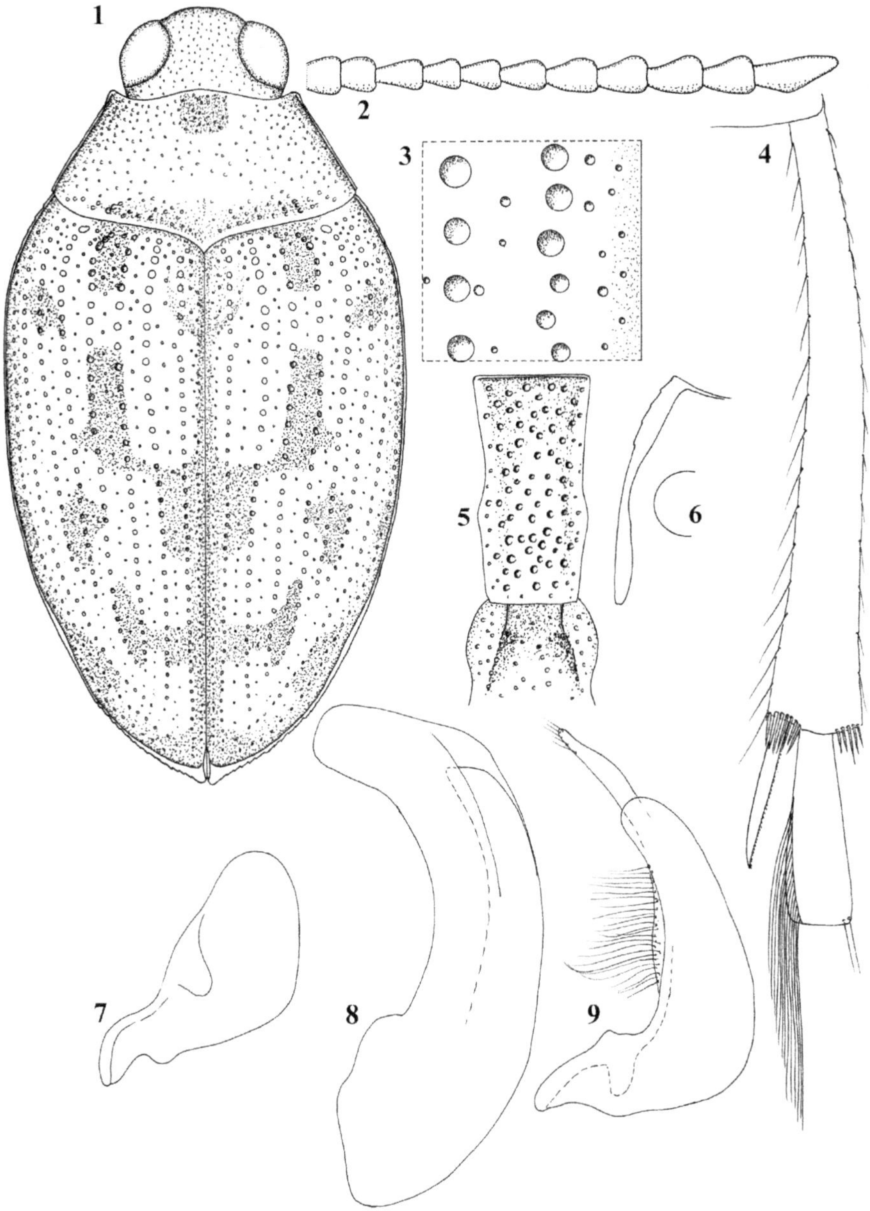

PLATE 181 *Haliplus* (*Paraliaphlus*) *grandis*. 1, dorsal view; 2, antenna; 3, elytral punctures near base and suture; 4, hind tibia in dorsal view; 5, prosternal and metaventral process; 6, prosternal process in lateral view; 7, left paramere; 8, penis; 9, right paramere

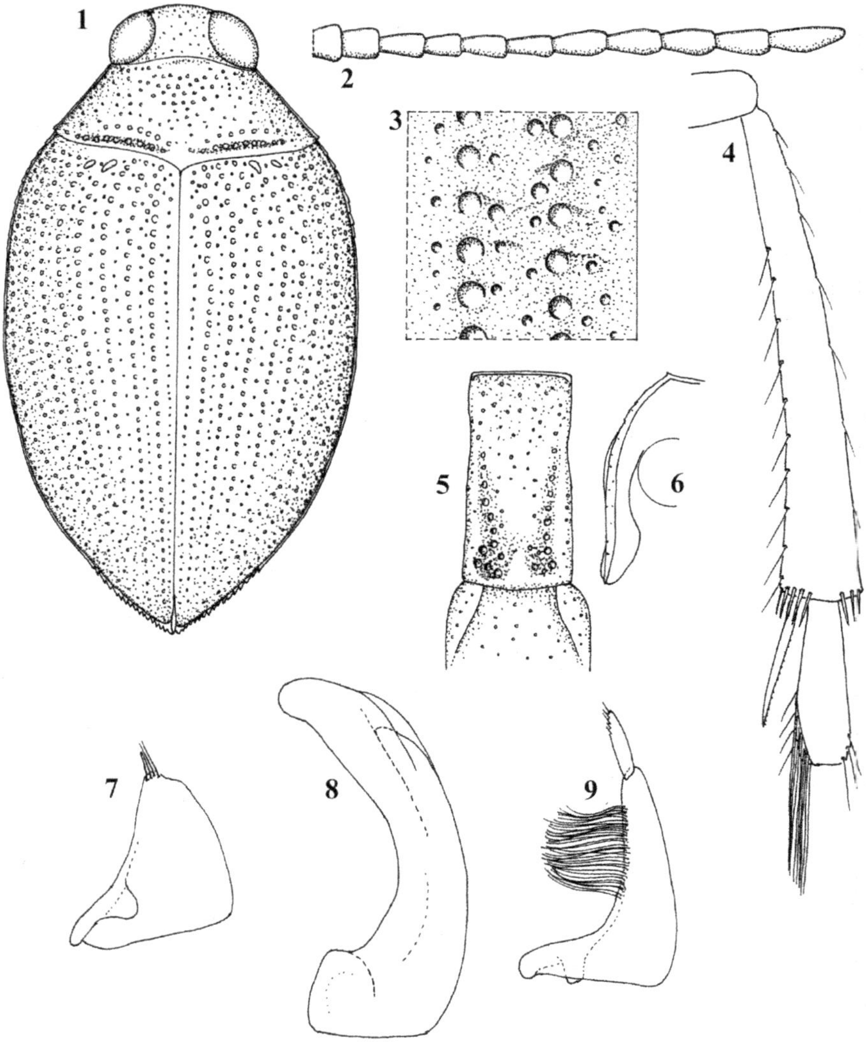

PLATE 182 *Haliplus* (*Paraliaphlus*) *kenneri*. 1, dorsal view; 2, antenna; 3, elytral punctures near base and suture; 4, hind tibia in dorsal view; 5, prosternal and metaventral process; 6, prosternal process in lateral view; 7, left paramere; 8, penis; 9, right paramere

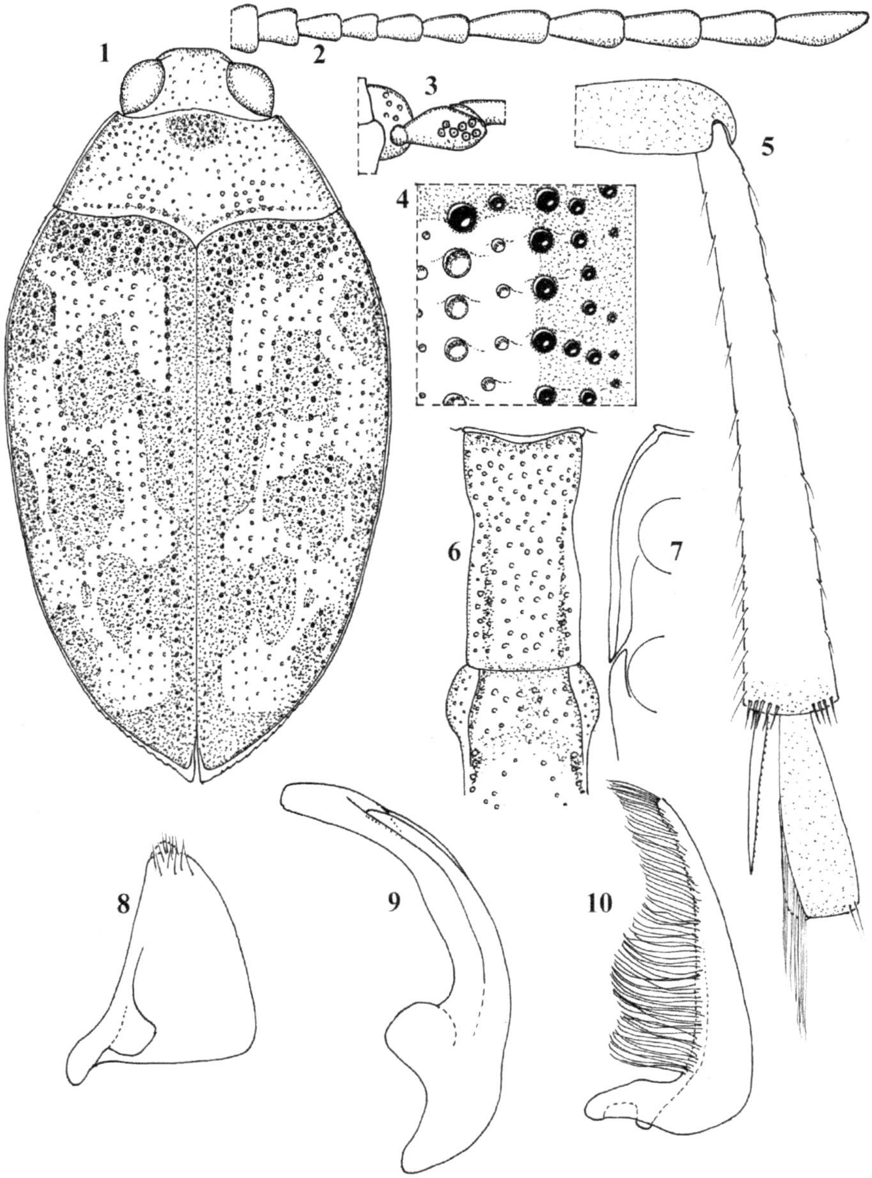

PLATE 183 *Haliplus* (*Paraliaphlus*) *leopardus*. 1, dorsal view; 2, antenna; 3, coxa of midleg; 4, elytral punctures near base and suture; 5, hind tibia in dorsal view; 6, prosternal and metaventral process; 7, prosternal process in lateral view; 8, left paramere; 9, penis; 10, right paramere

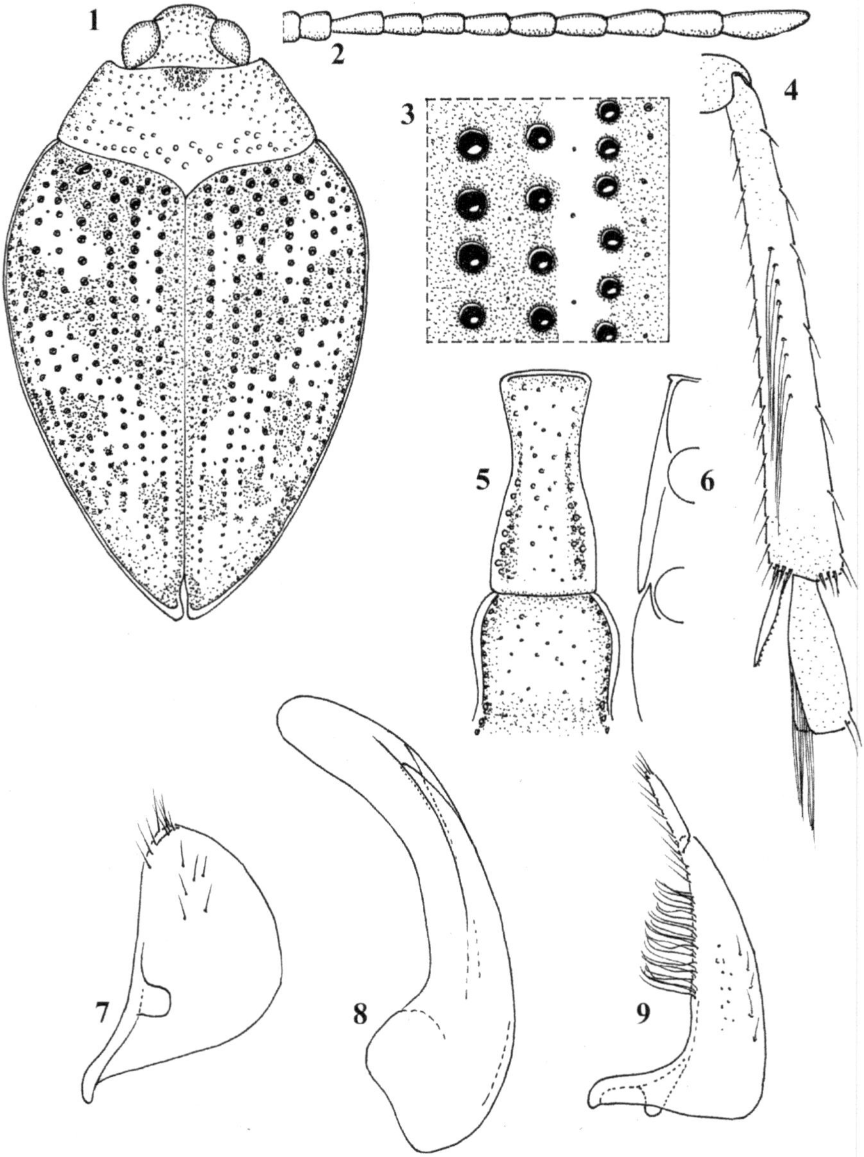

PLATE 184 *Haliplus* (*Paraliaphlus*) *lucigelinae*. 1, dorsal view; 2, antenna; 3, elytral punctures near base and suture; 4, hind tibia in dorsal view; 5, prosternal and metaventral process; 6, prosternal process in lateral view; 7, left paramere; 8, penis; 9, right paramere

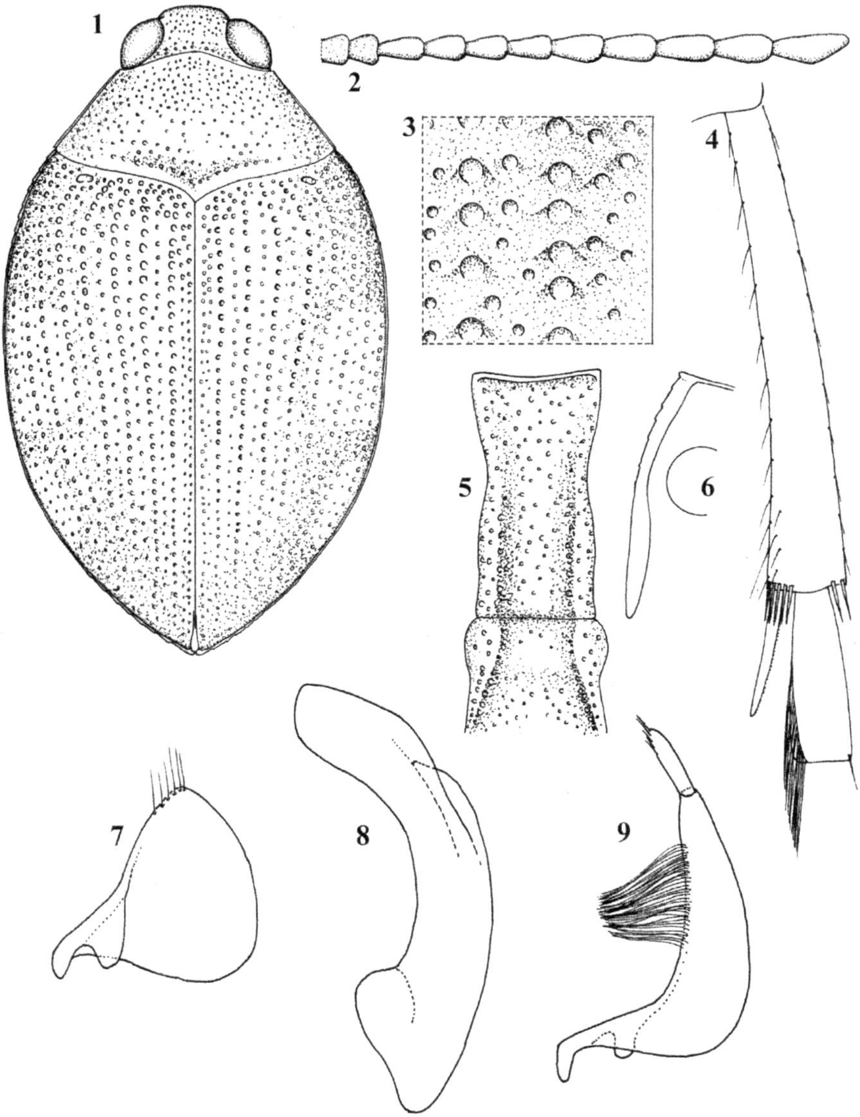

PLATE 185 *Haliplus* (*Paraliaphlus*) *mimeticus*. 1, dorsal view; 2, antenna; 3, elytral punctures near base and suture; 4, hind tibia in dorsal view; 5, prosternal and metaventral process; 6, prosternal process in lateral view; 7, left paramere; 8, penis; 9, right paramere

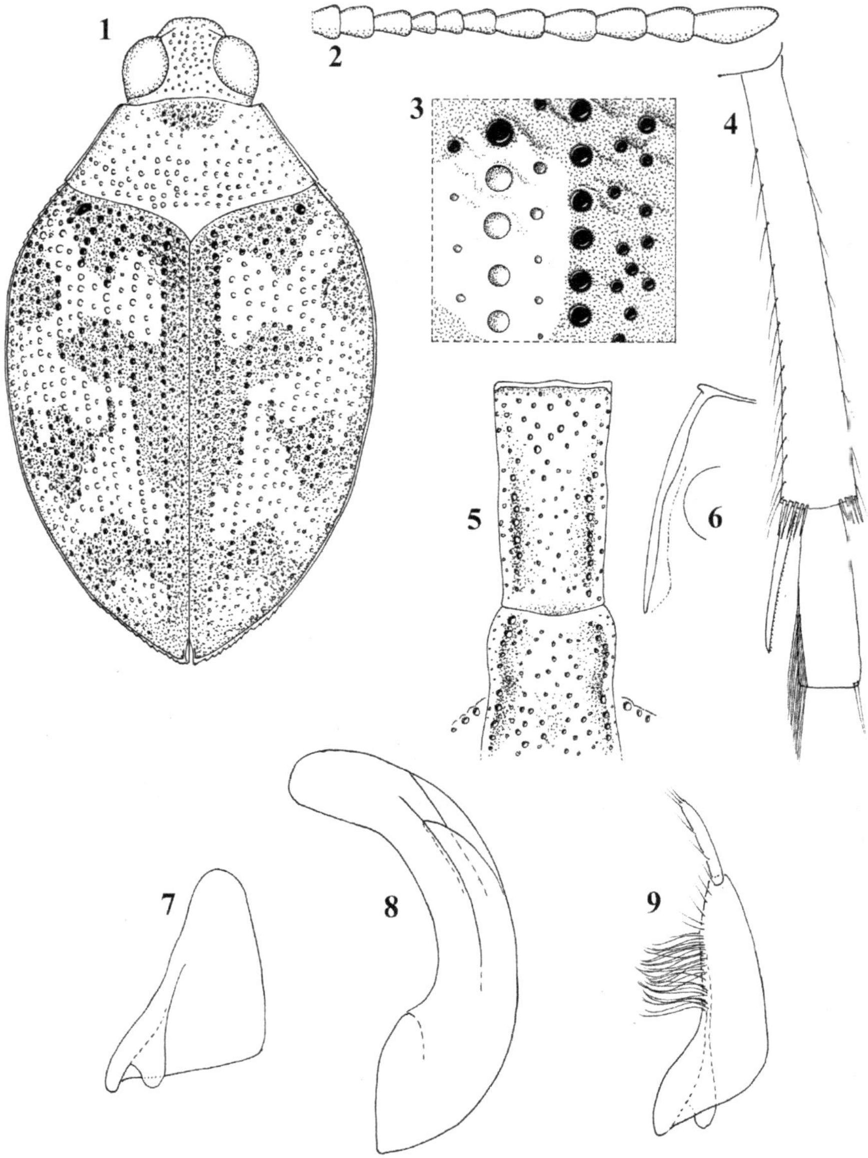

PLATE 186 *Haliplus* (*Paraliaphlus*) *mutchleri*. 1, dorsal view; 2, antenna; 3, elytral punctures near base and suture; 4, hind tibia in dorsal view; 5, prosternal and metaventral process; 6, prosternal process in lateral view; 7, left paramere; 8, penis; 9, right paramere

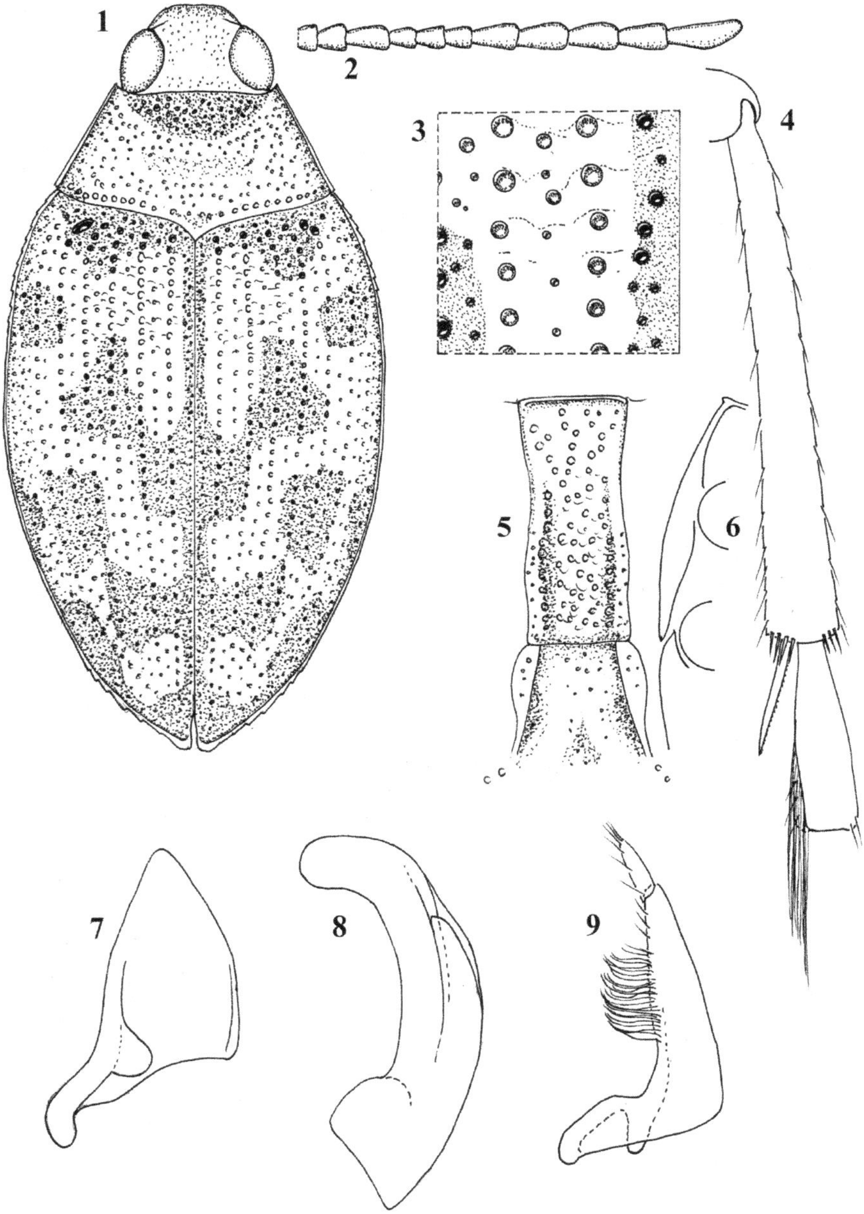

PLATE 187 *Haliplus* (*Paraliaphlus*) *pantherinus*. 1, dorsal view; 2, antenna; 3, elytral punctures near base and suture; 4, hind tibia in dorsal view; 5, prosternal and metaventral process; 6, prosternal process in lateral view; 7, left paramere; 8, penis; 9, right paramere

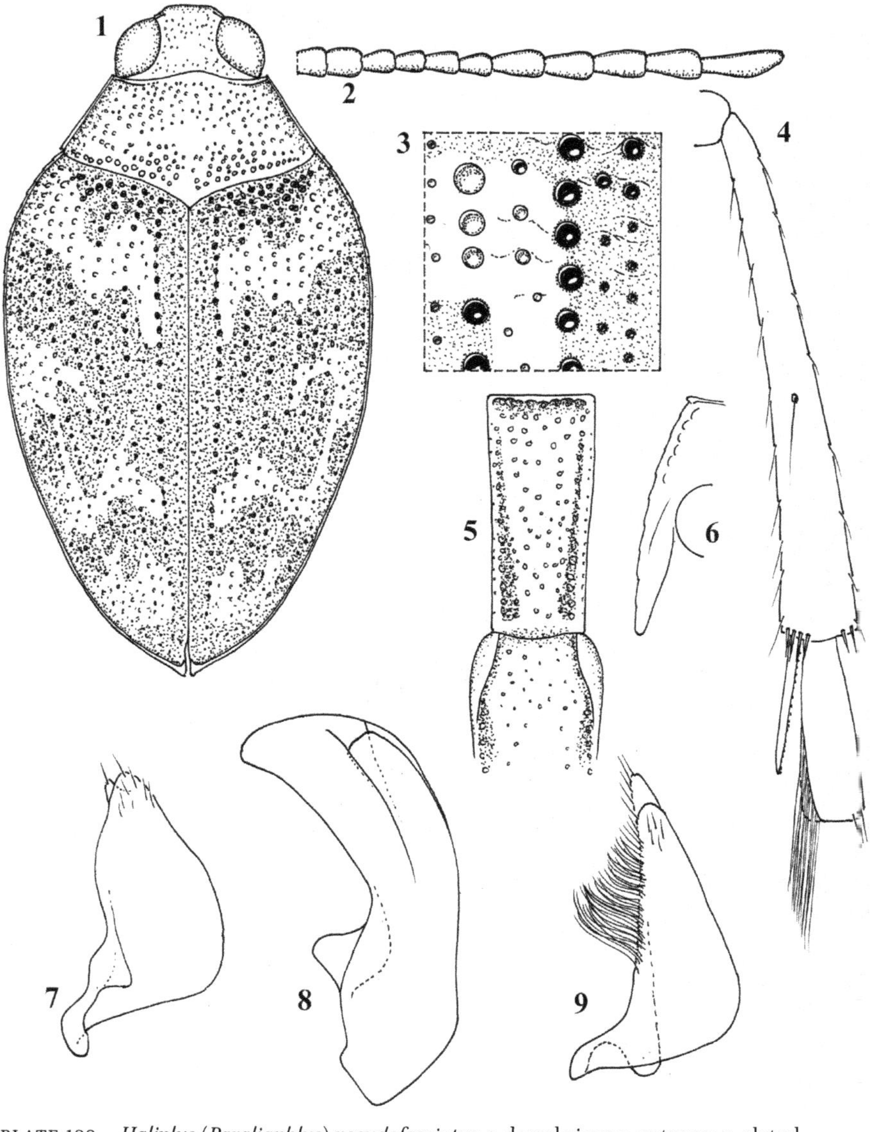

PLATE 188 *Haliplus* (*Paraliaphlus*) *pseudofasciatus*. 1, dorsal view; 2, antenna; 3, elytral punctures near base and suture; 4, hind tibia in dorsal view; 5, prosternal and metaventral process; 6, prosternal process in lateral view; 7, left paramere; 8, penis; 9, right paramere

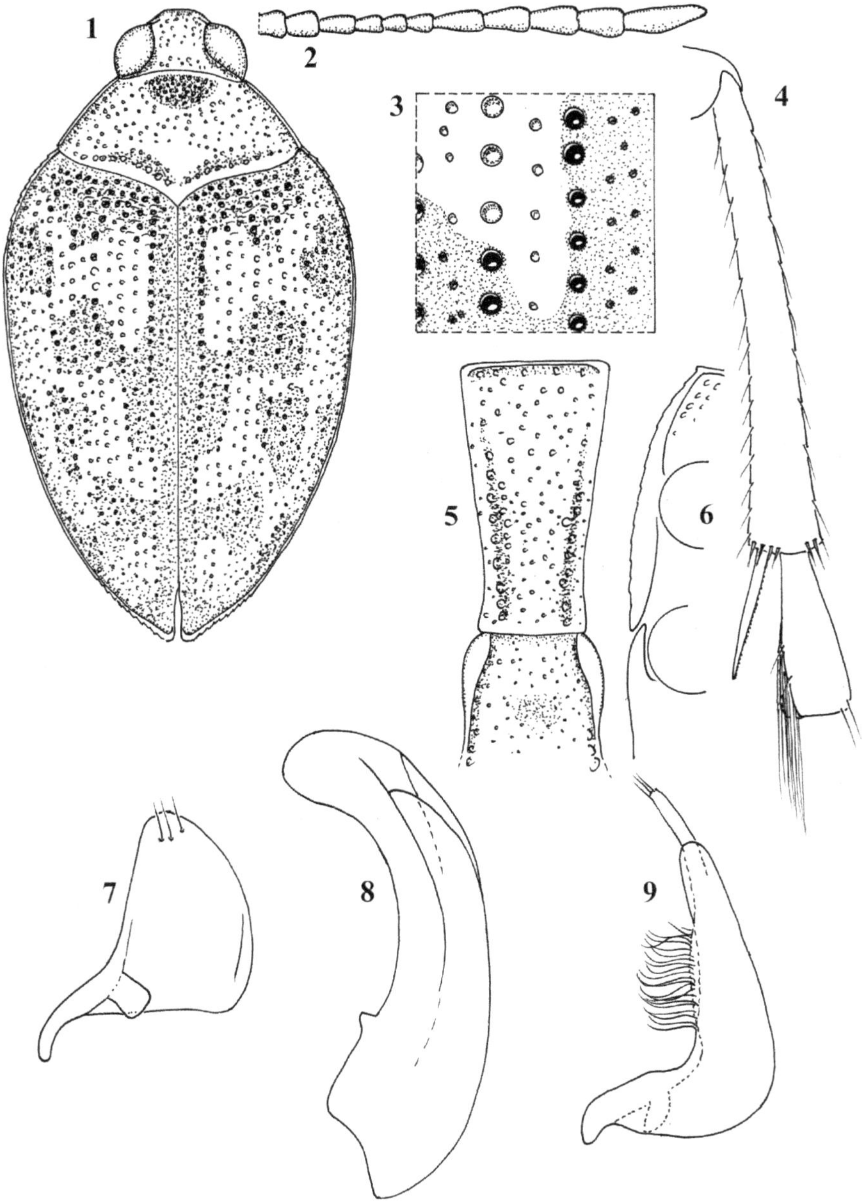

PLATE 189 *Haliplus* (*Paraliaphlus*) *punctatus*. 1, dorsal view; 2, antenna; 3, elytral punctures near base and suture; 4, hind tibia in dorsal view; 5, prosternal and metaventral process; 6, prosternal process in lateral view; 7, left paramere; 8, penis; 9, right paramere

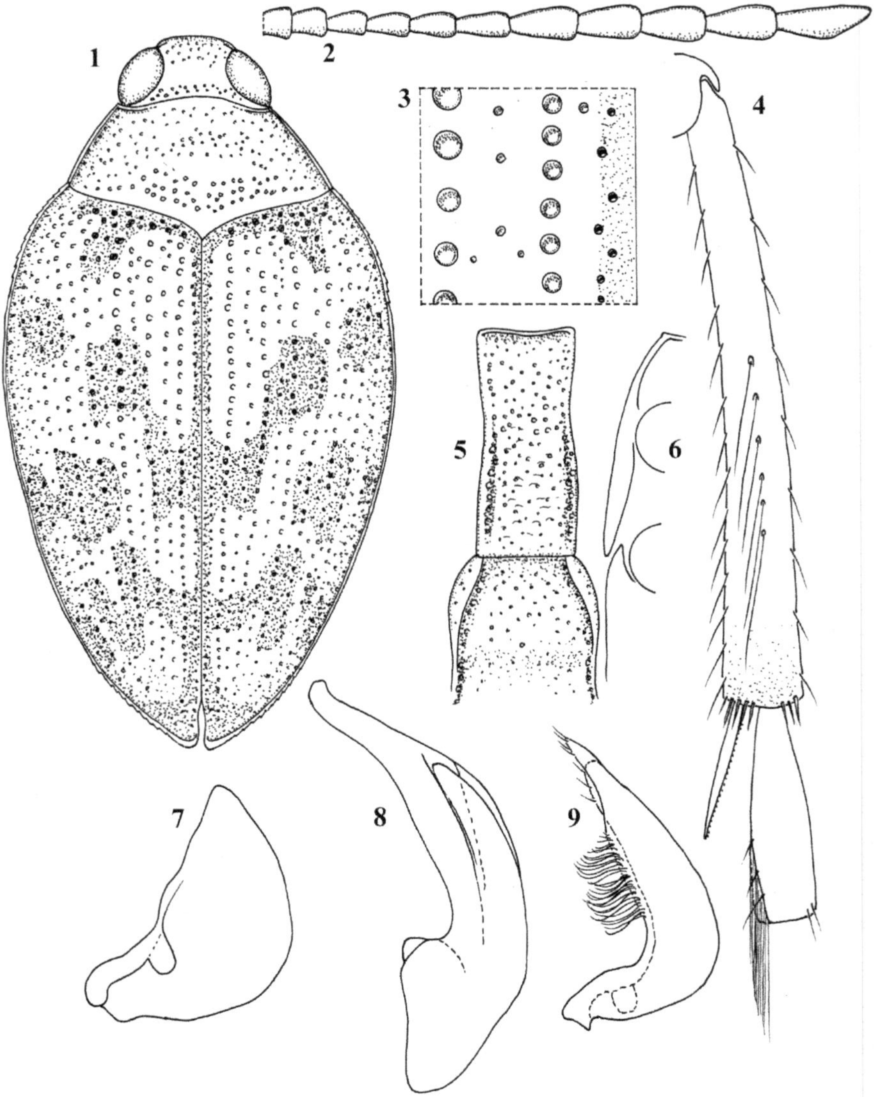

PLATE 190 *Haliplus* (*Paraliaphlus*) *tortilipenis*. 1, dorsal view; 2, antenna; 3, elytral punctures near base and suture; 4, hind tibia in dorsal view; 5, prosternal and metaventral process; 6, prosternal process in lateral view; 7, left paramere; 8, penis; 9, right paramere

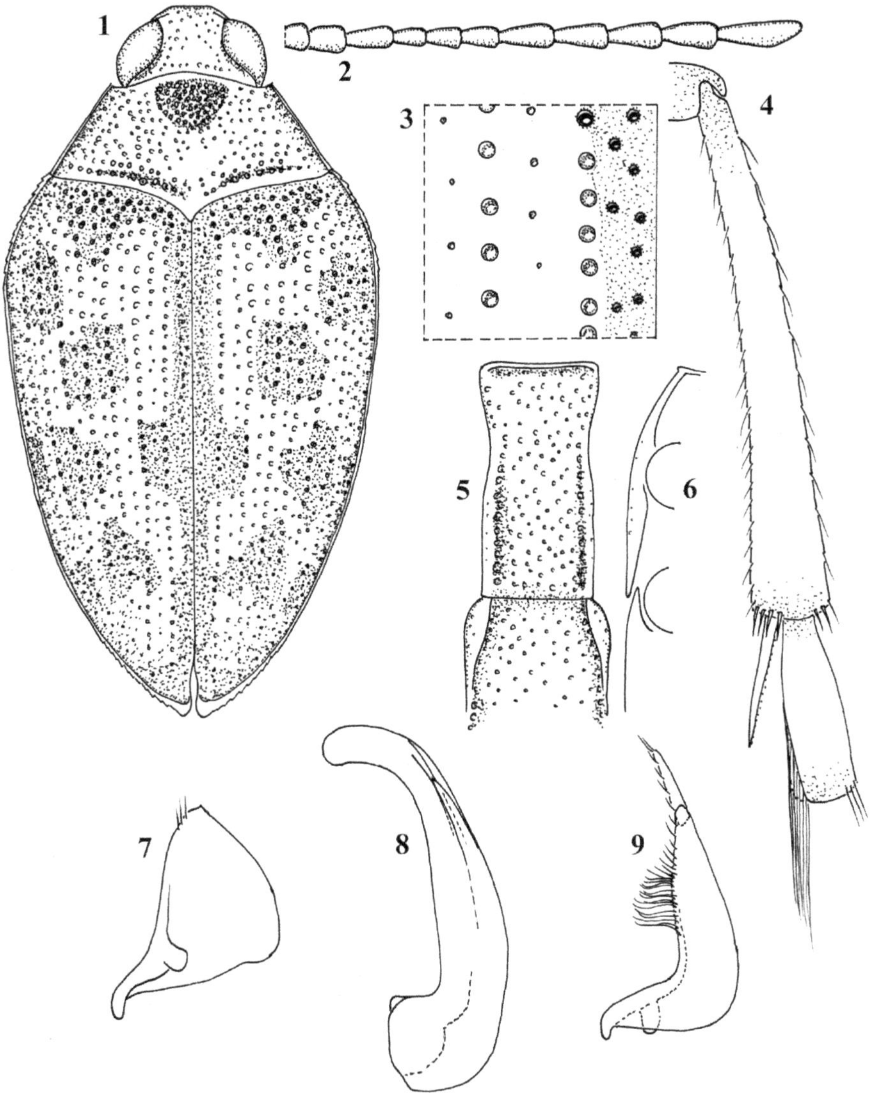

PLATE 191 *Haliplus* (*Paraliaphlus*) *triopsis*. 1, dorsal view; 2, antenna; 3, elytral punctures near base and suture; 4, hind tibia in dorsal view; 5, prosternal and metaventral process; 6, prosternal process in lateral view; 7, left paramere; 8, penis; 9, right paramere.

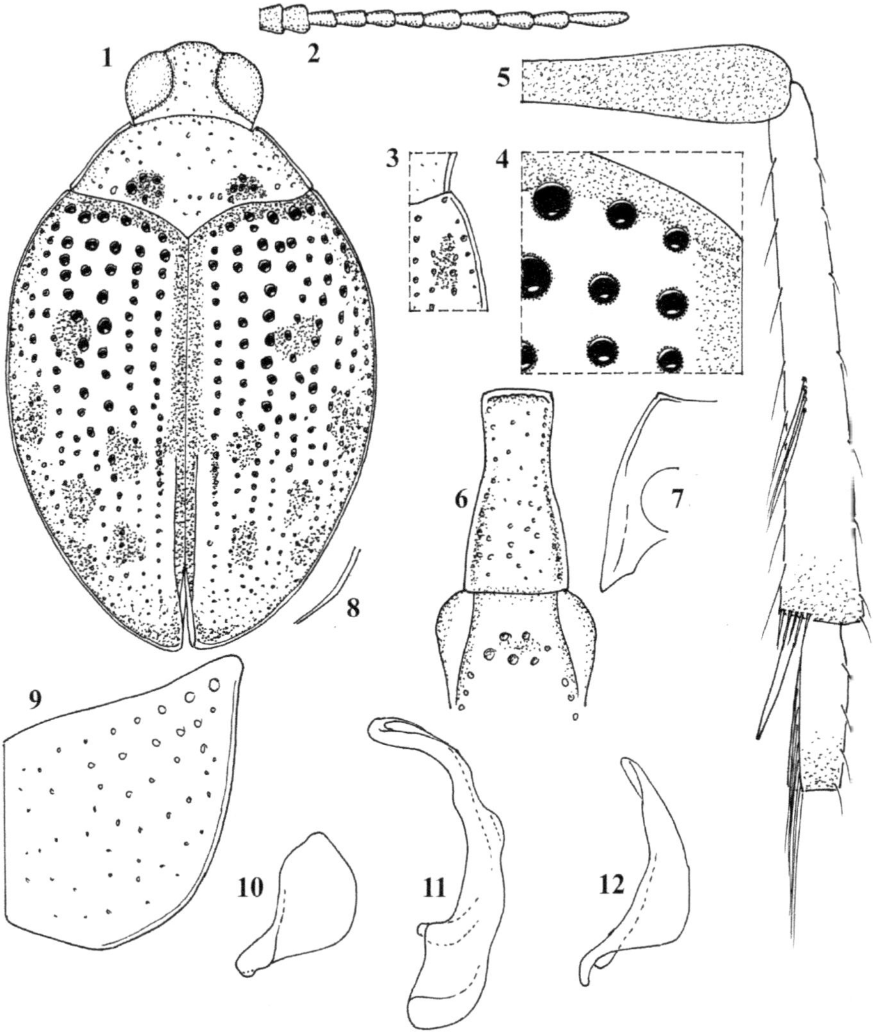

PLATE 192 *Peltodytes* (*Neopeltodytes*) *bradleyi*. 1, dorsal view; 2, antenna; 3, elytral shoulder in lateral view; 4, elytral punctures near base and suture; 5, hind tibia in dorsal view; 6, prosternal and metaventral process; 7, prosternal process in lateral view; 8, preapical tooth on elytral margin; 9, hind coxal plate; 10, left paramere; 11, penis; 12, right paramere

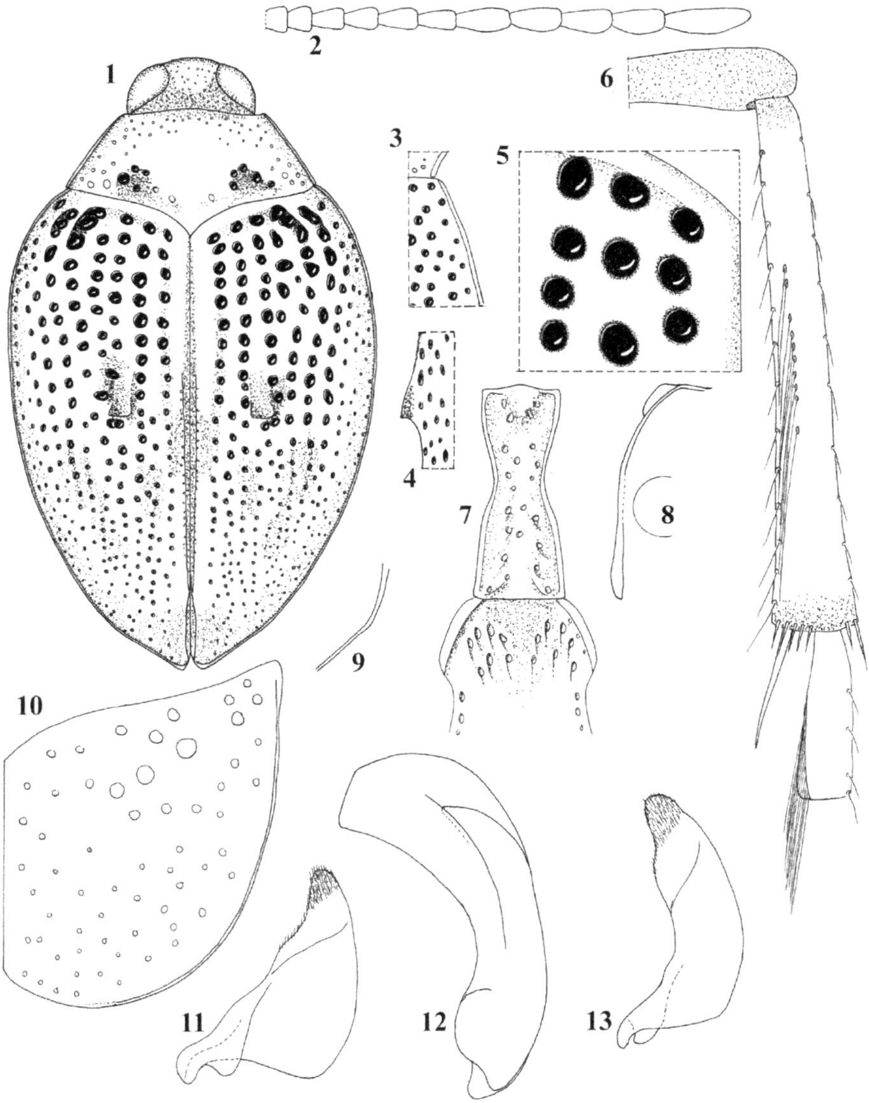

PLATE 193 *Peltodytes* (*Neopeltodytes*) *callosus*. 1, dorsal view; 2, antenna; 3, elytral shoulder in lateral view; 4, elytral tubercle in lateral view; 5, elytral punctures near base and suture; 6, hind tibia in dorsal view; 7, prosternal and metaventral process; 8, prosternal process in lateral view; 9, preapical tooth on elytral margin; 10, hind coxal plate; 11, left paramere; 12, penis; 13, right paramere

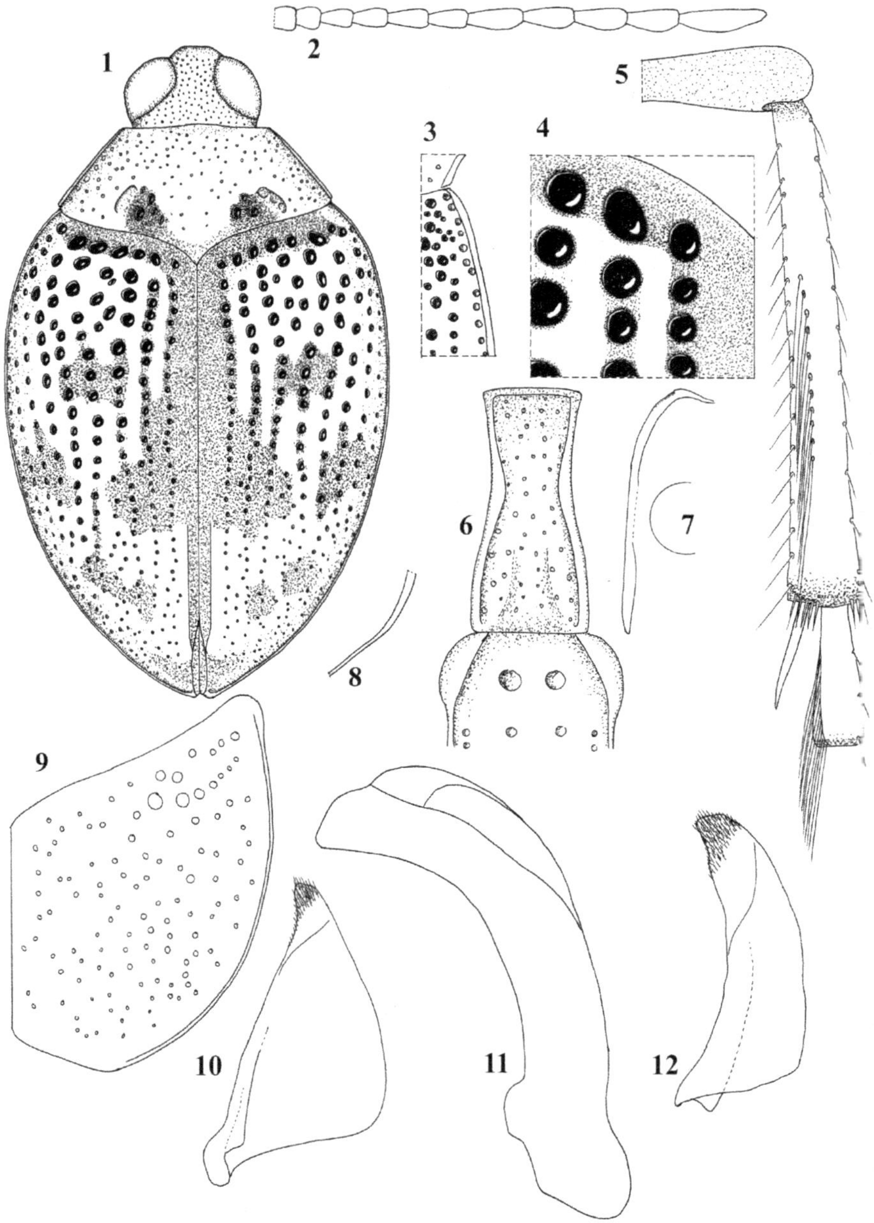

PLATE 194 *Peltodytes* (*Neopeltodytes*) *darlingtoni*. 1, dorsal view; 2, antenna; 3, elytral shoulder in lateral view; 4, elytral punctures near base and suture; 5, hind tibia in dorsal view; 6, prosternal and metaventral process; 7, prosternal process in lateral view; 8, preapical tooth on elytral margin; 9, hind coxal plate; 10, left paramere; 11, penis; 12, right paramere

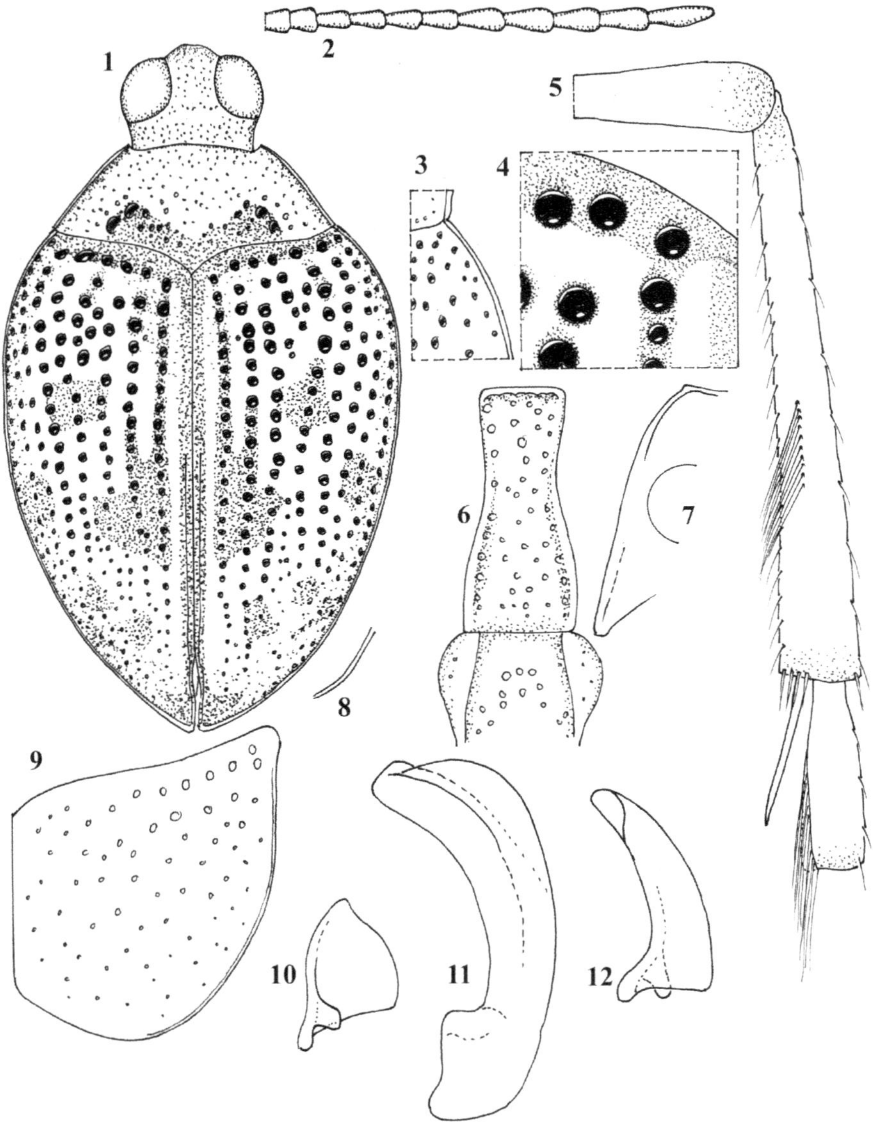

PLATE 195 *Peltodytes* (*Neopeltodytes*) *dietrichi*. 1, dorsal view; 2, antenna; 3, elytral shoulder in lateral view; 4, elytral punctures near base and suture; 5, hind tibia in dorsal view; 6, prosternal and metaventral process; 7, prosternal process in lateral view; 8, preapical tooth on elytral margin; 9, hind coxal plate; 10, left paramere; 11, penis; 12, right paramere

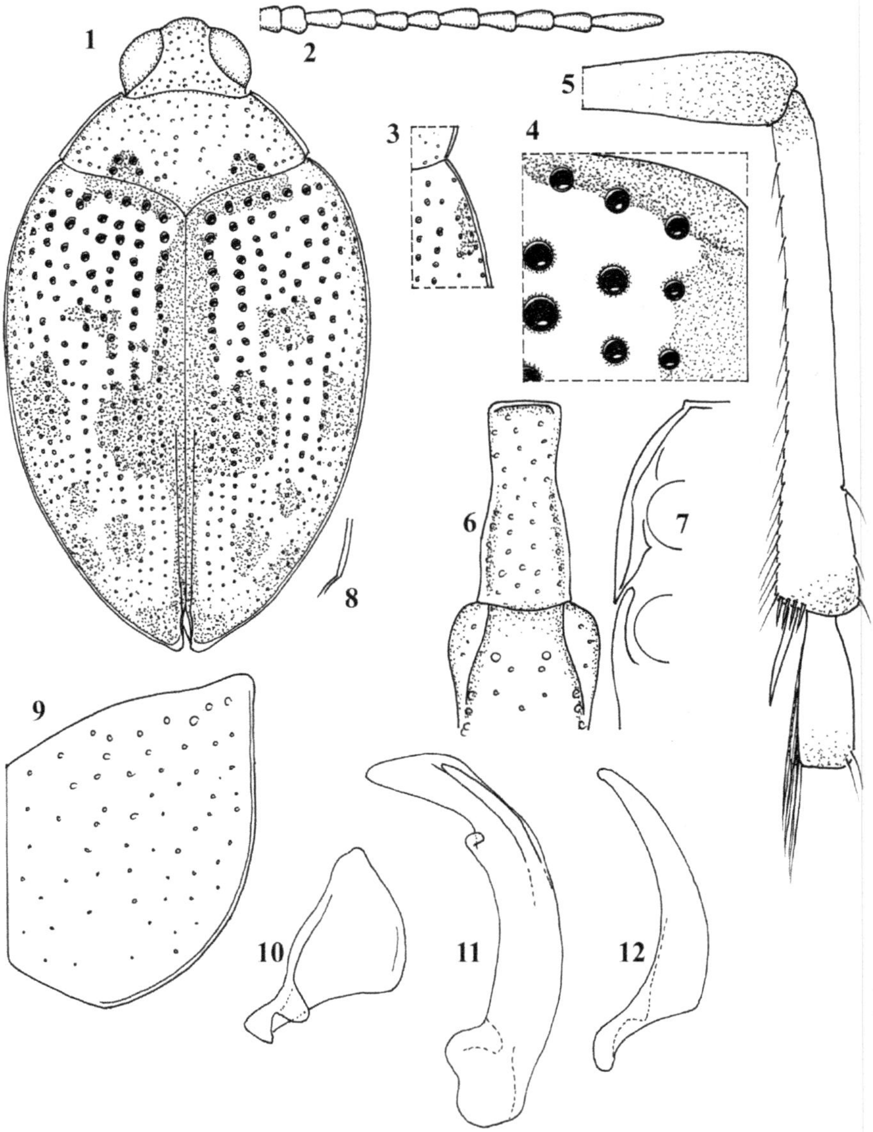

PLATE 196 *Peltodytes* (*Neopeltodytes*) *dunavani*. 1, dorsal view; 2, antenna; 3, 4, elytral shoulder in lateral view; 5, elytral punctures near base and suture; 6, hind tibia in dorsal view; 7, prosternal and metaventral process; 8, prosternal process in lateral view; 9, preapical tooth on elytral margin;10, hind coxal plate; 11, left paramere; 11, penis; 12, right paramere

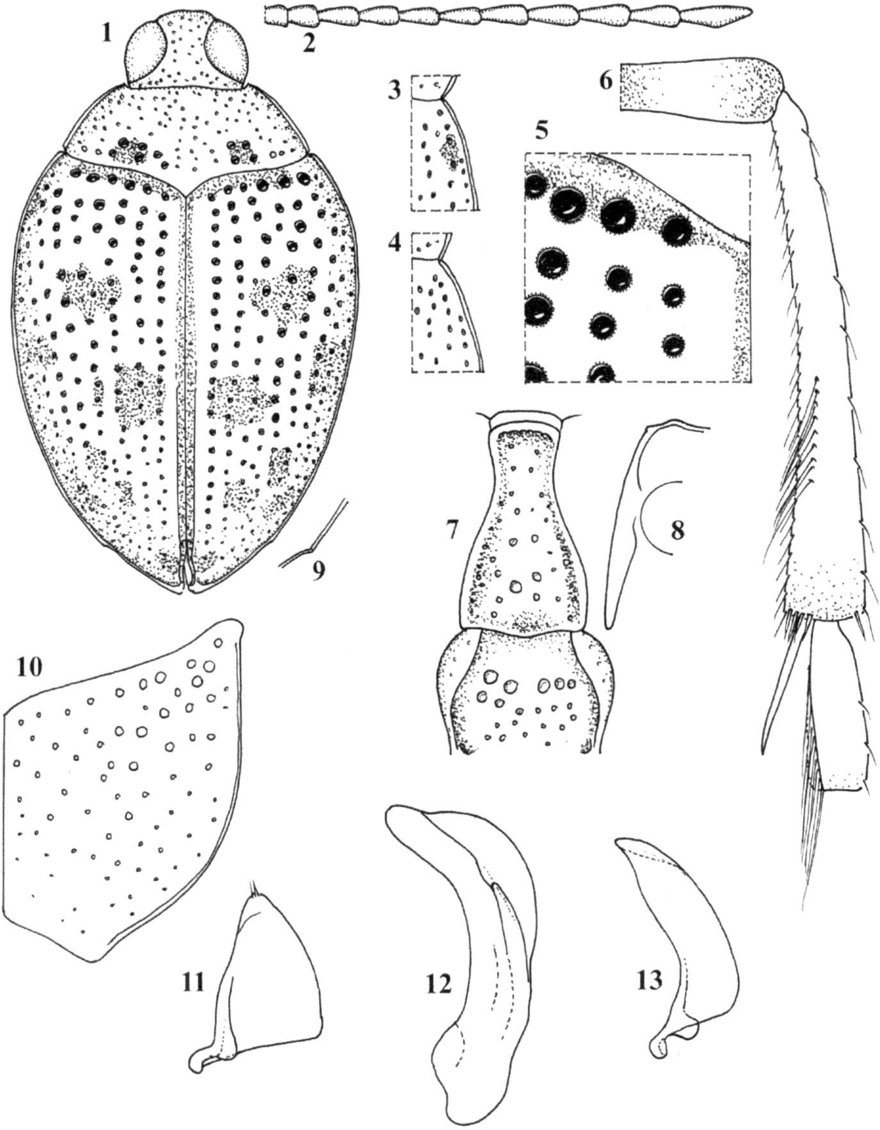

PLATE 197 *Peltodytes* (*Neopeltodytes*) *duodecimpunctatus*. 1, dorsal view; 2, antenna; 3, 4, elytral shoulder in lateral view; 5, elytral punctures near base and suture; 6, hind tibia in dorsal view; 7, prosternal and metaventral process; 8, prosternal process in lateral view; 9, preapical tooth on elytral margin; 10, hind coxal plate; 11, left paramere; 12, penis; 13, right paramere

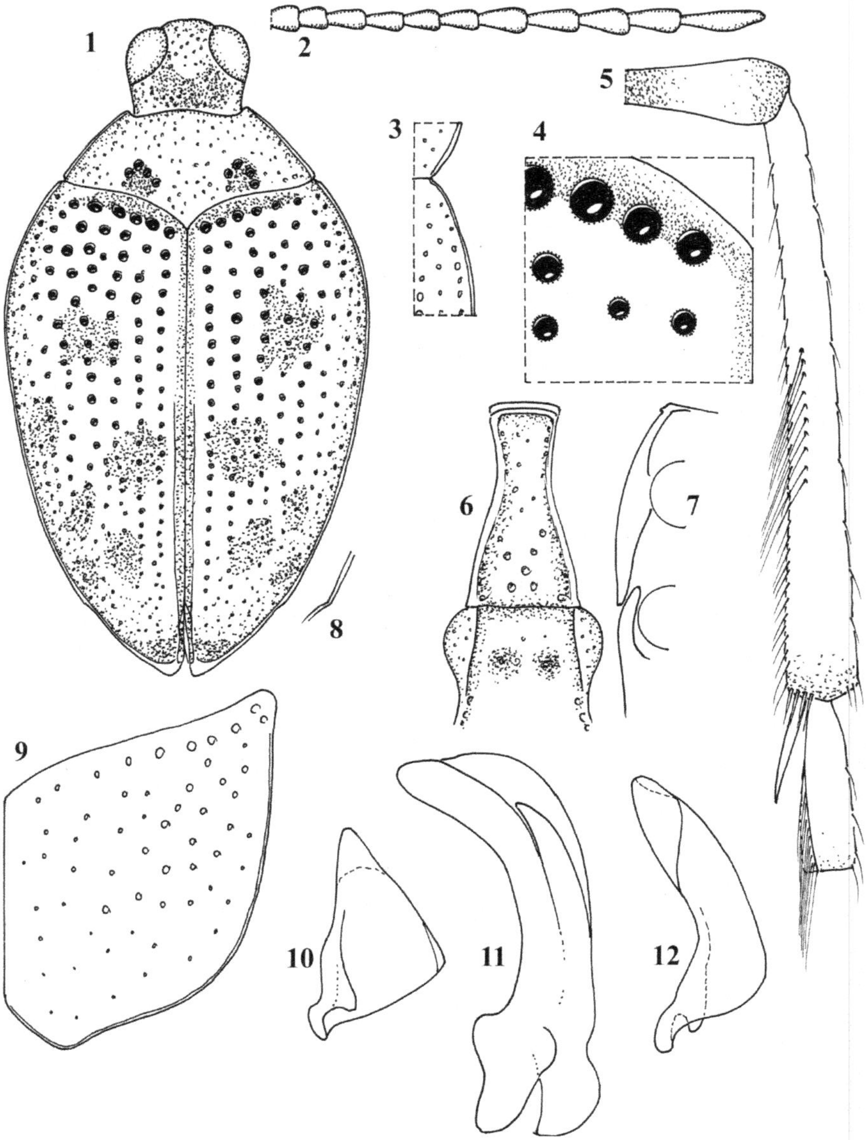

PLATE 198 *Peltodytes* (*Neopeltodytes*) *edentulus*. 1, dorsal view; 2, antenna; 3, elytral shoulder in lateral view; 4, elytral punctures near base and suture; 5, hind tibia in dorsal view; 6, prosternal and metaventral process; 7, prosternal process in lateral view; 8, preapical tooth on elytral margin; 9, hind coxal plate; 10, left paramere; 11, penis; 12, right paramere

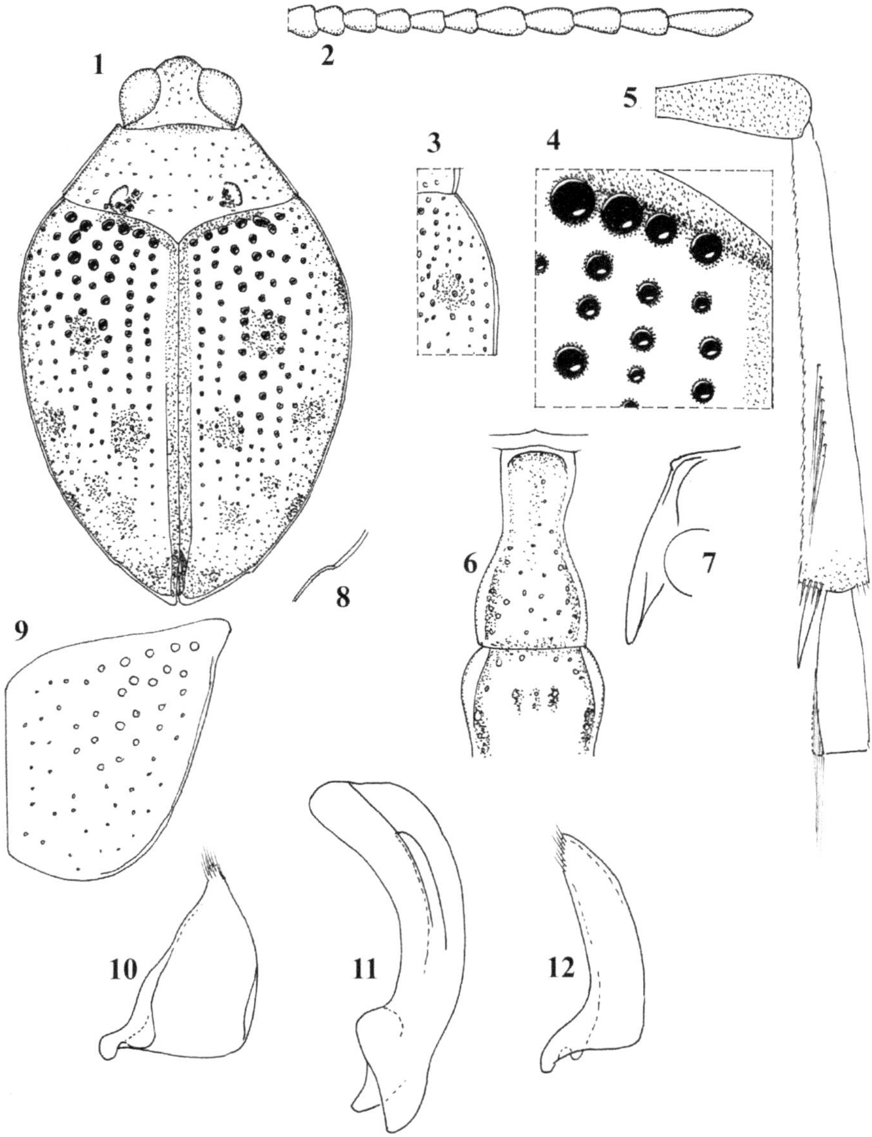

PLATE 199 *Peltodytes* (*Neopeltodytes*) *festivus*. 1, dorsal view; 2, antenna; 3, elytral shoulder in lateral view; 4, elytral punctures near base and suture; 5, hind tibia in dorsal view; 6, prosternal and metaventral process; 7, prosternal process in lateral view; 8, preapical tooth on elytral margin; 9, hind coxal plate; 10, left paramere; 11, penis; 12, right paramere

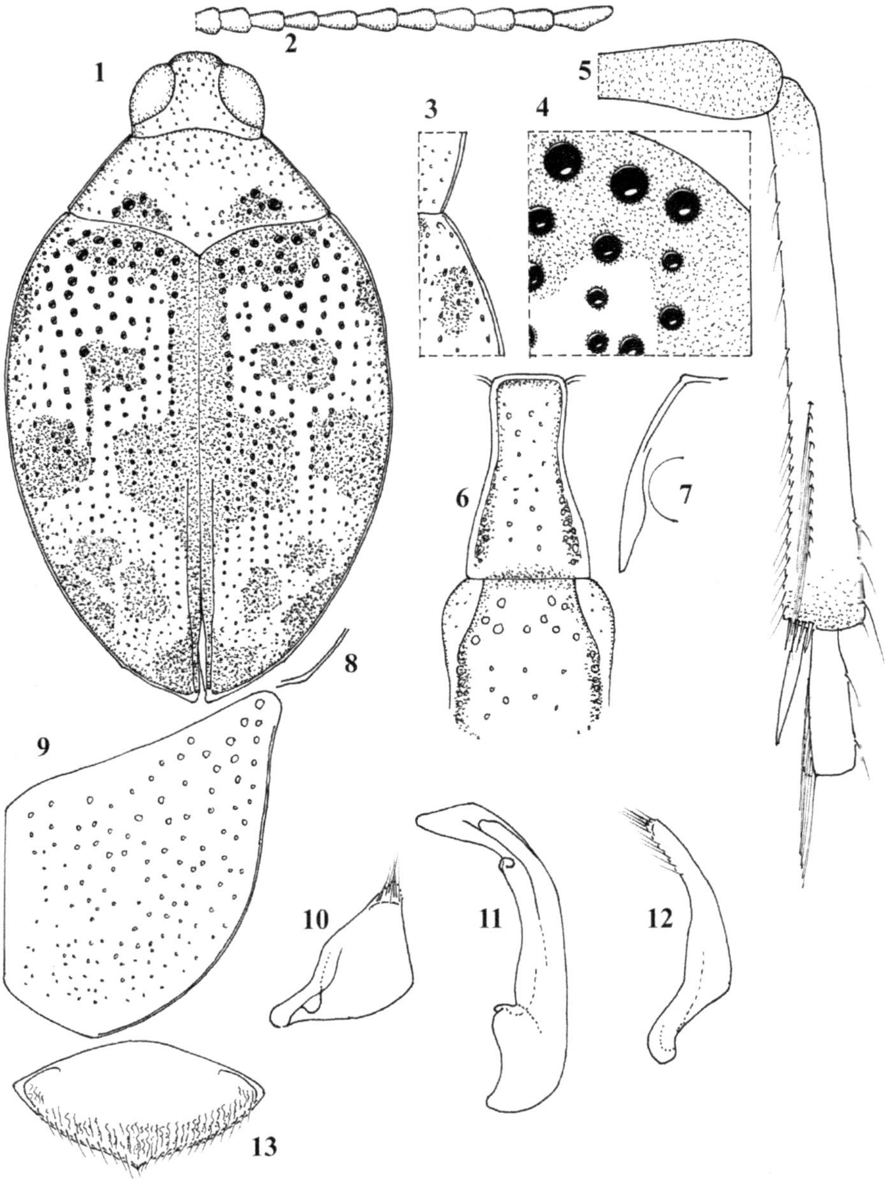

PLATE 200 *Peltodytes* (*Neopeltodytes*) *floridensis*. 1, dorsal view; 2, antenna; 3, elytral shoulder in lateral view; 4, elytral punctures near base and suture; 5, hind tibia in dorsal view; 6, prosternal and metaventral process; 7, prosternal process in lateral view; 8, preapical tooth on elytral margin; 9, hind coxal plate; 10, left paramere; 11, penis; 12, right paramere; 13, last sternite

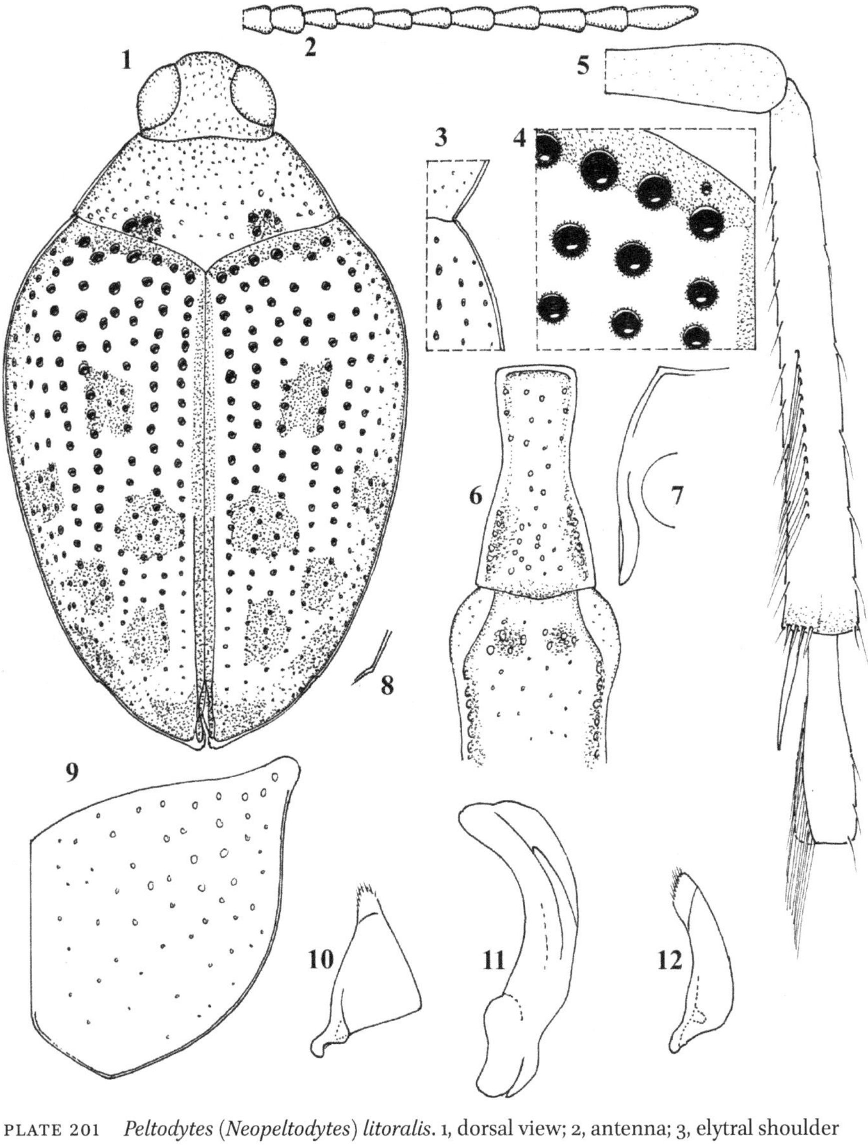

PLATE 201 *Peltodytes* (*Neopeltodytes*) *litoralis*. 1, dorsal view; 2, antenna; 3, elytral shoulder in lateral view; 4, elytral punctures near base and suture; 5, hind tibia in dorsal view; 6, prosternal and metaventral process; 7, prosternal process in lateral view; 8, preapical tooth on elytral margin; 9, hind coxal plate; 10, left paramere; 11, penis; 12, right paramere

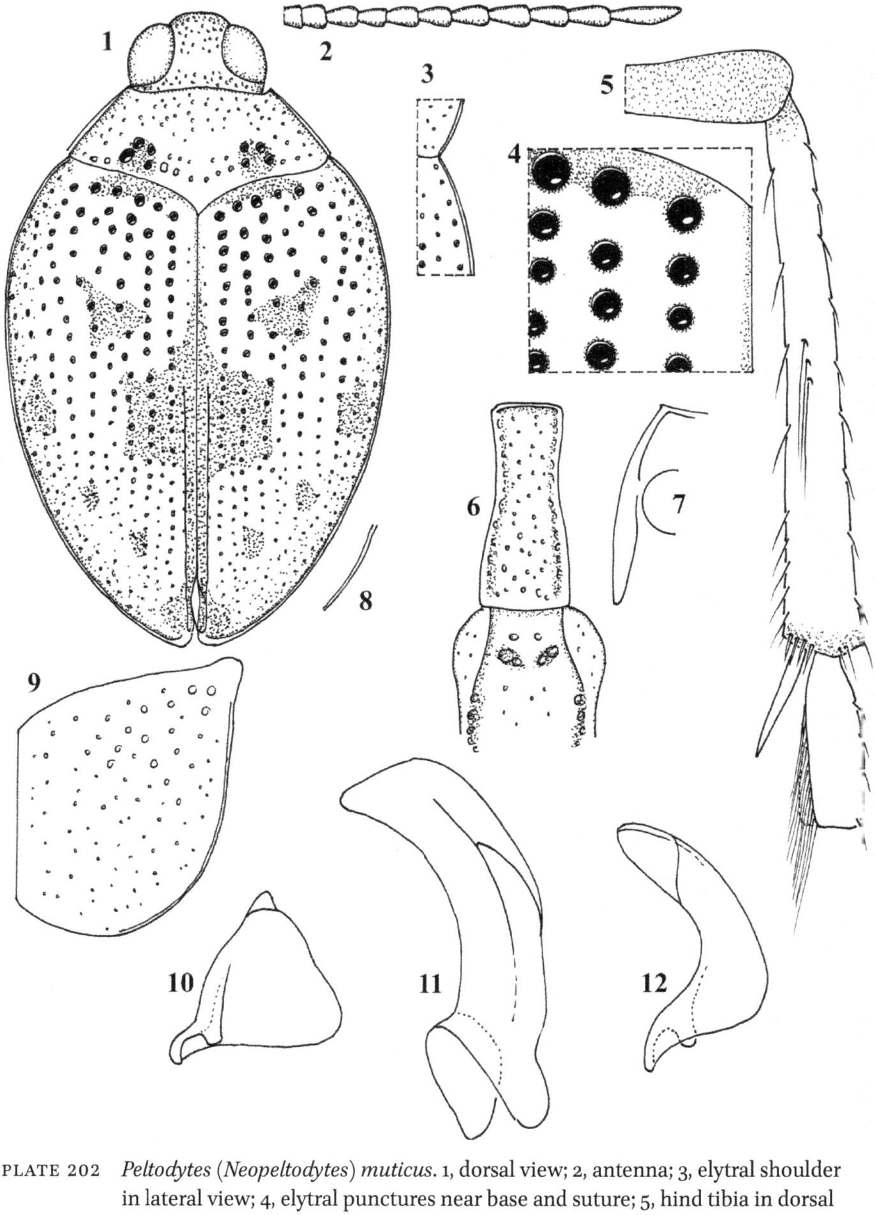

PLATE 202 *Peltodytes* (*Neopeltodytes*) *muticus*. 1, dorsal view; 2, antenna; 3, elytral shoulder in lateral view; 4, elytral punctures near base and suture; 5, hind tibia in dorsal view; 6, prosternal and metaventral process; 7, prosternal process in lateral view; 8, preapical tooth on elytral margin; 9, hind coxal plate; 10, left paramere; 11, penis; 12, right paramere

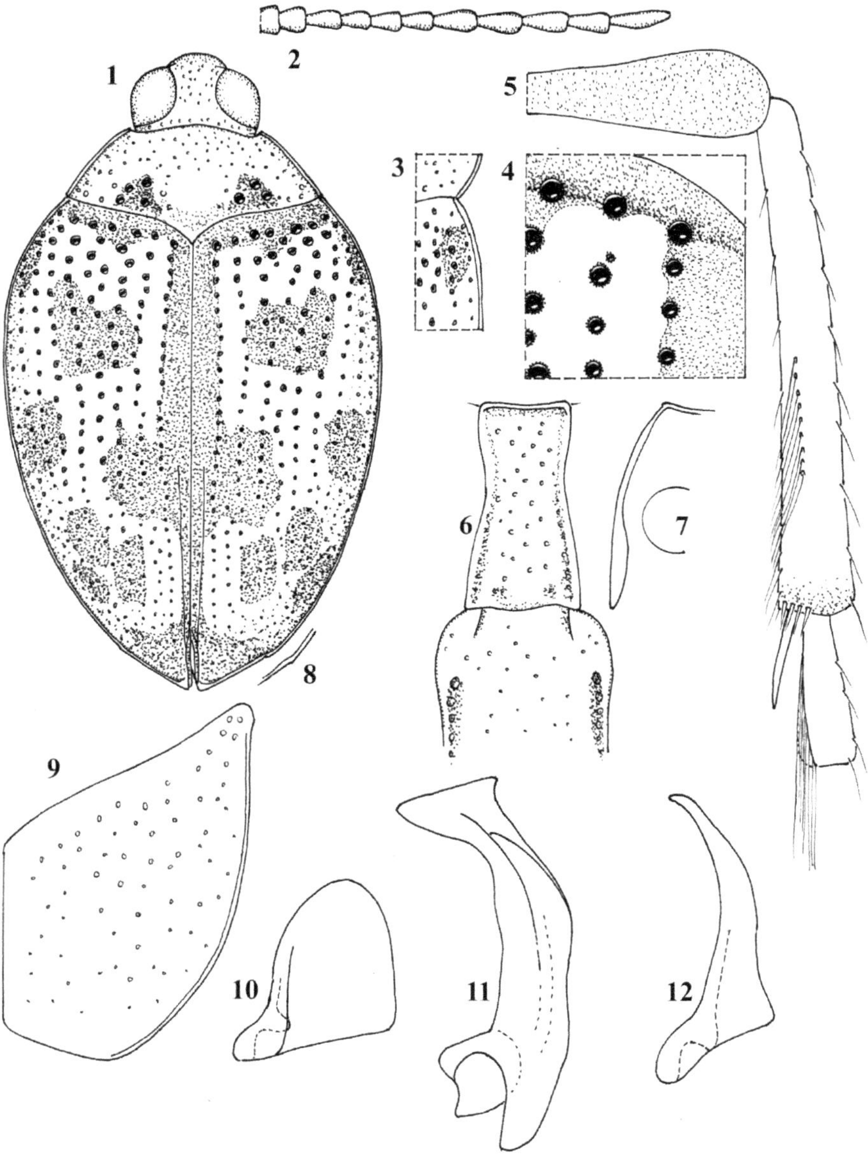

PLATE 203 *Peltodytes* (*Neopeltodytes*) *oppositus*. 1, dorsal view; 2, antenna; 3, elytral shoulder in lateral view; 4, elytral punctures near base and suture; 5, hind tibia in dorsal view; 6, prosternal and metaventral process; 7, prosternal process in lateral view; 8, preapical tooth on elytral margin; 9, hind coxal plate; 10, left paramere; 11, penis; 12, right paramere

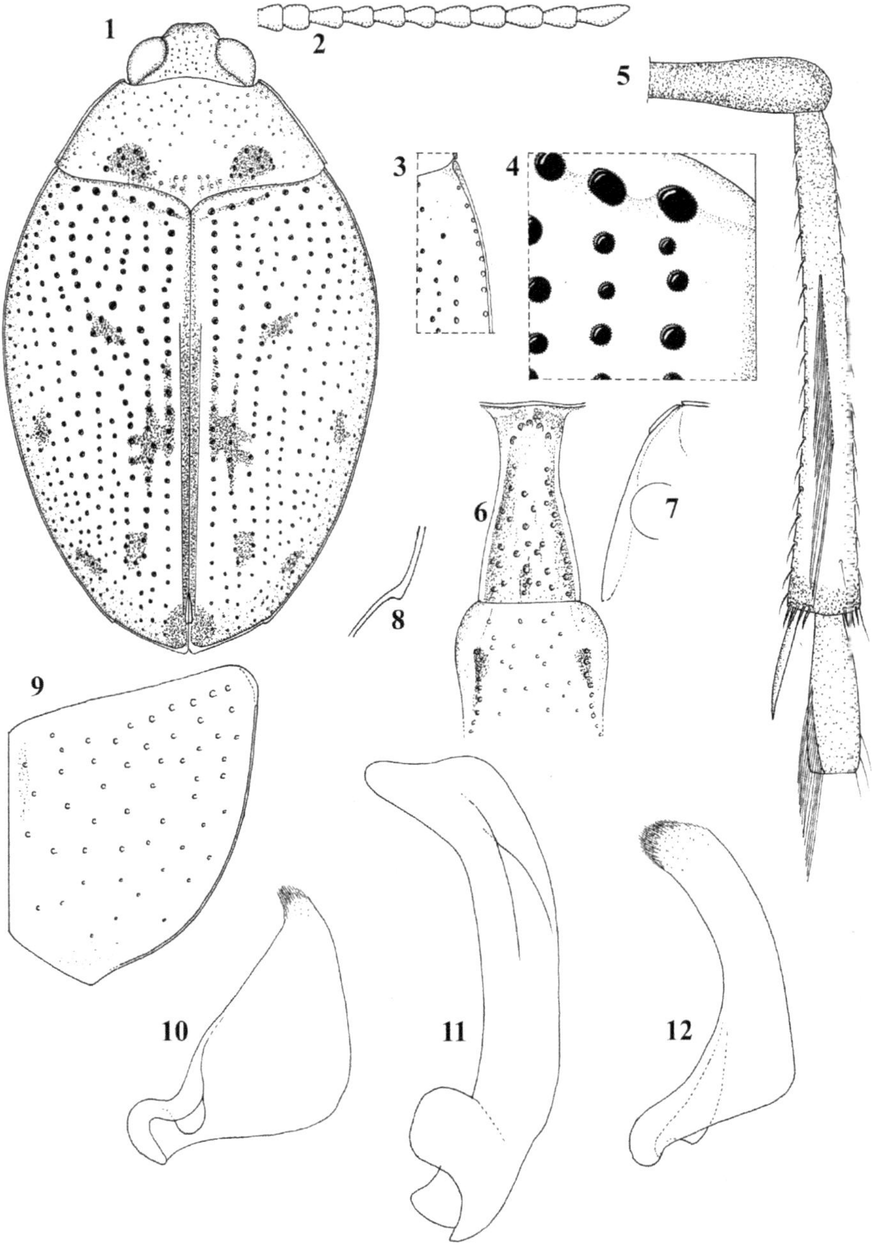

PLATE 204 *Peltodytes* (*Neopeltodytes*) *ovalis*. 1, dorsal view; 2, antenna; 3, elytral shoulder in lateral view; 4, elytral punctures near base and suture; 5, hind tibia in dorsal view; 6, prosternal and metaventral process; 7, prosternal process in lateral view; 8, preapical tooth on elytral margin; 9, hind coxal plate; 10, left paramere; 11, penis; 12, right paramere

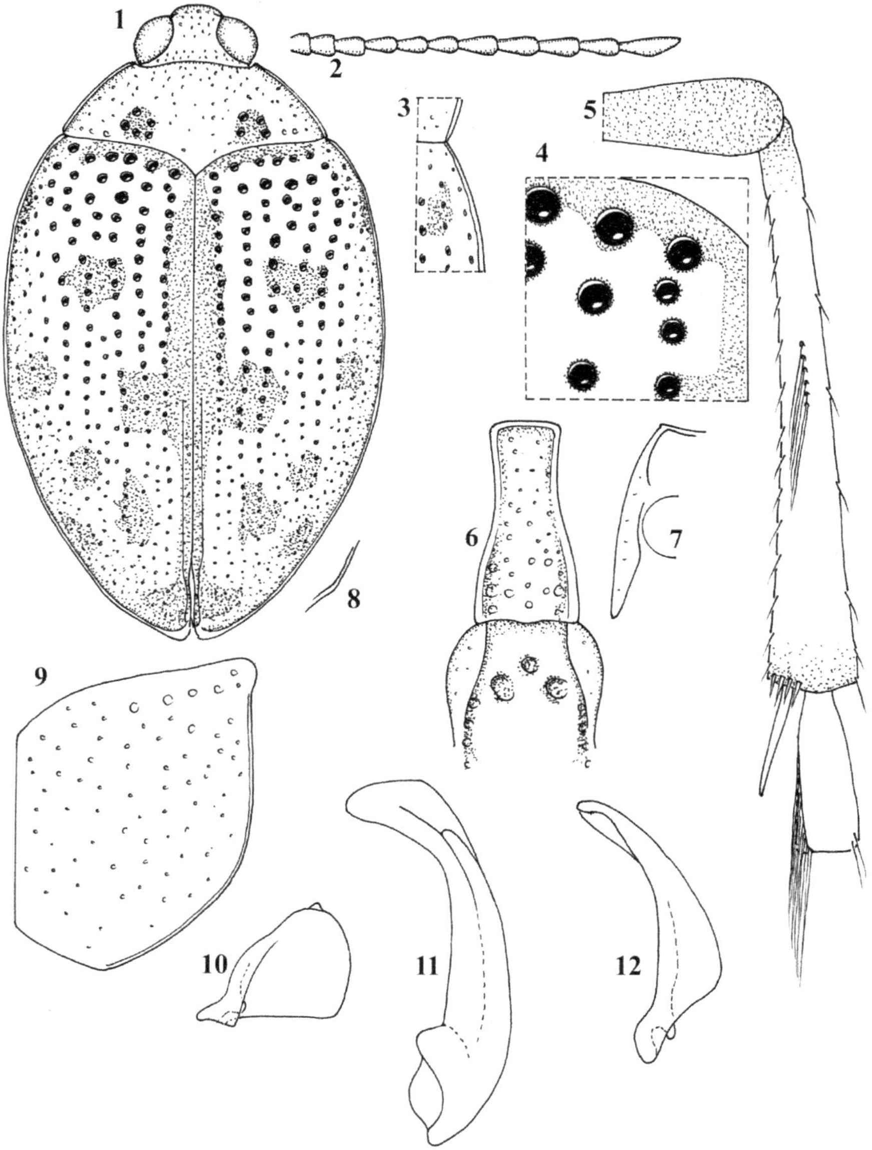

PLATE 205 *Peltodytes* (*Neopeltodytes*) *pedunculatus*. 1, dorsal view; 2, antenna; 3, elytral shoulder in lateral view; 4, elytral punctures near base and suture; 5, hind tibia in dorsal view; 6, prosternal and metaventral process; 7, prosternal process in lateral view; 8, preapical tooth on elytral margin; 9, hind coxal plate; 10, left paramere; 11, penis; 12, right paramere

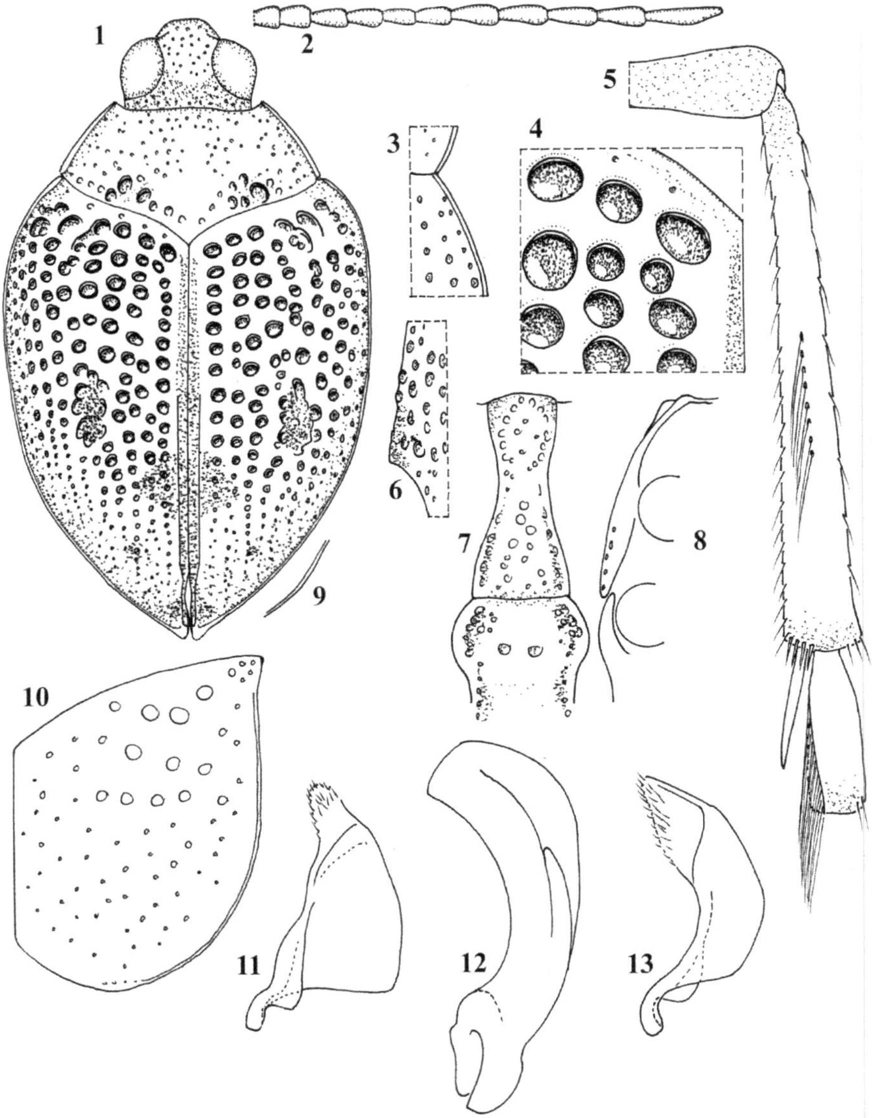

PLATE 206 *Peltodytes* (*Neopeltodytes*) *roughleyi*. 1, dorsal view; 2, antenna; 3, elytral shoulder in lateral view; 4, elytral punctures near base and suture; 5, hind tibia in dorsal view; 7, elytral tubercle in lateral view; 7, prosternal and metaventral process; 8, prosternal process in lateral view; 9, preapical tooth on elytral margin; 10, hind coxal plate; 11, left paramere; 12, penis; 13, right paramere

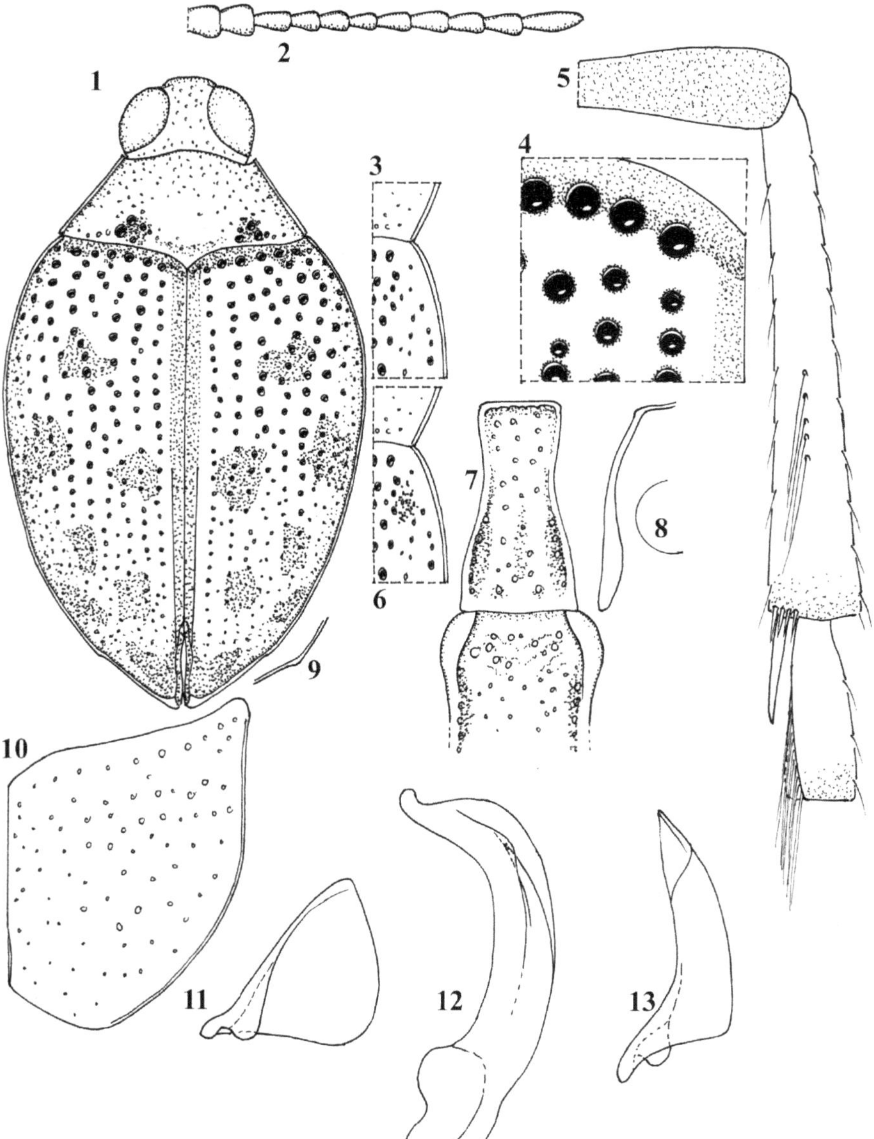

PLATE 207 *Peltodytes* (*Neopeltodytes*) *sexmaculatus*. 1, dorsal view; 2, antenna; 3, elytral shoulder in lateral view; 4, elytral punctures near base and suture; 5, hind tibia in dorsal view; 6, elytral shoulder in lateral view with dark mark; 7, prosternal and metaventral process; 8, prosternal process in lateral view; 9, preapical tooth on elytral margin; 10, hind coxal plate; 11, left paramere; 12, penis; 13, right paramere

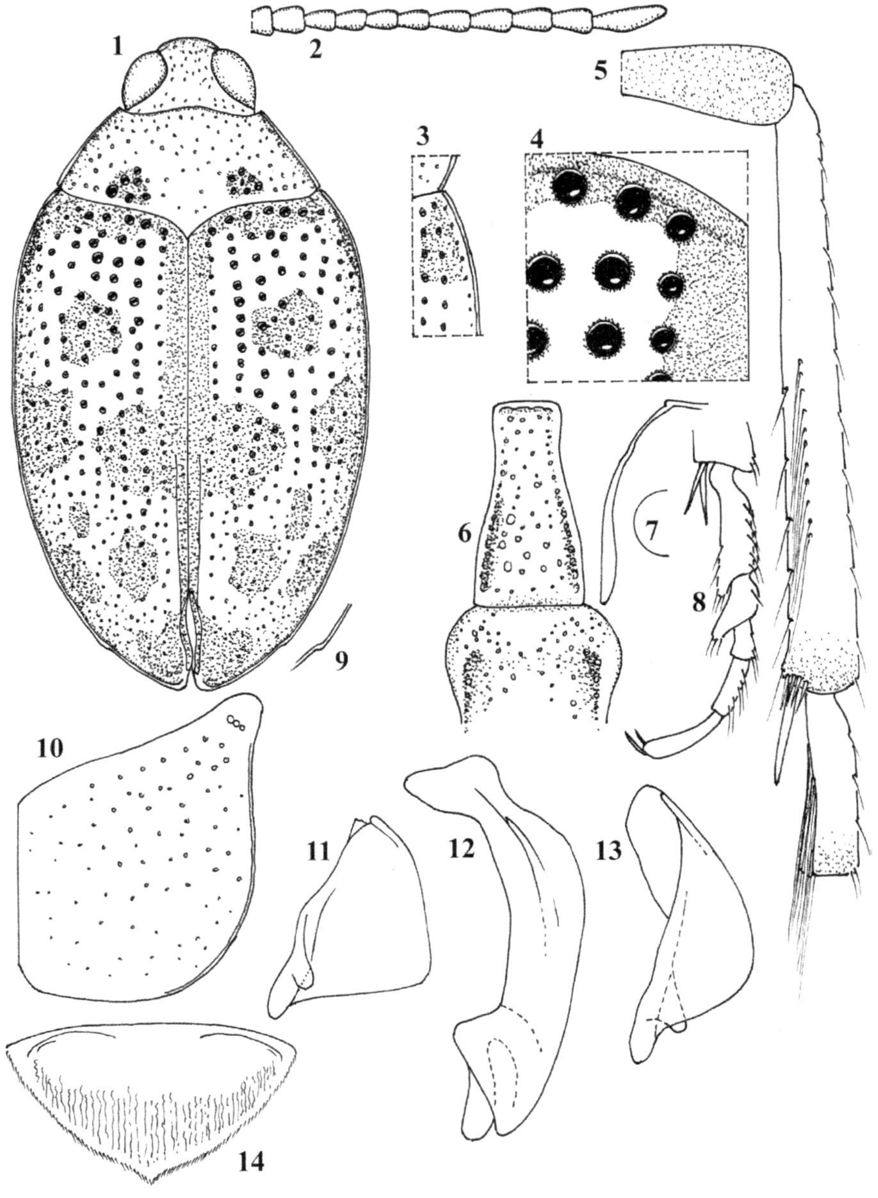

PLATE 208 *Peltodytes* (*Neopeltodytes*) *shermani*. 1, dorsal view; 2, antenna; 3, elytral shoulder in lateral view; 4, elytral punctures near base and suture; 5, hind tibia in dorsal view; 6, prosternal and metaventral process; 7, prosternal process in lateral view; 8, tarsus of midleg; 9, preapical tooth on elytral margin; 10, hind coxal plate; 11, left paramere; 12, penis; 13, right paramere; 14, last sternite

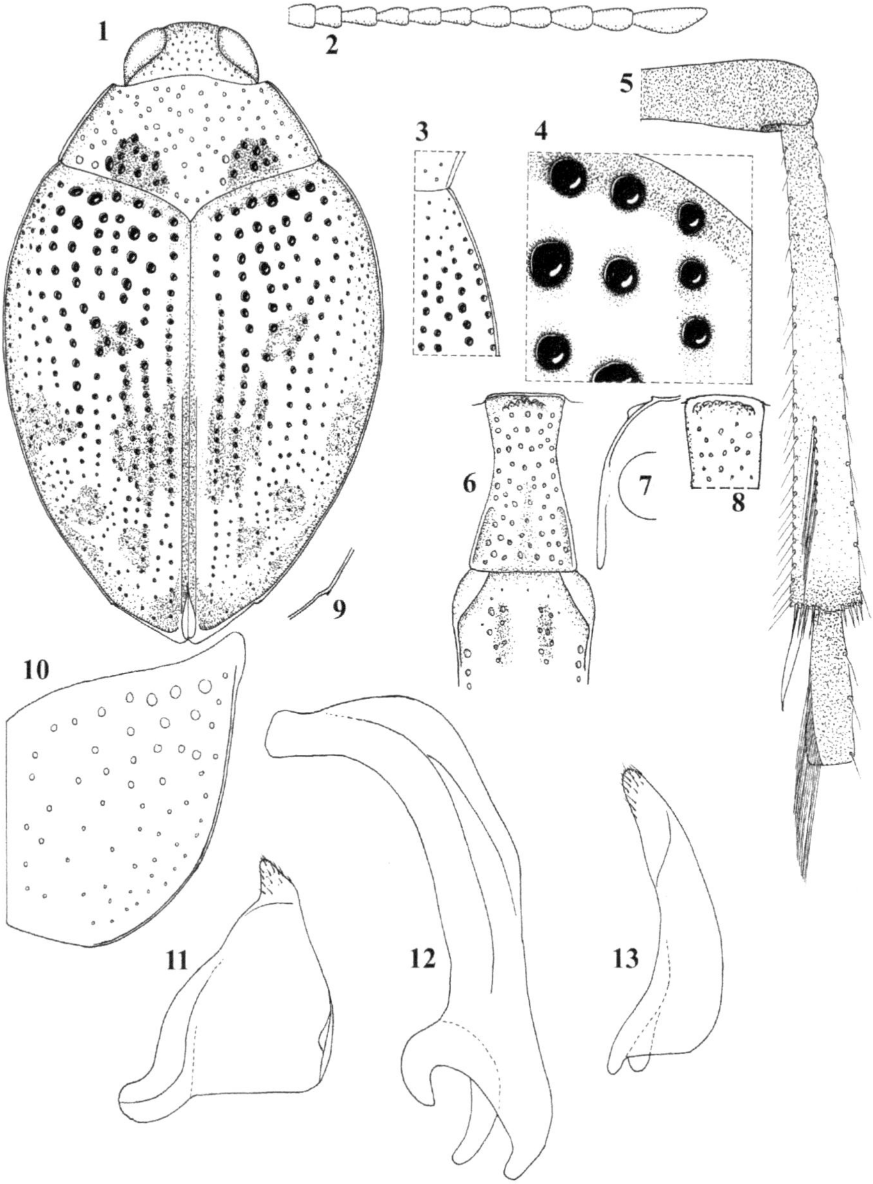

PLATE 209 *Peltodytes* (*Neopeltodytes*) *simplex*. 1, dorsal view; 2, antenna; 3, elytral shoulder in lateral view; 4, elytral punctures near base and suture; 5, hind tibia in dorsal view; 6, prosternal and metaventral process; 7, prosternal process in lateral view; 8, alternative shape of anterior edge of prosternal process; 9, preapical tooth on elytral margin; 10, hind coxal plate; 11, left paramere; 12, penis; 13, right paramere

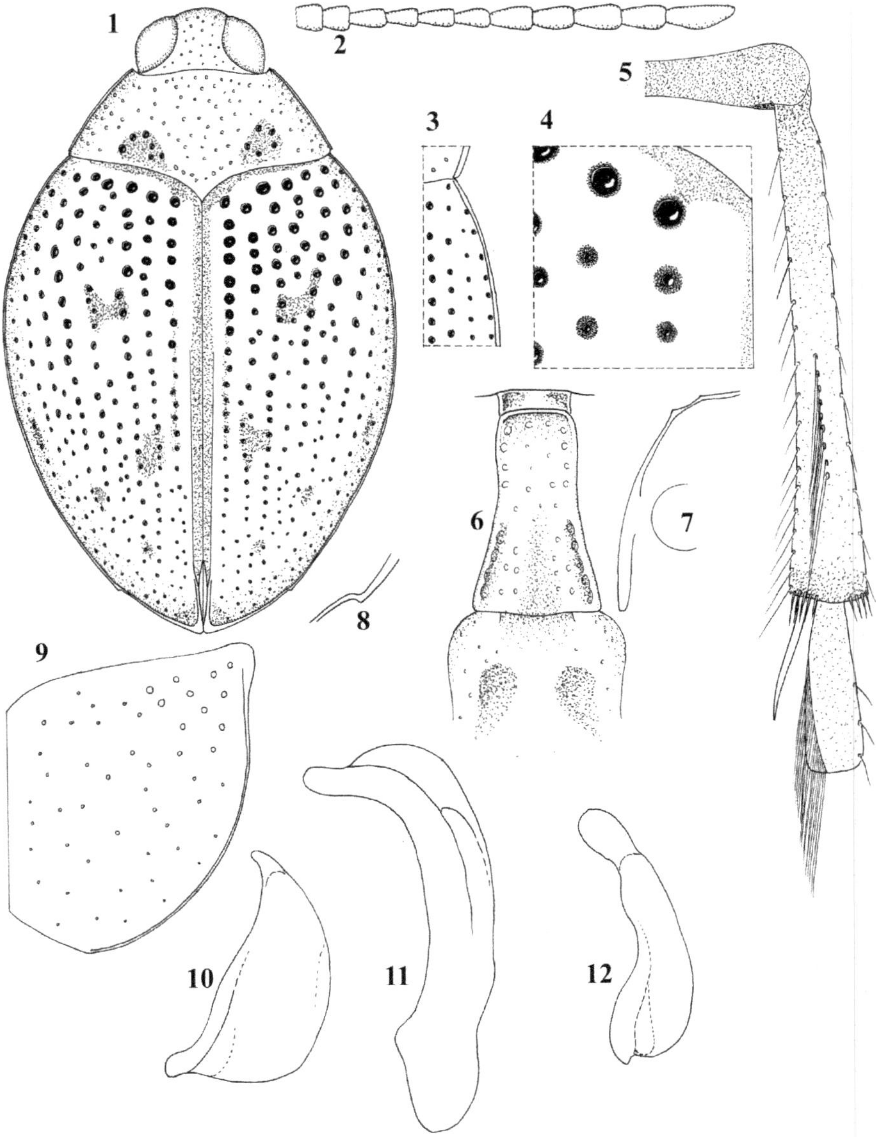

PLATE 210 *Peltodytes* (*Neopeltodytes*) *tamaulipensis*. 1, dorsal view; 2, antenna; 3, elytral shoulder in lateral view; 4, elytral punctures near base and suture; 5, hind tibia in dorsal view; 6, prosternal and metaventral process; 7, prosternal process in lateral view; 8, preapical tooth on elytral margin; 9, hind coxal plate; 10, left paramere; 11, penis; 12, right paramere

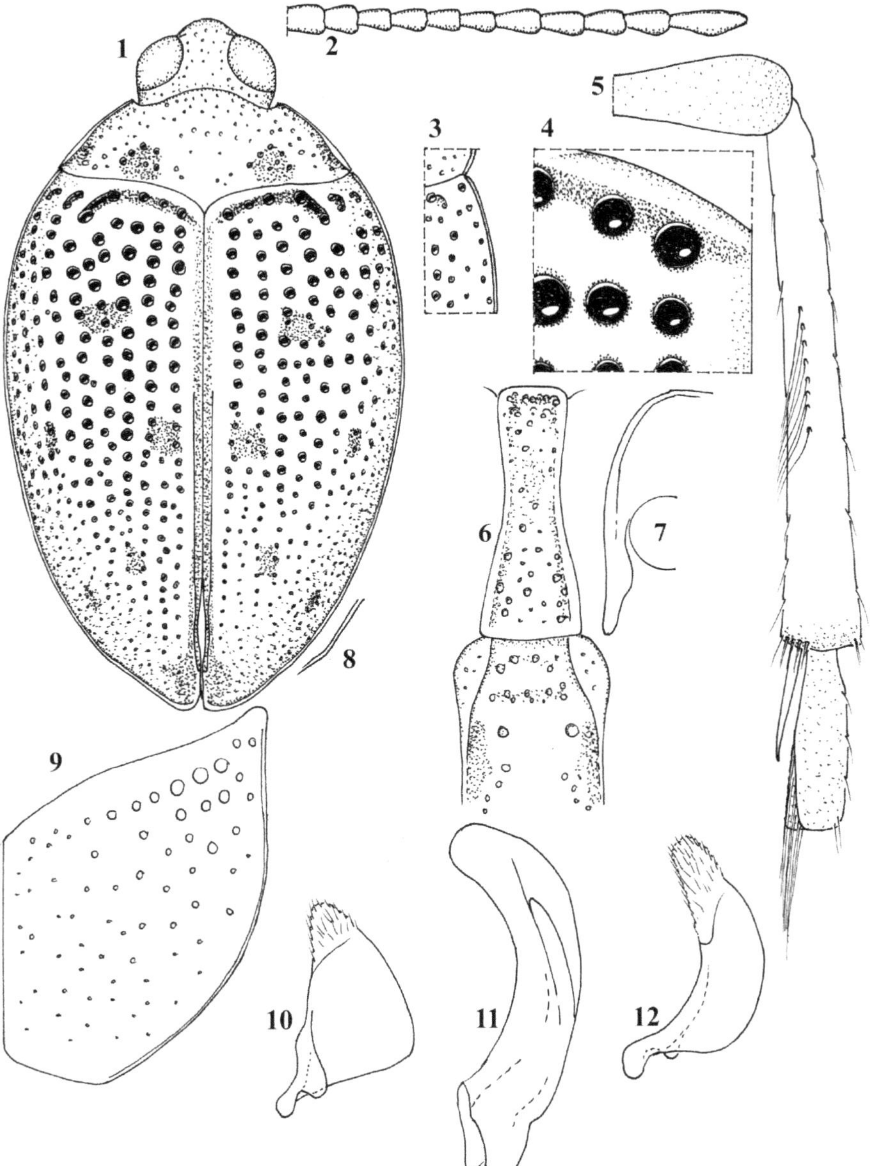

PLATE 211 *Peltodytes* (*Neopeltodytes*) *tortulosus*. 1, dorsal view; 2, antenna; 3, elytral shoulder in lateral view; 4, elytral punctures near base and suture; 5, hind tibia in dorsal view; 6, prosternal and metaventral process; 7, prosternal process in lateral view; 8, preapical tooth on elytral margin; 9, hind coxal plate; 10, left paramere; 11, penis; 12, right paramere

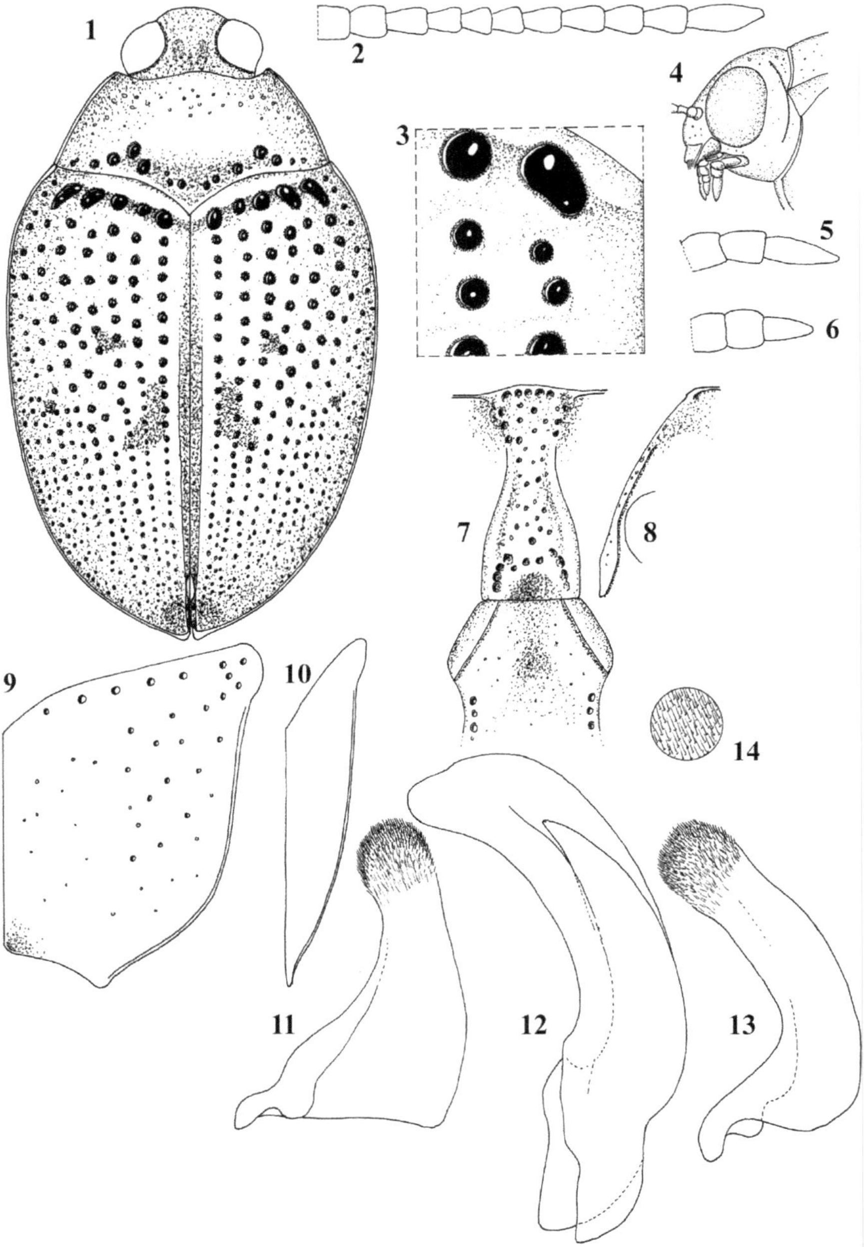

PLATE 212 *Peltodytes* (*Peltodytes*) *caesus*. 1, dorsal view; 2, antenna; 3, elytral punctures near base and suture; 4, head in lateral view; 5, maxillary palpus; 6, labial palpus; 7, prosternal and metaventral process; 8, prosternal process in lateral view; 9, hind coxal plate;10, hind coxal plate in lateral view; 11, left paramere; 12, penis; 13, right paramere; 14, structure of top of left and right paramere

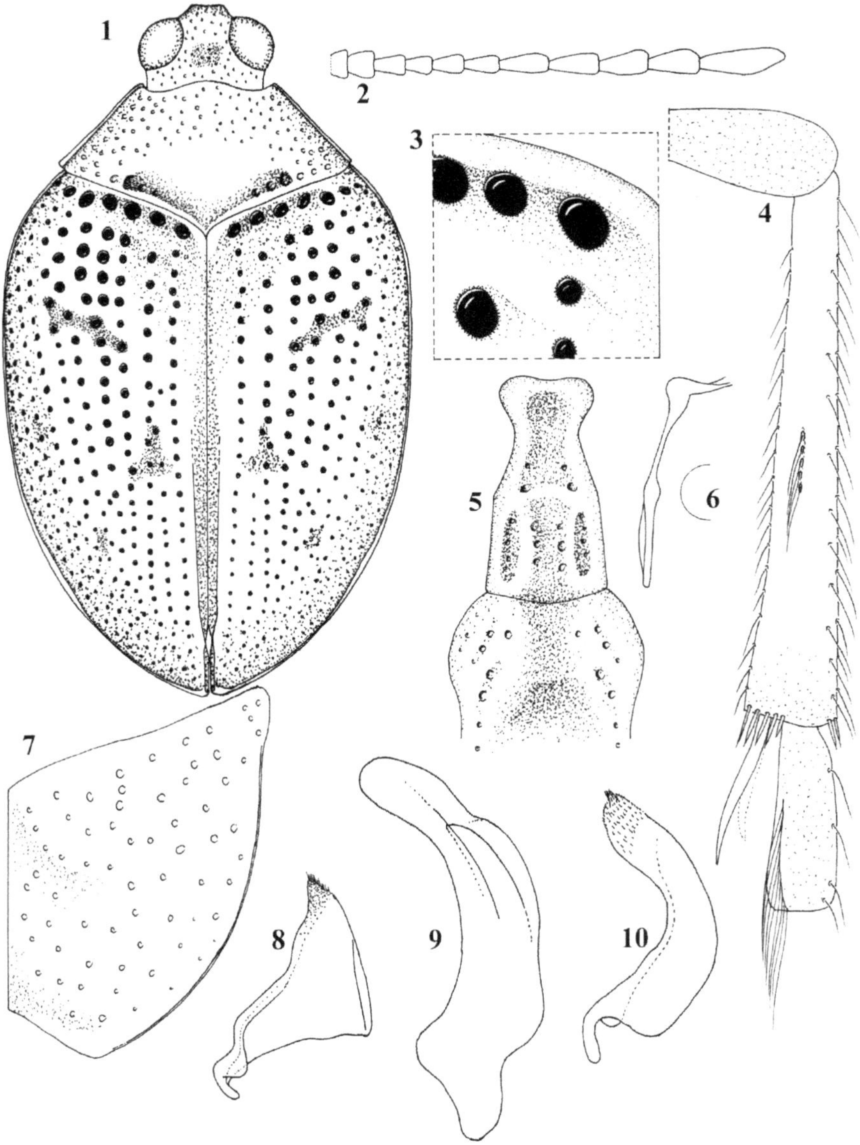

PLATE 213 *Peltodytes* (*Peltodytes*) *congoensis*. 1, dorsal view; 2, antenna; 3, elytral punctures near base and suture; 4, hind tibia in dorsal view; 5, prosternal and metaventral process; 6, prosternal process in lateral view; 7, hind coxal plate; 8, left paramere; 9, penis; 10, right paramere

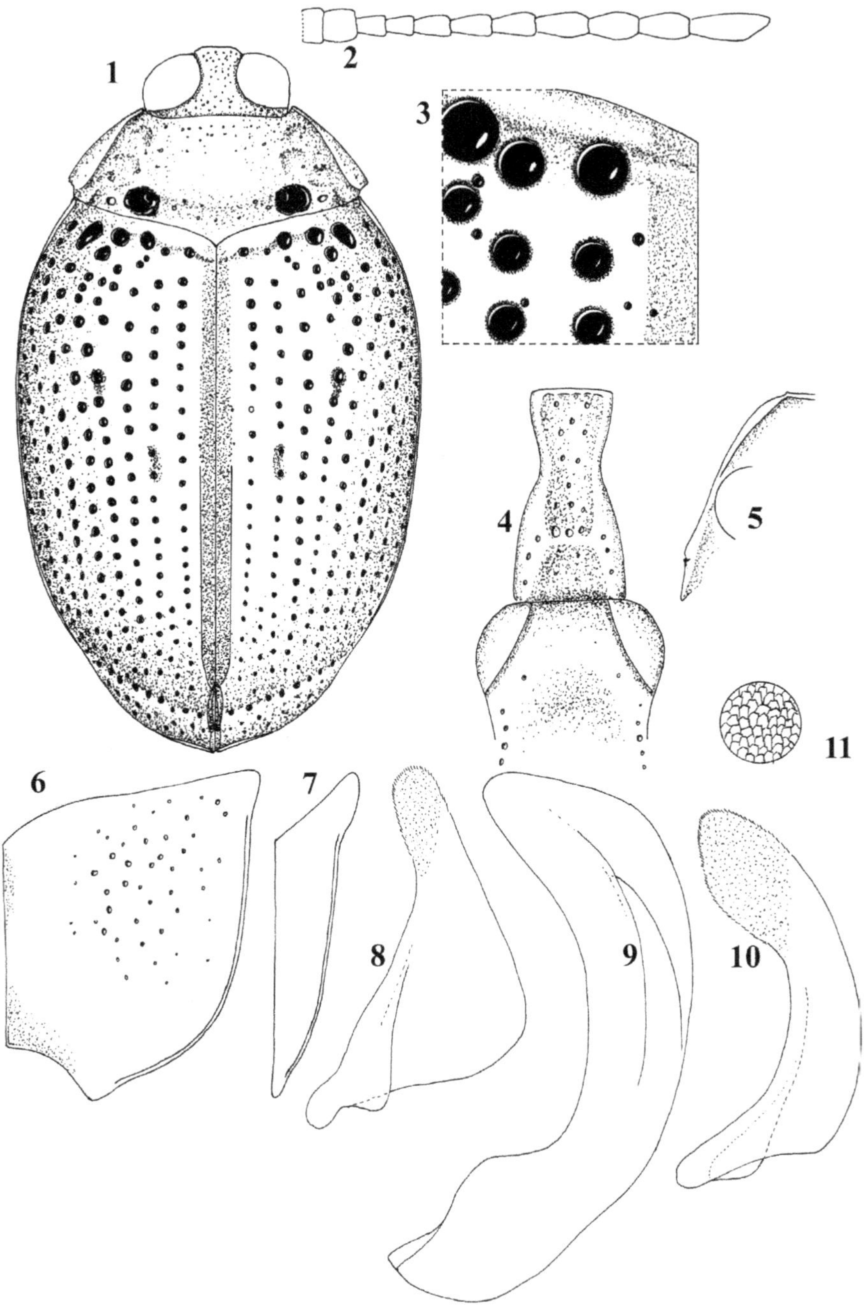

PLATE 214 *Peltodytes* (*Peltodytes*) *coomani*. 1, dorsal view; 2, antenna; 3, elytral punctures near base and suture; 4, prosternal and metaventral process; 5, prosternal process in lateral view; 6, hind coxal plate; 7, hind coxal plate in lateral view; 8, left paramere; 9, penis; 10, right paramere; 11, structure of top of left and right paramere

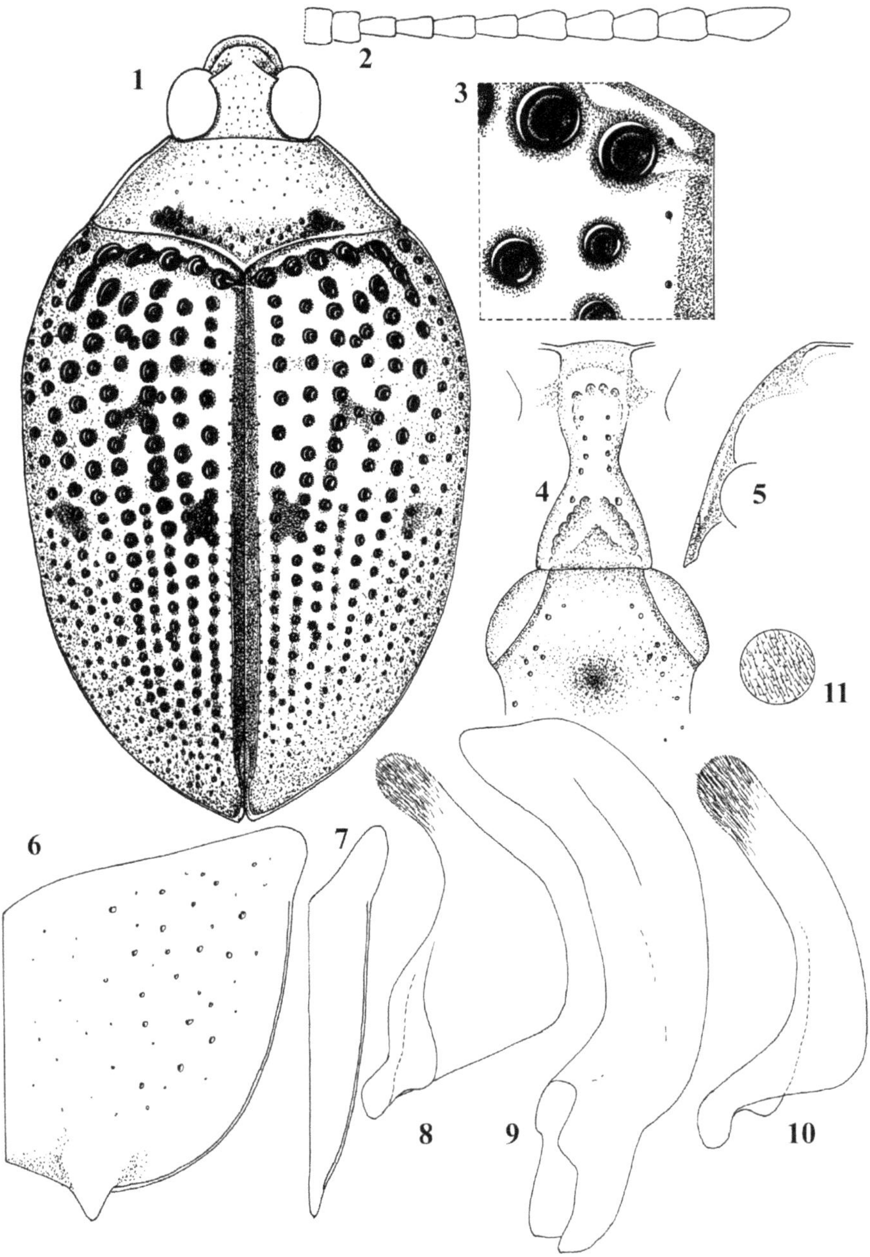

PLATE 215 *Peltodytes* (*Peltodytes*) *dauricus*. 1, dorsal view; 2, antenna; 3, elytral punctures near base and suture; 4, prosternal and metaventral process; 5, prosternal process in lateral view; 6, hind coxal plate; 7, hind coxal plate in lateral view; 8, left paramere; 9, penis; 10, right paramere; 11, structure of top of left and right paramere

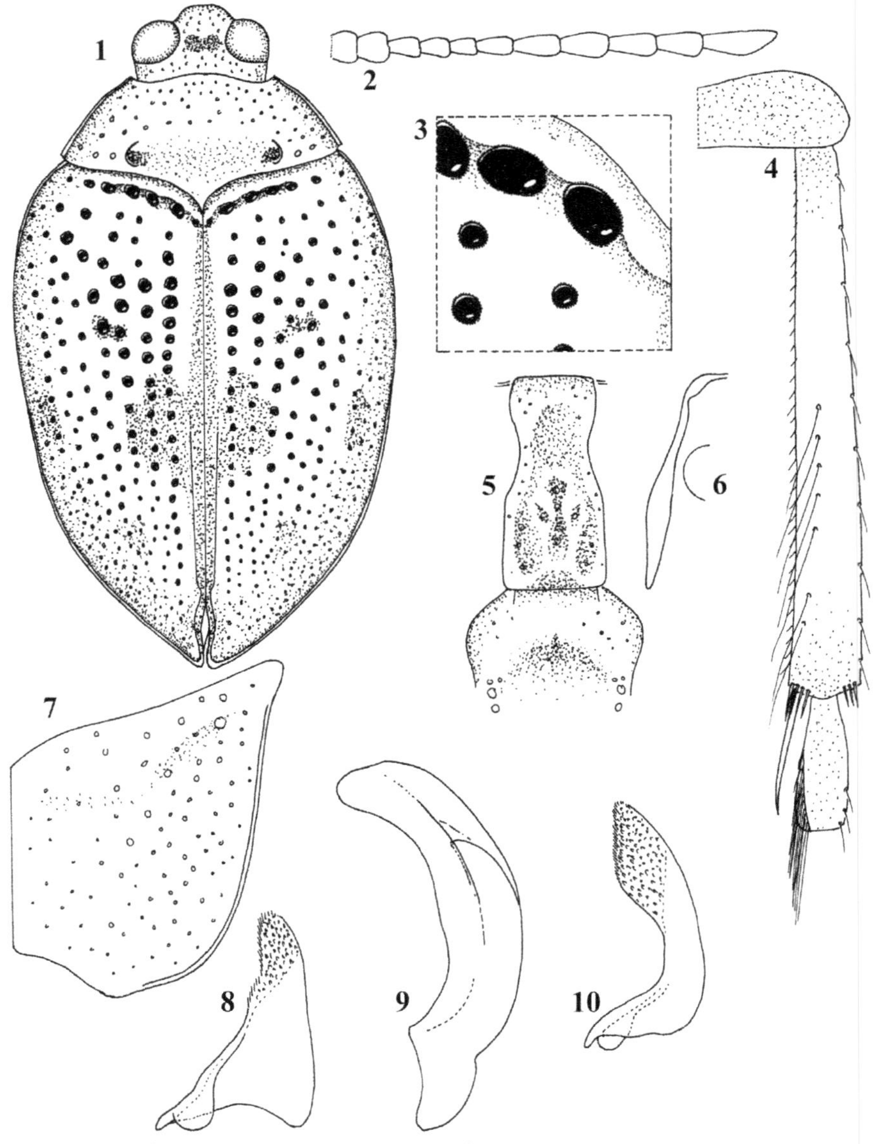

PLATE 216 *Peltodytes* (*Peltodytes*) *ghanaensis*. 1, dorsal view; 2, antenna; 3, elytral punctures near base and suture; 4, hind tibia in dorsal view; 5, prosternal and metaventral process; 6, prosternal process in lateral view; 7, hind coxal plate; 8, left paramere; 9, penis; 10, right paramere

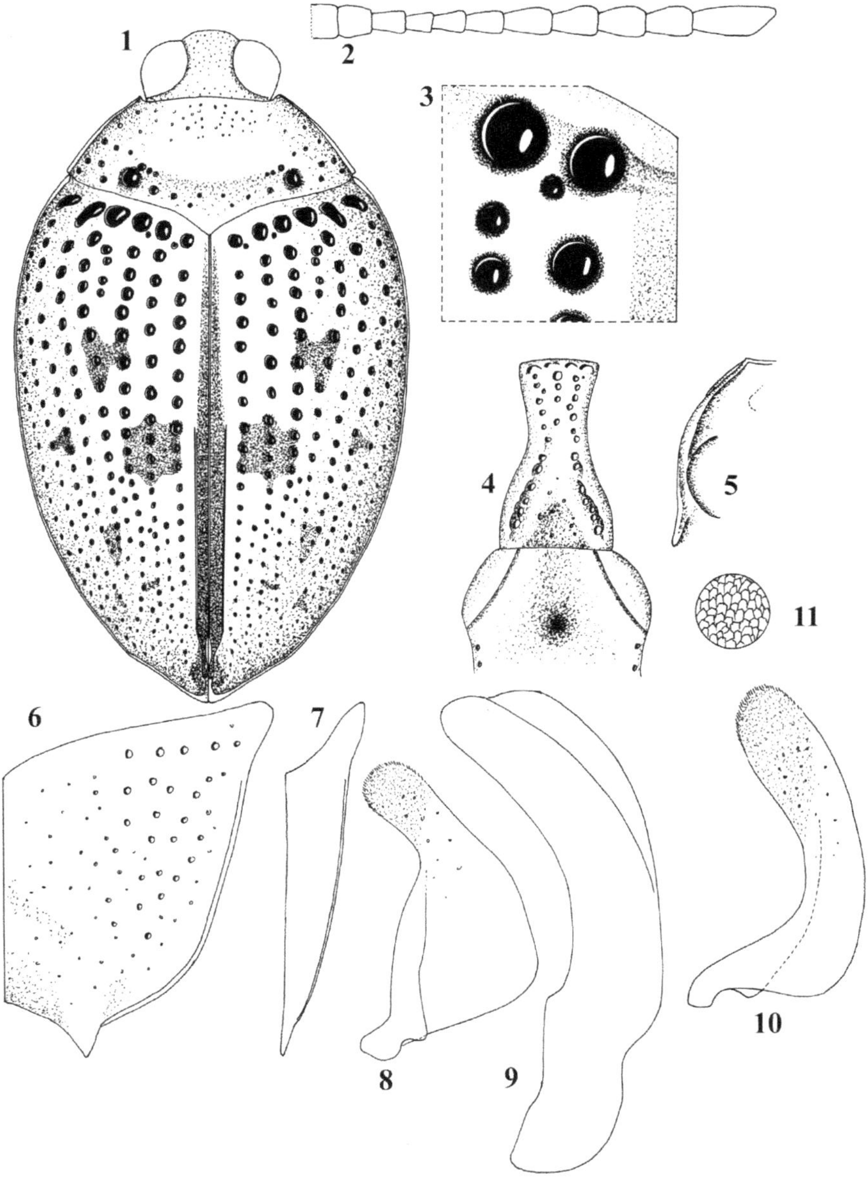

PLATE 217 *Peltodytes* (*Peltodytes*) *intermedius*. 1, dorsal view; 2, antenna; 3, elytral punctures near base and suture; 4, prosternal and metaventral process; 5, prosternal process in lateral view; 6, hind coxal plate; 7, hind coxal plate in lateral view; 8, left paramere; 9, penis; 10, right paramere; 11, structure of top of left and right paramere

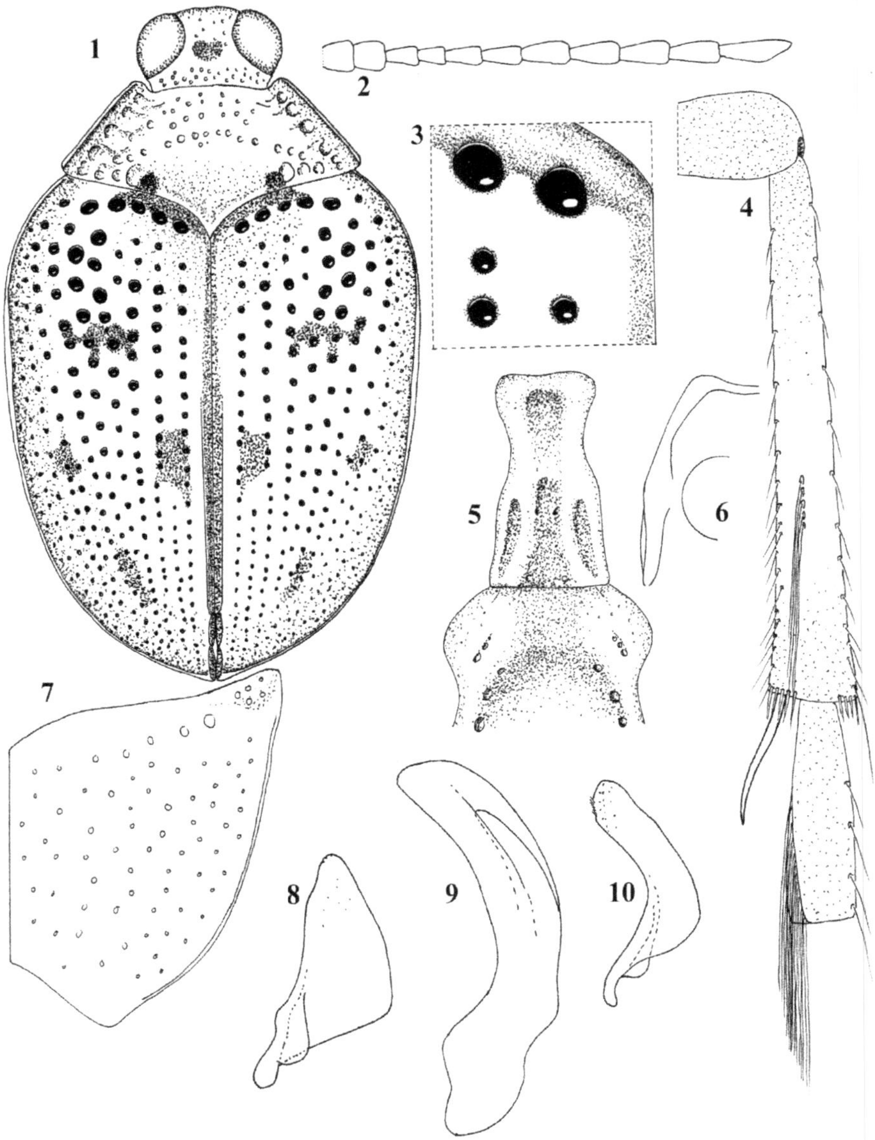

PLATE 218 *Peltodytes* (*Peltodytes*) *nodieri*. 1, dorsal view; 2, antenna; 3, elytral punctures near base and suture; 4, hind tibia in dorsal view; 5, prosternal and metaventral process; 6, prosternal process in lateral view; 7, hind coxal plate; 8, left paramere; 9, penis; 10, right paramere

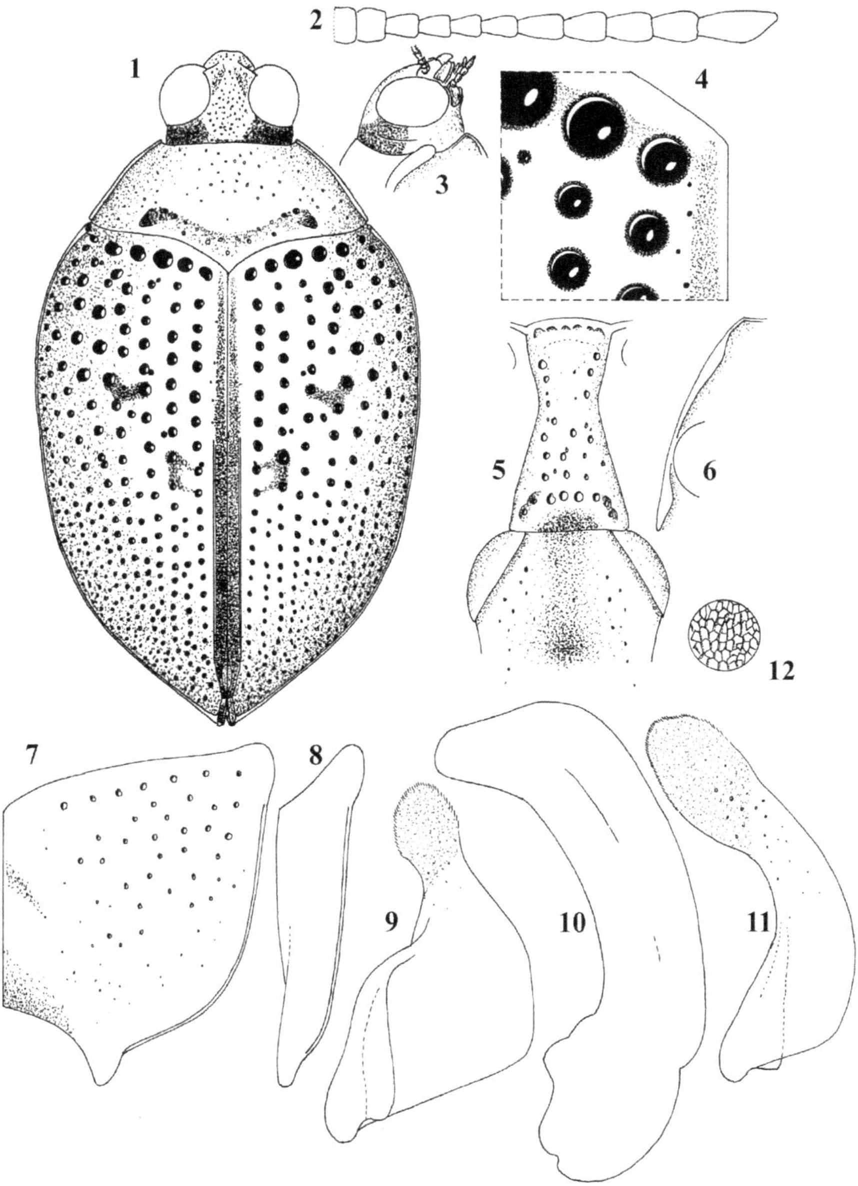

PLATE 219 *Peltodytes (Peltodytes) pekinensis*. 1, dorsal view; 2, antenna; 3, head in lateral view; 4, elytral punctures near base and suture; 5, prosternal and metaventral process; 6, prosternal process in lateral view; 7, hind coxal plate; 8, hind coxal plate in lateral view; 9, left paramere; 10, penis; 11, right paramere; 12, structure of top of left and right paramere

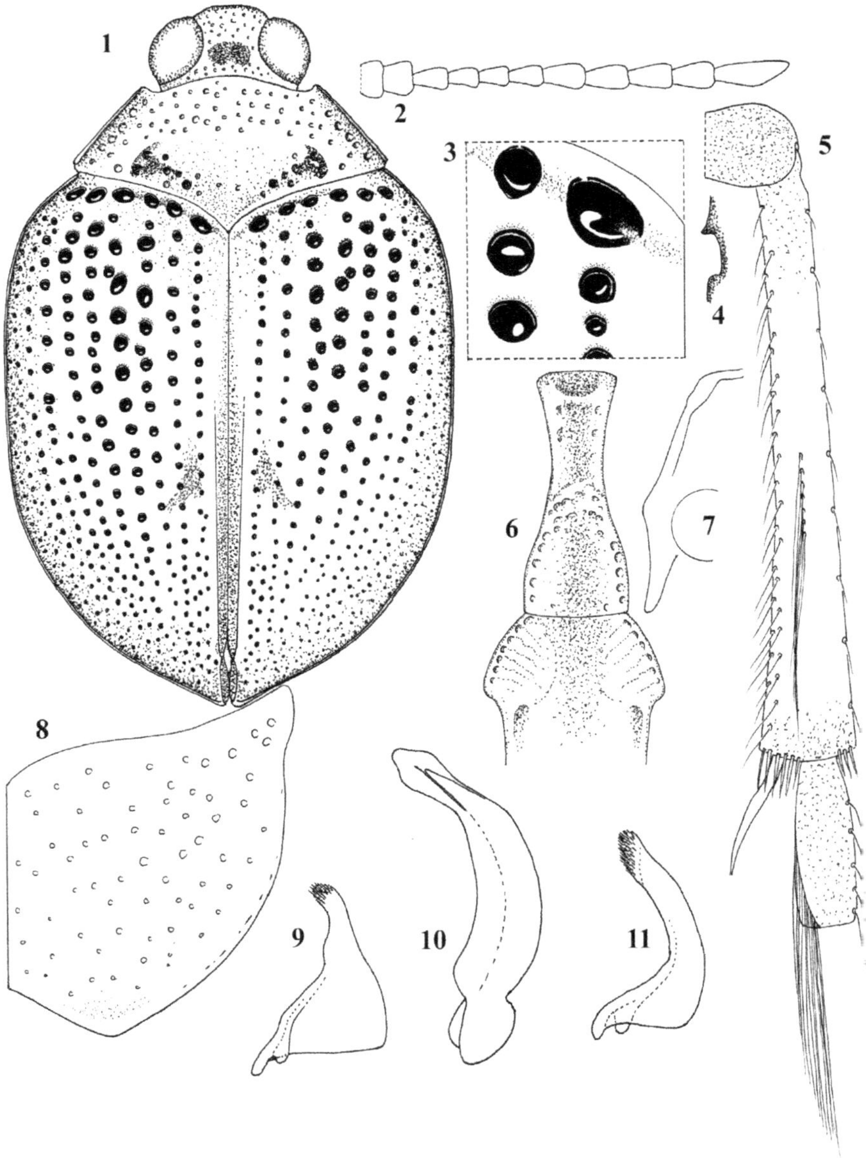

PLATE 220 *Peltodytes* (*Peltodytes*) *quadratus*. 1, dorsal view; 2, antenna; 3, elytral punctures near base and suture; 4, elytral puncture in cross section; 5, hind tibia in dorsal view; 6, prosternal and metaventral process; 7, prosternal process in lateral view; 8, hind coxal plate; 9, left paramere; 10, penis; 11, right paramere

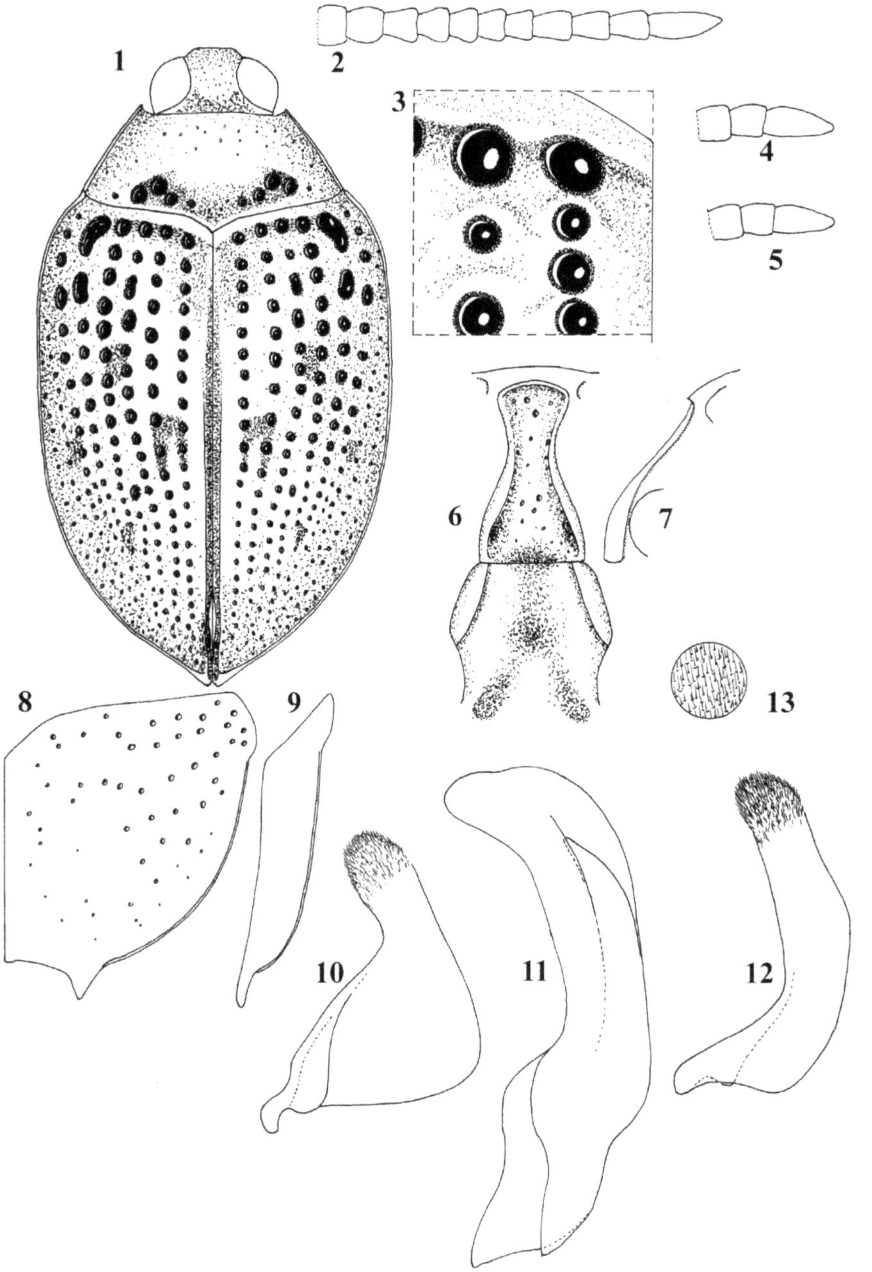

PLATE 221 *Peltodytes* (*Peltodytes*) *rotundatus*. 1, dorsal view; 2, antenna; 3, elytral punctures near base and suture; 4, maxillary palpus; 5, labial palpus; 6, prosternal and metaventral process; 7, prosternal process in lateral view; 8, hind coxal plate; 9, hind coxal plate in lateral view; 10, left paramere; 11, penis; 12, right paramere; 13, structure of top of left and right paramere

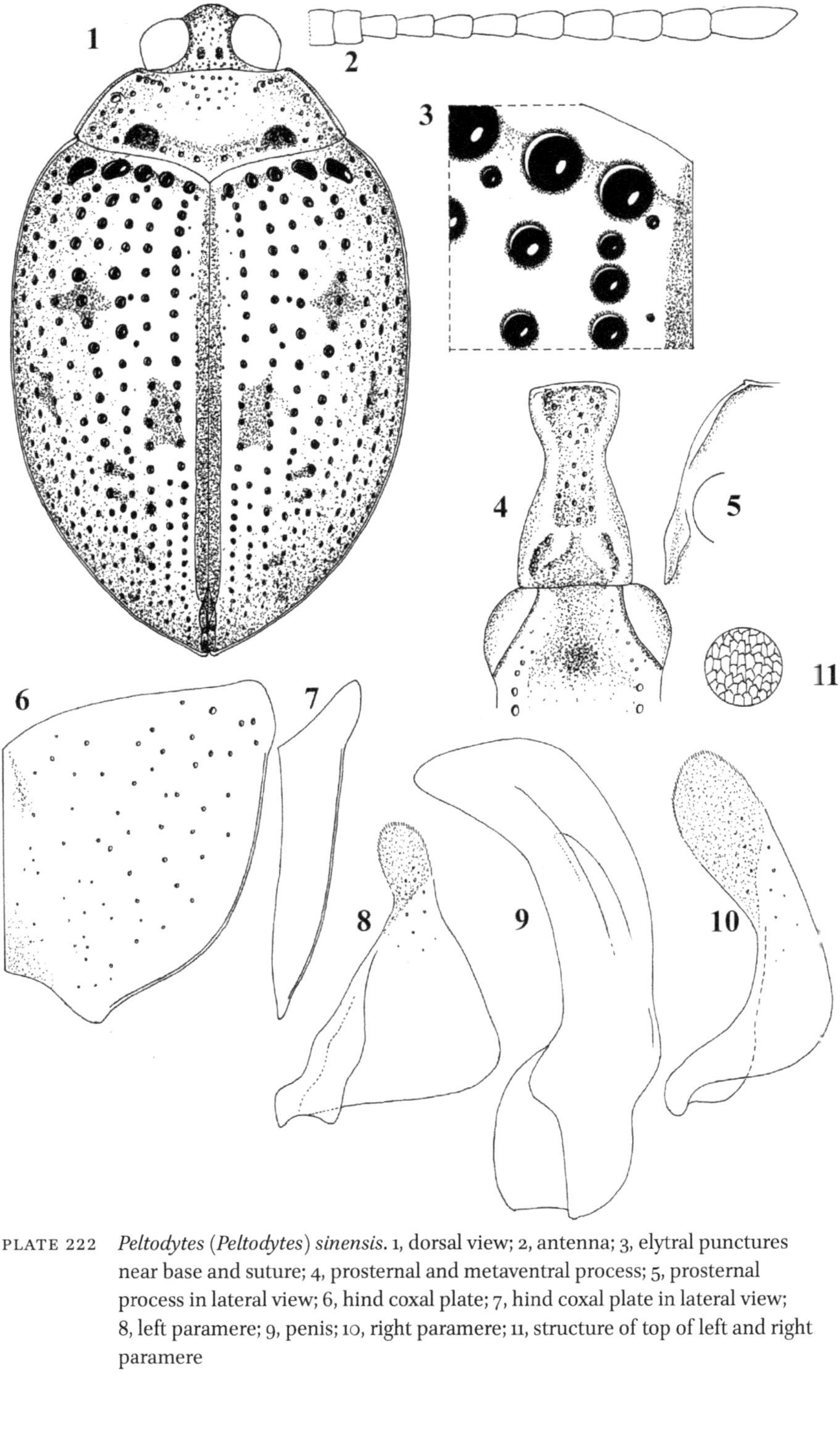

PLATE 222 *Peltodytes* (*Peltodytes*) *sinensis*. 1, dorsal view; 2, antenna; 3, elytral punctures near base and suture; 4, prosternal and metaventral process; 5, prosternal process in lateral view; 6, hind coxal plate; 7, hind coxal plate in lateral view; 8, left paramere; 9, penis; 10, right paramere; 11, structure of top of left and right paramere

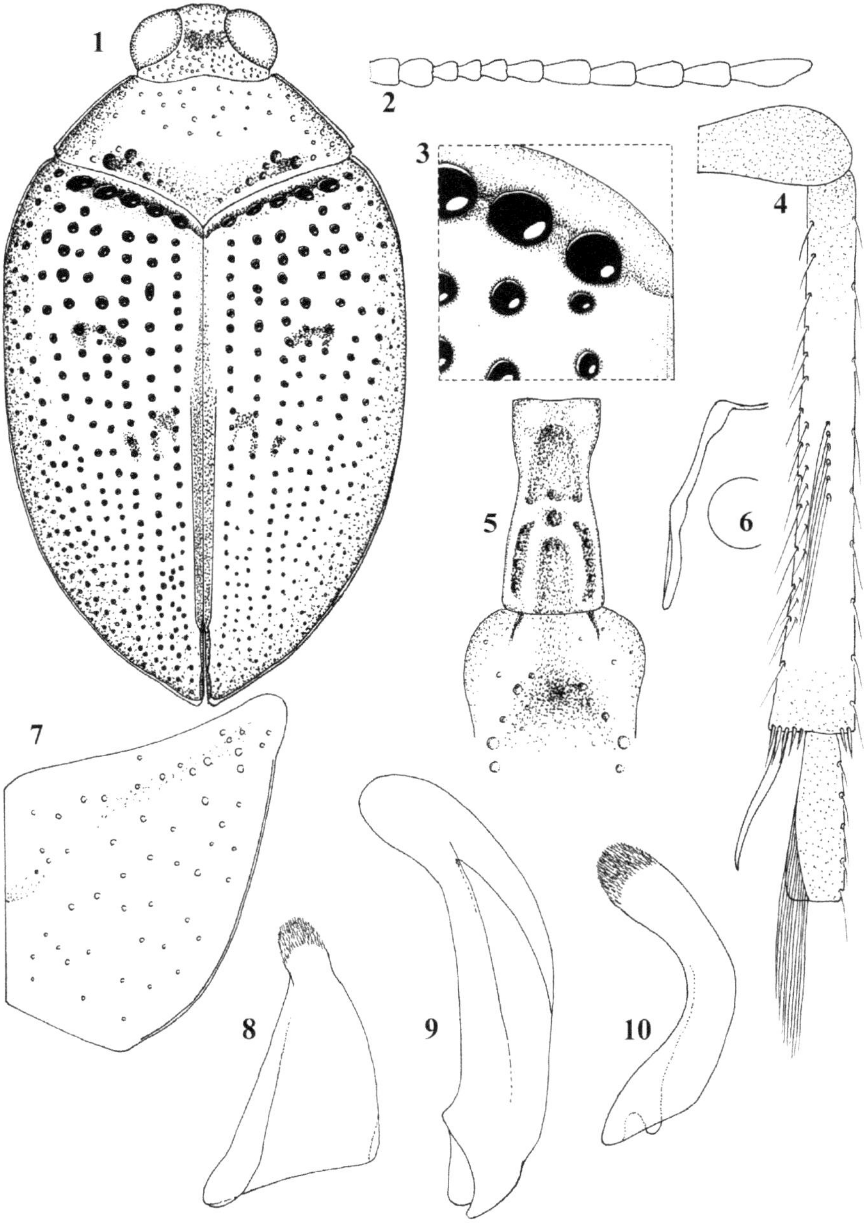

PLATE 223 *Peltodytes* (*Peltodytes*) *speratus*. 1, dorsal view; 2, antenna; 3, elytral punctures near base and suture; 4, hind tibia in dorsal view; 5, prosternal and metaventral process; 6, prosternal process in lateral view; 7, hind coxal plate; 8, left paramere; 9, penis; 10, right paramere

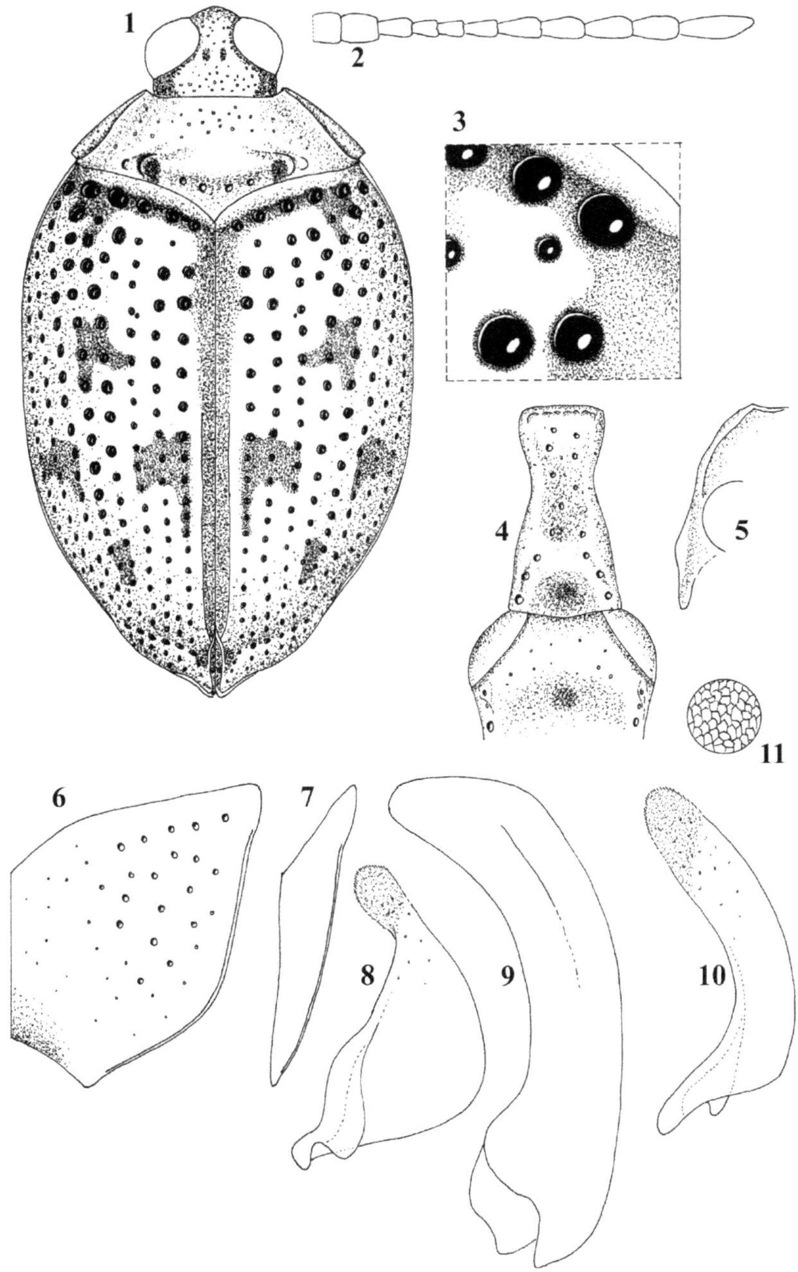

PLATE 224 *Peltodytes* (*Peltodytes*) *sumatrensis*. 1, dorsal view; 2, antenna; 3, elytral punctures near base and suture; 4, prosternal and metaventral process; 5, prosternal process in lateral view; 6, hind coxal plate; 7, hind coxal plate in lateral view; 8, left paramere; 9, penis; 10, right paramere; 11, structure of top of left and right paramere

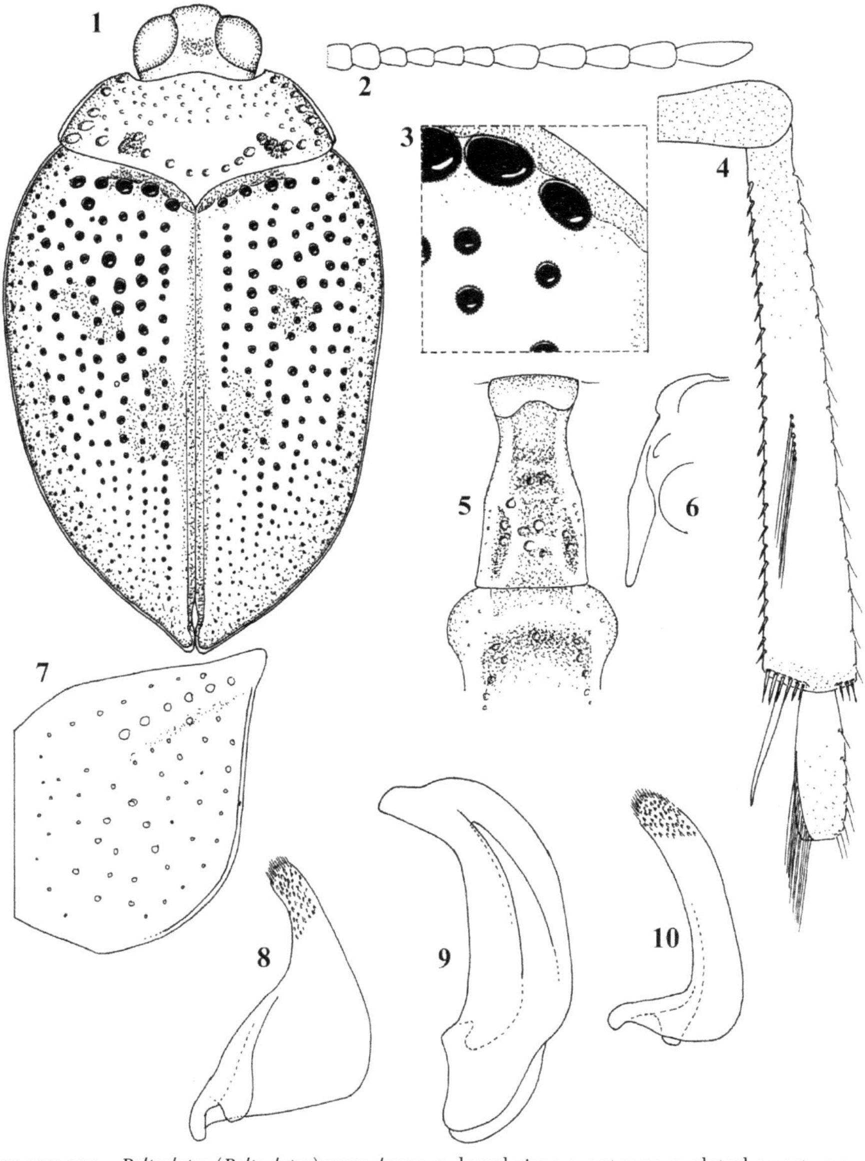

PLATE 225 *Peltodytes* (*Peltodytes*) *ugandanus*. 1, dorsal view; 2, antenna; 3, elytral punctures near base and suture; 4, hind tibia in dorsal view; 5, prosternal and metaventral process; 6, prosternal process in lateral view; 7, hind coxal plate; 8, left paramere; 9, penis; 10, right paramere

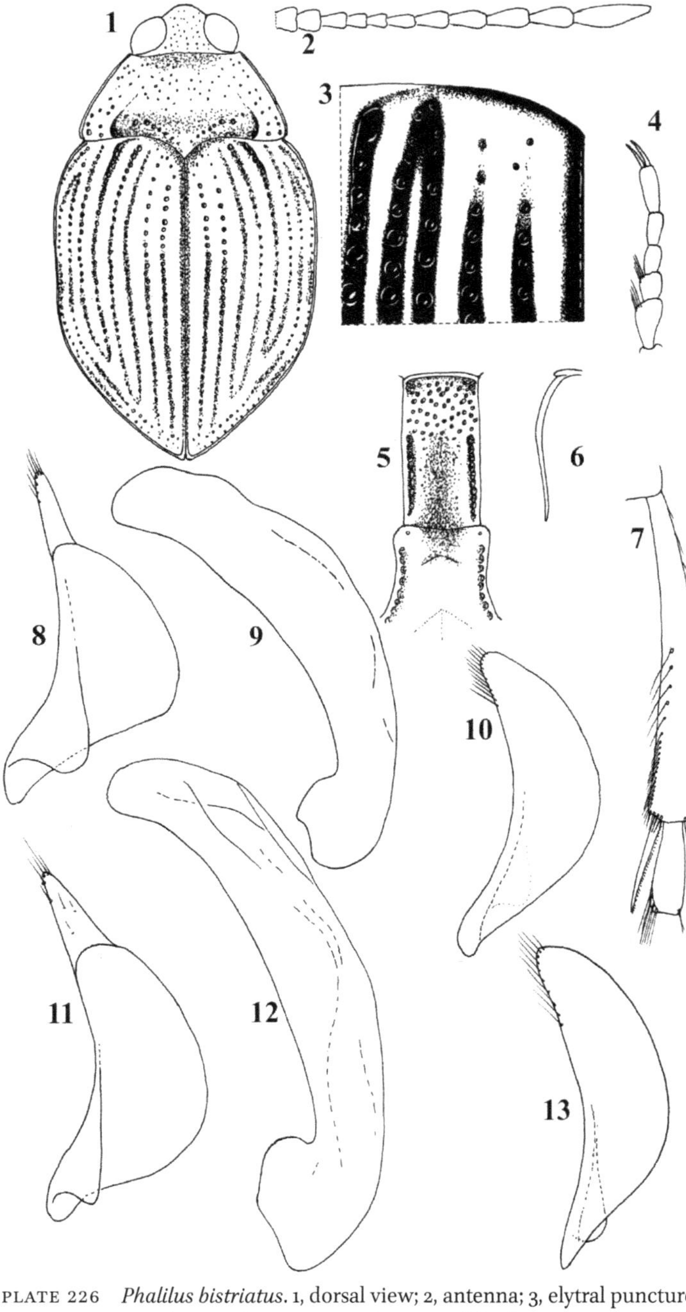

PLATE 226 *Phalilus bistriatus*. 1, dorsal view; 2, antenna; 3, elytral punctures near base and suture; 4, male tarsus of foreleg; 5, prosternal and metaventral process; 6, prosternal process in lateral view; 7, hind tibia in dorsal view; 8, 11, left paramere; 9, 12, penis; 10, 13, right paramere

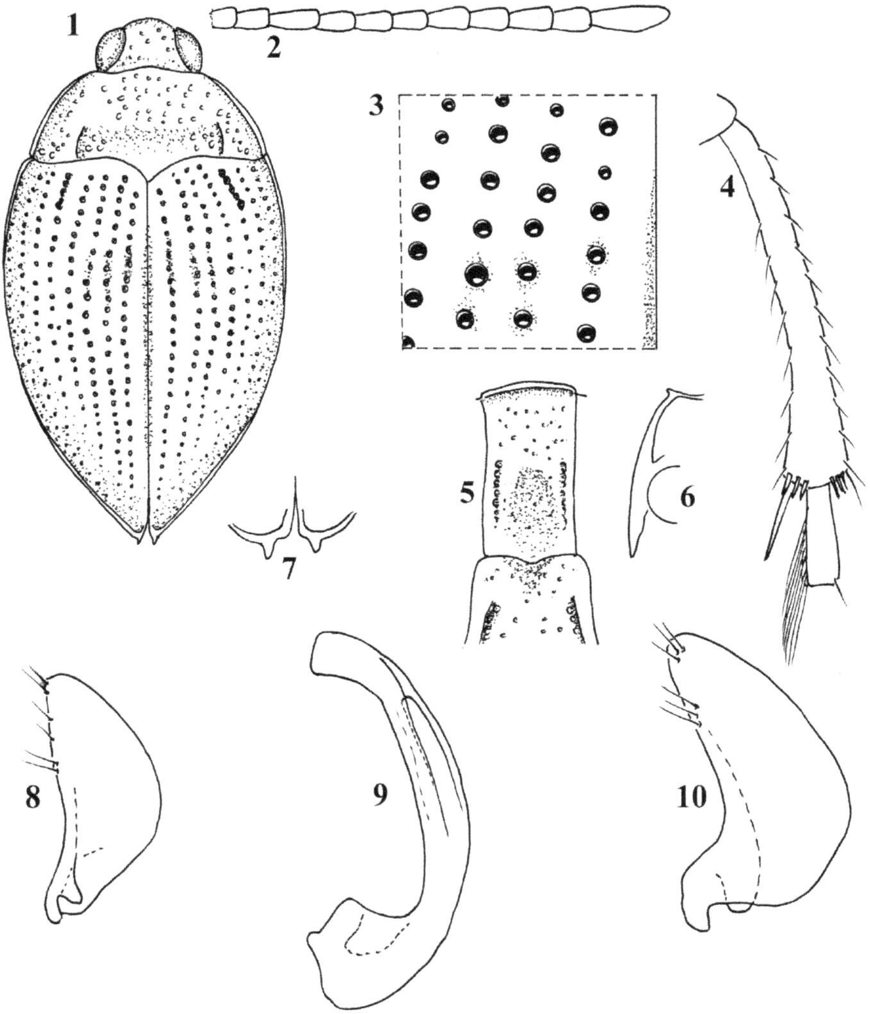

PLATE 227 *Phalilus fortescueensis*. 1, dorsal view; 2, antenna; 3, elytral punctures near base and suture; 4, hind tibia in dorsal view; 5, prosternal and metaventral process; 6, prosternal process in lateral view; 7, apical points of elytra; 8, left paramere; 9, penis; 10, right paramere

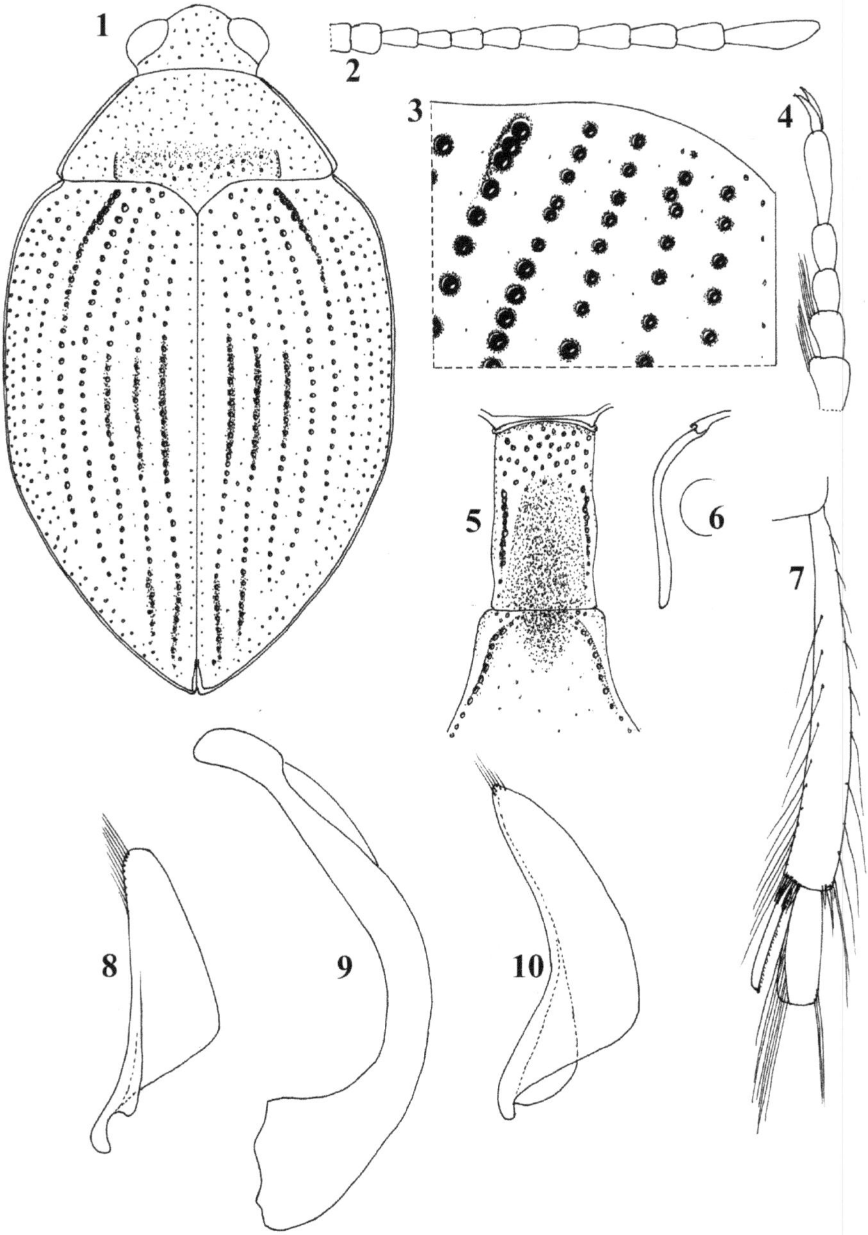

PLATE 228 *Phalilus fuscatus*. 1, dorsal view; 2, antenna; 3, elytral punctures near base and suture; 4, male tarsus of foreleg; 5, prosternal and metaventral process; 6, prosternal process in lateral view; 7, hind tibia in dorsal view; 8, left paramere; 9, penis; 10, right paramere

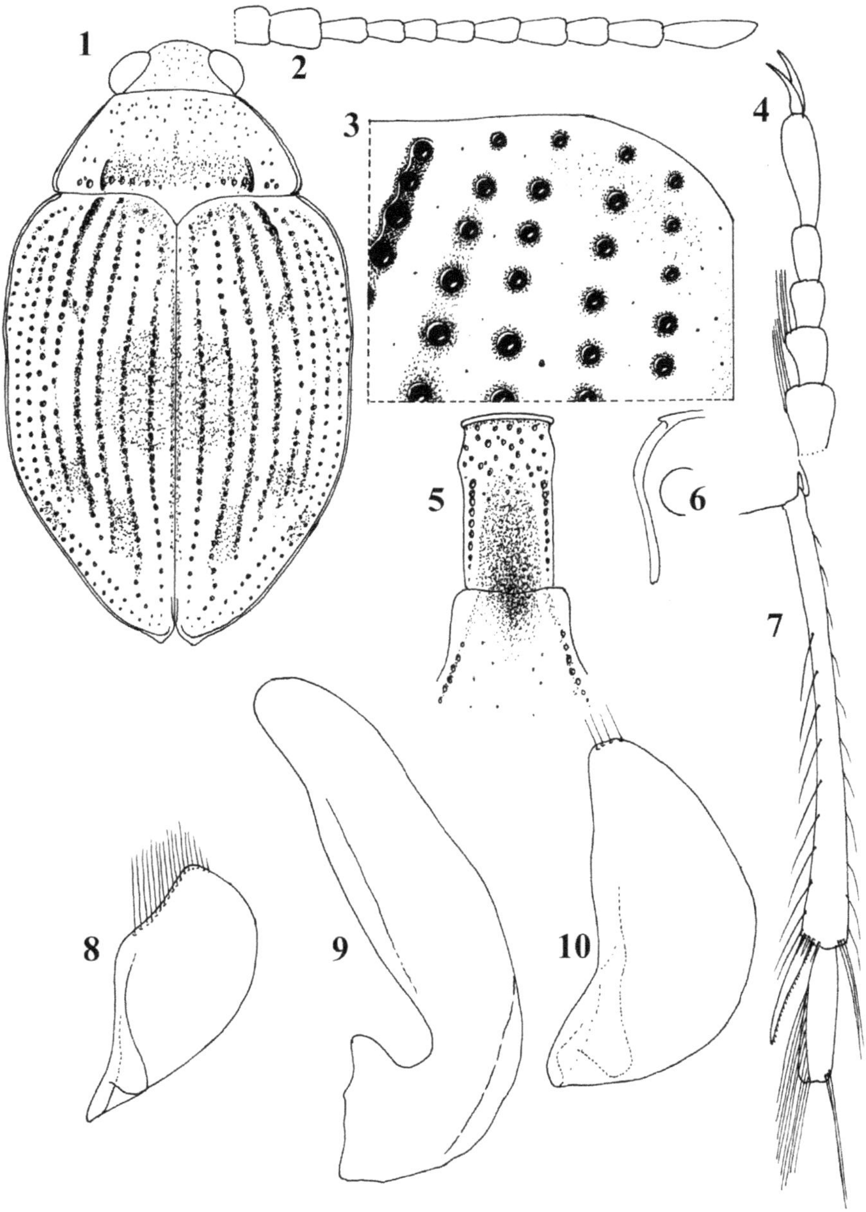

PLATE 229 *Phalilus gibbus*. 1, dorsal view; 2, antenna; 3, elytral punctures near base and suture; 4, male tarsus of foreleg; 5, prosternal and metaventral process; 6, prosternal process in lateral view; 7, hind tibia in dorsal view; 8, left paramere; 9, penis; 10, right paramere

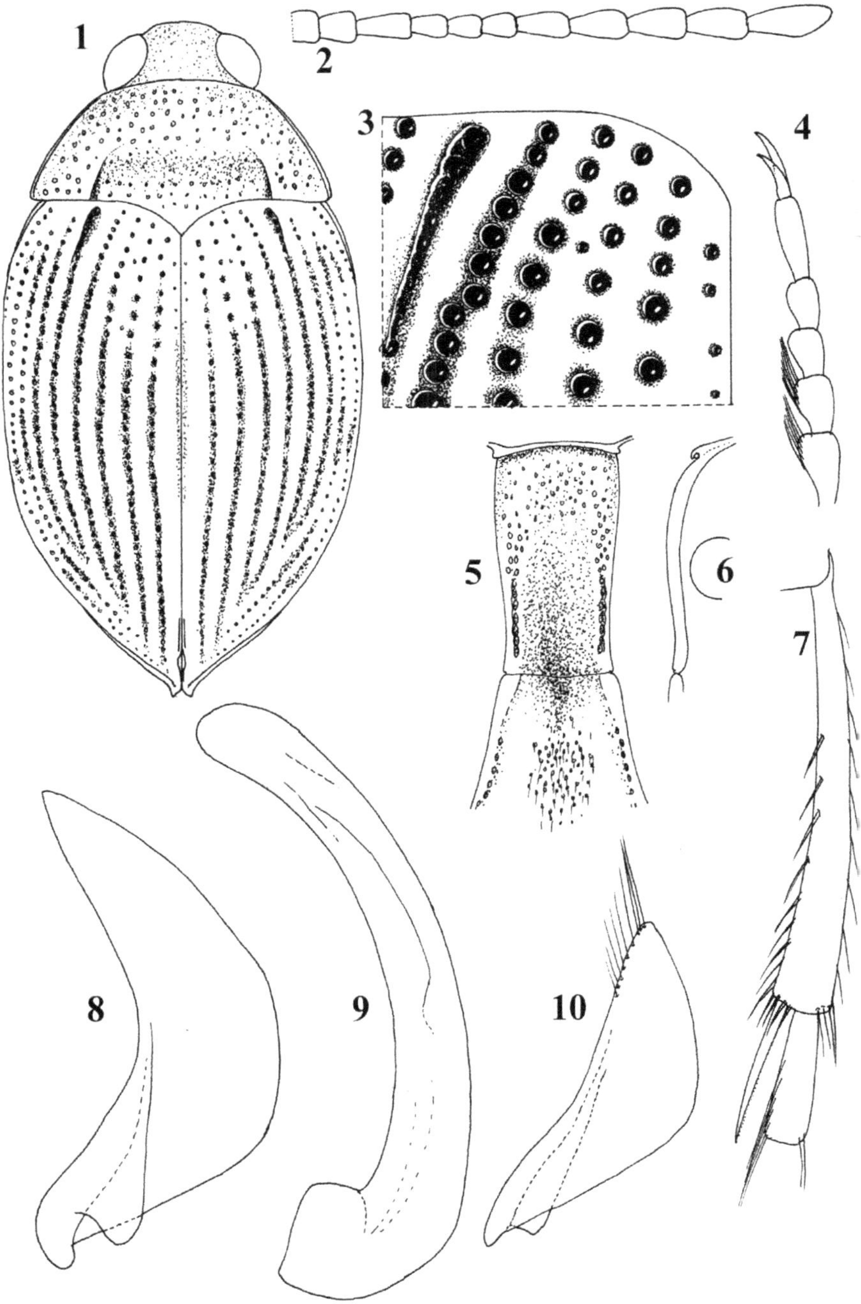

PLATE 230 *Phalilus hydei*. 1, dorsal view; 2, antenna; 3, elytral punctures near base and suture; 4, male tarsus of foreleg; 5, prosternal and metaventral process; 6, prosternal process in lateral view; 7, hind tibia in dorsal view; 8, left paramere; 9, penis; 10, right paramere

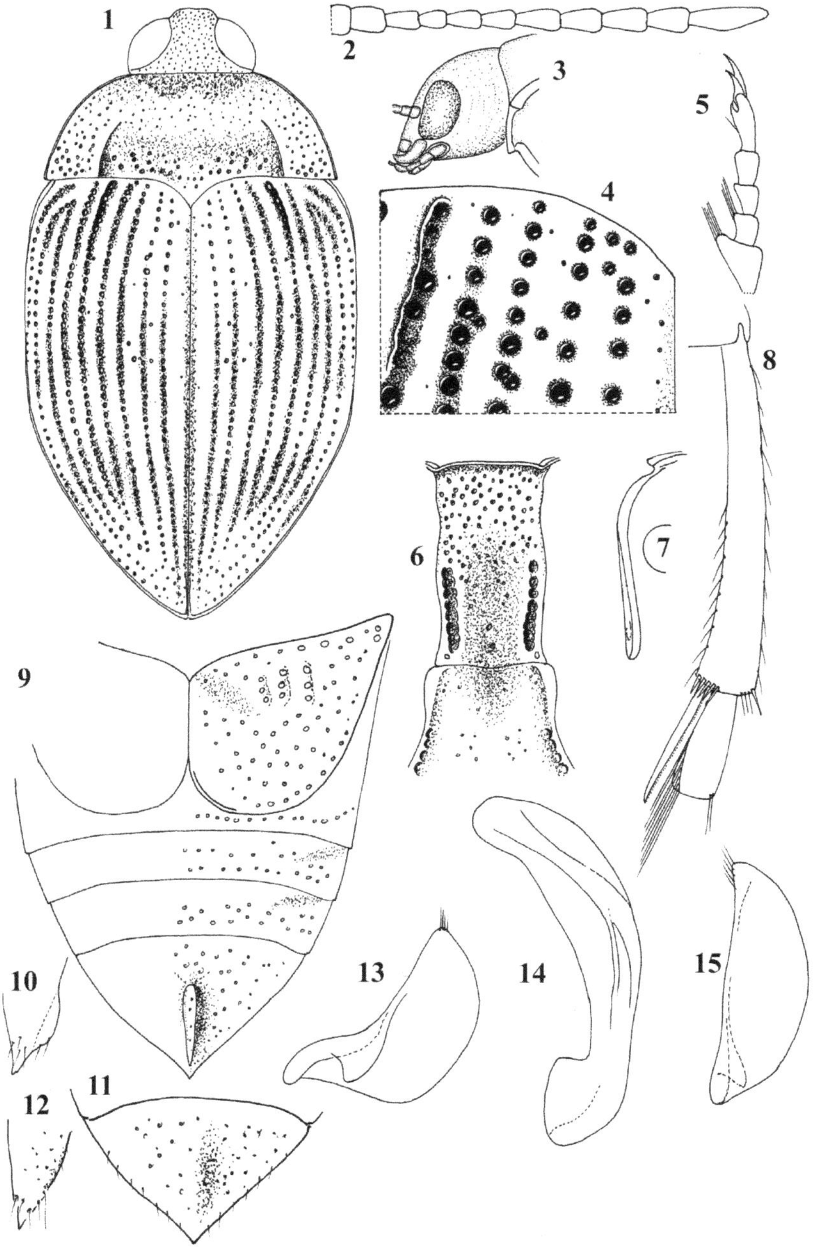

PLATE 231 *Phalilus oberthuri*. 1, dorsal view; 2, antenna; 3, head in lateral view; 4, elytral punctures near base and suture; 5, male tarsus of foreleg; 6, prosternal and metaventral process; 7, prosternal process in lateral view; 8, hind tibia in dorsal view; 9, coxal plates and sternites of male; 10, last male sternite in lateral view; 11, last female sternite; 12, last female sternite in lateral view; 13, left paramere; 14, penis; 15, right paramere

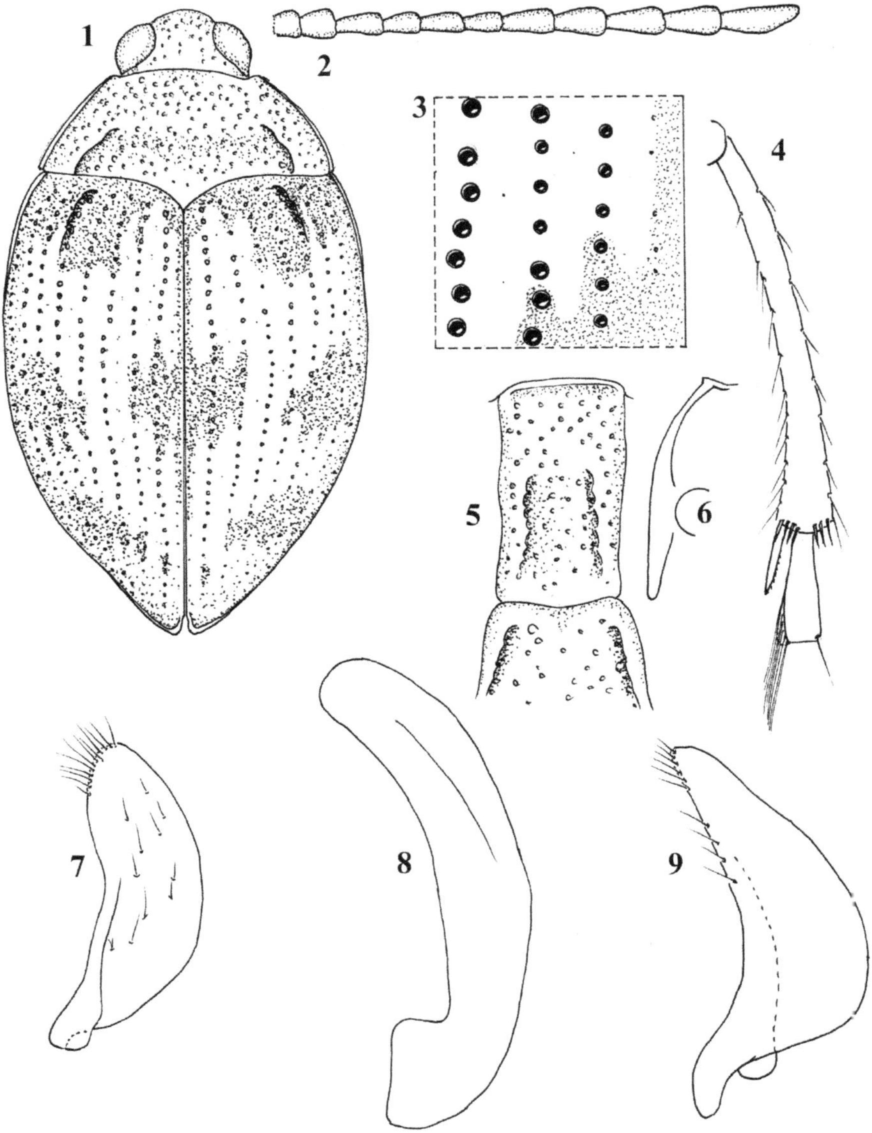

PLATE 232 *Phalilus pinderi*. 1, dorsal view; 2, antenna; 3, elytral punctures near base and suture; 4, hind tibia in dorsal view; 5, prosternal and metaventral process; 6, prosternal process in lateral view; 7, left paramere; 8, penis; 9, right paramere

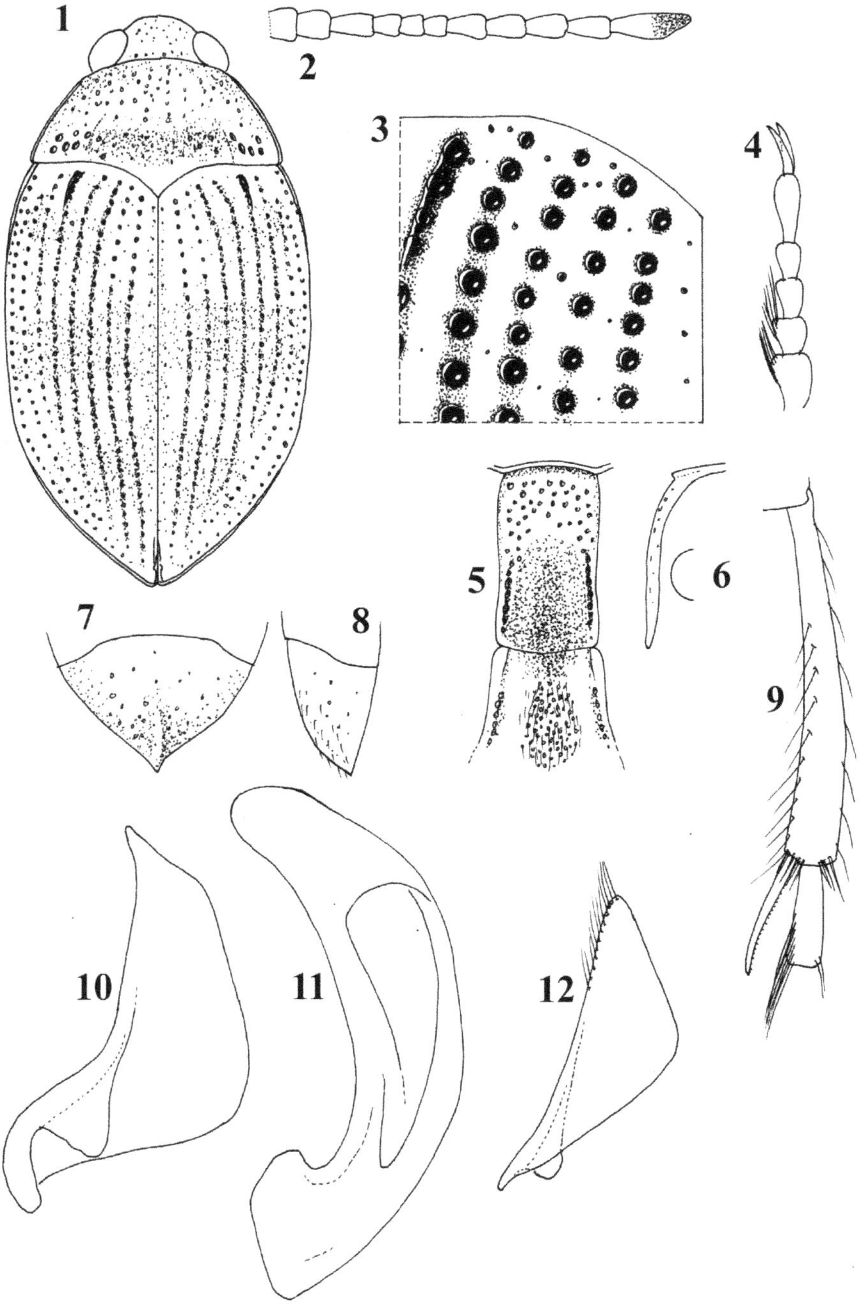

PLATE 233 *Phalilus storeyi*. 1, dorsal view; 2, antenna; 3, elytral punctures near base and suture; 4, male tarsus of foreleg; 5, prosternal and metaventral process; 6, prosternal process in lateral view; 7, last sternite; 8, last sternite in lateral view; 9, hind tibia in dorsal view; 10, left paramere; 11, penis; 12, right paramere

Plates 234–259. Photos of All Species in Alphabetical Order

PLATE 234 1, *Brychius elevatus*; 2, *B. glabratus*; 3, *B. hornii*; 4, *B. hungerfordi*; 5, *B. pacificus*; 6, *Haliplus* (*Algophilus*) *lathridioides*; 7, *Haliplus* (*Apteraliplus*) *parvulus*; 8, *Haliplus* (*Australiplus*) *sindus* (by Chris Watts); 9, *Haliplus* (*Borealiplus*) *borealis*. Pictures not on the same scale

 HTTPS://DOI.ORG/10.1163/9789004759671-018

PLATE 235 1, *Haliplus* (*Borealiplus*) *oblongus*; 2. *Haliplus* (*Haliaplus*) *laminatus*; 3; *Haliplus* (*Haliplidius*) *confinis*; 4, *H.* (*H.*) *obliquus*; 5, *H.* (*H.*) *rejseki*; 6. *H.* (*H.*) *subseriatus*; 7, *H.* (*H.*) *varius*; 8, *Haliplus* (*Haliplus*) *aliae*; 9, *H.* (*H.*) *apicalis*. Pictures not on the same scale

PLATE 236 1, *Haliplus* (*Haliplus*) *blanchardi*; 2, *H.* (*H.*) *distinctus*; 3, *H.* (*H.*) *dorsomaculatus*; 4, *H.* (*H.*) *falli*; 5, *H.* (*H.*) *fluviatilis*; 6, *H.* (*H.*) *fulvicollis*, 7, *H.* (*H.*) *furcatus*; 8, *H.* (*H.*) *fuscicornis*; 9, *H.* (*H.*) *harminae*. Pictures not on the same scale

PLATE 237 1, *Haliplus* (*Haliplus*) *heydeni*; 2, *H.* (*H.*) *immaculatus*; 3, *H.* (*H.*) *immaculicollis*; 4, *H.* (*H.*) *interjectus*; 5, *H.* (*H.*) *kamiyai*; 6, *H.* (*H.*) *kirgisiensis* (by Jaakko Mattila & Pekka Malinen); 7, *H.* (*H.*) *lineolatus*; 8, *H.* (*H.*) *longulus*; 9, *H.* (*H.*) *morii* (by Hiroyuki Yoshitomi). Pictures not on the same scale

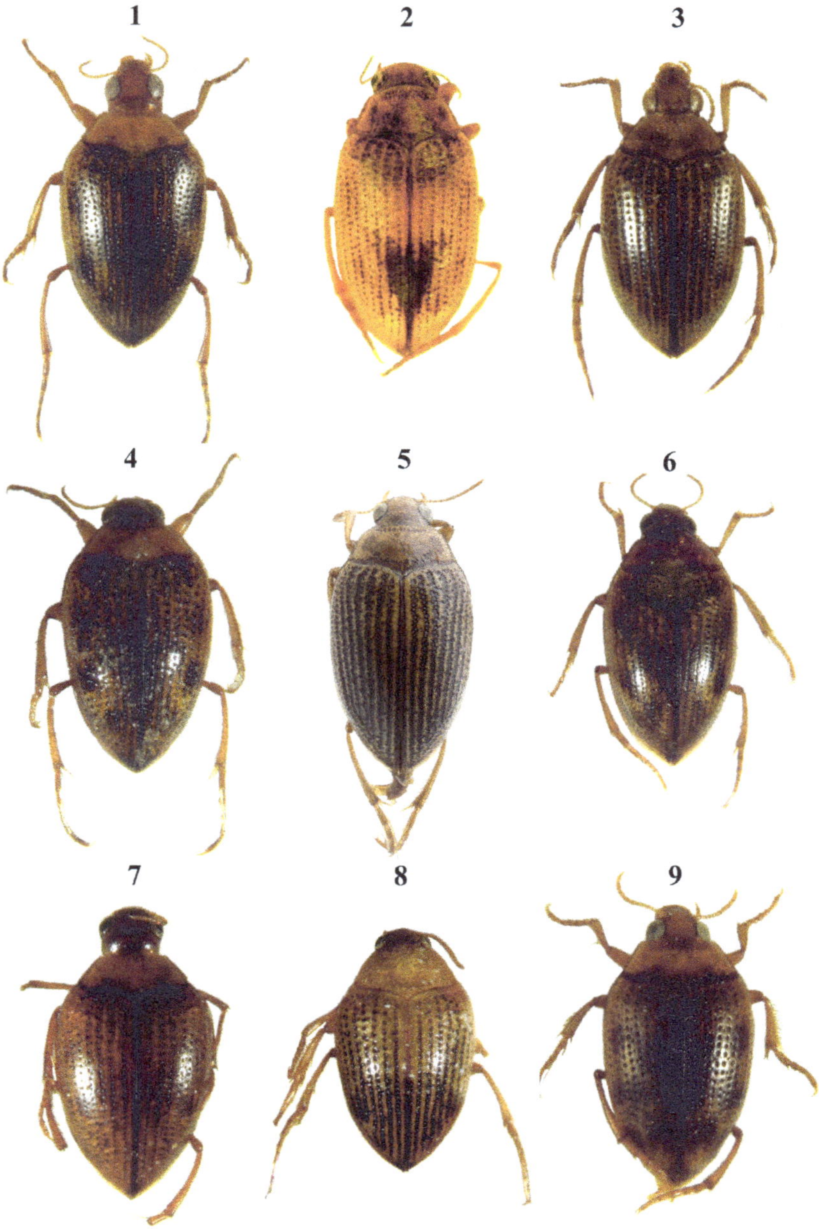

PLATE 238 1, *Haliplus* (*Haliplus*) *ruficollis*; 2, *H.* (*H.*) *samojedorum*; 3. *H.* (*H.*) *sibiricus*; 4, *H.* (*H.*) *simplex*; 5, *H.* (*H.*) *stagninus*; 6, *H.* (*H.*) *steppensis*; 7, *H.* (*H.*) *turkmenicus*; 8, *H.* (*H.*) *zacharenkoi*; 9, *Haliplus* (*Liaphlus*) *abbreviatus*. Pictures not on the same scale

PLATE 239 1, *Haliplus* (*Liaphlus*) *africanus*; 2, *H.* (*L.*) *alastairi*; 3, *H.* (*L.*) *alluaudi*; 4, *H.* (*L.*) *andalusicus*; 5, *H.* (*L.*) *angusi*; 6, *H.* (*L.*) *angustifrons*; 7, *H.* (*L.*) *annulatus*; 8, *H.* (*L.*) *arrowi*; 9, *H.* (*L.*) *astrakhanus*. Pictures not on the same scale

PLATE 240 1, *Haliplus* (*Liaphlus*) *australis*; 2, *H.* (*L.*) *bachmanni*; 3, *H.* (*L.*) *basinotatus*; 4, *H.* (*L.*) *bonariensis*; 5, *H.* (*L.*) *brasiliensis* (by Michael Balke & Katja Neven); 6, *H.* (*L.*) *camposi*; 7, *H.* (*L.*) *carinatus*; 8, *H.* (*L.*) *chinensis*; 9, *H.* (*L.*) *colombiensis*. Pictures not on the same scale

PLATE 241 1, *Haliplus* (*Liaphlus*) *columbiensis*; 2, *H.* (*L.*) *concolor*; 3, *H.* (*L.*) *confluentus*; 4, *H.* (*L.*) *costaricanus*; 5, *H.* (*L.*) *coxaspinipes*; 6, *H.* (*L.*) *crassus*; 7, *H.* (*L.*) *cribrarius*; 8, *H.* (*L.*) *cubensis*; 9, *H.* (*L.*) *cylindricus*. Pictures not on the same scale

PLATE 242 1, *Haliplus* (*Liaphlus*) *dalmatinus*; 2, *H.* (*L.*) *diopus* (by Christophe Rivier & Antoine Mantilleri); 3, *H.* (*L.*) *discessus*; 4, *H.* (*L.*) *drechseli*; 5, *H.* (*L.*) *ebolovensis*; 6, *H.* (*L.*) *elsaltous*; 7, *H.* (*L.*) *excoffieri*; 8, *H.* (*L.*) *eximius*; 9, *H.* (*L.*) *exsecratus*. Pictures not on the same scale

PLATE 243 1, *Haliplus* (*Liaphlus*) *ferruginipes* (by Chris Watts); 2, *H.* (*L.*) *figuratus* (by Jaakko Mattlia & Pekka Malinen); 3, *H.* (*L.*) *fiorentini* (by Cesar Benetti); 4, *H.* (*L.*) *flavicollis*; 5, *H.* (*L.*) *fulvus*; 6, *H.* (*L.*) *fuscipennis*; 7, *H.* (*L.*) *gamma*; 8, *H.* (*L.*) *garambanus*; 9, *H.* (*L.*) *genieri*. Pictures not on the same scale

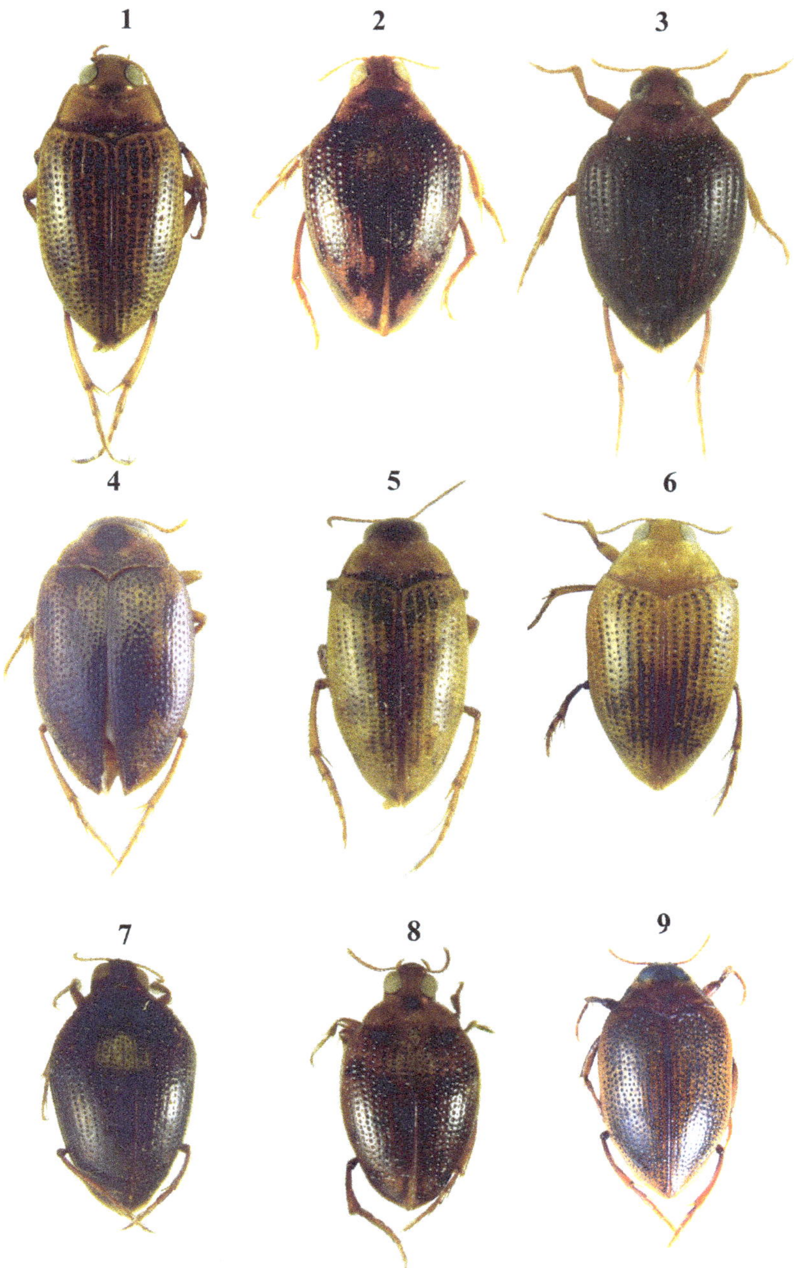

PLATE 244 1, *Haliplus* (*Liaphlus*) *gracilis*; 2, *H.* (*L.*) *gravidoides*; 3, *H.* (*L.*) *gravidus*; 4, *H.* (*L.*) *guignoti*; 5, *H.* (*L.*) *guttatus*; 6, *H.* (*L.*) *halsei*; 7, *H.* (*L.*) *havaniensis*; 8, *H.* (*L.*) *heppneri*; 9, *H.* (*L.*) *incrassatus*. Pictures not on the same scale.

PLATE 245 1, *Haliplus* (*Liaphlus*) *indicus*; 2, *H.* (*L.*) *indistinctus*; 3, *H.* (*L.*) *insularis*; 4, *H.* (*L.*) *javanicus*; 5, *H.* (*L.*) *kantneri*; 6, *H.* (*L.*) *kapuri*; 7, *H.* (*L.*) *kotoshonis*; 8, *H.* (*L.*) *lamottei*; 9, *H.* (*L.*) *langleyi*. Pictures not on the same scale

PLATE 246 1. *Haliplus* (*Liaphlus*) *leechi*; 2, *H.* (*L.*) *lewisii*; 3, *H.* (*L.*) *maculatus*; 4, *H.* (*L.*) *maculicollis*; 5, *H.* (*L.*) *maculipennis*; 6, *H.* (*L.*) *madagascariensis*; 7, *H.* (*L.*) *megapunctatus*; 8, *H.* (*L.*) *mesoamericanus*; 9; *H.* (*L.*) *methneri*. Pictures not on the same scale

PLATE 247 1, *Haliplus* (*Liaphlus*) *mexicanus*; 2, *H.* (*L.*) *mimulus*; 3, *H.* (*L.*) *minimus*; 4, *H.* (*L.*) *mucronatus*; 5, *H.* (*L.*) *nanus*; 6, *H.* (*L.*) *napolovi*; 7, *H.* (*L.*) *natalensis*; 8. *H.* (*L.*) *nedungaduensis*; 9, *H.* (*L.*) *nieseri*. Pictures not on the same scale

PLATE 248 1. *Haliplus* (*Liaphlus*) *nigerianus*; 2, *H.* (*L.*) *nitens*; 3, *H.* (*L.*) *novateutonianus*; 4, *H.* (*L.*) *oklahomensis*; 5, *H.* (*L.*) *ornatipennis*; 6, *H.* (*L.*) *ovalis*; 7, *H.* (*L.*) *panamanus*; 8, *H.* (*L.*) *perroti* (by Christophe Rivier & Antoine Mantilleri); 9, *H.* (*L.*) *peruanus* (by Michael Balke & Katja Neven). Pictures not on the same scale

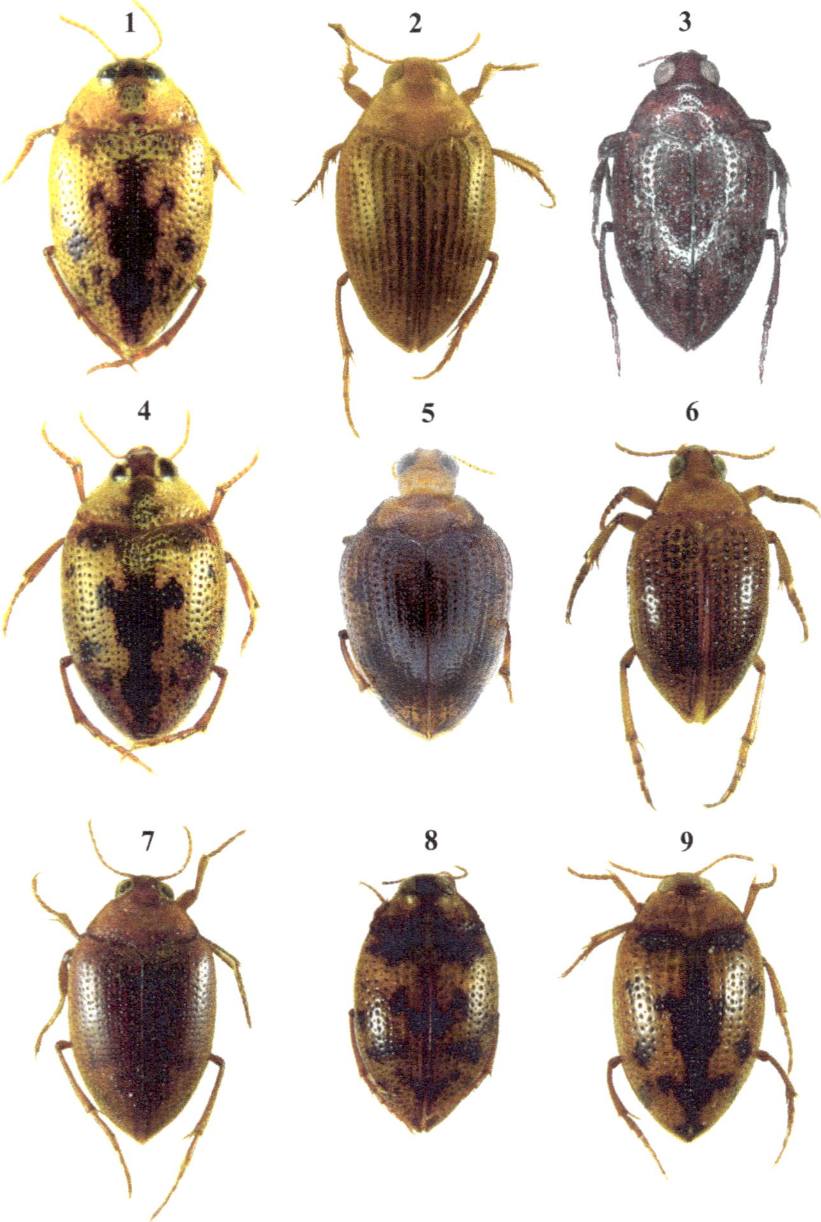

PLATE 249 1, *Haliplus* (*Liaphlus*) *philippinus*; 2, *H.* (*L.*) *pilbaraensis*; 3, *H.* (*L.*) *pruthii*; 4, *H.* (*L.*) *pulchellus*; 5, *H.* (*L.*) *regili* (by Cesar Benetti); 6, *H.* (*L.*) *rubidus*; 7, *H.* (*L.*) *rufescens*; 8, *H.* (*L.*) *samosirensis*; 9, *H.* (*L.*) *sharpi*. Pictures not on the same scale

PLATE 250 1, *Haliplus* (*Liaphlus*) *signatipennis*; 2, *H.* (*L.*) *signatus*; 3, *H.* (*L.*) *solitarius*; 4, *H.* (*L.*) *srilankanus*; 5, *H.* (*L.*) *stepheni*; 6, *H.* (*L.*) *sulawesicus*; 7, *H.* (*L.*) *tantoyucanus*; 8, *H.* (*L.*) *tanzanianus*; 9, *H.* (*L.*) *testaceus*. Pictures not on the same scale

PLATE 251 1, *Haliplus* (*Liaphlus*) *testudo*; 2, *H.* (*L.*) *thoracicus*; 3, *H.* (*L.*) *timmsi*; 4, *H.* (*L.*) *tocumenus*; 5, *H.* (*L.*) *triplehorni*; 6, *H.* (*L.*) *tumidus*; 7, *H.* (*L.*) *unicarinatus*; 8, *H.* (*L.*) *uniformis*; 9, *H.* (*L.*) *valdiviensis*. Pictures not on the same scale

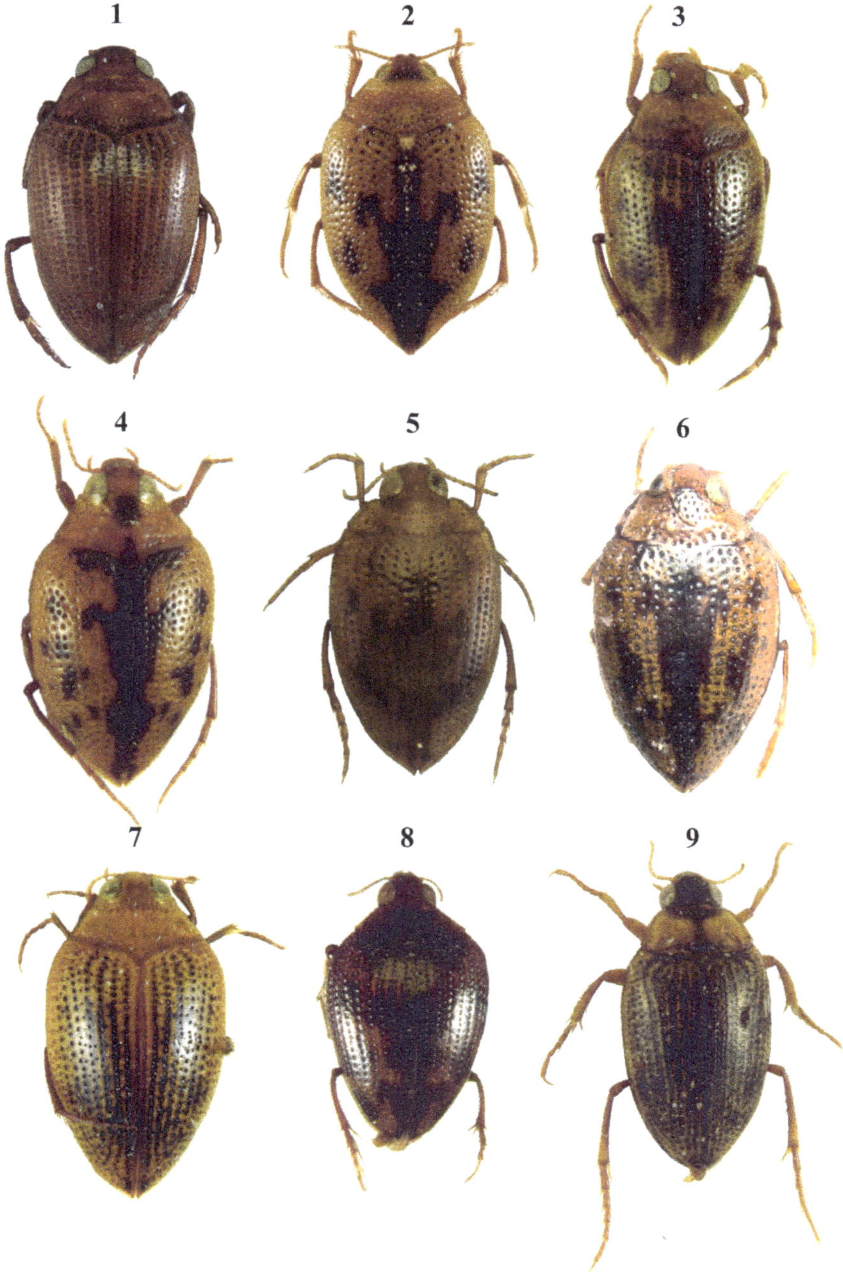

PLATE 252 1, *Haliplus* (*Liaphlus*) *vancouverensis*; 2, *H.* (*L.*) *varicator*; 3, *H.* (*L.*) *variegatus*; 4, *H.* (*L.*) *venustus*; 5, *H.* (*L.*) *volpei*; 6, *H.* (*L.*) *watsoni*; 7, *H.* (*L.*) *wattsi*; 8, *H.* (*L.*) *youngi*; 9, *Haliplus* (*Neohaliplus*) *lineatocollis*. Pictures not on the same scale

PLATE 253 1, *Haliplus* (*Neohaliplus*) *ruficeps*; 2, *Haliplus* (*Nipponiplus*) *japonicus*; 3. *H.* (*N.*) *latreillei*; 4, *H.* (*N.*) *regimbarti*; 5, *Haliplus* (*Paraliaphlus*) *alberti* (by Yoandri Megna); 6, *H.* (*P.*) *apostolicus*; 7, *H.* (*P.*) *connexus*; 8, *H.* (*P.*) *deceptus*; 9, *H.* (*P.*) *fasciatus*. Pictures not on the same scale

PLATE 254 1, *Haliplus* (*Paraliaphlus*) *grandis*; 2, *H.* (*P.*) *kenneri*; 3, *H.* (*P.*) *leopardus*; 4, *H.* (*P.*) *lucigelinae*; 5, *H.* (*P.*) *mimeticus*; 6. *H.* (*P.*) *mutchleri*; 7, *H.* (*P.*) *pantherinus*; 8, *H.* (*P.*) *pseudofasciatus*; 9, *H.* (*P.*) *punctatus*. Pictures not on the same scale

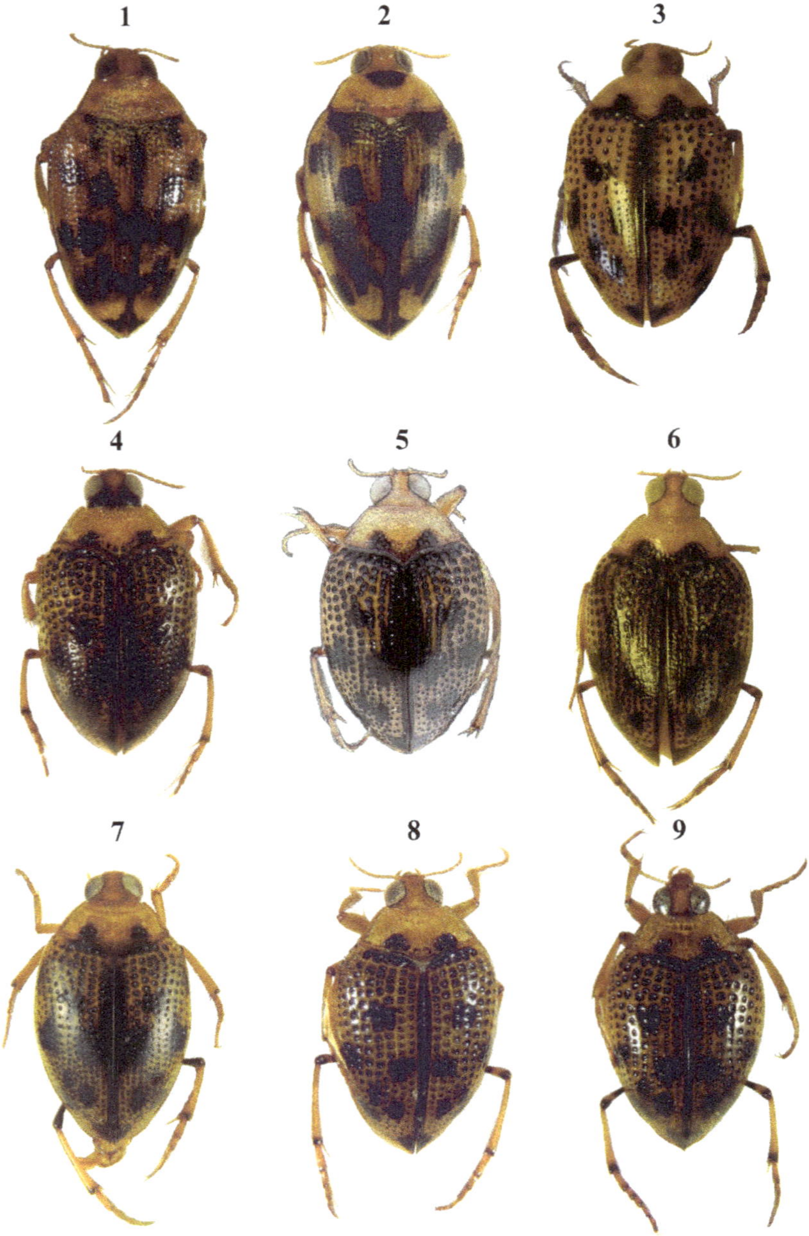

PLATE 255 1, *Haliplus* (*Paraliaphlus*) *tortilipenis*; 2, *H.* (*P.*) *triopsis*; 3, *Peltodytes* (*Neopeltodytes*) *bradleyi*; 4, *P.* (*N.*) *callosus*; 5, *P.* (*N.*) *darlingtoni*; 6, *P.* (*N.*) *dietrichi*; 7, *P.* (*N.*) *dunavani*; 8, *P.* (*N.*) *duodecimpunctatus*; 9, *P.* (*N.*) *edentulus*. Pictures not on the same scale

PLATE 256 1, *Peltodytes (Neopeltodytes) festivus*; 2, *P.* (*N.*) *floridensis*; 3, *P.* (*N.*) *litoralis*; 4, *P.* (*N.*) *muticus*; 5, *P.* (*N.*) *oppositus*; 6, *P.* (*N.*) *ovalis*; 7, *P.* (*N.*) *pedunculatus*; 8, *P.* (*N.*) *roughleyi*; 9, *P.* (*N.*) *sexmaculatus*. Pictures not on the same scale

PLATE 257 1, *Peltodytes* (*Neopeltodytes*) *shermani*; 2, *P.* (*N.*) *simplex*; 3, *P.* (*N.*) *tamaulipensis*; 4, *P.* (*N.*) *tortulosus*; 5, *Peltodytes* (*Peltodytes*) *caesus*; 6, *P.* (*P.*) *congoensis*; 7, *P.* (*P.*) *coomani*; 8, *P.* (*P.*) *dauricus*; 9, *P.* (*P.*) *ghanaensis*. Pictures not on the same scale

PLATE 258 1. *Peltodytes* (*Peltodytes*) *intermedius*; 2, *P.* (*P.*) *nodieri*; 3, *P.* (*P.*) *pekinensis*; 4, *P.* (*P.*) *quadratus*; 5, *P.* (*P.*) *rotundatus*; 6, *P.* (*P.*) *sinensis*; 7, *P.* (*P.*) *speratus*; 8, *P.* (*P.*) *sumatrensis*; 9, *P.* (*P.*) *ugandanus*. Pictures not on the same scale

PLATE 259 1, *Phalilus bistriatus*; 2, *P. fortescueensis*; 3, *P. fuscatus*; 4, *P. gibbus*; 5, *P. hydei*; 6, *P. oberthuri*; 7, *P. pinderi*; 8, *P. storey*. Pictures not on the same scale

Plates 260–281. Distribution Maps of All Species, Arranged by Faunal Region

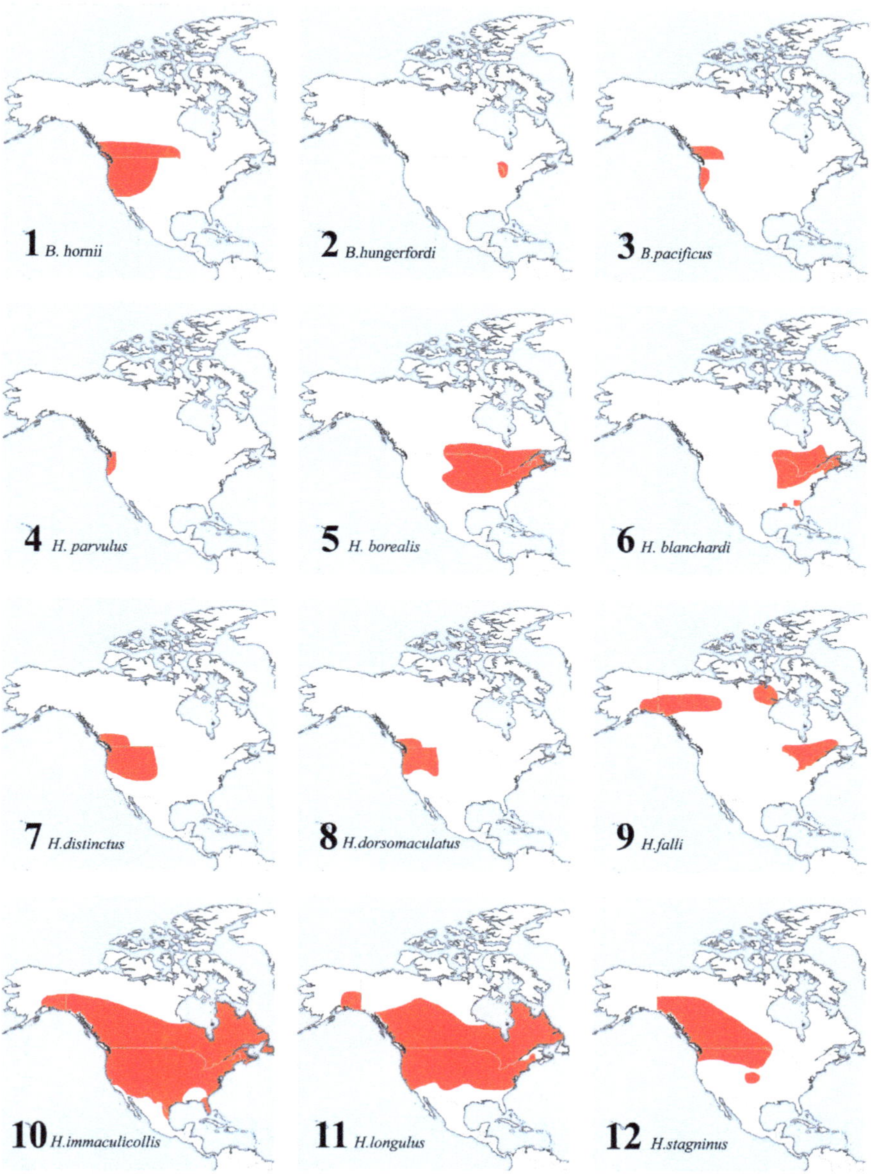

PLATE 260 Distribution of Nearctic Haliplidae

 | HTTPS://DOI.ORG/10.1163/9789004759671-019

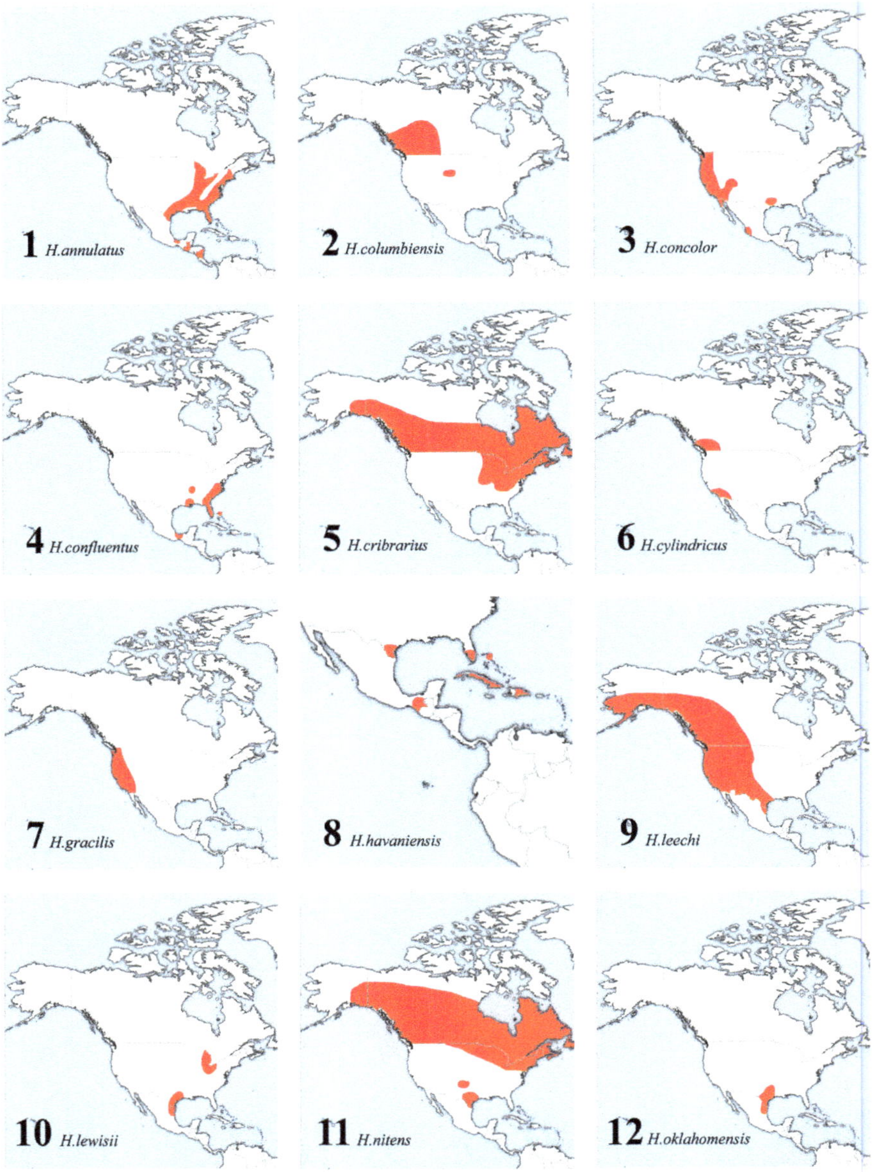

PLATE 261 Distribution of Nearctic Haliplidae

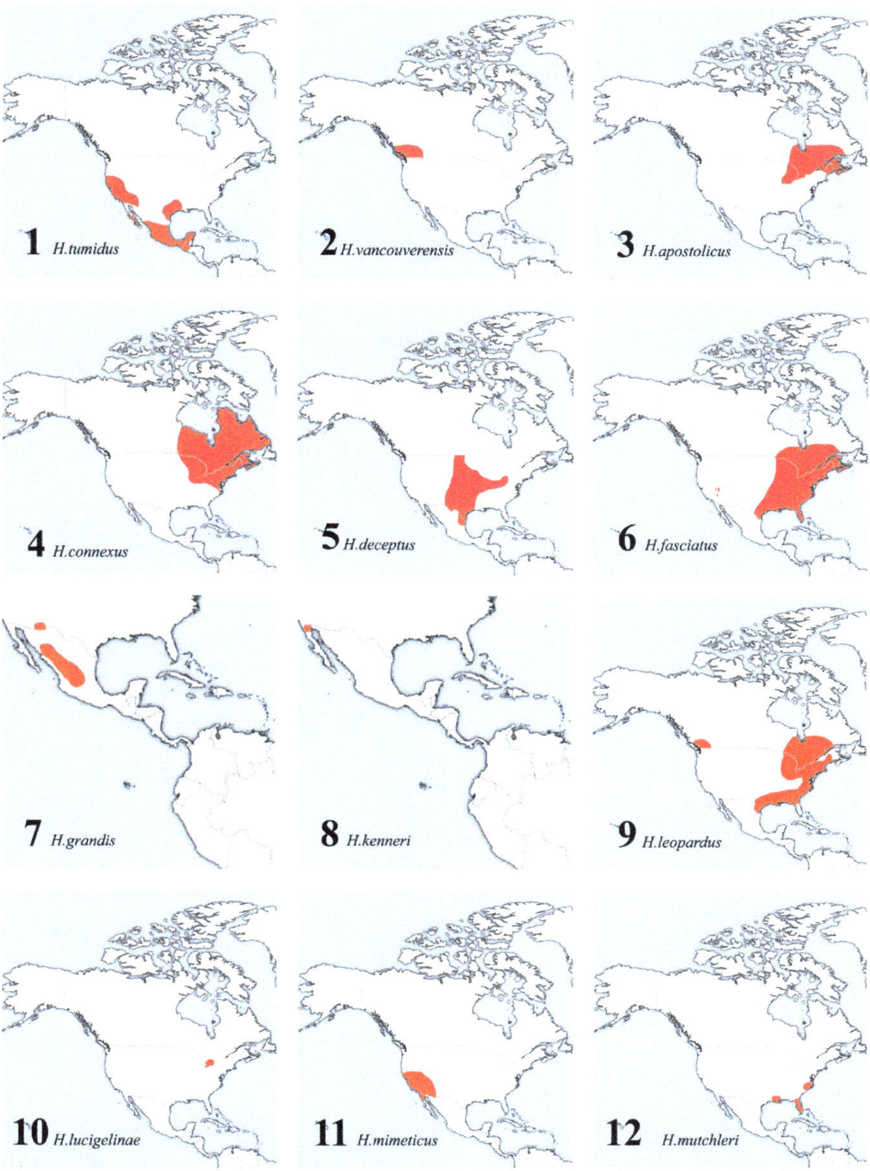

PLATE 262 Distribution of Nearctic and Neotropical Haliplidae

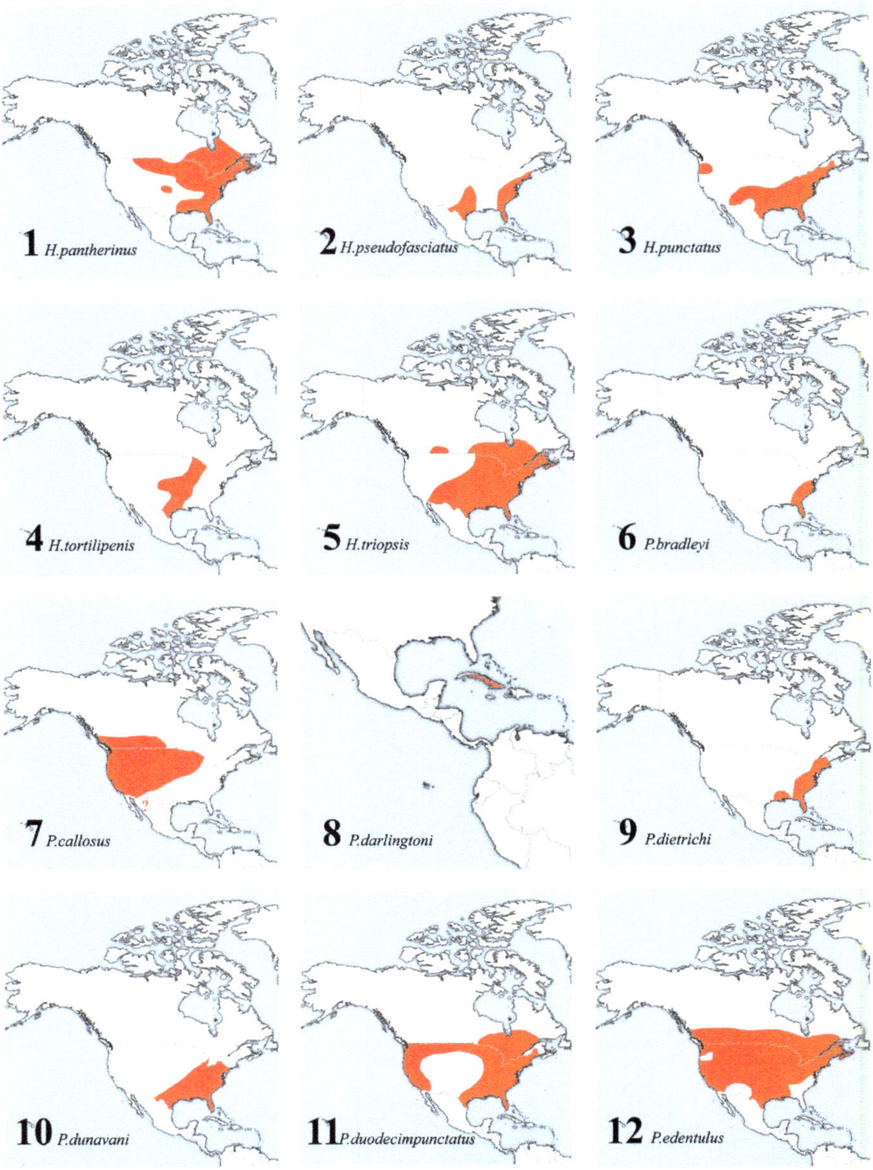

PLATE 263 Distribution of Nearctic and Neotropical Haliplidae

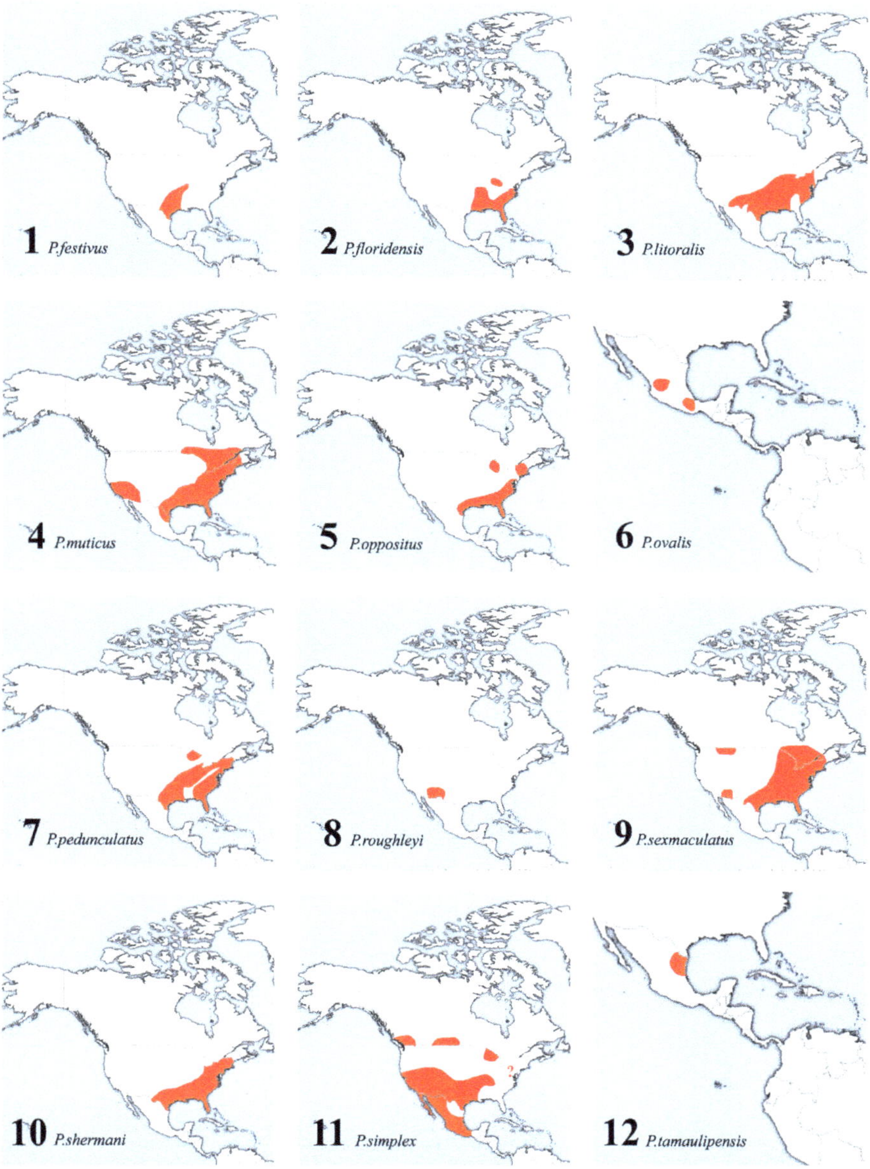

PLATE 264 Distribution of Nearctic and Neotropical Haliplidae

PLATE 265 Distribution of Nearctic and Neotropical Haliplidae

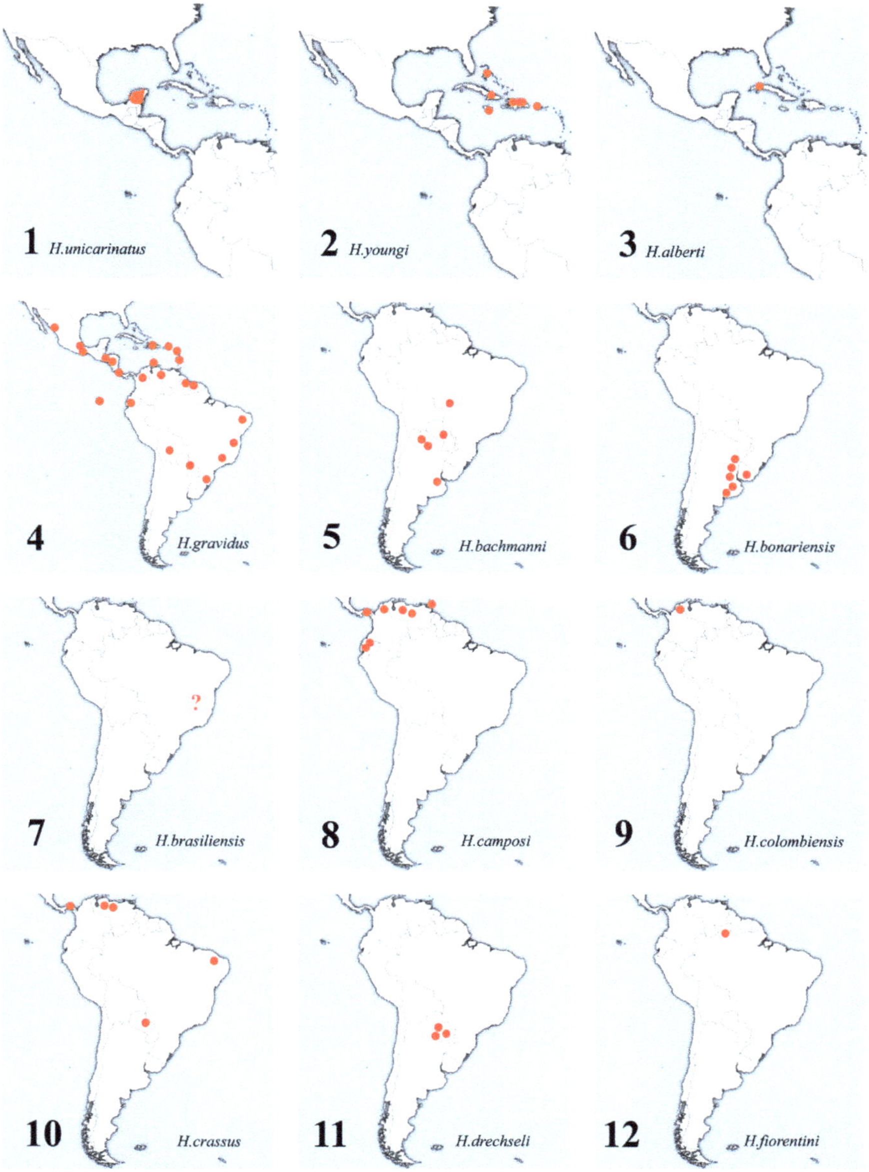

PLATE 266 Distribution of Neotropical Haliplidae

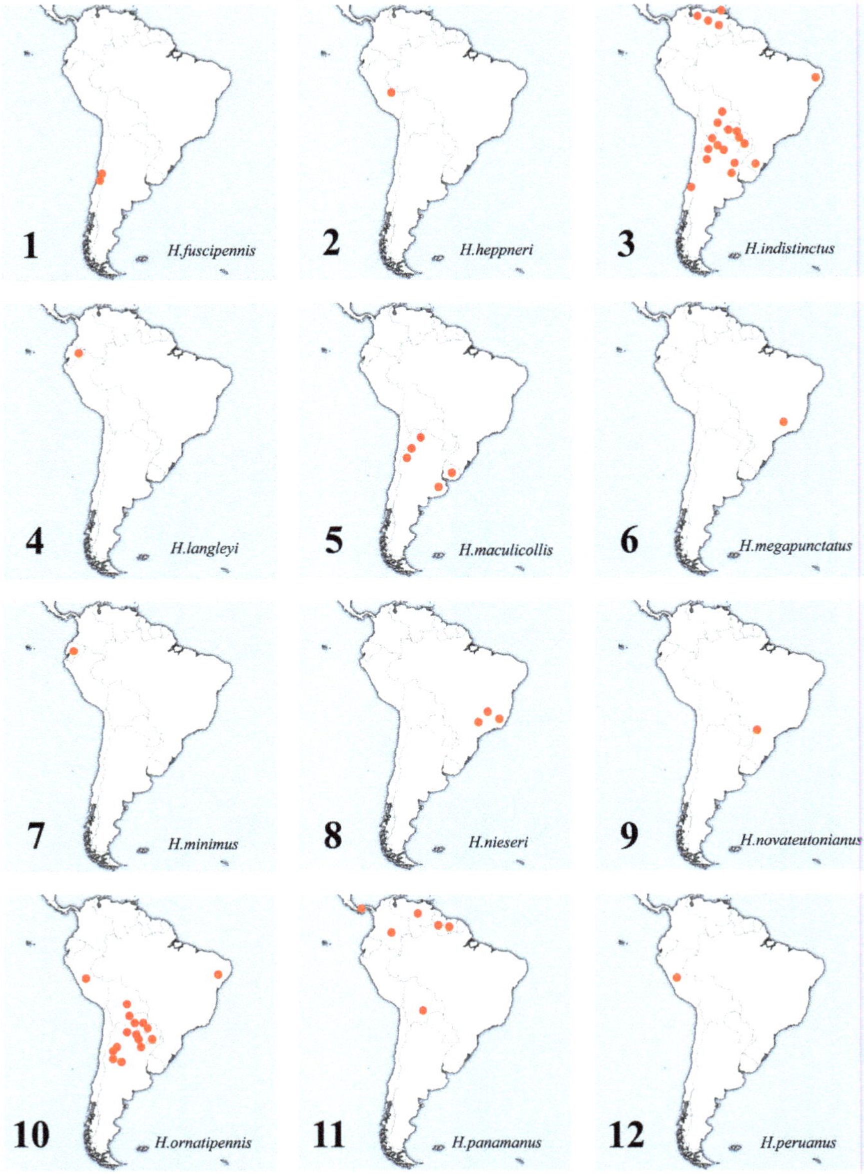

PLATE 267 Distribution of Neotropical Haliplidae

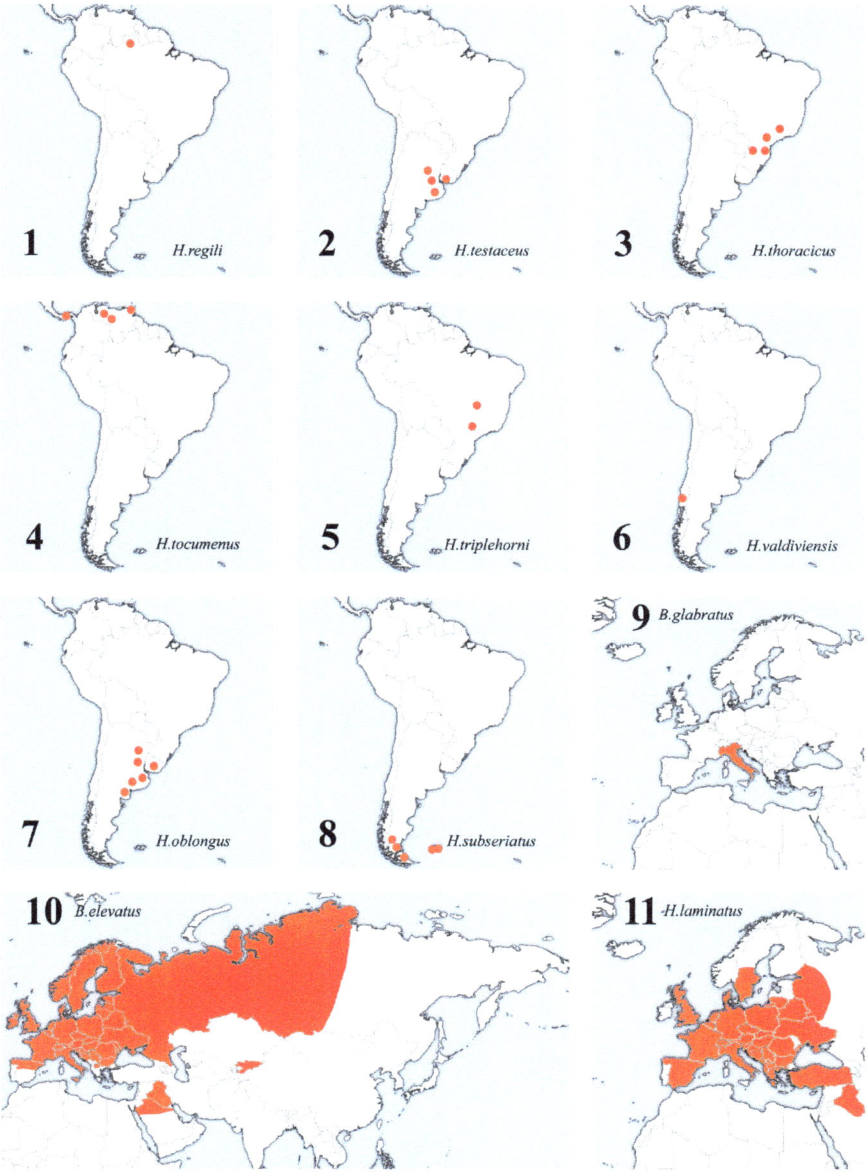

PLATE 268 Distribution of Neotropical and Palaearctic Haliplidae

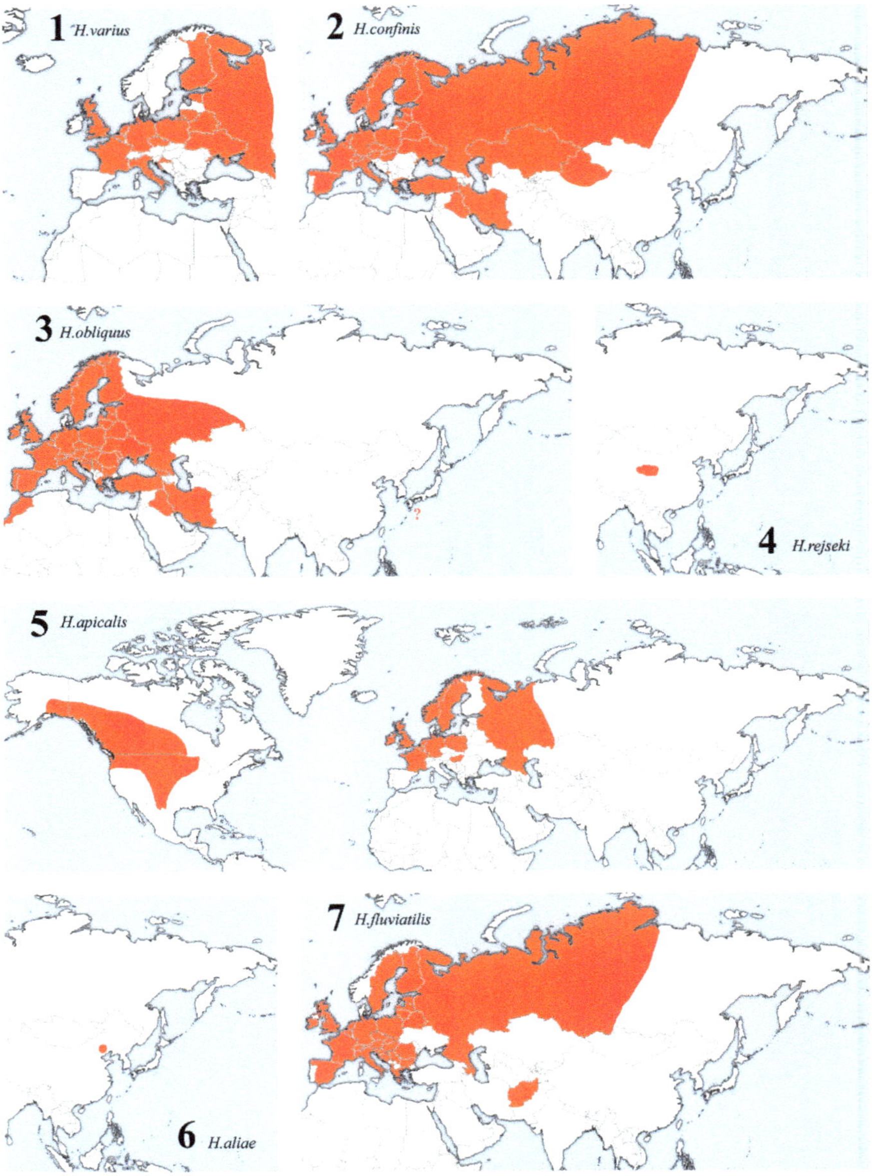

PLATE 269 Distribution of Nearctic and Palaearctic Haliplidae

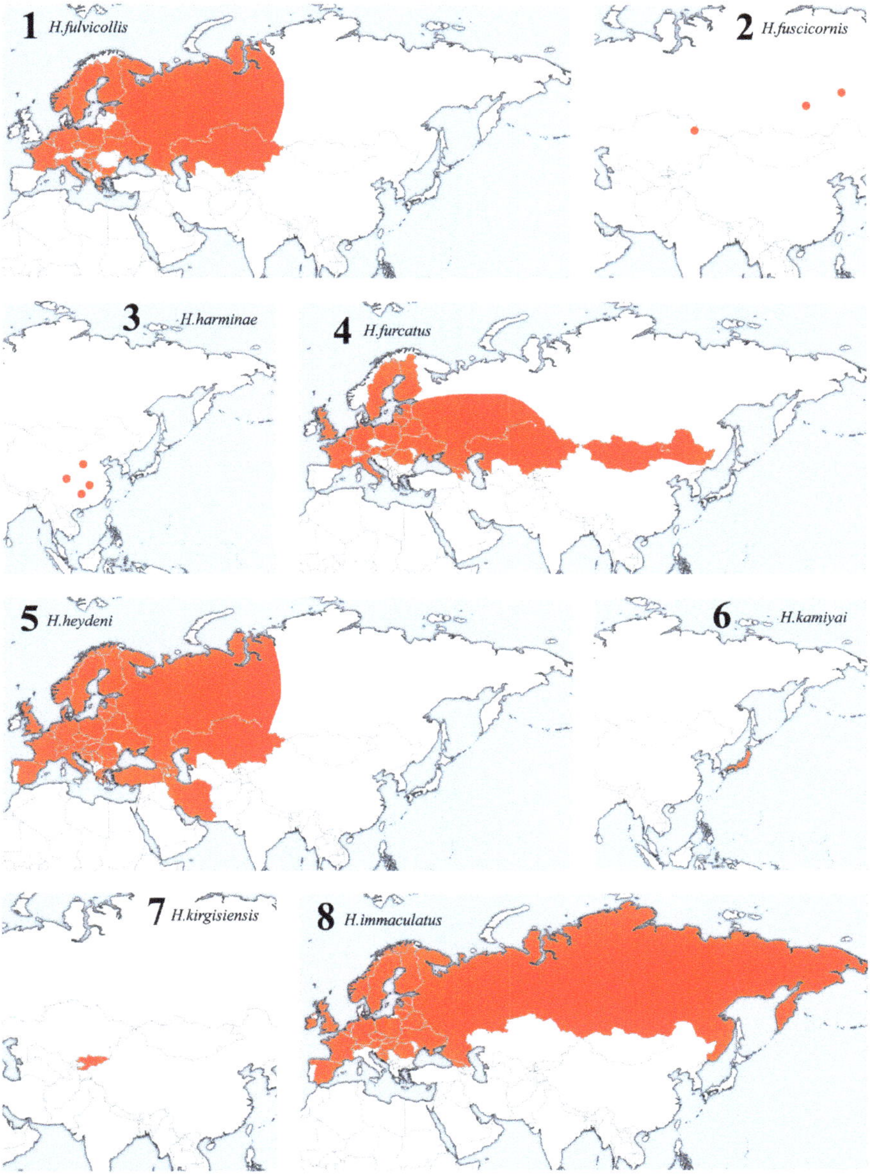

PLATE 270 Distribution of Palaearctic Haliplidae

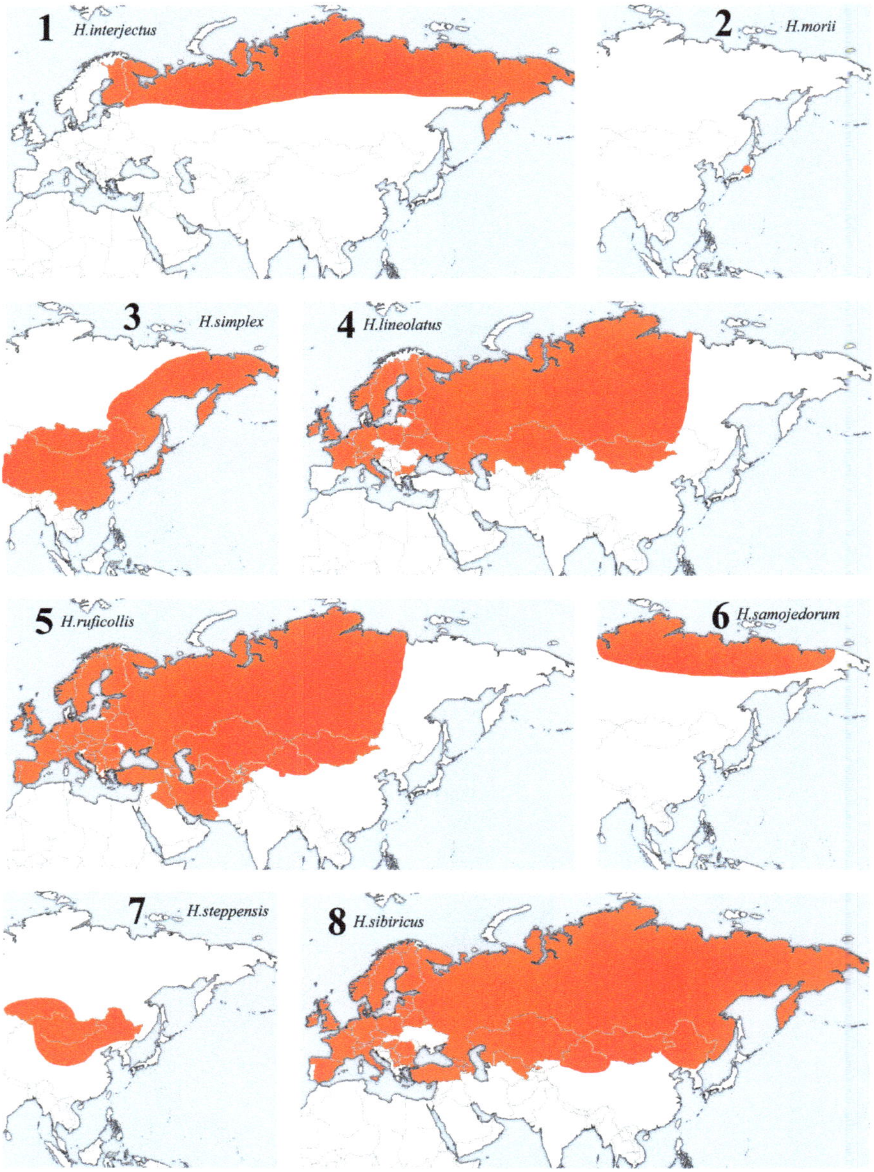

PLATE 271 Distribution of Palaearctic Haliplidae

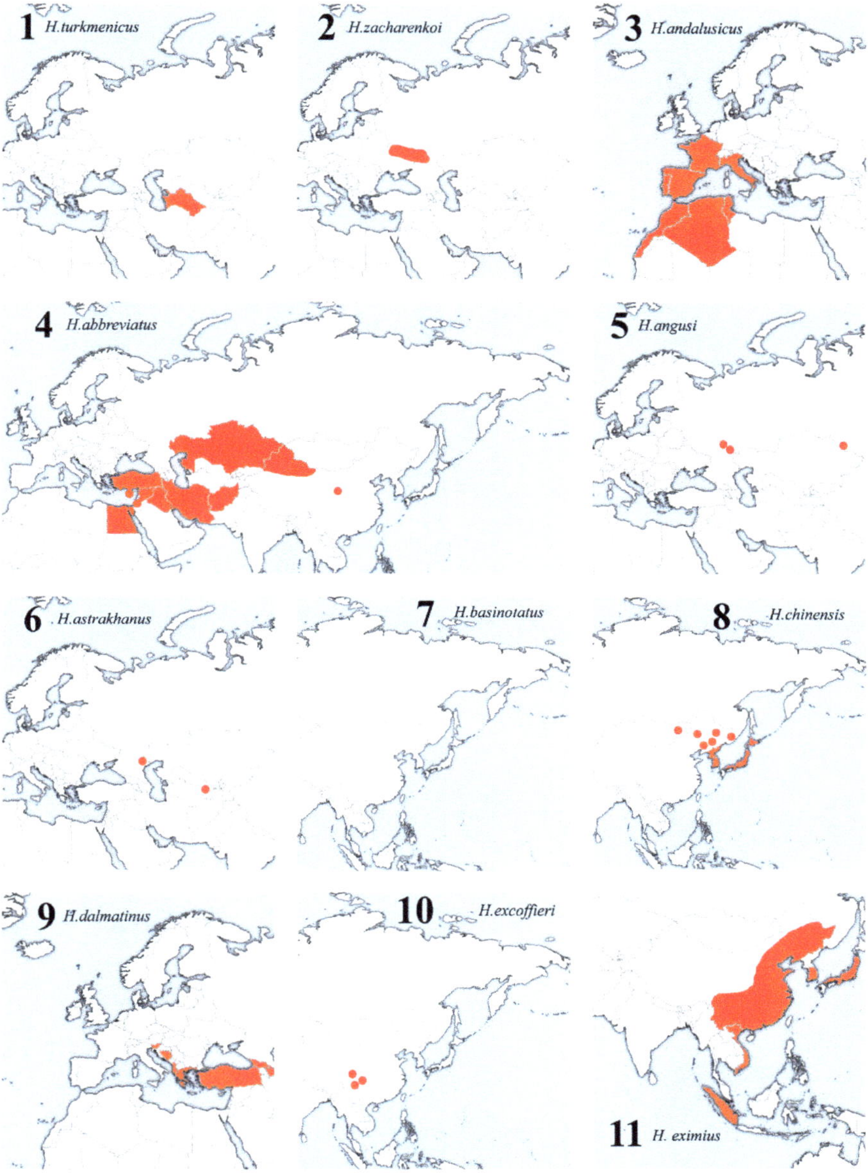

PLATE 272 Distribution of Palaearctic Haliplidae

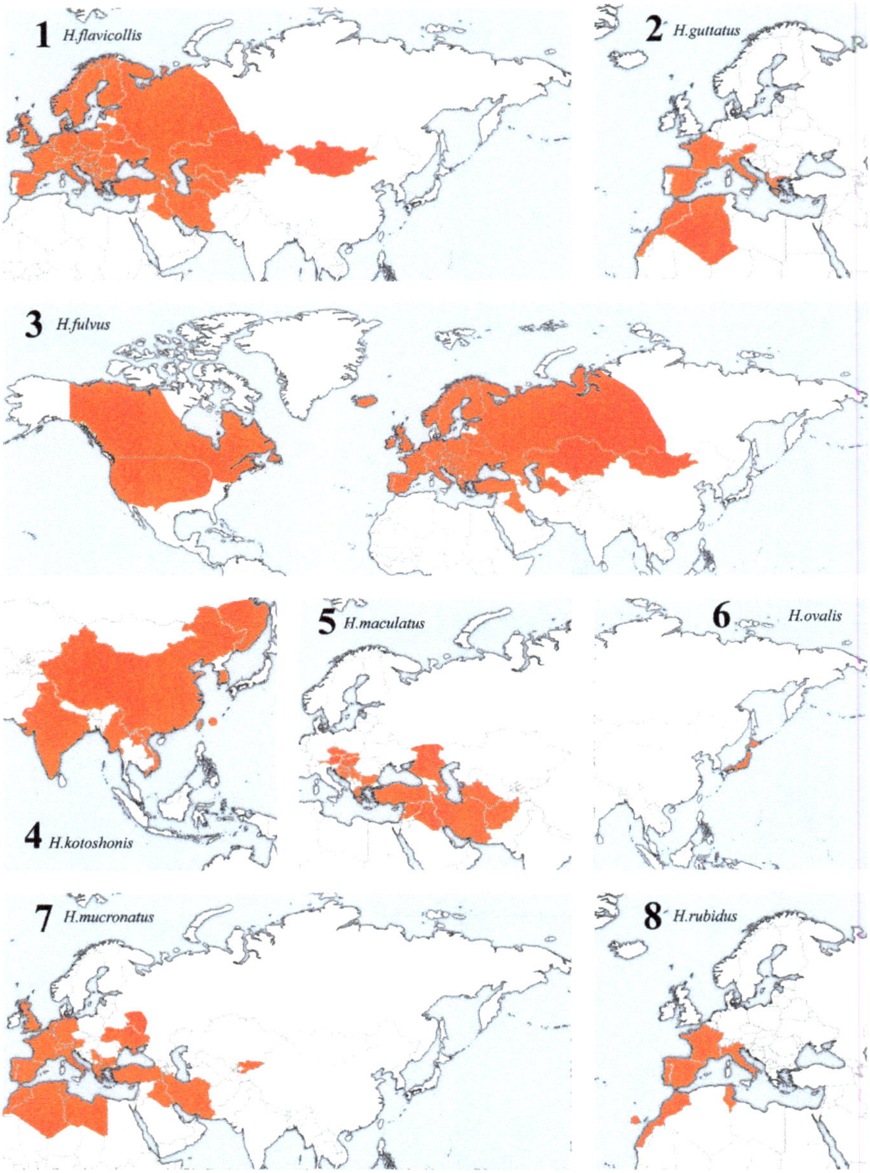

PLATE 273 Distribution of Nearctic, Palaearctic and Oriental Haliplidae

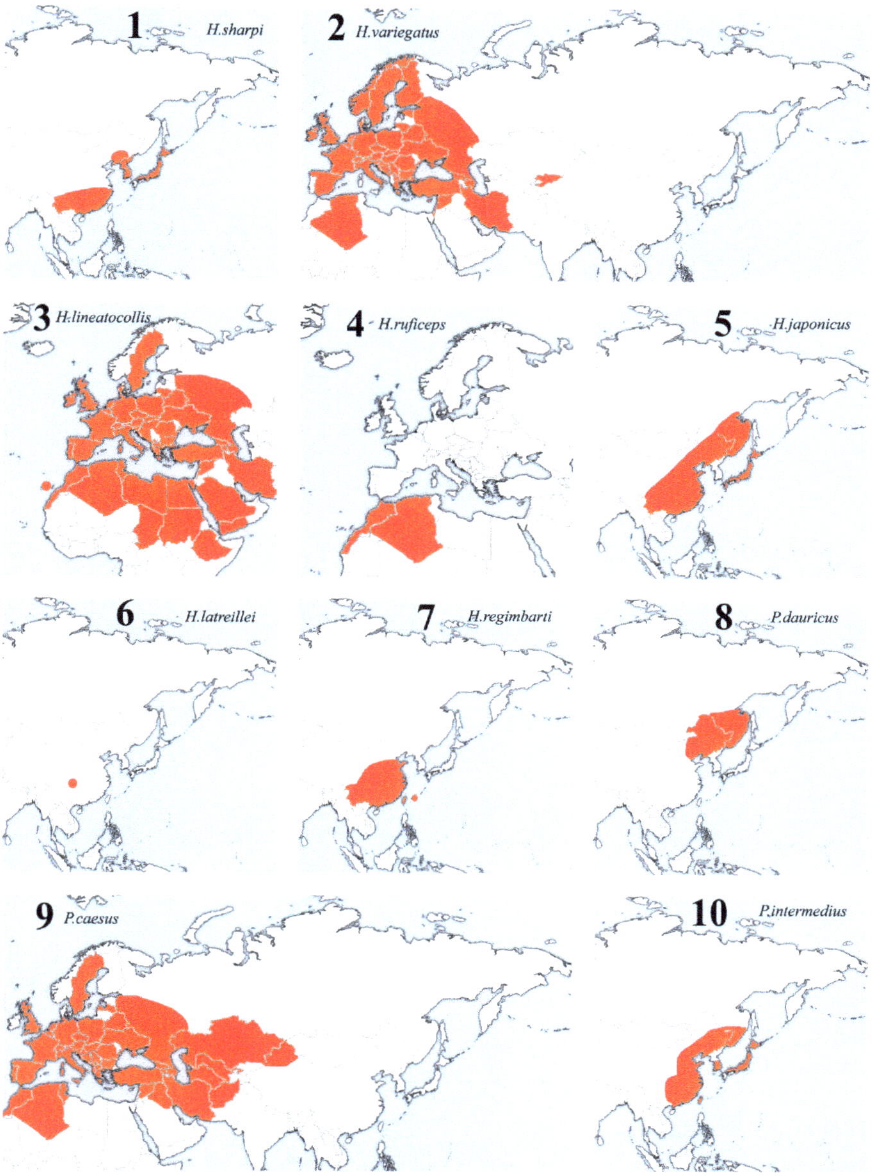

PLATE 274 Distribution of Palaearctic Haliplidae

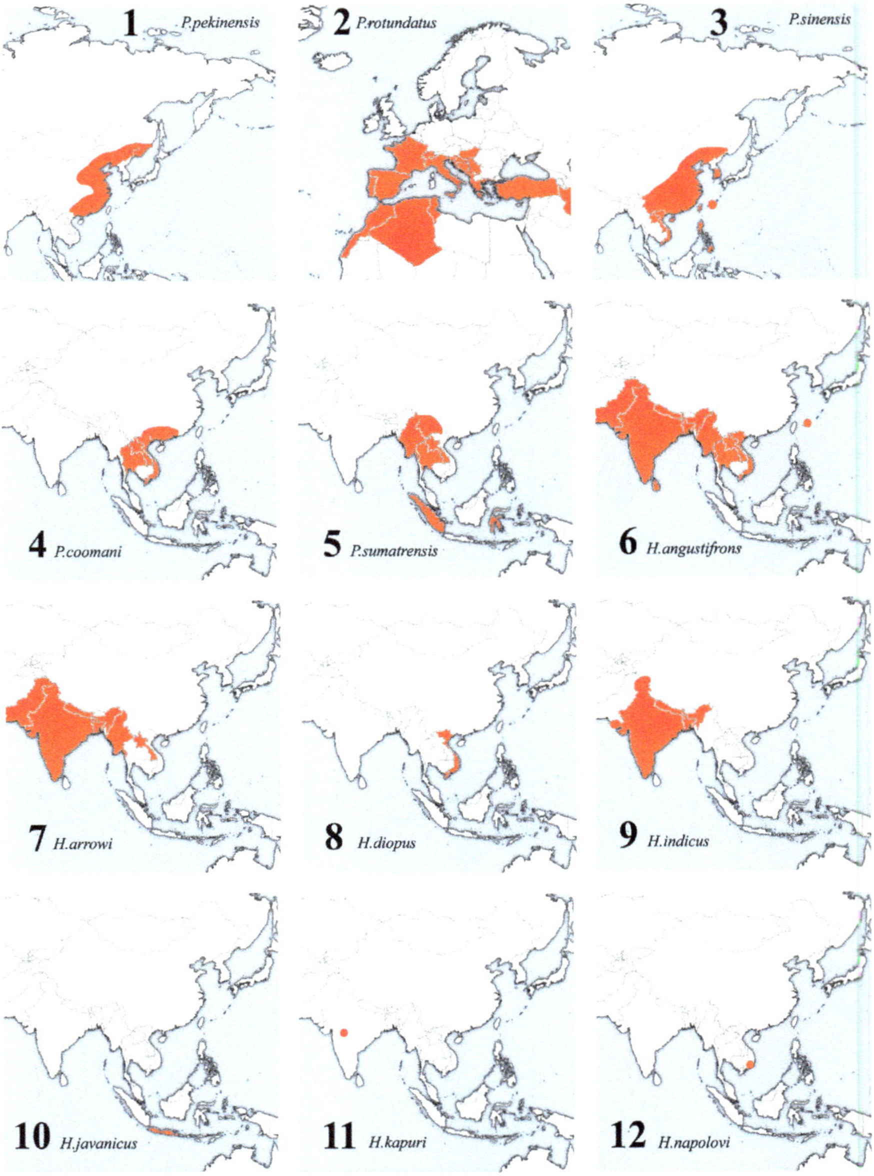

PLATE 275 Distribution of Palaearctic and Oriental Haliplidae

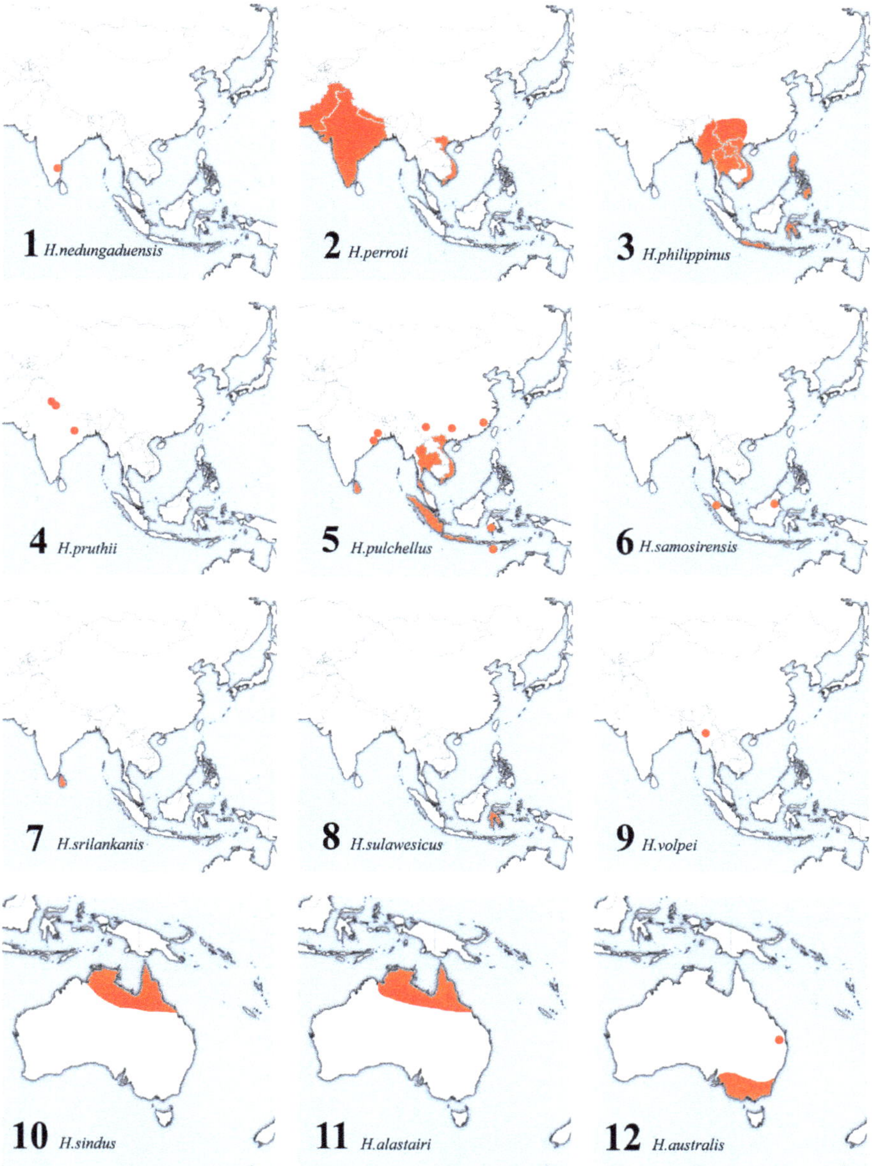

PLATE 276 Distribution of Oriental and Australasian Haliplidae

PLATE 277 Distribution of Australasian Haliplidae

PLATE 278 Distribution of Australasian and Afrotropical Haliplidae

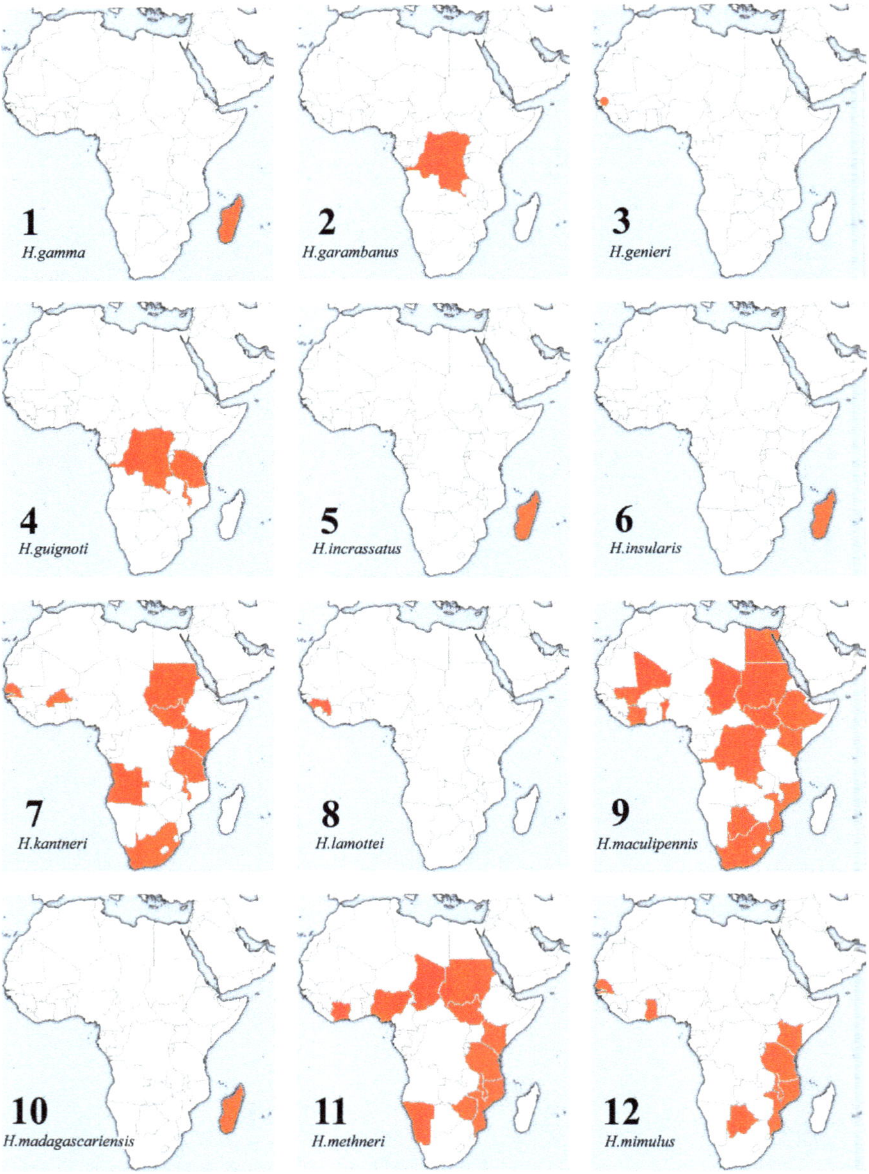

PLATE 279 Distribution of Afrotropical Haliplidae

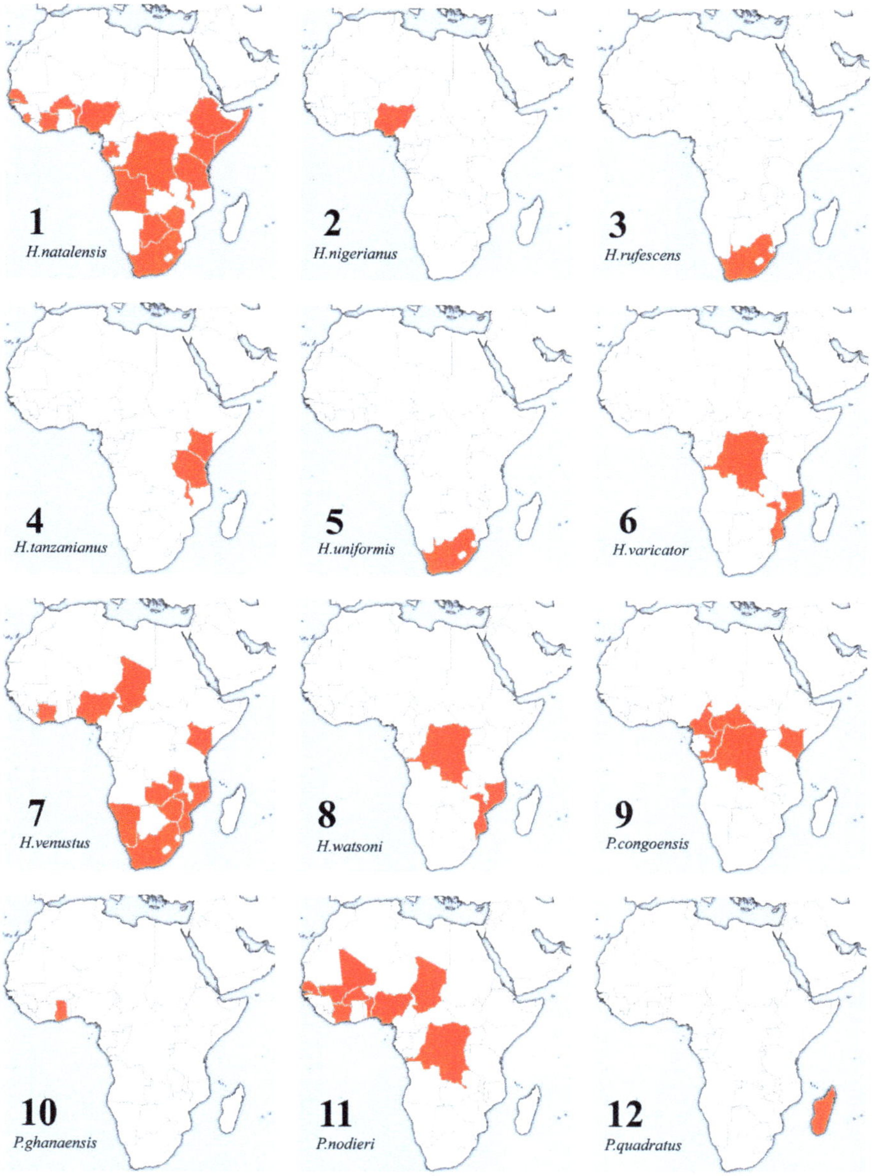

PLATE 280 Distribution of Afrotropical Haliplidae

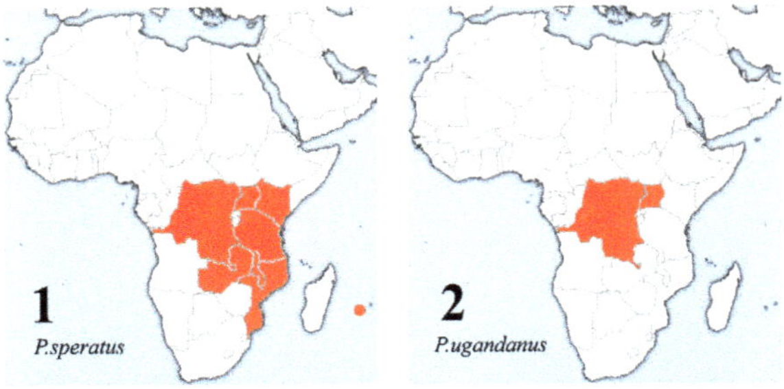

PLATE 281 Distribution of Afrotropical Haliplidae

Index

Synonyms and not available names are in italics.

Page numbers are in regular style, while plate numbers are in ***bold italics***

Family-group names

Genus-group and subgenus-group names

Species-group names

 | HTTPS://DOI.ORG/10.1163/9789004759671-020

www.ingramcontent.com/pod-product-compliance
Ingram Content Group UK Ltd.
Pitfield, Milton Keynes, MK11 3LW, UK
UKHW022228051026
15624UKWH00002B/410

9 789004 759664